T0300737

# Introduction to Probability, Statistics & R

Sujit K. Sahu

# Introduction to Probability, Statistics & R

Foundations for Data-Based Sciences

 Springer

Sujit K. Sahu (iD)
School of Mathematical Sciences
University of Southampton
Southampton, UK

ISBN 978-3-031-37864-5      ISBN 978-3-031-37865-2   (eBook)
https://doi.org/10.1007/978-3-031-37865-2

Mathematics Subject Classification: 62-01, 60-01

This Springer imprint is published by the registered company Springer Nature Switzerland AG
The registered company address is: Gewerbestrasse 11, 6330 Cham, Switzerland

If disposing of this product, please recycle the paper.

*To my family, parents, teachers and
professors who taught me everything.*

# Preface

It is a daunting task to contemplate writing an introductory textbook in Statistics and Probability since there are so many excellent text books already available in the market. However there seems to be a lack of textbooks/literature detailing statistical software R which contain sufficient levels of mathematical rigour whilst also remaining unambigious and comprehensible to the reader. There are even fewer introductory books which can be adopted as entry level texts for degree programmes with a large mathematical content, such as mathematics with statistics, operations research, economics, actuarial science and data science.

The market is saturated with a vast number of books on introductory mathematical statistics. Among the most prominent books in this area are the ones written by Professor Robert Hogg and his co-authors, e.g. the book *Probability and Statistical Inference*, 8th Edition (2010) by Hogg and Tanis (Pearson). Laid out in 11 chapters, this book is one of the best out there for learning mathematical statistics. Competitive books include: (i) *Probability and Statistics* by M. H. Degroot and M. J. Schervish (Pearson), (iii) *Statistical Inference* by G. Casella and R. Berger (Duxbury), (iii) *A First Course in Probability* by S. A. Ross (Pearson), (iv) *Mathematical statistics with applications* by D. D. Wackerly, W. Mendenhall and R. L. Scheaffer (Duxbury). These books assume a higher level of preparedness in mathematical methods that the typical first year undergraduate students do not have. Also these texts typically do not provide a plethora of examples that can help put the target audience of the first year undergraduate students at ease. Such students, fresh out of secondary school, are used to seeing lots of examples in each topic they studied. Universities in USA mostly adopt such text books in their masters level statistics courses. Lastly, none of these books integrate R in their presentation of the topics.

Apart from the above list, the book most relevant to the current textbook is *Introductory Statistics with R* by Peter Dalgaard published by Springer in 2008. Dalgaard's book is more targeted for a biometric/medical science audience whereas the current textbook targets students in a wider field of data-based and mathematical sciences. For example, Dalgaard's book include multiple regression and survival analysis. Multiple regression is too advanced for first year and survival analysis is too advanced for even second year students, who still are in the process acquiring skills in statistical inference and modelling. Also, unlike the Dalgaard's book, the current textbook does not assume knowledge of basic statistics to start with and hence is appealing to a wider audience of mathematics students who have not learned probability and statistics in their previous studies. The book *Teaching Statistics: A Bag of Tricks* by Andrew Gelman and Deborah Nolan (2nd Edition), published by the Oxford University Press, discusses many excellent methods for teaching practical statistics. However, this book is concerned about teaching statistics to aspiring applied scientists as well as mathematicians.

The current book aims to fill this gap in the market by taking a more direct targeted approach in providing an authentic text for introducing both **mathematical and applied statistics** with R. The book aims to provide a gentle introduction by keeping in mind the knowledge gap created by previous, ether none or non-rigorous, studies of statistics without the proper and rigorous use of mathematical symbols and proofs. This self-contained introductory book is also designed to appeal to first time students who were not previously exposed to the ideas of probability and statistics but have some background in mathematics. Many worked examples in the book are likely to be attractive to them, and those examples will build a transition bridge from their previous studies to university level mathematics. Moreover, integration of R throughout is designed to make learning statistics easy to understand, fun and exciting.

The book is presented in five parts. Part I (Chaps. 1 and 2) introducing basic statistics and R does not assume knowledge and skills in higher level mathematics such as multivariate calculus and matrix algebra. Part II (Chaps. 3 to 8) introduces standard probability distributions and the central theorem. Part III (Chaps. 9 to 12) introduces basic ideas of statistical inference. As a result, and quite deliberately, this part presents statistical inference methods such as the $t$-tests and confidence intervals without first deriving the necessary $t$ and $\chi^2$ distributions. Such derivations are delayed until the later chapters in Part IV. In this part (Chaps. 13 to 16), we present materials for typical second year courses in statistical distribution theory discussing advanced concepts of moment generating functions, univariate and bivariate transformation, multivariate distributions and concepts of convergence. Both this and the final Part V (Chaps. 17, 18 and 19) assume familiarity of results in multivariate calculus and matrix algebra. Part V of the book is devoted to introducing ideas in statistical modelling, including simple and multiple linear regression and one way analysis of variance. Several data sets are used as running examples, and dedicated R code blocks are provided to illustrate many key concepts such as the Central Limit Theorem and the weak law of large numbers. The reader can access those by installing the accompanying R package `ipsRdbs` in their computer.

I am highly indebted to all of my current and past mathematics and statistics colleagues in the Universities of Cardiff and Southampton, especially: Brian Bailey, Stefanie Biedermann, Dankmar Böhning, Russell Cheng, Jon Cockayne, Frank Dunstan, Jon Forster, Steven Gilmour, Terence Iles, Gerard Kennedy, Alan Kimber, Susan Lewis, Wei Liu, Zudi Lu, John W. McDonald, Robin Mitra, Barry Nix, Helen Ogden, Antony Overstall, Vesna Perisic, Philip Prescott, Dasha Semochkina, T. M. Fred Smith, Peter W. Smith, Alan Welsh, Dave Woods, Chieh-Hsi Wu, and Chao Zheng, whose lecture notes for various statistics courses inspired me to put together this manuscript. Often I have used excerpts from their lecture notes, included their data sets, examples, exercises and illustrations, without their full acknowledgement and explicit attribution. However, instead of them, I acknowledge responsibility for the full content of this book.

I also thank all my bachelor's and master's degree students who read and gave feedback on earlier versions of my lecture notes leading to drafting of this book. Specifically I thank three Southampton BSc students: Mr Minh Nguyen, Mr Ali Aziz and Mr Luke Brooke who read and corrected a preliminary draft of this book. I also thank PhD students Mr Indrajit Paul (University of Calcutta), who introduced me to use the latex package tikz for drawing several illustrations, and Ms Joanne Ellison (Southampton), who helped me typeset and proofread Parts I–III of the book. Lastly, I thank two anonymous reviewers whose suggestions I incorporated to improve various aspects including coverage and presentation.

Winchester, UK

Sujit K. Sahu

# Contents

# Part I

# Introduction to Basic Statistics and R

# Introduction to Basic Statistics

**1**

**Abstract**

Chapter 1: This chapter introduces basic statistics such as the mean, median and mode and standard deviation. It also provides introduction to many motivating data sets which are used as running examples throughout the book. An accessible discussion is also provided to debate issues like: "Lies, damned lies and statistics" and "Figures don't lie but liars can figure."

## 1.1 What Is Statistics?

### 1.1.1 Early and Modern Definitions

The word *statistics* has its roots in the Latin word *status* which means the state, and in the middle of the eighteenth century was intended to mean:

*collection, processing and use of data by the state.*

With the rapid industrialisation of Europe in the first half of the nineteenth century, statistics became established as a discipline. This led to the formation of the Royal Statistical Society, the premier professional association of statisticians in the UK and also world-wide, in 1834. During this nineteenth century growth period, statistics acquired a new meaning as the interpretation of data or methods of extracting information from data for decision making. Thus statistics has its modern meaning as the methods for:

*collection, analysis and interpretation of data.*

Indeed, the Oxford English Dictionary defines *statistics* as: "*The practice or science of collecting and analysing numerical data in large quantities, especially for the*

*purpose of inferring proportions in a whole from those in a representative sample.*"
Note that the word 'state' has been dropped from its definition. Dropping of the word
'state' reflects the wide spread use of statistics in everyday life and in industry—not
only the government. The ways to compile interesting statistics, more appropriately
termed as statistical methods, are now essential for every decision maker wanting to
answer questions and make predictions by observing data.

Example questions from everyday life may include: will it rain tomorrow? Will
the stock market crash tomorrow? Does eating red meat make us live longer? Is
smoking harmful during pregnancy? Is the new shampoo better than the old as
claimed by its manufacturer? How do I invest my money to maximise the return?
How long will I live for? A student joining university may want to ask: Given my
background, what degree classification will I get at graduation? What prospects do
I have in my future career?

### 1.1.2   Uncertainty: The Main Obstacle to Decision Making

The main obstacle to answering the types of questions above is *uncertainty*,
which means **lack of one-to-one correspondence between cause and effect**. For
example, having a diet of (hopefully well-cooked!) red meat for a period of time
is not going to kill someone immediately. The effect of smoking during pregnancy
is difficult to judge because of the presence of other factors, e.g. diet and lifestyle;
such effects will not be known for a long time, e.g. at least until the birth. Thus,
according to a famous quote:

> **"Uncertainty is the only certainty there is, ..."**

Yet another quote claims, "In statistics, there is uncertainty over the past, present
and future." This again emphasises the importance of presence of uncertainty in the
data for the purposes of drawing conclusions with certainty.

### 1.1.3   Statistics Tames Uncertainty

It[1] is clear that we may never be able to get to the bottom of every case to learn the
full truth and so will have to make a decision under uncertainty; thus we conclude
that mistakes cannot be avoided when making decisions based on uncertain data. If
mistakes cannot be avoided, it is better to know how often we make mistakes (which
provides knowledge of the amount of uncertainty) by following a particular rule of

---

[1] This section is based on Section 2.2 of the book *Statistics and Truth* by C. R. Rao cited as Rao
[15].

decision making. Such knowledge could be put to use in finding a rule of decision making which does not betray us too often, or which minimises the frequency of wrong decisions, or which minimises the loss due to wrong decisions. Thus we have the following equation due to Rao [15]:

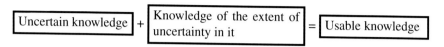

$$\boxed{\text{Uncertain knowledge}} + \boxed{\begin{array}{l}\text{Knowledge of the extent of}\\ \text{uncertainty in it}\end{array}} = \boxed{\text{Usable knowledge}}$$

In the above equation uncertain data are noted as uncertain knowledge and the decisions we make based on data are denoted by the phrase usable knowledge. The amount of uncertainty, as alluded to in the middle box, is evaluated by applying appropriate statistical methods. Without an explicit assessment of uncertainty, conclusions (or decisions) are often meaningless guesses with vast amounts of uncertainty. Although such conclusions may turn out to be correct just by sheer chance, or luck, in a given situation, the methods used to draw such conclusions cannot *always* be guaranteed to yield sound decisions. A carefully crafted statistical method, with its explicit assessment of uncertainty, will allow us to make better decisions on average, although it is to be understood that it is not possible to guess exactly always correctly in the presence of uncertainty.

How does statistics tame uncertainty? The short answer to this question is by evaluating it. Uncertainty, once evaluated, can be reduced by eliminating the causes and contributors of uncertainty as far as possible and then by hunting for better statistical methods which have lower levels of uncertainty. Explicit statistical model based methods may help in reducing uncertainties. This book will illustrate such uncertainty reduction in later chapters. *Uncertainty reduction* is often the most important task left to the statisticians as any experimenter is free to guess about any aspects of their experiments. Indeed, in many practical situations results (and conclusions) are quoted without any mention (and assessment) of uncertainty. Such cases are dangerous as those may give a false sense of security implied by the drawn conclusions. The associated, perhaps un-evaluated, levels of uncertainty may completely overwhelm the drawn conclusions.

### 1.1.4 Place of Statistics Among Other Disciplines

Studying statistics equips the learner with the basic skills in data analysis and doing science with data in any scientific discipline. Statistical methods are to be used wherever there is any presence of uncertainty in the drawn conclusions and decisions. Basic statistics, probability theory, and statistical modelling provide the solid foundation required to learn and use advanced methods in modern data science, machine learning and artificial intelligence. Students studying mathematics as their major subject may soon discover that learning of statistical theories gives them the opportunity to practice their deductive mathematical skills on real life problems.

In this way, they will be able to improve at mathematical methods while studying statistical methods. This book will illustrate these ideas repeatedly.

The following quote by Prof. C. R. Rao, see Rao [15], sums up the place of statistics among other disciplines.

> "All *knowledge* is, in final analysis, *history*.
> All *sciences* are, in the abstract, *mathematics*.
> All *judgements* are, in their rationale, *statistics*."

Application of statistics and statistical methods require dealing with uncertainty which one can never be sure about. Hence the mention of the word 'judgements' in the above quote. Making judgements requires a lot of common sense. Hence common sense thinking together with applications of mathematical and inductive logic is very much required in any decision making using statistics.

### 1.1.5 Lies, Damned Lies and Statistics?

Statistics and statistical methods are often attacked by statements such as the famous quotation in the title of this section. Some people also say, "you can prove anything in statistics!" and many such jokes. Such remarks bear testimony to the fact that often statistics and statistical methods are miss-quoted without proper verification and robust justification. It is clear that some people may intentionally miss-use statistics to serve their own purposes while some other people may be incompetent in statistics to draw sound conclusions, and hence decisions in practical applications. Thus, admittedly and regretfully, statistics can be very much miss-used and miss-interpreted especially by dis-honest individuals.

However, we statisticians argue:

* "Figures won't lie, but liars can figure!"
* "Data does not speak for itself"
* "Every number is guilty unless proved innocent."

Hence, although people may miss-use the tools of statistics, it is our duty to learn, question and sharpen those tools to develop scientifically robust and strong arguments. As discussed before, statistical methods are the only viable tool whenever there is uncertainty in decision making. It will be wrong to feel certainty where no certainty exists in the presence of uncertainty. In scientific investigations, statistics is an inevitable instrument in search of truth when uncertainty cannot be totally removed from decision making. Of-course, a statistical method may not yield the best predictions in every practical situation, but a systematic and robust application of statistical methods will eventually win over pure guesses. For example, statistical methods are the only definitive proof that cigarette smoking is bad for human health.

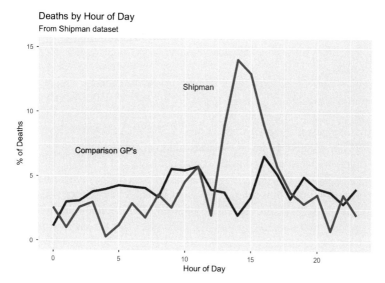

**Fig. 1.1** The time at which Shipman's patients died, compared to the times at which patients of other local family doctors died. This is Figure 0.2 (reproduced here with permission) in the book "The Art of Statistics" by David Spiegelhalter

### 1.1.6   Example: Harold Shipman Murder Enquiry

To illustrate and motivate the study of statistics consider the Harold Shipman murder enquiry data example as discussed in the book, *The Art of Statistics* by Spiegelhalter [20]. Shipman was a British family doctor and serial killer who killed at least 215 of his most elderly patients by injecting opium between 1975 and 1998. His killing spree went undetected until an enquiry was launched during 1998–1999. He was finally convicted in January 2000. Figure 1.1 provides a graph of the percentages of patients dying in each of the 24 hours in a day. No sophisticated statistical analysis is required to detect the obvious pattern in the data, which shows that 2PM is the very unusual peak time of death for Shipman's patients. Further background and details are provided in Chapter 1 of the book by Prof David Spiegelhalter [20].

   This example illustrates the importance of studying probability and statistics for data based sciences, although it did not require any sophisticated statistical methods about to be presented in the book. However, it is essential to learn a plethora of statistical methods to fully appreciate the strength of the evidence present in Fig. 1.1.

### 1.1.7   Summary

In summary, we note that statistical methods are, often, the only and essential tools to be applied whenever there is uncertainty and complete enumeration is not possible. Any analysis of empirically collected data must stand up to scientific (read as statistical) scrutiny and hence statistical knowledge is essential for conducting any scientific investigation. As a result, it is enormously advantageous to have good statistical data analysis skills to advance in most career paths in academia, industry and government.

This section has also discussed the main purpose of statistics—mainly to assess and to reduce uncertainty in the drawn conclusions. It also noted the place of statistics among different scientific disciplines and mathematics. Statistics and mathematics are best studied together as statistical applications provide rich training grounds for learning mathematical methods and mathematical theories and logic, on the other hand, help develop and justify complex statistical methods.

This section also tackles the often discussed misconceptions regarding the use of statistics in everyday life. It is often argued that statistics and statistical methods can be used to both prove or disprove a single assertion. This section has put forward the counter argument that data, being pure numbers, does not lie but users of statistics are liable to make mistakes either un-knowingly or knowingly through deceptions. A robust use of statistics is recommended so that the drawn conclusions can stand up to scientific scrutiny. Unfortunately, this task is to be taken by the producers of statistics so that only sound conclusions are reported in the first place.

For further reading, we note two accessible books: *Statistics and Truth* by Rao [15] and (ii) *The Art of Statistics* by David Spiegelhalter [20]. In addition, there are many online resources that discuss the joy of statistics. For example, we recommend the reader to watch the *YouTube* video **Joy of Statistics**.[2]

To acknowledge the references used for this book, we note the excellent textbooks written by Goon et al. [7], Casella and Berger [3], DeGroot and Schervish [4] and Ross [17]. We also acknowledge two books of worked examples in probability and statistics by Dunstan et al. [5, 6]. We also borrowed example exercises from the *Cambridge International AS and A-level Mathematics Statistics* (2012) book published by Hodder Education (ISBN-9781444146509) and written by Sophie Goldie and Roger Porkess.

## 1.2    Example Data Sets

Before introducing the example data sets it may be pertinent to ask the question, "How does one collect data in statistics?" Recall the definition of statistics in Sect. 1.1.1 where it states that statistics uses a representative sample to infer proportions in a whole, which we call the population. To collect a representative sample from a population the experimenter must select individuals randomly or haphazardly using a lottery for example. Otherwise we may introduce bias. For example, in order to gauge student opinion in a university, an investigator should not only survey the international students. However, there are cases when systematic sampling, e.g., selecting every third caller in a radio phone-in show for a prize, or

---

[2] https://www.youtube.com/watch?v=cdf0k545yDA.

sampling air pollution hourly or daily, may be preferable. This discussion regarding random sample collection and designed experiments, where the investigator controls the values of certain experimental variables and then measures a corresponding output or response variable, is deferred to Sects. 9.1 and 12.7. Until then we assume that we have data from $n$ randomly selected sampling units.

For each of the $n$ sampling units, we may collect information regarding a single or multiple characteristics. In the first case we conveniently denote the data by $x_1, x_2, \ldots, x_n$, so that these values are numeric, either discrete counts, e.g. number of road accidents, or continuous, e.g. heights of 18-year old girls, marks obtained in an examination. In case multiple measurements are taken, we often introduce more elaborate notations. For example, the variable of interest for the $i$th individual is denoted by $y_i$ and the other variables, assuming $p$ many, for individual $i$ may be denoted by $x_{i1}, \ldots, x_{ip}$. The billionaires data set introduced below provides an example of this.

We now introduce several data sets that will be used as running examples throughout this book. Later chapters of this book may use the same data set to illustrate different statistical concepts and theories. All these data sets are downloadable from the online supplement of this book and also included in the R package `ipsRdbs` accompanying this book. Some of these data sets are taken from the data and story library.[3]

### Example 1.1 (Fast Food Service Time)

The table (data obtained from the online Data and Story library in 2018) below provides the service times (in seconds) of customers at a fast-food restaurant (Fig. 1.2). The first row is for customers who were served from 9–10AM and the second row is for customers who were served from 2–3PM on the same day. The data set is `ffood` in the R package `ipsRdbs` with a help file obtained using the command `?ffood`.

| AM | 38, | 100, | 64, | 43, | 63, | 59, | 107, | 52, | 86, | 77 |
|----|-----|------|-----|-----|-----|-----|------|-----|-----|-----|
| PM | 45, | 62, | 52, | 72, | 81, | 88, | 64, | 75, | 59, | 70 |

Note that the service times are not paired, i.e. the times in the first column, 38 and 45, are not related. Those are time in seconds for two different customers possibly served by different workers in two different shifts. Issues that we would like to investigate include analyses of the AM and PM service times and comparison of differences between the times. ◄

---

[3] https://dasl.datadescription.com/.

**Fig. 1.2**  A fast food restaurant in Kyiv, Ukraine 2012. A photo by Sharon Hahn Darlin, source: Wikimedia Commons. https://www.flickr.com/photos/sharonhahndarlin/8088905486/. License: CC-BY-2.0

**Table 1.1**  Number of weekly computer failures over two years

| 4 | 0 | 0 | 0 | 3 | 2 | 0 | 0 | 6 | 7 |
|---|---|---|---|---|---|---|---|---|---|
| 6 | 2 | 1 | 11 | 6 | 1 | 2 | 1 | 1 | 2 |
| 0 | 2 | 2 | 1 | 0 | 12 | 8 | 4 | 5 | 0 |
| 5 | 4 | 1 | 0 | 8 | 2 | 5 | 2 | 1 | 12 |
| 8 | 9 | 10 | 17 | 2 | 3 | 4 | 8 | 1 | 2 |
| 5 | 1 | 2 | 2 | 3 | 1 | 2 | 0 | 2 | 1 |
| 6 | 3 | 3 | 6 | 11 | 10 | 4 | 3 | 0 | 2 |
| 4 | 2 | 1 | 5 | 3 | 3 | 2 | 5 | 3 | 4 |
| 1 | 3 | 6 | 4 | 4 | 5 | 2 | 10 | 4 | 1 |
| 5 | 6 | 9 | 7 | 3 | 1 | 3 | 0 | 2 | 2 |
| 1 | 4 | 2 | 13 | | | | | | |

**Example 1.2 (Computer Failures)**

This data set (Table 1.1) contains weekly failures of a university computer system (See Fig. 1.3) over a period of two years. The source of the data set is the book 'A Handbook of Small Data Sets' by Hand et al. [8], thanks to Prof Jon Forster (author of Kendall et al. [9]) for sharing this. We will use this data set to illustrate commands in the R software package and also in statistical modelling. The data set is cfail in the R package ipsRdbs and there is a help file obtained by issuing the command ?cfail. ◀

**Example 1.3 (Number of Bomb Hits in London During World War II (See Fig. 1.4))**

This data set is taken from the research article Shaw and Shaw [19] via the book by Hand et al. [8] and Prof Dankmar Böhning, (author of Böhning et al. [2]). The city of Greater London is divided into 576 small areas of one-quarter square

**Fig. 1.3** PCs running Windows. Photo by Project Manhattan. License: CC BY-SA 3.0

**Fig. 1.4** London Blitz. Photo by H. F. Davis

kilometre each. The number of bomb hits during World War II in each of the 576 areas was recorded. The table below provides the frequencies of the numbers of hits.

| Number of hits | 0 | 1 | 2 | 3 | 4 | 5 | Total |
|---|---|---|---|---|---|---|---|
| Frequency | 229 | 211 | 93 | 35 | 7 | 1 | 576 |

Thus, 229 areas were not hit at all, 211 areas had exactly one hit and so on. Like the previous computer failure data example, we will use this to illustrate and compare statistical modelling methods. The data set is `bombhits` in the R package `ipsRdbs` and there is a help file which is accessed by issuing the command `?bombhits`. ◀

## Example 1.4 (Weight Gain of Students)

This data set was collected to investigate if students (see Fig. 1.5) tend to gain weight during their first year in college/university. In order to test this, David Levitsky, a Professor of Nutrition in the Cornell University (USA), recruited students from two large sections of an introductory course in health care, see the article Levitsky et al. [11]. Although they were volunteers, they appeared to match the rest of the freshman class in terms of demographic variables such as sex and ethnicity. Sixty-eight students were weighed during the first week of the semester, then again 12 weeks later. The table below provides the first and

**Fig. 1.5** College students. Source: https://www.freestock.com/free-photos/college-university-students-smiling-classroom-23344615 Image used under license from Freestock.com

last three rows of the data set in kilograms, which was converted from imperial measurement in pounds and ounces.

| Student number | Initial weight (kg) | Final weight (kg) |
| --- | --- | --- |
| 1 | 77.6 | 76.2 |
| 2 | 49.9 | 50.3 |
| 3 | 60.8 | 61.7 |
| ⋮ | ⋮ | ⋮ |
| 66 | 52.2 | 54.0 |
| 67 | 75.7 | 77.1 |
| 68 | 59.4 | 59.4 |

This data set will be used to illustrate simple exploratory charts in R and also to demonstrate what is known as statistical hypothesis testing. The data set is wgain in the R package ipsRdbs, with an associated help file ?wgain. ◄

---

**Example 1.5 (Body Fat Percentage)**

Knowledge of the fat content of the human body is physiologically and medically important. The fat content may influence susceptibility to disease, the outcome of disease, the effectiveness of drugs (especially anaesthetics) and the ability to withstand adverse conditions including exposure to cold and starvation. In practice, fat content is difficult to measure directly—one way is by measuring body density which requires subjects to be weighed underwater! For this reason, it is useful to try to relate simpler measures such as skin-fold thicknesses (which

**Fig. 1.6**  Womens 5000 m start at the 2012 Olympics by Nick Webb. Source: Wikipedia. License: CC BY 2.0

are readily measured using calipers) to body fat content and then use these to estimate the body fat content.

Dr R. Telford, working for the Australian Institute of Sport (AIS), collected skin-fold (the sum of four skin-fold measurements) and percent body fat measurements on 102 elite athletes training at the AIS (see Fig. 1.6). Obtained from Prof Alan H. Welsh (author of Welsh [21]), the data set `bodyfat` is made available from the R package `ipsRdbs` and the R command `?bodyfat` provides further information and code for exploring and modelling the data, which has been perfprmed in Chap. 17. ◄

| Athlete | Skin-fold | Body-fat (%) |
|---------|-----------|--------------|
| 1       | 44.5      | 8.47         |
| 2       | 41.8      | 7.68         |
| 3       | 33.7      | 6.16         |
| ⋮       | ⋮         | ⋮            |
| 100     | 47.6      | 8.51         |
| 101     | 60.4      | 11.50        |
| 102     | 34.9      | 6.26         |

**Example 1.6 (Wealth of Billionaires)**

Fortune magazine publishes a list of the world's billionaires each year. The 1992 list includes 225 individuals from five regions: **Asia, Europe, Middle East,**

**Fig. 1.7** Stack of 100 dollar bills. Source: Wikimedia Commons. License: CC BY-SA 3.0

United States, and Other. For these 225 individuals we also have their wealth (in billions of dollars, see Fig. 1.7) and age (in years). The first and last two rows of the data set are given in the table below.

| Wealth | Age | Region |
|--------|-----|--------|
| 37.0 | 50 | M |
| 24.0 | 88 | U |
| ⋮ | ⋮ | ⋮ |
| 1 | 9 | M |
| 1 | 59 | E |

This example will investigate differences in wealth of billionaires due to age and region using many exploratory graphical tools and statistical methods. The data set is `bill` in the R package `ipsRdbs` and a help file is obtained by issuing the command `?bill`. ◄

## 1.3　Basic Statistics

Having motivated to study statistics and introduced the data sets, our mission in this section is to learn some basic summary statistics and graphical tools through the use of the R software package. This section also aims to explore the summary statistics using basic mathematical tools such as the summation symbol $\sum$ and minimisation methods, which will be used repeatedly in the later chapters.

Suitable summaries of data are used to describe the data sets. The needs and motivation behind data collection often dictate what particular statistical summaries to report in the data description. Usually the non-numeric categorical variables are summarised by frequency tables. For example, we may report the number of billionaires in each of the five regions. For numeric variables we would like to know the centre of the data, i.e., measures of location or central tendency, and the spread or variability. The two sections below introduce these measures.

## 1.3.1  Measures of Location

This section defines three most common measures of location: mean, median and mode. It also justifies appropriateness of their use as a representative value for the data through theoretical arguments. The section ends with a discussion to decide the most suitable measure in practical applications.

### 1.3.1.1 Mean, Median and Mode

Suppose that the we have the data $x_1, x_2, \ldots, x_n$ for which we are seeking a representative value, which will be a function of the data. The sample mean denoted by $\bar{x}$ and defined by

$$\bar{x} = \frac{1}{n}(x_1 + x_2 + \cdots + x_n) = \frac{1}{n}\sum_{i=1}^{n} x_i,$$

is a candidate for that representative value. Two other popular measures are the sample median and sample mode which we define below.

The sample median is the middle value in the ordered list of values $x_1, x_2, \ldots, x_n$. Consider the AM service time data in Example 1.1 where the values are: 38, 100, 64, 43, 63, 59, 107, 52, 86, 77. Obviously, we first write these values in order:

$$38 < 43 < 52 < 59 < 63 < 64 < 77 < 86 < 100 < 107.$$

There does not exist a unique middle value. But it is clear that the middle value must lie between 63 and 64. Hence, median is defined to be any value between 63 and 64. For the sake of definiteness, we may chose the mid-point 63.5 as the median.

In general, how do we find the middle value of the numbers $x_1, x_2, \ldots, x_n$? Mimicking the above example, we first write these values in order:

$$x_{(1)} \leq x_{(2)} \leq \cdots \leq x_{(n)},$$

where $x_{(1)}$ denotes the minimum and $x_{(n)}$ denotes the maximum of the $n$ data values $x_1, x_2, \ldots, x_n$. Note that we simply do not write $x_1 \leq x_2 \leq \cdots \leq x_n$ since that would be wrong when data are not arranged in increasing order. Hence the new notation $x_{(i)}, i = 1, \ldots, n$ has been introduced here. For example, if 10 randomly observed values are: 9, 1, 5, 6, 8, 2, 10, 3, 4, and 7, then $x_1 = 9$ but $x_{(1)} = 1$.

If $n$ is odd then $x_{(\frac{n+1}{2})}$ is the median value. For example, if $n = 11$ then the 6th value in the ordering of 11 sample values is the sample median. If $n$ is even then sample median is defined to be any value in the interval $\left(x_{(\frac{n}{2})}, x_{(\frac{n}{2}+1)}\right)$. For convenience, we often take the mid-point of the interval as the sample median. Thus, if $n = 10$ then the sample median is defined to be the mean of the 5th and 6th ordered values out of these $n = 10$ sample values. Thus, for the 10 observed values 9, 1, 5, 6, 8, 2, 10, 3, 4, and 7, the sample median is 5.5.

To recap, the sample median is defined as the observation ranked $\frac{1}{2}(n + 1)$ in the ordered list if $n$ is odd. If $n$ is even, the median is any value between $\frac{n}{2}$th and $(\frac{n}{2} + 1)$th in the ordered list. For example, for the AM service times, $n = 10$ and $38 < 43 < 52 < 59 < 63 < 64 < 77 < 86 < 100 < 107$. So the median is any value between 63 and 64. For convenience, we often take the mean of these. So the median is 63.5 seconds. Note that we use the unit of the observations when reporting any measure of location.

The third measure of location is the sample mode which is the most frequent value in the sample. In the London bomb hits Example 1.3, 0 is the mode of the number of bomb hits. If all sample values are distinct, as in the AM service time data example, then there is no unique mode. In such cases, especially when $n$ is large, sample data may be presented in groups and the modal class may be found leading to an approximation for the mode. We, however, do not discuss frequency data in any further detail.

### 1.3.1.2  Which of the Three Measures to Use?

Which one of the three candidate representative values, sample mean, median, and mode shall we choose in a given situation? This question can be answered by considering a possibly fictitious imaginary idea of loss incurred in choosing a particular value, i.e., either the sample mean or median, as the representative value for all the observations. (Here we are thinking that we are guessing all the sample values by the representative value and there will be a loss, i.e. a penalty to be paid for incorrect guessing.)

Suppose a particular number $a$, e.g. the sample mean, is chosen to be the value representing the numbers $x_1, x_2, \ldots, x_n$. The loss we may incur in choosing $a$ to represent any $x_i$ could be a function of the error $x_i - a$. Hence the total loss is $\sum_{i=1}^{n}(x_i - a)$. But note that the total loss is not a going to be a good measure since some of the individual losses will be negative and some will be positive resulting in a small value or even a negative value of total loss. Hence, we often assume the squared-error loss, $(x_i - a)^2$ or the absolute error loss $|x_i - a|$ for representing the observation $x_i$ so that the errors in ether direction (positive or negative) attracts similar amount of losses. Then we may choose the $a$ that minimises the total error given by $\sum_{i=1}^{n}(x_i - a)^2$ in the case of squared-error loss. In case we assume absolute error loss we will have to find the $a$ that minimises the sum of the absolute losses, $\sum_{i=1}^{n}|x_i - a|$.

It turns out that the sample mean is the $a$ that minimises $\sum_{i=1}^{n}(x_i - a)^2$ and sample median is the $a$ that minimises $\sum_{i=1}^{n}|x_i - a|$. The sample mode minimises a third type of loss obtained be considering the 0–1 loss function. In 0–1 loss, the loss is defined to be zero if $x_i = a$ and 1 if $x_i \neq a$ for $i = 1, \ldots, n$. That is, the loss is zero if $a$ is the correct guess for $x_i$ and 1 if $a$ is an incorrect guess. It is now intuitively clear that the sample mode will minimise the total of the 0–1 loss function since if $a$ =sample mode then the loss is going to be 0 for most of the observations, $x_1, \ldots, x_n$, resulting in the smallest value for total of the 0–1 loss.

Here now prove that *the sample mean minimises the sum of squares of the errors*, denoted by:

$$\text{SSE} = \sum_{i=1}^{n} (x_i - a)^2.$$

To establish this we can use the derivative method in Calculus, see the exercises. Here is an important alternative derivative free proof which will be used in other similar circumstances in later chapters.

***Proof*** Here the trick is to subtract and then add the sample mean $\bar{x}$ inside the square $(x_i - a)^2$. Then the task is to simplify after expanding the square as follows:

$$
\begin{aligned}
\sum_{i=1}^{n}(x_i - a)^2 &= \sum_{i=1}^{n}(x_i - \bar{x} + \bar{x} - a)^2 \quad \text{[subtract and add } \bar{x}\text{]} \\
&= \sum_{i=1}^{n}\left\{(x_i - \bar{x})^2 + 2(x_i - \bar{x})(\bar{x} - a) + (\bar{x} - a)^2\right\} \\
&= \sum_{i=1}^{n}(x_i - \bar{x})^2 + 2(\bar{x} - a)\sum_{i=1}^{n}(x_i - \bar{x}) + \sum_{i=1}^{n}(\bar{x} - a)^2 \\
&= \sum_{i=1}^{n}(x_i - \bar{x})^2 + n(\bar{x} - a)^2,
\end{aligned}
$$

since $\sum_{i=1}^{n}(x_i - \bar{x}) = n\bar{x} - n\bar{x} = 0$. Now note that the first term is free of $a$; the second term is non-negative for any value of $a$. Hence the minimum occurs when the second term is zero, i.e. when $a = \bar{x}$. This completes the proof. $\square$

The trick of adding and subtracting the mean, expanding the square and then showing that the cross-product term is zero will be used several times in this book in the later chapters. Hence it is important to learn this proof. The result is a very important in statistics, and this will be used several times in this book. Note that, $\text{SSE}= \sum_{i=1}^{n}(x_1 - a)^2$ is the sum of squares of the deviations of $x_1, x_2, \ldots, x_n$ from any number $a$. The established identity states that:

The sum of (or mean) squares of the deviations of $x_1, x_2, \ldots, x_n$ from any number $a$ is minimised when $a$ is the sample mean of $x_1, x_2, \ldots, x_n$.

In the proof we also noted that $\sum_{i=1}^{n}(x_i - \bar{x}) = 0$. This is stated as:

The sum of the deviations of a set of numbers from their mean is zero.

(1.1)

The sum of the deviations of a set of numbers from their mean is zero.

We now prove that *the sample median minimises the sum of the absolute deviations*, defined by:

$$\text{SAD} = \sum_{i=1}^{n} |x_i - a|.$$

Here the derivative approach, stated in the exercises to prove the previous result for the sample mean, does not work since the derivative does not exist for the absolute function. Instead we use the following argument.

*Proof* First, order the observations (see Fig. 1.8):

$$x_{(1)} \le x_{(2)} \le \cdots \le x_{(n)}.$$

Now note that:

$$\begin{aligned}
\text{SAD} &= \sum_{i=1}^{n} |x_i - a| \\
&= \sum_{i=1}^{n} |x_{(i)} - a| \\
&= |x_{(1)} - a| + |x_{(n)} - a| + |x_{(2)} - a| + |x_{(n-1)} - a| + \cdots
\end{aligned}$$

Now see Fig. 1.9 for a visualisation of the following argument. From the top line in the figure it is clear that $S_1(a) = |x_{(1)} - a| + |x_{(n)} - a|$ is minimised when $a$ is such that $x_{(1)} \le a \le x_{(n)}$, in which case $S_1(a)$ takes the value $|x_{(1)} - x_{(n)}|$. Otherwise, suppose $a$ lies outside of the interval $\left(x_{(1)}, x_{(n)}\right)$. For example, if $a < x_{(1)}$ then

**Fig. 1.8**  Illustration of ordered observations

**Fig. 1.9**  Visualisation of the proof that the sample median minimises the sum of the absolute deviations

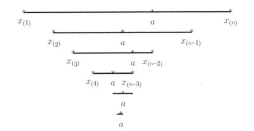

$S_1(a)$ will take the value $|x_{(1)} - x_{(n)}|$ plus $2 \times |x_{(1)} - a|$. Thus to minimise $S_1(a)$ we must put $a$ somewhere in the interval $(x_{(1)}, x_{(n)})$.

Continuing this argument we conclude that, $|x_{(2)} - a| + |x_{(n-1)} - a|$ is minimised when $a$ is such that $x_{(2)} \le a \le x_{(n-1)}$. Finally, when $n$ is odd, the last term $|x_{(\frac{n+1}{2})} - a|$ is minimised when $a = x_{(\frac{n+1}{2})}$ or the middle value in the ordered list. In this case $a$ has been defined as the sample median above. If, however, $n$ is even, the last pair of terms will be $|x_{(\frac{n}{2})} - a| + |x_{(\frac{n}{2}+1)} - a|$. This will be minimised when $a$ is any value between $x_{(\frac{n}{2})}$ and $x_{(\frac{n}{2}+1)}$, which has been defined as the sample median in case $n$ is even. Hence this completes the proof.  $\square$

This establishes the fact that:

> the sum (or mean) of the absolute deviations of $x_1, \ldots, x_n$ from any number $a$ is minimised when $a$ is the sample median of $x_1, \ldots, x_n$.

There is the concept of third type of loss, called a 0-1 loss, when we are searching for a measure of central tendency. In this case, it is intuitive that the best guess $a$ will be the mode of the data, which is the most frequent value.

**Which of the Three (Mean, Median and Mode) Should We Prefer?** Obviously, the answer will depend on the type of loss we may assume for the particular problem. The decision may also be guided by the fact the sample mean gets more affected by extreme observations while the sample median does not. For example for the AM service times, suppose the next observation is 190. The median will be 64 instead of 63.5 but the mean will shoot up to 79.9.

## 1.3.2   Measures of Spread

The measures of central tendency defined in the previous section does not convey anything regarding the variability or spread of the data. Often, it is of interest to know how tightly packed the data are around the chosen centre, one of sample mean, median or mode. This section discusses three measures of spread or variability.

A quick measure of the spread is the *range*, which is defined as the difference between the maximum and minimum observations. For example, for the AM service times in the Fast Food Example 1.1 the range is 69 ($= 107 - 38$) seconds. The range, however, is not a very useful measure of spread, as it is extremely sensitive to the values of the two extreme observations. Furthermore, it gives little information about the distribution of the observations between the two extremes.

A better measure of spread is given by the sample standard deviation, denoted by $s$, which the square-root of the *sample variance*, $s^2$, defined by

$$\text{Var}(x) = s^2 = \tfrac{1}{n-1} \sum_{i=1}^{n}(x_i - \bar{x})^2.$$

The sample variance is defined with the divisor $n - 1$ since there are some advantages which will be discussed in Sect. 9.4.1. The divisor $n - 1$ is the default in R. for the command `var`.

The population variance is defined with the divisor $n$ instead of $n-1$ in the above. Although $s^2$ above is defined as the mean of sum of squares of the deviations from the mean, we do not normally calculate it using that formula. Instead, we use the following fundamental identity:

$$
\begin{aligned}
\sum_{i=1}^{n}(x_i - \bar{x})^2 &= \sum_{i=1}^{n}\left(x_i^2 - 2x_i\bar{x} + \bar{x}^2\right) \\
&= \sum_{i=1}^{n} x_i^2 - 2\bar{x}(n\bar{x}) + n\bar{x}^2 \\
&= \sum_{i=1}^{n} x_i^2 - n\bar{x}^2.
\end{aligned}
$$

Hence we prefer to calculate variance by the formula:

$$\text{Var}(x) = s^2 = \tfrac{1}{n-1}\left(\sum_{i=1}^{n} x_i^2 - n\bar{x}^2\right)$$

and the *standard deviation* is taken as the square-root of the variance. For example, the standard deviation of the AM service times is 23.2 seconds. Note that standard deviation has the same unit as the observations.

A third measure of spread is what is known as the inter-quartile range (IQR). The IQR is the difference between the third, $Q_3$ and first, $Q_1$ quartiles, which are respectively the observations ranked $\frac{1}{4}(3n + 1)$ and $\frac{1}{4}(n + 3)$ in the ordered list, $x_{(1)} \leq x_{(2)} \leq \cdots \leq x_{(n)}$. Note that the sample median is the second quartile, $Q_2$. When $n$ is even, definitions of $Q_3$ and $Q_1$ are similar to that of the median, $Q_2$. The lower and upper quartiles, together with the median, divide the observations up into four sets of equal size. For the AM service times

$$38 < 43 < 52 < 59 < 63 < 64 < 77 < 86 < 100 < 107$$

$Q_1$ lies between 52 and 59, while $Q_3$ lies between 77 and 86. Some linear interpolation methods are used to find approximate values in R. We, however, do not discuss this any further.

Usually the three measures: range, sd and IQR are not used interchangeably. The range is often used in data description, the most popular measure, standard

**Fig. 1.10**  A sketch of a boxplot diagram

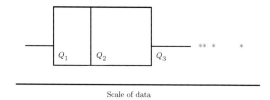

deviation, is used as a measure of variability or concentration around the sample mean and the IQR is most often used in graphical summaries of the data such as the boxplot which is described in the next section.

### 1.3.3   Boxplot

A boxplot of sample data, e.g. computer failure data, plots the three quartiles and also provides valuable information regarding the shape and concentration of the data. From a boxplot, we can immediately gain information concerning the centre, spread, and extremes of the distribution of the observations (Fig. 1.10).

Constructing a boxplot involves the following steps:

1. Draw a vertical (or horizontal) axis representing the interval scale on which the observations are made.
2. Calculate the median, and upper and lower quartiles ($Q_1$, $Q_3$) as described above. Calculate the inter-quartile range (or 'midspread') $H = Q_3 - Q_1$.
3. Draw a rectangular box alongside the axis, the ends of which are positioned at $Q_1$ and $Q_3$. Hence, the box covers the 'middle half' of the observations). $Q_1$ and $Q_3$ are referred to as the 'hinges'.
4. Divide the box into two by drawing a line across it at the median.
5. The whiskers are lines which extend from the hinges as far as the most extreme observation which lies within a distance $1.5 \times H$, of the hinges.
6. Any observations beyond the ends of the whiskers (further than $1.5 \times H$ from the hinges) are suspected outliers and are each marked on the plot as individual points at the appropriate values. (Sometimes a different style of marking is used for any outliers which are at a distance greater than $H$ from the end of the whiskers).

Figure 1.11 shows a boxplot of the number of weekly computer failure data introduced in Example 1.2. The two quartiles $Q_1$ and $Q_3$ are drawn as vertical lines at the two edges of the box and the median is the bold vertical line drawn through the middle of the box. The whiskers are the horizontal lines drawn at the two edges of the box. There are three extreme observations plotted as individual points beyond the whisker at the right of the plot. Comparing the lengths of the two whiskers, we see that the data shows an un-even distribution where there is a larger spread of values in the right hand side of the median, or the right tail. Such a distribution of

**Fig. 1.11** A boxplot of
computer failure data

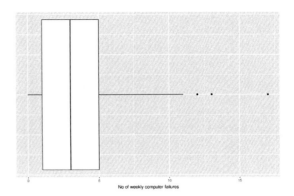

No of weekly computer failures

the data is often called to be positively (or right) skewed. We will learn how to draw
such a plot using R in Chap. 2.

Exploration of statistical data uses many other plots such as the stem and leaf
plot, histogram, barplot and pie chart. However, this textbook does not provide
discussion of such plots for brevity. Instead, the interested reader is referred to
school level elementary statistics textbooks for detailed discussions on such topics.

### 1.3.4   Summary

This section has introduced three measures of location: mean, median and mode,
each of which is optimal under a different consideration. We have also introduced
three measures of variability range, sd and the IQR, each of which has the same unit
as the original data.

## 1.4    Exercises

**1.1 (Addition with the Summation Symbol $\sum$)**

- Assume $x_1, \ldots, x_n$ and $y_1, \ldots, y_n$ are real numbers not all zero. Also, $a, b, k$ are
  real numbers and $n > 0$ is an integer.
- We write $\sum_{i=1}^{n} x_i$ to denote the sum $x_1 + x_2 + \cdots + x_n$. We should always include the
  limits and the dummy (e.g. $i$), i.e., $\sum_{i=1}^{n} x_i$, and we do not encourage the notation
  $\sum x$ since it does not make it clear what numbers are being added up. Also note
  that $\sum_{i=1}^{n} x_i = \sum_{j=1}^{n} x_j$, i.e, the letter $i$ or $j$ we write for the dummy does not
  matter.

1. Prove that $\sum_{i=1}^{n} k\,x_i = k \sum_{i=1}^{n} x_i$.
2. Prove that $\sum_{i=1}^{n} (k + x_i) = n\,k + \sum_{i=1}^{n} x_i$.
3. Prove that $\sum_{i=1}^{n} (x_i - \bar{x}) = 0$, where $\bar{x} = \frac{1}{n} \sum_{i=1}^{n} x_i$.
4. Prove that $\sum_{i=1}^{n} (x_i - \bar{x})^2 = \sum_{i=1}^{n} x_i^2 - n\bar{x}^2$.
5. Suppose that the values of $x_1, \ldots, x_n$ are known and we want to minimise the sum $\sum_{i=1}^{n} (x_i - a)^2$ with respect to the variable $a$. Prove that $\sum_{i=1}^{n} (x_i - a)^2$ is minimised when $a = \bar{x}$ by using the derivative method described below.

   To optimise $f(a)$, we first solve the equation $f'(a) = 0$. We then see if $f''(a)$, evaluated at the solution, is positive or negative. The function $f(a)$ attains a local **minimum** at the solution if the sign is positive. The function $f(a)$ attains a local **maximum** at the solution if the sign is negative. There is neither a minima nor a maxima if the second derivative is zero at the solution. Such a point is called a *point of inflection*.

### 1.2 (Mean-Variance)

1. Suppose we have the data: $x_1 = 1$, $x_2 = 2, \ldots, x_n = n$. Find the mean and the variance. For variance use the divisor $n$ instead of $n - 1$.
2. Suppose $y_i = ax_i + b$ for $i = 1, \ldots, n$ where $a$ and $b$ are real numbers. Show that:
   (a) $\bar{y} \equiv \frac{1}{n} \sum_{i=1}^{n} y_i = a\bar{x} + b$ and
   (b) $\mathrm{Var}(y) = a^2 \mathrm{Var}(x)$ where $\bar{x} = \frac{1}{n} \sum_{i=1}^{n} x_i$
   and for variance it is possible to use either the divisor $n$, i.e. $\mathrm{Var}(x) = \frac{1}{n} \sum_{i=1}^{n} (x_i - \bar{x})^2$ or $n - 1$, i.e. $\mathrm{Var}(x) = \frac{1}{n-1} \sum_{i=1}^{n} (x_i - \bar{x})^2$. The divisor does not matter as the results hold regardless. **Hint:** For the second part, start with the left hand side, $\mathrm{Var}(y) = \frac{1}{n} \sum_{i=1}^{n} (y_i - \bar{y})^2$ and substitute $y_i$ and $\bar{y}$ in terms of $x_i$ and $\bar{x}$.
3. Suppose $a \le x_i \le b$ for $i = 1, \ldots, n$. Show that $a \le \bar{x} \le b$.

### 1.3 (Variance Inequality)

1. Prove that for any set of numbers $x_1, x_2, \ldots, x_n$,

$$\left( x_1^2 + x_2^2 + \cdots + x_n^2 \right) \ge \frac{(x_1 + x_2 + \ldots x_n)^2}{n},$$

i.e. sum of squares of $n$ numbers is greater than equal to the square of the sum divided by $n$. **Hint:** You may start by assuming $\sum_{i=1}^{n} (x_i - \bar{x})^2 \ge 0$ and then expand the square within the summation symbol.

## 1.4 (Additional Data)

1. Assume that $x_1, \ldots, x_n, x_{n+1}$ are given real numbers. Prove that:
   (a) $\bar{x}_{n+1} = \frac{x_{n+1} + n\bar{x}_n}{n+1}$
   (b) $ns_{n+1}^2 = (n-1)s_n^2 + \frac{n}{n+1}(x_{n+1} - \bar{x}_n)^2$.
2. Assume that $x_1, \ldots, x_m$ and $y_1, \ldots, y_n$ are real numbers not all zero, where $m$ and $n$ are positive integers. Let $z_1, \ldots, z_{n+m}$ denote the combined $m + n$ observations, i.e. $\mathbf{z} = (x_1, \ldots, x_m, y_1, \ldots, y_n)$ without loss of generality.
   Let $\bar{x}, s_x^2, \bar{y}, s_y^2, \bar{z}, s_z^2$ denote the sample mean and variance pair of the $x$'s, $y$'s and $z$'s respectively.
   (a) Prove that $\bar{z}$ is given by:

$$\bar{z} = \frac{m\,\bar{x} + n\,\bar{y}}{m + n}.$$

   (b) Prove that the sample variance of the $z$ values is given by:

$$s_z^2 = \frac{(m-1)s_x^2 + (n-1)s_y^2}{m + n - 1} + \frac{m(\bar{x} - \bar{z})^2 + n(\bar{y} - \bar{z})^2}{m + n - 1}.$$

- These two formulae allow us to calculate the mean and variance of the combined data easily.

**1.5 (Two Variables and the Cauchy-Schwarz Inequality)** Suppose that $(x_1, y_1), (x_2, y_2). \ldots, (x_n, y_n)$ are given pairs of numbers.

1. Prove that

$$\sum_{i=1}^{n}(y_i - \bar{y})(x_i - \bar{x}) = \sum_{i=1}^{n}(y_i - \bar{y})x_i = \sum_{i=1}^{n} y_i(x_i - \bar{x}) = \sum_{i=1}^{n} y_i x_i - n y x.$$

2. Prove the Cauchy-Schwarz Inequality.

$$\left(x_1^2 + x_2^2 + \cdots + x_n^2\right)\left(y_1^2 + y_2^2 + \cdots + y_n^2\right) \geq (x_1 y_1 + x_2 y_2 + \cdots + x_n y_n)^2.$$

**Hint**: You can either try the induction method or use the fact that for any set of numbers $a_1, \ldots a_n$ and $b_1, \ldots b_n$.:

$$\sum_{i=1}^{n}(a_i - b_i)^2 \geq 0, \quad \text{and then substitute } a_i = \frac{x_i}{\sqrt{\sum_{i=1}^{n} x_i^2}}, \; b_i = \frac{y_i}{\sqrt{\sum_{i=1}^{n} y_i^2}}.$$

# Getting Started with R

# 2

**Abstract**

Chapter 2: This chapter introduces the R software package and discusses how to get started with many examples. It revisits some of the data sets already mentioned in Chap. 1 by drawing simple graphs and obtaining summary statistics.

## 2.1  What Is R?

The R language provides many facilities for analysis and data handling, graphical display, statistical modelling and programming. It also allows the user extreme flexibility in manipulating and analysis of large data sets. R is freely available from the web, developed through the leadership of Ross Ihaka and Robert Gentleman. The CRAN[1] (Comprehensive R Archive Network) provides a range of information including downloading and how to getting started. The CRAN website also links many tutorial pages written by many authors. To get used to the common R commands, the reader may find it helpful to download a R language Cheatsheet from the webpage: https://iqss.github.io/dss-workshops/R/Rintro/base-r-cheat-sheet.pdf or elsewhere on the internet.

R is an *object-oriented* language, which means that everything is stored as a particular type of object, with different operations being appropriate for different types of object. For example, *vectors* and *matrices* are both types of object in R. Data are usually stored in a *data frame* object, and results of statistical analyses are

---

The original version of this chapter has been revised. A correction to this chapter can be found at https://doi.org/10.1007/978-3-031-37865-2_21

---

[1] https://cran.r-project.org/.

stored in an object of the appropriate type. Summarising and exploratory methods are then applied to such objects.

R has an extensive on-line help system. You can access this using the **Help** menu. The help system is particularly useful for looking up commands or functions that you know exist but whose name or whose syntax you have forgotten. An alternative way of obtaining information about a function is to type `help`(<function name>) or ?<function name>, for example `help`(`plot`) or ?`plot`.

In this book we will code in R using a freely available front-end called Rstudio which can be freely downloaded from the Rstudio website.[2] Both R and Rstudio can work with any of the three computer operating systems: Mac, Windows and Linux. Rstudio is the preferred choice for working with R since it has nicer operational functionality with more menu driven options.

### 2.1.1  R Basics

R can be started by just launching the Rstudio programme. In Rstudio there is the *R console* that allows us to type in commands at the prompt > directly. To exit R we need to either type in the command > `q`() in the *R console* then hit the Enter key in the keyboard or click the Run button in the menu. We may also exit R and Rstudio by following the menu **File→Exit**.

In R , just typing a command will not produce anything. We will have to execute the command either by hitting the Enter key on the keyboard or by clicking the Run button in the Rstudio menu.

R commands are always of the form <function>(<arguments>). For example, **mean**(x) obtains the mean of values contained in the vector x. In the case where there are no options, e.g. the command `q`(), we still need to add the brackets. This is because R treats all of its commands as functions. If the brackets are omitted, then R thinks that we do not want to execute the function but we would simply like to see the R code which the function executes. For example, the reader can type **plot** in the *R console* and then hit the Enter button to see what happens.

All the data sets introduced in Sect. 1.2 are included in the dedicated R contributed package ipsRdbs. It is recommended that the reader installs this package by issuing the commands:

```
install.packages("ipsRdbs")
library(ipsRdbs)
ls("package:ipsRdbs")
```

The installation commands need to be run only once. It is not necessary to install the packages again in subsequent R sessions. However, the library command **library**(ipsRdbs) is required to invoke the package whenever the reader wants to access data and specific code from the package.

---

[2] https://rstudio.com/products/rstudio/download/.

The assignment operator in R is `<-`, *i.e.* a 'less than' symbol immediately followed by a hyphen. For example,

```
x <- 2 + 2 # The output should be 4!
```

We can also use the `"="` symbol for assignment but in this book we shall only use the `<-` symbol for assignment. Note that an assignment does not produce any output (unless we have made an error, in which case an error message will appear). To see the result of an assignment, we need to examine the contents of the object we have assigned the result of the command to. For example, typing x and then hitting Enter, should now give the output `[1] 4`. The `[1]` indicates that 4 is the first component of x. Of-course, x only has one component here, but this helps to keep track when the output is a vector of many components.

Note that anything we type after # sign is a comment and R will ignore. This is used for documentation.

When calling a function, the arguments can be placed in any order provided that they are explicitly named. Any unnamed argument passed to a function is assigned to the first variable which has not yet been assigned. Any arguments which have defaults, may be omitted. For example, consider the function `qnorm` which gives the quantiles of the normal distribution which we will learn in Sect. 7.2. The help file `?qnorm` reveals that the order of arguments is p, **mean** and **sd**. However, the commands

```
qnorm(0.95, mean=-2.0, sd=3.0)
qnorm(0.95, sd=3.0, mean=-2.0)
qnorm(mean=-2.0, sd=3.0, p=0.95)
```

all will have the same effect and they all will produce the same result.

## 2.1.2   Script Files

Typing R commands directly in the *R console* is not recommended since it is cumbersome to type-in long commands correctly in one go and the commands are all forgotten as soon as one quits R. To avoid such problems we normally put the R commands in a file that we save in our computer. R script files are saved with the file extension `.R` for example, `Rfile1.R` in a suitable folder in the computer. We can open a saved file by following: **File** → **Open File** in Rstudio. The **File** menu also allows the user to open a new script file where commands can be written and then executed as follows.

To run a bunch of commands in the opened script file we highlight the bunch and then press the `Run` button in Rstudio (towards the top right corner of the script Window with a green colour arrow) or the `Run line or selection` menu button in R.

## 2.1.3 Working Directory in R

The most important task in R is to set the working directory in R. The working directory is the sub-folder in the computer where we would like to save our data and R programme files. There are essentially two steps that we will have to follow: (i) create a dedicated folder in the computer and (ii) let R know of the folder location. In the discussion below we shall assume that we are working in a Windows computer and we can create a folder called `ProbStats` in the `C:` drive. This folder can also be located in the sub-folder `C:/Users/username/ProbStats` where `username` is the computer login-name of the user. However, for convenience, we will simply refer to this as the `C:/ProbStats` folder. Note that we avoid folder names with spaces, e.g. we do not recommend folder names such as `C:/Prob Stats`.

Once a dedicated folder, e.g. `C:/ProbStats` has been created, we issue the command

```
setwd("C:/ProbStats/")
```

to set the working directory in R. To print the current working directory we can issue the command `getwd()`. In `Rstudio`, a more convenient way to set the working directory is: by following the menu **Session** → **Set Working Directory**. It then gives us a dialogue box to navigate to the folder we want.

## 2.1.4 Reading Data into R

All the data sets used in the book are included in the accompanying R package `ipsRdbs` and hence the reader can just install that R package to obtain those data sets. The data reading methods discussed in this section are nevertheless included so that the reader can learn to work with their own data sets as well.

The data sets used in this book are available for download from the link.[3] It is recommended that the user first downloads this zip file and extracts the files in the current working directory, say `C:/ProbStats`. The following data reading commands will not work unless the data sets are saved in the current working directory itself.

R allows many different ways to read data. To read just a vector of numbers separated by tab or space we use `scan("filename.txt")` where the data are stored in the file called `"filename.txt"` in the current working directory. To read a tab-delimited text file of data with the first row giving the column headers, the command is: `read.table("filename.txt", head=TRUE)` where again `"filename.txt"` is a file containing data. To read comma-separated files (such as the ones exported by EXCEL), the command is `read.table("filename.csv", head=TRUE, sep=",`

---

[3] `https://www.sujitsahu.com/ipsRdbsdata.zip`.

") or simply **read.csv**("filename.csv", **head**=TRUE). The option **head**=TRUE instructs R that the first row of the data file contains the column headers.

Assuming that the working directory has been set correctly, we may use the following commands to read the data sets:

```
cfail <- scan("cfail.txt") # Reads the computer failure data
ffood <- read.csv("ffood.csv", head=T) # Reads the fast food
    service time data
wgain <- read.table("wtgain.txt", head=T) # Reads the weight
    gain data set
bill <- read.table("billionaires.txt", head=T) # Reads the
    billionaire data set
```

Reading data into R is perhaps the most difficult task for beginner learners of R. If the above commands did not work, the user is able to read the data sets directly from the book web-page by issuing the following commands:

```
path <- "https://www.sujitsahu.com/ipsRdbs/"
cfail <- scan(paste0(path, "cfail.txt"))
ffood <- read.csv(paste0(path, "ffood.csv"), head=T)
wgain <- read.table(paste0(path, "wtgain.txt"), head=T) # Reads
    the weight gain data set
bill <- read.table(paste0(path, "billionaires.txt"), head=T) #
    Reads the billionaire data set
```

These and other data sets discussed in the book can also be found in the R package ipsRdbs.

R does not automatically show the data after reading. To see the data we need to issue a command like: cfail, **head**(ffood); **tail**(bill) etc. after reading in the data. The **head** and **tail** commands print the top and bottom rows of the input data set. The data sets ffood, wgain and bill are data frames, which are like spread sheets. The columns of such a data frame are accessed by using the $ symbol. For example, ffood$AM gives the 10 service times observed during the morning.

## 2.1.5   Summary Statistics from R

There are many commands to extract summary statistics in R. The four main ones that we use in this book are **summary**, **var**, **sd** and **table**. The **summary** command obtains various summaries depending on the input argument. For example, the reader is asked to examine the outputs of the commands:

```
summary(cfail)
summary(ffood)
```

```
summary(wgain)
summary(bill)
```

To calculate the sample variance and the sample standard deviations the commands are **var** and **sd**. For example, **var**(cfail), **var**(ffood$AM), **var**(**c**(ffood $AM, ffoood$PM)) are commands to obtain the sample variances of the inputs. In the last command, **c**(ffood$AM, ffoood$PM) combines the two vector of morning and afternoon service times into a single vector of 20 service times. For example, x <-**c**(1, 5) puts the numbers 1 and 5 in the object x. The standard deviations are obtained by using the **sd** command, e.g. **sd**(cfail).

The **table** command tabulates the frequencies of unique values in the input. For example, **table**(cfail) is a valid command. We can obtain a frequency distribution of region in bill by issuing the command **table**(bill$region).

## 2.2   R Data Types

In this section we aim to learn a bit more of the R language so that we can manipulate and query data sets in R . We also learn to create new columns of data by applying transformation and data manipulation. The reader's task is to understand the commands by examining the output in each case.

The most common data types in R are vectors, matrices and data frames. The first two of these are exactly the same as we learn in linear algebra. (As an aside, all the matrix manipulations e.g. addition, multiplication and inversion, can be done numerically in R .) The third type, data frame, are rectangular arrays where columns can be of different types, e.g. the ffood and bill we saw previously. The main difference between a data frame and a matrix is that the columns of a data frame can contain different types of data, e.g. numbers (weight) and characters (race, sex). A matrix data type will not allow mixing of data types and hence the data frame type is more useful in analysing large practical data sets.

### 2.2.1   Vectors and Matrices

• **Vectors** are ordered strings of data values. A vector can be one of numeric, character, logical or complex types. For example: x <-**c**(1, 4, 7, 10, 13) puts the five numbers in the vector x. We can access parts of x by calling things like:
```
x[1] #gives the first element of x.
x[2:4] #gives the elements x[2], x[3], x[4].
x[-(2:4)] #gives all but x[2], x[3], x[4].
```
There are various commands for creating vectors. For example, y <-5:15 puts the numbers 5, 6, . . . , 15 in the vector y. Hence the : operator generates a simple sequence of successive numbers (with increment 1) between the two endpoints.

Investigate the vectors produced by the following commands, i.e. issue the
commands one by one and then print them by just typing their names and hitting
Run:

```
x <- seq(from=1, to=13, by =3) } # a better way of inputting
   the x above.
?seq # prints out the help file.
a1 <- c(1,3,5,6,8,21) # if we have to input irregular data.
a2 <- seq(5,25, length=5)
a3 <- c(a1,a2)
a4 <- seq(from=min(a1), to=max(a1), length=10)
a5 <- rep(2, 5)
a6 <- c(1, 3, 9)
a7 <- rep(a6, times=2)
a8 <- rep(a6, each=2)
a9 <- rep(a6, c(2, 3, 1))
cbind(a7, a8, a9) # Can we see the differences between a7, a8
   and a9?
```

In addition, we can add, subtract and multiply vectors. For example, examine the
output of `2*a6`, `a7+a8` etc. R performs these operations element-wise.
- **Matrices** are rectangular arrays consisting of rows and columns. All data must be
of the same mode. For example, `y <-matrix(1:6, nrow=3,ncol=2)` creates a
$3 \times 2$ matrix, called `y`. We can access parts of `y` by calling things like:

```
y[1,2] #gives the first row second column entry of y
y[1,] #gives the first row of y
y[,2] #gives the second column of y
```
and so on.
Individual elements of vectors or matrices, or whole rows or columns of matrices
may be updated by assigning them new values, e.g.

```
a1[1] <- 3
y[1,2] <- 3
y[,2] <- c(2,2, 2)
```

We can do arithmetic with the matrices, for example suppose
`x <-2*matrix (1:6, nrow=3,ncol=2)`
Now we can simply write `z <-x+y` to get the sum. However, `x*y` will get us a
new matrix whose elements are the simple products of corresponding elements
of `x` and `y`.

## 2.2.2    Data Frames and Lists

- **Data frames** are rectangular arrays where columns could be of different types. Columns of data frames are vectors and are denoted by `<data frame name>` `$<variable name>`. Data frames are also indexed like matrices, so elements, rows and columns of data, can all be accessed as for matrices above. Create a data frame called `dframe` by issuing the command:

  `dframe <-data.frame(x=1:10, y=rnorm(10))`

  We can add a new column to a data frame, `dframe` say, by issuing:

  `dframe$xy <-dframe$x *dframe$y`

  Note that most operations on vectors are performed component-wise, so for example `dframe$x *dframe$y` results in a vector of the same length as `dframe$x` and `dframe$y`, containing the component-wise products. Similarly, the commands `dframe$x^2 -1, 3*sqrt(0.5*dframe$x)` and `log(dframe$x)/2` all create vectors of the same length as `dframe$x`, with the relevant operation performed component by component.

  However, certain statistical operations on vectors result in scalars, for example the functions **mean, median, var, min, max, sum, prod** *etc.* Try, for example, **mean**`(dframe$x)` and **var**`(dframe$x)`.

- The `View` command lets us see its data frame argument like a spreadsheet. For example, type `View(dframe)`. In `Rstudio` the `View` command is invoked by double clicking the name of the particular object in the 'Environment Window'. We can print the list of all the objects in the current environment by issuing the `ls` `()` command. The command for deleting (removing) objects is **rm**`(name)` where `name` is the object to be removed.

- **Lists** are used to collect objects of different types. For example, a list may consist of two matrices and three vectors of different size and modes. The components of a list have individual names and are accessed using `<list name>$<component name>`, similar to data frames (which are themselves lists, of a particular form). For example,

  `myresults <-list(mean=10, sd=3.32, values=5:15)`

  Now `myresults$mean` will print the value of the member mean in the list `myresults`.

## 2.2.3    Factors and Logical Vectors

- **Factor** There is a data type called `factor` which is normally used to hold a categorical variable, for example the `region` column in `bill` is a factor. Here are some further examples:

  `citizen <-factor(c("uk", "us", "no", "in", "es", "in"))`

  Some functions to use with factors are **levels, table**, etc. For example, type these lines:

```
table(citizen)
levels(citizen)
levels(bill$region)  # Assuming we read the billionaire data
    set already.
levels(bill$region) <- c("Asia", "Europe", "Mid-East", "Other"
    , "USA")
```

- **Logical vectors**
  We can select a set of components of a vector by indicating the relevant components in square brackets. For example, to select the first element of `a1 <-c(1,3,5,6,8,21)` we just type in `a1[1]`. However, we often want to select components, based on their values, or on the values of another vector. For example, how can we select all the values in `a1` which are greater than 5? For the `bill` data set we may be interested in all the rows of `bill` which have wealth greater than 5, or all the rows for region A.
  Typing a *condition* involving a vector returns a logical vector of the same length containing T (true) for those components which satisfy the condition and F (false) otherwise. For example, try:

```
a1[a1>5]
bill$wealth > 5
bill$region == "A"
```

  (note the use of `==` in a logical operation, to distinguish it from the assignment `=`). A logical vector may be used to select a set of components of any other vector. Try

```
bill.wealth.ge5 <- bill[bill$wealth>5, ]
bill.wealth.ge5
bill.region.A <- bill[ bill$region == "A", ]
bill.region.A
```

  Note that the comma in the above two commands instructs R to get all the columns of the data frame `bill`.
  The operations **&** (and) and | (or) operate on pairs of logical vectors. For example if `x <-1:10`, then `x>3 &x<7` returns
  `[1] F F F T T T F F F F`
  and `x<3 | x>7` returns
  `[1] T T F F F F F T T T`
- The functions **any** and **all** take a logical vector as their argument, and return a single logical value. For example, **any**`(x>3 &x<7)` returns T, because at least one component of its argument is T, whereas **all**`(x>3 &x<7)` returns F, because not every component of its argument is T.

- A little exercise. How can we choose subsets of a data frame? For example, how can we pick only the odd numbered rows? Hint: We can use the `seq` or `rep` command learned before. For example, a `<-seq(1, 10, by =2)` and `oddrows <-bill[a, ]`

## 2.3    Plotting in R

The main commands to draw exploratory graphs using R are: `stem,barplot, hist, plot, pie` and `boxplot`. Below we illustrate these commands using the data sets previously introduced in Chap. 1. We assume that the data sets, `cfail`, `food`, and `bill` are present in the current R workspace.

The reader can type in the commands and then press `Run` after each completed line. The comments after the # sign can be ignored—those are used to document code. Many arguments in the below commands are not explained in detail so that the reader can produce the plots speedily without getting bogged down in the details. Those arguments and options can be studied in more detail later.

Various parameters of the plot are controlled by the `par` command. To learn about these type `?par`. Graphs can be saved or copied to Clipboard from the menus on the graphics device.

### 2.3.1    Stem and Leaf Diagrams

A stem and leaf diagram is produced by the command `stem`, e.g.,

```
stem(ffood$AM)
```

### 2.3.2    Bar Plot

A bar plot is obtained by using the `barplot` command. For example,

```
barplot(table(bill$region), col=2:6)
```

provides a coloured bar plot of the frequencies of different regions. The `table` command calculates the frequencies and the colours are chosen by the `col` argument.

### 2.3.3 Histograms

Histograms are produced by using the `hist` command, e.g., `hist(cfail)`. This command can be modified

```
hist(cfail, xlab="Number of weekly computer failures")
```

to change the x-axis label. Here are some other illustrations for the billionaires data example.

```
hist(bill$wealth) # produces a dull looking plot
hist(bill$wealth, nclass=20) # produces a more detailed plot.
hist(bill$wealth, nclass=20, xlab="Wealth", main="Histogram of
    wealth of billionaires") # produces a more informative
    plot.
```

### 2.3.4 Scatter Plot

To obtain a scatter plot of the before and after weights of the students in the weight gain data set, `wgain`, we issue the command

```
plot(wgain$initial, wgain$final)
```

We can add a 45° degree line by the command:

```
abline(0, 1, col="red")
```

A nicer and more informative plot can be obtained by:

```
plot(wgain$initial, wgain$final, xlab="Wt in Week 1", ylab="Wt
    in Week 12", pch="*", las=1)
abline(0, 1, col="red")
title("A scatter plot of the weights in Week 12 against the
    weights in Week 1")
```

Here are some further illustrations of the `plot` command for the billionaires data example.

```
plot(bill$age, bill$wealth) # A very dull plot.
plot(bill$age, bill$wealth, xlab="Age", ylab="Wealth", pch="*") #
    A bit better.
```

The following modifications improve the plot a great deal.

```
plot(bill$age, bill$wealth, xlab="Age", ylab="Wealth", type="n
    ")
# Lays the plot area but does not plot.
text(bill$age, bill$wealth, labels=bill$region, cex=0.7, col
    =2:6)
# Adds the points to the empty plot.
# Provides a better looking plot with more information.
```

### 2.3.5  Boxplots

To draw boxplots use the boxplot command, e.g., `boxplot(cfail)`. The default boxplot shows the median and whiskers drawn to the nearest observation from the first and third quartiles but not beyond the distance 1.5 times the inter-quartile range. Points beyond the two whiskers are suspected outliers and are plotted individually. `boxplot(ffood)` generates two boxplots side-by-side: one for the AM service times and the other for the PM service times.

Side-by-side boxplots for the billionaires example are produced by:

```
boxplot(data=bill, wealth ~region, col=2:6)}  # Side by side
    box plots of wealth by region.
```

The $\sim$ notation, used in the above command, in R has a left hand side and a right hand side. In the left hand side we put the variable which goes in the y-axis and the right hand side may contain the formula terms which go in the x-axis, y $\sim$ x, y $\sim$ x1 + x2. These two are examples of what is called a `formula` in R , which we use in linear regression modelling in Chap. 17.

## 2.4    Appendix: Advanced Materials

### 2.4.1  The Functions `apply` and `tapply`

In this section we learn some more advanced but essential R commands that are often used in statistical data analysis. For example, we may want to find out the mean and variance of the billionaires categorised by region. This will help us answer questions like are US billionaires richer than Asian billionaires? We will also explore a few advanced plotting ideas which allow us to learn much more from the data.

The reader is warned that the materials presented here are advanced and it is not necessary to learn these to read the chapters ahead The materials presented here are a bit difficult in nature but these do not require any knowledge from the chapters presented subsequently. Hence the reader can come back to learning these materials whenever they are ready.

It is often desirable, in data analysis to carry out the same statistical operation separately on different segments of a data frame, matrix or list. The function `apply` allows us to do this when we want to perform the same function on each row or each column of a matrix or data frame. For example,

```
x <- matrix(1:12, byrow=T, ncol=4) # type x to see what matrix we
    have got.
apply(x, 2, mean) # produces four column means of x
apply(x, 1, mean) # produces three row means of x
```

The function `tapply` allows us to carry out a statistical operation on subsets of a given vector, defined according to the values of a specified vector. For example, to calculate the mean value of `wealth` for each region A, E, M, O, U separately we use

```
tapply(X=bill$wealth, INDEX=bill$region, FUN=mean)
tapply(X=bill$wealth, INDEX=bill$region, FUN=sd)
```

We can round the numbers for nicer printing using:

```
round(tapply(X=bill$wealth, INDEX=bill$region, FUN=mean), 2)
```

Read the help files `?tapply`, `?tapply` and `?round`.

### 2.4.2 Graphing Using `ggplot2`

All the above graphics are obtained using the base commands in R and this is sufficient for the purposes of reading this book. These graphics plots can be made a lot nicer by using a more advanced graphics package called `ggplot2`, although these advanced materials can be skipped at the first reading if it is so desired by the reader. Using `ggplot2` we obtain graphics objects which can be saved, re-called and then edited as the user wishes. Below we provide some instructions to get started.

- The package is installed by issuing the command `install.packages("ggplot2")` in R.
- If installation is successful, then we invoke the library by issuing the command `library(ggplot2)` in R.
- Before plotting the billionaires data set we re-name the levels of the region:

```
levels(bill$region) <- c("Asia", "Europe", "Mid-East", "Other
    ", "USA")
```

- Now we obtain a basic ggplot first:

```
g1 <- ggplot(data=bill, aes(x=age, y=wealth)) +
geom_point(aes(col=region, size=wealth))
g1 # Plots the object g1
```

- Observe that the `ggplot` command takes a data frame argument and `aes` is the short form for aesthetics. The arguments of `aes` can be varied depending on the desired plot. Here we just have the `x` and `y`'s required to draw a scatter plot.
- We can add a smooth curve as follows.

```
g2 <- g1 + geom_smooth(method="loess", se=F) # Hit Run
g2 # Hit Run to see what has been added.
```

- We can add some informative title and axis labels.

```
g3 <- g2 +
labs(subtitle="Wealth vs Age of Billionaires", x="Age", y="
    Wealth (Billion US $)", caption = "Source: Fortune
    Magazine, 1992.")
g3
```

The resulting plot is provided in Fig. 2.1. The above code lines are included in the R package `ipsRdbs`. The code can be seen by issuing the commands

```
library(ipsRdbs)
?bill
```

### 2.4.3   Writing Functions: Drawing a Butterfly

This section shows a fun exercise of drawing a nice graph by writing a function in R. Functions in R are a bunch of code lines put together so that they may complete a more complicated task. User written functions, like the built-in functions, may depend on some inputs, e.g. data and parameters and may result in some desired output or graphics. To illustrate, we may type in the following statements inside an empty R script file. We then highlight from below

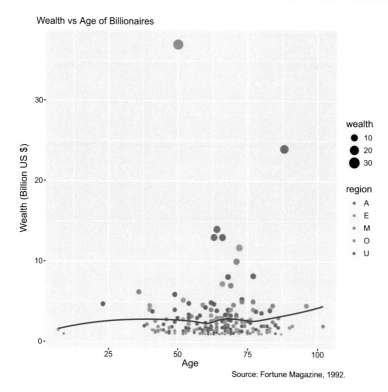

**Fig. 2.1**  Wealth of billionaires in different continents

```
butterfly <- function(color = 2, p1=2, p2=4) {
  theta <- seq(from=0.0, to=24 * pi, len = 2000)
  radius <- exp(cos(theta)) - p1 * cos(p2 * theta)
  radius <- radius + sin(theta/12)
  x <- radius * sin(theta)
  y <- - radius * cos(theta)
  plot(x, y, type = "l", axes = F, xlab = "", ylab = "", col =
      color)
}
```

upto the end curly brace above. We then press the Run button. If there are no error messages, we may issue the following commands to obtain graphics output shown in Fig. 2.2.

**Fig. 2.2** Different shapes using the butterfly programme

```
butterfly(p1=20, p2=4)
butterfly(color = 6)
par(mfrow=c(2, 2))
butterfly(color = 6)
butterfly(p1=5, p2=5, color=2)
butterfly(p1=10, p2=1.5, color = "seagreen")
butterfly(p1=20, p2=4, color = "blue")
```

The above `butterfly` programme is inluded in the R package `ipsRdbs`. The pgragramme help file can be accessed by the command `?butterfly`.

## 2.5    Summary

This chapter shows how to get started with R . It is expected that the reader will install R and Rstudio on their computer. The readers are also asked to make themselves familiar with the data sets discussed above. These data sets will be used throughout this book.

R is intuitive and easy to learn, and there are a wealth of community resources online. Using such resources it is possible to draw beautiful publication quality graphics that we may see in scientific books.

R is not a spreadsheet programme like EXCEL and it is excellent for advanced statistical methods where EXCEL may struggle. There is a R cheatsheet that we can download from the internet for more help with getting started.

## 2.6    Exercises

### 2.1 (R Exercises)

1. Read the computer failure data by issuing the command `cfail <-scan("
   cfail.txt")`. [Before issuing the command, make sure that you have set the
   correct working directory so that R can find this data file.]
   (a) Describe the data using the commands **summary**, **var** and **table**. Also
       obtain an histogram (**hist**) plot and a boxplot (**boxplot**) of the data and
       write a brief comment on the shape of the distribution.
   (b) Use **plot**(cfail) to produce a plot of the failure numbers (y-axis) against
       the order in which the data were collected (x-axis). Do you detect any
       pattern, e.g. an upward or downward trend? **Hint:** You may modify the
       command to **plot**(cfail, type="l") to perhaps get a better looking plot.
       The argument is lower case letter "l" not the number 1.
   (c) The object cfail is a vector of length 104. The first 52 values are for the
       weeks in the first year and the last 52 values are for the second year. Find
       the two annual means and variances.
       **Hint:** To do this you need to create an index corresponding to each of
       the 104 values. The index should take the value 1 for the first 52 values
       (in the first year) and 2 for the remaining values in the second year. Use
       the **rep** (for replicate) command to generate this. Issue year **<-rep**(c
       (1, 2), each=52). This will replicate 1, 52 times followed by 2, 52
       times and put the resulting vector in year. Type **cbind**(year, cfail)
       and see what happens. Now you can calculate the year-wise means by
       issuing the command: **tapply**(X=cfail, INDEX=year, FUN=**mean**). The
       function **tapply** applies the supplied FUN argument to the X argument
       separately for each unique index supplied by the INDEX argument. That
       is, the command returns the mean value for each group of observations
       corresponding to distinct values of the INDEX variable year.
2. Fortune magazine publishes a list of the world's billionaires each year. The
   1992 list includes 225 individuals. Their wealth, age, and geographic location
   ( Asia, Europe, Middle East, United States, and Other) are reported. Variables
   are: wealth: Wealth of family or individual in billions of dollars; age: Age in
   years (for families it is the maximum age of family members); region: Region
   of the World (Asia, Europe, Middle East, United States and Other). The head
   and tail values of the data set are given below.

| Wealth | Age | Region |
|--------|-----|--------|
| 37.0 | 50 | M |
| 24.0 | 88 | U |
| ⋮ | ⋮ | ⋮ |
| 1 | 9 | M |
| 1 | 59 | E |

Read the data by issuing the command:
`bill <-read.table("billionaires.txt", head=T)`.

(a) Obtain the summary of the data set and comment on each of the three columns of data.

(b) Obtain the mean and variance of the wealth for each of the 5 regions of the world and comment.
**Hint:** You can use the `tapply`(X=bill$wealth, INDEX=bill$region, FUN=`mean`) command to do this.

(c) Produce a set of boxplots to demonstrate the difference in distribution of wealth according to different parts of the world and comment on it.
**Hint:** You may issue the command: `boxplot`(wealth∼region, **data** = bill)

(d) Produce a scatter plot of the wealth against age and comment on the relationship between them. Do we see any outlying observations? Do you think a linear relationship between age and wealth will be sensible?
Ensure that all your plots have informative titles and axis labelling—The function `title` and the function arguments xlab and ylab, for example ylab="wealth in billions of US dollars" are helpful for this.

**2.2 (A Data Analysis Exercise)** We will use the age guessing data collected by the students during a lecture given to all the first year mathematics students. In that class students who sat next to each other were formed into 55 groups of sizes 2 and 3. Each group guessed the ages of 10 Southampton mathematicians of different races. The resulting data set contains the error committed by each group of students for each of the 10 photographs. The results for each age guess by each group of students form a row of the data set. The column headings and their descriptions are provided below.

1. group: This is the group number of the group of students who sat together for the age guessing exercise. There were 55 groups in total.
2. size: Number of students in the group.
3. females: Number of female students in the group. Hence the number of males in each group is: size—females. There were no other gender type of students. This can be used to investigate if female students are on average better at guessing ages from photographs.
4. photo: photograph number guessed, can take value 1 to 10 for 10 photographs.

5. sex: Gender of the photographed person.
6. race: Race of the photographed person.
7. est_age: Estimated age of the person in the photograph.
8. tru_age: True age of the person in the photograph.
9. error: Error in age estimation: est_age—tru_age
10. abs_error: absolute value of the error: |est_age—tru_age|

Use the `err_age` data set in the R package `ipsRdbs` to answer the following questions.

1. How many rows and columns are there in the data set?
2. How many students were there in the age guessing exercise on that day? We may use the built-in `sum` command. (Think, it is not 1500!) How many of the students were male and how many were female?
3. Looking at the column tru_age (e.g. by obtaining a frequency table), find the number of photographed mathematicians for each unique value of age. Remember there are only 10 photographed mathematicians!
4. The `table` command can take multiple arguments for cross-tabulation. Use the `table` command to obtain a two-way table providing the distribution of 10 photographed mathematicians in different categories of race and gender.
5. What are the minimum and maximum true ages of the photographed mathematicians?
6. Obtain a barplot of the true age distribution. This is the unknown population distribution of the true ages of photographed mathematicians.
7. Obtain a histogram of the estimated age column and compare this with the true age distribution seen in the barplot drawn above.
8. What is the command for plotting estimated age (on the y-axis and) against true age?
9. What are the means and standard deviations for the columns: size, females, est_age, tru_age, error and abs_error?
10. What is the mean number of males in each group? What is the mean number of females in each group?
11. How many of the photographs were of each race?
12. Note down the frequency table of the sign of the errors. That is, obtain the numbers of negative, zero and positive errors. We may use the built-in `sign` function for this.
13. Obtain a histogram for the errors and another for the absolute errors. Which one is bell shaped and why?
14. Obtain a histogram for the square-root of the absolute errors. Does it look more bell shaped than the histogram of just the absolute errors?
15. Draw a boxplot of the absolute errors and comment on its shape.
16. Is it easier to guess the ages of female mathematicians?
17. Draw a side by side boxplot of the absolute errors for the two groups of mathematicians: males and females.

18. Is it easier to guess the ages of black mathematicians? How would we order the mean absolute error by race?
19. Is it easier to guess the ages of younger mathematicians?
20. Which person's age is the most difficult to guess?

# Part II

# Introduction to Probability

This is Part II of this book where we introduce the ideas of probability from first definitions.

# Introduction to Probability

# 3

**Abstract**

Chapter 3: The basic concepts of probability are introduced in this chapter. Elementary methods of counting, the number of permutations and the number of combinations are introduced and illustrated. Elementary methods for calculating probabilities are discussed and the general urn problem in probability is defined.

Probabilities are often used to express the uncertainty of events of interest happening. For example, we may say that: (i) it is highly likely that last year's winning club will retain the English premiership football league title this season. One can also put a numerical value quantifying this 'highly likely' chance, e.g. 80%. By way of another example, we may say that the probability of a tossed fair coin landing heads is 0.5. So, it is clear that probabilities mean different things to different people. As we have commented previously in Chap. 1, there is uncertainty everywhere. Hence, probabilities are used as tools to quantify the associated uncertainty. The theory of statistics has its basis in the mathematical theory of probability. A statistician must be fully aware of what probability means to him/her and what it means to other people. In this chapter we will learn the basic definitions of probability and how to find them.

## 3.1 Two Types of Probabilities: Subjective and Objective

The two examples above, premiership title winner and tossing of a coin, convey two different interpretations of probability. The premiership winning probability is the commentator's own subjective belief, isn't it? The commentator certainly has not performed a large experiment involving all the 20 teams over the whole (future) season under all playing conditions, players, managers and transfers. This notion is known as subjective probability. *Subjective probability* gives a measure of the plausibility of the proposition, to the person making it, in the light of past experience (e.g. the last year's winners are the current champions) and other evidence (e.g. they may have spent a very large amount of money buying players for the new season). There are plenty of other examples, e.g. I think there is a 70% chance that the FTSE 100 stock market index will rise tomorrow, or according to the Met Office there is a 40% chance that we will have a white Christmas this year in London. Subjective probabilities are increasingly being used cleverly in a statistical framework called Bayesian inference. Such methods allow one to combine expert opinion and evidence from data to make the best possible inferences and prediction. We will return to introducing the Bayes Theorem in Sect. 4.5 and discussing Bayesian inference in Sect. 10.3.

The second definition of probability comes from the long-term relative frequency of a result of a random experiment (e.g. coin tossing) which can be repeated an infinite number of times under *essentially similar conditions*. First we give some essential definitions.

**Random Experiments**  An experiment is said to be random if we cannot predict its exact outcome in advance, e.g. tossing of a fair coin, even though we can write down all the possible outcomes which together are called the **sample space**, which we denote by $S$. For example, in a coin tossing experiment, $S = \{H, T\}$ where $H$ and $T$ denote respectively the outcome head and tail. If we toss two coins together, $S = \{HH, HT, TH, TT\}$.

**Event**  An event is defined as a particular result of the random experiment. For example, HH (two heads) is an event when we toss two coins together. Similarly, at least one head e.g. {HH, HT, TH} is an event as well. Events are denoted by capital letters $A, B, C, \ldots$ or $A_1, B_1, A_2$ etc., and a single outcome is called an *elementary event*, e.g. HH. An event which is a group of elementary events is called a *composite event*, e.g. at least one head. How to determine the probability of a given event $A$, denoted by $P(A)$, is the focus of probability theory.

**Probability as Relative Frequency**  Imagine we are able to repeat a random experiment under identical conditions and count how many of those repetitions result in the event $A$. The relative frequency of $A$, i.e. the ratio

$$\frac{\text{the number of repetitions resulting in } A}{\text{total number of repetitions}},$$

approaches a fixed limit value as the number of repetitions increases. This limit value is defined as $P(A)$.

As a simple example, in the experiment of tossing a particular coin, suppose we are interested in the event $A$ of getting a 'head'. We can toss the coin 1000 times (i.e. do 1000 replications of the experiment) and record the number of heads out of the 1000 replications. Then the relative frequency of $A$ out of the 1000 replications is the proportion of heads observed.

Sometimes, however, it is much easier to find $P(A)$ by using some 'common knowledge' about probability. For example, if the coin in the example above is fair (i.e. $P(H) = P(T)$), then this information and the common knowledge that $P(H) + P(T) = 1$ immediately implies that $P(H) = 0.5$ and $P(T) = 0.5$. Next, the essential 'common knowledge' about probability will be formalised as the *axioms of probability* in Sect. 3.3, which form the foundation of probability theory. But before that, we need to learn a bit more about the event space, the collection of all possible events in the following section.

## 3.2   Union, Intersection, Mutually Exclusive and Complementary Events

In order to proceed further we draw the following parallels between set theory and probability theory. The sample space $S$ in probability is called the whole set in set theory and it is composed of all possible elementary events, i.e. outcomes from a single experiment.

|     | Set theory       | Probability theory |
|-----|------------------|--------------------|
| (1) | Space            | Sample space       |
| (2) | Element or point | Elementary event   |
| (3) | Set              | Event              |

---

**Example 3.1 (Dice Throw)**

Roll a six-faced dice and observe the score on the uppermost face. Here $S = (1, 2, 3, 4, 5, 6)$, which is composed of six elementary events. ◄

The *union of two given events A and B*, denoted as $(A$ or $B)$ or $A \cup B$, consists of the outcomes that are either in $A$ or $B$ or both. 'Event $A \cup B$ occurs' means 'either $A$ or B occurs or both occur'. Figure 3.1 shows $A \cup B$ as the total area covered by the two circles $A$ and $B$, which are thought to be as two events in the sample space denoted by the enclosing rectangle.

For example, in Example 3.1, suppose $A$ is the event that *an even number is observed*. This event consists of the set of outcomes 2, 4 and 6, i.e. $A = \{x : x$ is even$\} = \{2, 4, 6\}$. Suppose $B$ is the event that *a number larger than 3 is observed*. This event consists of the outcomes 4, 5 and 6, i.e. $B = \{x : x$ is greater than3$\} = \{4, 5, 6\}$. Hence the event

$$A \cup B = \{x : x \text{ is an even number or a number larger than } 3\} = \{2, 4, 5, 6\}.$$

**Fig. 3.1** The union and
intersection of two events $A$
and $B$

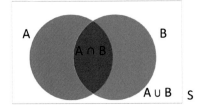

**Fig. 3.2** Two mutually
exclusive events

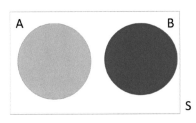

Clearly, when a 6 is observed, both $A$ and $B$ have occurred. The *intersection of two given events A and B*, denoted as $(A$ and $B)$ or $A \cap B$, consists of the outcomes that are common to both $A$ and $B$. 'Event $A \cap B$ occurs' means 'both A and B occur'. In the left panel of Fig. 3.1, $A \cap B$ is seen to be the zone covered by *both* the circles $A$ and $B$.

For example, in Example 3.1, $A \cap B = \{4, 6\}$. Additionally, if

$$C = \{x : x \text{ is a number less than } 6\} = \{1, 2, 3, 4, 5\},$$

the intersection of events $A$ and $C$ is the event

$$A \cap C = \{x : x \text{ is an even number less than } 6\} = \{2, 4\}.$$

The union and intersection of two events can be generalised in an obvious way to the union and intersection of more than two events.

Two events $A$ and $D$ are said to be *mutually exclusive* (also termed as *disjoint*) if $A \cap D = \emptyset$, where $\emptyset$ denotes the empty set or the null set, i.e. $A$ and $D$ have no outcomes in common. Intuitively, '$A$ and $D$ are mutually exclusive' means '$A$ and $D$ cannot occur simultaneously in the experiment'. Figure 3.2 shows two mutually exclusive events $A$ and $B$ as there is no common area covered by both the circles $A$ and $B$.

In Example 3.1, if $D = \{x : x \text{ is an odd number}\} = \{1, 3, 5\}$, then $A \cap D = \emptyset$ and so $A$ and $D$ are mutually exclusive. As expected, $A$ and $D$ cannot occur simultaneously in the experiment.

For a given event $A$, *the complement of* $A$ is the event that consists of all the outcomes not in $A$ and is denoted by $A'$. Note that $A \cup A' = S$ and $A \cap A' = \emptyset$.

## 3.3 Axioms of Probability

The long term limiting relative frequency definition and interpretation of probability in Sect. 3.1 cannot be used universally because of several reasons, e.g. the limit may not exist, the repetitions may not be feasible etc. Hence early pioneers, such as Russian mathematician A. N. Kolmogorov, defined probability using the following three axioms which are assumed to be true universally. Moreover, this axiomatic approach does not cause any problem in interpretation of probability. The three axioms defining probability are:

**A1:** $P(S) = 1$ where $S$ is the sample space.
**A2:** $0 \leq P(A) \leq 1$ for any event $A$,
**A3:** $P(A \cup B) = P(A) + P(B)$ provided that $A$ and $B$ are mutually exclusive events.

Thus any measure or definition of probability of any event $A$ in a sample space $S$ must obey the above three axioms. The axioms **A1** says that the probability of the whole sample space $S$ is 1, the maximum possible value of probability according to axiom **A2**. Axiom **A2** also guarantees that probabilities are always non-negative. Axiom **A3** defines the probability of union of mutually exclusive (or disjoint) events as the sum of the probabilities of the constituent events. The axiom states this rule for two events which can be generalised for any number of events using the method of induction as noted below. Here are some of the consequences of the axioms of probability:

(C1) For any event $A$, $P(A') = 1 - P(A)$, since $A \cup A' = S$. Thus, the probability of the complimentary event is one minus the probability of the given event.
(C2) From the (C1) and Axiom **A1**, $P(\emptyset) = 1 - P(S) = 0$. Hence if $A$ and $B$ are mutually exclusive events, i.e. $A \cap B = \emptyset$, then $P(A \cap B) = 0$.
(C3) If $D$ is a subset of $E$, $D \subset E$, then $E = (E \cap D') \cup D$ see left panel of Fig. 3.3 where the two events $E \cap D'$ and $D$ are mutually exclusive. Hence $P(E) = P(E \cap D') + P(D)$ by axiom **A3** and thus:

$$P(E \cap D') = P(E) - P(D)$$

when $D$ is a subset of $E$.
(C4) By looking at the right panel of Fig. 3.3, we can write

$$A = (A \cap B) \cup (A \cap B')$$

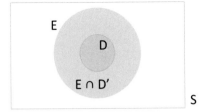 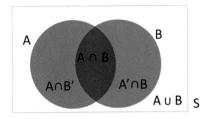

**Fig. 3.3** The left panel shows $D$ is a subset of $E$. The right panel shows the events $A \cap B'$, $A \cap B$, and $A' \cap B$ that may define the union $A \cup B$

for any two arbitrary events $A$ and $B$. By noting that the two events $(A \cap B)$ and $(A \cap B')$ are mutually exclusive, we have

$$P(A) = P(A \cap B) + P(A \cap B'),$$

which in turn implies for arbitrary events $A$ and $B$,

$$P(A \cap B') = P(A) - P(A \cap B). \tag{3.1}$$

(C5) It can be shown by mathematical induction that Axiom **A3** holds for more than two mutually exclusive events:

$$P(A_1 \cup A_2 \cup \cdots \cup A_k) = P(A_1) + P(A_2) + \ldots + P(A_k)$$

provided that $A_1, \ldots, A_k$ are mutually exclusive events.

Hence, the probability of an event A is the sum of the probabilities of the individual outcomes that make up the event.

(C6) For the union of two arbitrary events, we have the *General addition rule*: For any two events $A$ and $B$,

$$P(A \cup B) = P(A) + P(B) - P(A \cap B).$$

***Proof*** We can write $A \cup B = (A \cap B') \cup (A \cap B) \cup (A' \cap B)$, see the right panel of Fig. 3.3. All three of these are mutually exclusive events. From (3.1) we have,

$$P(A \cap B') = P(A) - P(A \cap B), P(A' \cap B) = P(B) - P(A \cap B).$$

Hence, using Axiom (A3),

$$\begin{aligned}
P(A \cup B) &= P(A \cap B') + P(A \cap B) + P(A' \cap B) \\
&= P(A) - P(A \cap B) + P(A \cap B) + P(B) - P(A \cap B) \\
&= P(A) + P(B) - P(A \cap B).
\end{aligned}$$

**Example 3.2 (Application to an Experiment with Equally Likely Outcomes)**

For an experiment with $N$ equally likely possible outcomes, the axioms (and the consequences above) can be used to find $P(A)$ of any event $A$ in the following way. From consequence (C5), we assign probability $1/N$ to each outcome since there are $N$ equally likely and exhaustive events $A_1, \ldots, A_N$ in the sample space $S$. The events $A_1, \ldots, A_N$ are said to be exhaustive if

$$A_1 \cup \ldots \cup A_N = S.$$

For any event $A$, we find $P(A)$ by adding up $1/N$ for each of the outcomes in event $A$:

$$P(A) = \frac{\text{number of outcomes in } A}{\text{total number of possible outcomes of the experiment}}.$$

Return to Example 3.1 where a six-faced dice is rolled. Suppose that one wins a bet if a 6 is rolled. Then the probability of winning the bet is $1/6$ as there are six possible outcomes in the sample space and exactly one of those, 6, wins the bet. Suppose $A$ denotes the event that an even-numbered face is rolled. Then $P(A) = 3/6 = 1/2$ as we can expect. ◄

**Example 3.3 (Dice Throw)**

Roll 2 distinguishable dice and observe the scores. Here

$$S = \{(1, 1), (1, 2), (1, 3), \ldots, (6, 3), (6, 4), (6, 5), (6, 6)\}$$

which consists of 36 possible outcomes or elementary events, $A_1, \ldots, A_{36}$. What is the probability of the outcome 6 in both the dice? The required probability is $1/36$. By complete enumeration of the sample space and assigning probability

1/36 to each outcome, it is possible to answer questions like, "What is the probability that the sum of the two dice is greater than 6?" ◄

## 3.4    Exercises

### 3.1 (Basic Probability)

1. I select two cards from a pack of 52 cards and observe the colour of each. Which of the following is an appropriate sample space S for the possible outcomes?
   (a)  S = (red, black)
   (b)  S = ((red, red), (red, black), (black, red), (black, black))
   (c)  S = (0, 1, 2)
   (d)  None of the above
2. In a particular game, a fair dice is rolled. If the number of spots showing is either 4 or 5 you win $1, if the number of spots showing is 6 you win $4, and if the number of spots showing is 1, 2 or 3 you win nothing. If it costs you $1 to play the game, find the probability that you win more than the cost of playing.
3. A fair dice is thrown. What is the probability that it shows a number which is even or greater than 4?
4. The probability that a man reads The Sun is 0.6, while the probability that he reads both The Sun and The Guardian is 0.1 and the probability that he reads neither paper is 0.2. What is the probability that he reads The Guardian? Draw a Venn diagram to illustrate your answer.
5. In a population of A-level students, a proportion 0.25 is left handed and a proportion 0.45 studies mathematics. However, half of the students are not left handed and do not study mathematics. Find the proportion that is left-handed and studies mathematics.
6. In a certain Scandinavian city 50% of females have blue eyes, 55% have blonde hair but 25% have neither characteristic. Find the probability that a randomly selected female will be a blue eyed blonde.

## 3.5    Using Combinatorics to Find Probability

In this section we find probabilities using specialist counting techniques called permutation and combination. This will allow us to find probabilities in a number of practical situations.

For example, suppose there are 4 boys and 6 girls available for a committee membership, but there are only 3 posts. How many possible committees can be formed? How many of those will be girls only? The UK National Lottery selects 6 numbers at random from 1 to 49. I buy one ticket—what is the probability that I will win the jackpot?

**Fig. 3.4**  A schematic
diagram of 6 possible routes

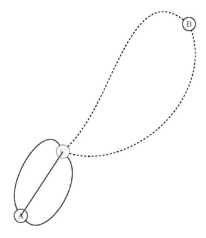

## 3.5.1  Multiplication Rule of Counting

Suppose that to complete a specific task, one has to complete $k(\geq 1)$ sub-tasks
sequentially. If there are $n_i$ different ways to complete the $i$-th sub-task ($i =
1, \ldots, k$) then there are $n_1 \times n_2 \times \ldots \times n_k$ different ways to complete the task.

### Example 3.4 (Counting)

Suppose there are 3 routes to London (marked C in Fig. 3.4) from Southampton
(marked A in Fig. 3.4) and then there are 2 routes to Cambridge (marked B
in Fig. 3.4) out of London. How many ways can I travel to Cambridge from
Southampton via London. The answer is obviously 6, see Fig. 3.4. ◀

## 3.5.2  The Number of Permutations of $k$ from $n$: $^{n}P_k$

The task is to select $k(\geq 1)$ from the $n$ ($n \geq k$) available people and sit the $k$ selected
people in $k$ (different) chairs. By considering the $i$-th sub-task as selecting a person
to sit in the $i$-th chair ($i = 1, \ldots, k$), it follows directly from the multiplication
rule above that there are $n(n-1) \cdots (n-[k-1])$ ways to complete the task. The
number $n(n-1) \cdots (n-[k-1])$ is called the number of permutations of $k$ from $n$
and denoted by

$$^{n}P_k = n(n-1) \cdots (n-[k-1]).$$

In particular, when $k = n$ we have $^nP_n = n(n-1)\cdots 1$, which is called '$n$ factorial' and denoted as $n!$. Note that $0!$ is defined to be 1. It is clear that

$$^nP_k = n(n-1)\cdots(n-[k-1]) = \frac{n(n-1)\cdots(n-[k-1]) \times (n-k)!}{(n-k)!}$$

$$= \frac{n!}{(n-k)!}.$$

---

**Example 3.5 (Football)**

How many possible rankings are there for the 20 football teams in the premier league at the end of a season? This number is given by $^{20}P_{20} = 20!$, which is a huge number! How many possible permutations are there for the top 4 positions who will qualify to play in Europe in the next season? This number is given by $^{20}P_4 = 20 \times 19 \times 18 \times 17$. ◄

---

### 3.5.3 The Number of Combinations of $k$ from $n$: $^nC_k$ or $\binom{n}{k}$

The task is to select $k(\geq 1)$ from the $n$ $(n \geq k)$ available people. Note that this task does NOT involve sitting the $k$ selected people in $k$ (different) chairs. We want to find the number of possible ways to complete this task, which is denoted as $^nC_k$ or $\binom{n}{k}$.

For this, let us reconsider the task of "selecting $k(\geq 1)$ from the $n$ $(n \geq k)$ available people and sitting the $k$ selected people in $k$ (different) chairs", which we already know from the discussion above has $^nP_k$ ways to complete.

Alternatively, to complete this task, one has to complete two sub-tasks sequentially. The first sub-task is to select $k(\geq 1)$ from the $n$ $(n \geq k)$ available people, which has $^nC_k$ ways. The second sub-task is to sit the $k$ selected people in $k$ (different) chairs, which has $k!$ ways. It follows directly from the multiplication rule that there are $^nC_k \times k!$ to complete the task. Hence we have

$$^nP_k = {}^nC_k \times k!, \quad i.e., \quad {}^nC_k = \frac{^nP_k}{k!} = \frac{n!}{(n-k)!k!}$$

---

**Example 3.6 (Football)**

How many possible ways are there to choose 3 teams for the bottom positions of the premier league table at the end of a season? This number is given by $^{20}C_3 = 20 \times 19 \times 18/3! = 1140$, which does not take into consideration the rankings of the three bottom teams! Similarly, there are $^{20}C_4$ ways of choosing 4 teams at the top who may play in the European championship in the next season. ◄

**Example 3.7 (Microchip)**

A box contains 12 microchips of which 4 are faulty. A sample of size 3 is drawn from the box *without replacement*.

- How many selections of 3 can be made?  Answer: $^{12}C_3$.
- How many samples have all 3 chips faulty?  Answer: $^4C_3$.
- How many selections have exactly 2 faulty chips?  Answer: $^8C_1\,^4C_2$. Note that in order to have exactly two faulty chips we must have exactly one non-faulty chip.
- How many samples have 2 or more faulty chips?  Answer: $^8C_1\,^4C_2 + {}^4C_3$. There can be exactly two or exactly 3 faulty chips.

◄

### 3.5.3.1 Calculating $^nC_k$

For non-negative integers $n$ and $k$ such that $k \le n$, the number of combinations

$$^nC_k \equiv \binom{n}{k} = {}^n C_{n-k} = \frac{n!}{k!\,(n-k)!}$$

since both $^nC_k$ and $^nC_{n-k}$ are equal to $\frac{n!}{k!\,(n-k)!}$. This is also easily argued by the fact that each time one chooses $k$ items from $n$ items, one leaves behind $n-k$ items. Hence the number of ways choosing $k$ items from $n$ items must be same as the number of ways choosing $n-k$ items from $n$ items. We now prove that:

$$^nC_k \equiv \binom{n}{k} = \frac{n \times (n-1) \times \cdots \times (n-k+1)}{1 \times 2 \times 3 \times \cdots \times k}.$$

**Proof** We have:

$$\begin{aligned}
\binom{n}{k} &= \tfrac{n!}{k!\,(n-k)!} \\
&= \tfrac{1}{k!} \tfrac{1 \times 2 \times 3 \times \cdots \times (n-k) \times (n-k+1) \times \cdots \times (n-1) \times n}{1 \times 2 \times 3 \times \cdots \times (n-k)} \\
&= \tfrac{1}{k!} \left[ (n-k+1) \times \cdots \times (n-1) \times n \right].
\end{aligned}$$

Hence the proof is complete.  □

This enables us to calculate $\binom{6}{2} = \frac{6 \times 5}{1 \times 2} = 15$. In general, $\binom{n}{k}$ is the ratio where:

the numerator is the multiplication of $k$ terms starting with $n$ and counting down, and the denominator is the multiplication of the first $k$ positive integers.

### 3.5.4   Calculation of Probabilities of Events Under Sampling 'at Random'

For the experiment of 'selecting a sample of size $n$ from a box of $N$ items without replacement', a sample is said to be selected *at random* if all the possible samples of size $n$ are equally likely to be selected. All the possible samples are then equally likely outcomes of the experiment and so assigned equal probabilities.

---

**Example 3.8 (Microchip Continued)**

In Example 3.7 assume that 3 microchips are selected at random without replacement. Then each outcome (a sample of size 3) has equal probability $1/^{12}C_3$. Now:

- $P$(all 3 selected microchips are faulty) $= {}^4C_3/^{12}C_3$.
- $P$(2 chips are faulty) $= {}^8C_1\,{}^4C_2/^{12}C_3$.
- $P$(2 or more chips are faulty) $= ({}^8C_1\,{}^4C_2 + {}^4C_3)/^{12}C_3$.

◄

### 3.5.5   A General 'urn Problem'

Example 3.7 is one particular case of the following general urn problem (see a schematic diagram in Fig. 3.5) which can be solved by the same technique.

A sample of size $n$ is drawn at random without replacement from a box of $N$ items containing a proportion $p$ of defective items. Hence there are $Np$ defective items are in the box and we assume this to be an integer. Now there are $N(1 - p)$ non-defective items in the box. For example, if $N = 100$ and $p = 0.25$, then $Np = 25$ and $N(1 - p) = 75$.

Note that we can easily remove the notation $p$ and instead use $M$ as the number of defective items, and so that $N - M$ is the number of non-defective items. But it is convenient to use the proportion $p$ of defective items so that we can easily distinguish between the binomial and hypergeometric distributions in Sects. 6.2 and 6.4.

Let $x$ denote a non-negative integer. The probability of exactly $x$ number of defectives in the sample of $n$ is

$$\frac{{}^{Np}C_x\ {}^{N(1-p)}C_{n-x}}{{}^{N}C_n}.$$

Note that for this to be meaningful we need to have $0 \le x \le n$, $0 \le x \le Np$ and $0 \le x \le N(1 - p)$. These values of $x$ and the corresponding probabilities make up what is called the *hyper-geometric distribution*, which will be discussed later in Sect. 6.4.

**Fig. 3.5**  A schematic of the urn problem, where $n$ items are chosen out of $N$ items of which $Np$ are defective and $N(1-p)$ are non-defective

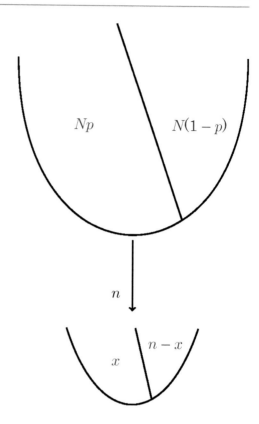

### Example 3.9 (Selecting a Committee)

There are 10 students available for a committee of which 4 are boys and 6 are girls. A random sample of 3 students are chosen to form the committee—what is the probability that exactly one is a boy?

The total number of possible outcomes of the experiment is equal to the number of ways of selecting 3 students from 10 and given by $^{10}C_3$. The number of outcomes in the event 'exactly one is a boy' is equal to the number of ways of selecting 3 students from 10 with exactly one boy, and given by $^{4}C_1\,^{6}C_2$. Hence

$$P(\text{exactly one boy}) = \frac{\text{number of ways of selecting one boy and two girls}}{\text{number of ways of selecting 3 students}}$$

$$= \frac{^{4}C_1\,^{6}C_2}{^{10}C_3} = \frac{4 \times 15}{120} = \frac{1}{2}.$$

Similarly,

$$P(\text{exactly two boys}) = \frac{^4C_2\,^6C_1}{^{10}C_3} = \frac{6 \times 6}{120} = \frac{3}{10}.$$

◀

---

**Example 3.10 (The UK National Lottery)**

In Lotto, a winning ticket has six numbers from 1 to 49 matching those on the balls drawn on a Wednesday or Saturday evening. The 'experiment' consists of drawing the balls from a box containing 49 balls. The 'randomness', the equal chance of any set of six numbers being drawn, is ensured by the spinning machine, which rotates the balls during the selection process. What is the probability of winning the jackpot?

Total number of possible selections of six balls/numbers is given by $^{49}C_6$. There is only 1 selection for winning the jackpot. Hence

$$P(\text{jackpot}) = \frac{1}{^{49}C_6} = 7.15 \times 10^{-8}.$$

which is roughly 1 in 13.98 million.

Other prizes are given for fewer matches. The corresponding probabilities are:

$$P(5 \text{ matches}) = \frac{^6C_5\,^{43}C_1}{^{49}C_6} = 1.84 \times 10^{-5}.$$

$$P(4 \text{ matches}) = \frac{^6C_4\,^{43}C_2}{^{49}C_6} = 0.0009686197$$

$$P(3 \text{ matches}) = \frac{^6C_3\,^{43}C_3}{^{49}C_6} = 0.0176504$$

There is one other way of winning by using the bonus ball—matching 5 of the selected 6 balls plus matching the bonus ball. The probability of this is given by

$$P(5 \text{ matches } + \text{ bonus}) = \frac{6}{^{49}C_6} = 4.29 \times 10^{-7}.$$

Adding all these probabilities of winning some kind of prize (i.e. at least one match) together gives

$$P(\text{winning}) \approx 0.0186 \approx 1/53.7.$$

So a player buying one ticket each week would expect to win a prize, (most likely a £10 prize for matching three numbers) about once a year ◀

## 3.5.6 Section Summary

We have learned the multiplication rule of counting and the number of permutations and combinations. We have applied the rules to find probabilities of interesting events, e.g. the jackpot in the UK National Lottery.

## 3.6 Exercises

**3.2 (Using Permutation and Combination)**

1. Calculate:

$$^6C_2, \quad ^5C_4, \quad ^nC_0, \quad ^nC_n, \quad ^{20}C_{18}.$$

2. Find how many arrangements there are of the letters in each of these words:

(i) PASS (ii) STATISTICS (iii) EXAM

3. A builder requires the services of both a plumber and an electrician. If there are 12 plumbers and 9 electricians available in the area, find the number of choices of the pair of contractors.

4. A rental car company has 10 Japanese and 15 European cars waiting to be serviced on a particular Saturday morning. Because there are so few mechanics available, only 6 cars can be serviced. Calculate
   (a) the number of outcomes in the sample space (i.e. the number of possible selections of 6 cars).
   (b) the number of outcomes in the event "3 of the selected cars are Japanese and the other 3 are European".
   (c) the probability (to 3 decimal places) that the event in (b) occurs, when the 6 cars are chosen at random.

5. Shortly after being put into service, some buses manufactured by a certain company have developed cracks on the underside of the main frame. Suppose a particular city has 25 of these buses, and cracks have actually appeared in 8 of them.
   (a) How many ways are there to select a sample of 5 buses from the 25 for a thorough inspection?
   (b) How many of these samples of 5 buses contain exactly 4 with visible cracks?
   (c) If a sample of 5 buses is chosen at random, what is the probability that exactly 4 of the 5 will have visible cracks (to 3 dp)?
   (c) If buses are selected as in part (c), what is the probability that at least 4 of those selected will have visible cracks (to 3 dp)?

6. A hand of 7 cards is dealt from a pack of 52. Find the probability that the pack contains:

(a) 4 spades ♠ and 3 hearts ♡.
(b) 3 spades ♠ and 1 heart ♡.
(c) 2 Aces.
(d) exactly one Ace and one King of the same suit.

7. A football team is to be formed of 4 defenders, 4 mid-fielders, 2 strikers and 1 goal-keeper. The team manager can choose from a group of 22 players consisting of 8 defenders, 8 mid-fielders, 4 strikers and 2 goal-keepers. How many different teams can be selected?

8. A cricket team consisting of 6 batsmen, 4 bowlers and 1 wicket-keeper is to be selected from a group of 18 cricketers comprising 9 batsmen, 7 bowlers and 2 wicket-keepers. How many different teams can be selected?

9. Simplify: (i) $\frac{(n+1)!}{(n-1)!}$  (ii) $\frac{(n-2)!}{n!}$.

10. Factorise:  $n! + (n+1)!$.

11. Prove the following by expanding the factorials:
$$^{(n+1)}C_k = {}^nC_k + {}^nC_{k-1}.$$
State the meaning of this identity so that even a layman (like some non-mathematically qualified person in your family!) can understand.

12. A bag contains 8 items, 3 red and 5 blue. Three items are randomly selected from the bag without replacement. Find the probability that the second selected item is red.

13. A committee of four is to be selected from five boys and four girls. The members are selected at random. What is the probability that the committee will be made of more boys than girls?

# Conditional Probability and Independence

**4**

**Abstract**

Chapter 4: This chapter introduces many advanced laws of probability such as the total probability theorem, conditional probability and the Bayes theorem. The famous Monty Python problem is discussed and illustrated using a simulation tool in R. The concept of independence is discussed and illustrated with many examples such system reliability and randomised response methods.

## 4.1 Definition of Conditional Probability

This chapter is all about using additional information, i.e. things that have already happened, in the calculation of probabilities. For example, a person may have a certain disease, e.g. diabetes or HIV/AIDS, whether or not they show any symptoms of it. Suppose a randomly selected person is found to have the symptom. Given this additional information, what is the probability that they have the disease? Note that having the symptom does not fully guarantee that the person has the disease.

Applications of conditional probability occur naturally in all sciences, e.g., actuarial science and medical studies, where conditional probabilities such as "what is the probability that a person will survive for another 20 years given that they are still alive at the age of 40?" are calculated.

In many real life problems, one has to determine the probability of an event $A$ when one already has some partial knowledge of the outcome of an experiment, i.e. another event $B$ has already occurred. For this, one needs to find the conditional probability.

**Example 4.1 (Dice Throw Continued)**

Return to the rolling of a fair dice (Example 3.1). Let

$$A = \{\text{a number greater than } 3\} = \{4, 5, 6\},$$

$$B = \{\text{an even number}\} = \{2, 4, 6\}.$$

It is clear that $P(B) = 3/6 = 1/2$. This is the unconditional probability of the event $B$. It is sometimes called the *prior* probability of $B$ without any knowledge about occurrence of $A$. However, suppose that we are told that the event $A$ has already occurred. What is the probability of $B$ now given that $A$ has already happened?

The sample space of the experiment is $S = \{1, 2, 3, 4, 5, 6\}$, which contains $n = 6$ equally likely outcomes, see the left panel of Fig. 4.1. Given the partial knowledge that event $A$ has occurred, only the $n_A = 3$ outcomes in $A = \{4, 5, 6\}$ could have occurred. Thus the sample space $S$ containing six elementary events shrinks to $A$, which contains only three elementary events, see the right panel of Fig. 4.1. Now only some of the outcomes in $B$ among these $n_A$ outcomes in $A$ will make event $B$ occur; the number of such outcomes is given by the number of outcomes $n_{A \cap B}$ in both $A$ and $B$, i.e., $A \cap B$, and equal to 2. Hence the probability of $B$, given the partial knowledge that event $A$ has occurred, is equal to

$$\frac{2}{3} = \frac{n_{A \cap B}}{n_A} = \frac{n_{A \cap B}/n}{n_A/n} = \frac{P(A \cap B)}{P(A)}.$$

Hence we say that $P(B|A) = 2/3$, which is often interpreted as the *posterior* probability of $B$ given $A$. The additional knowledge that $A$ has already occurred

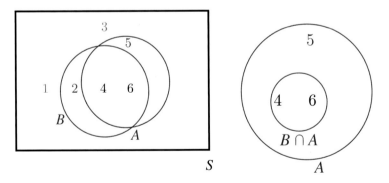

**Fig. 4.1** The left figure shows the two events $A$ and $B$ in the sample space $S$. The right panel shows the event $A$, which contains three possibilities, two of which are also in the event $B$. To calculate $P(B|A)$, we consider the right panel, where the three events in $A$ $\{4, 5, 6\}$ constitute the whole sample space since we have to assume that $A$ has alreday occurred—the other events 1, 2 and 3 are no longer possible

has helped us to revise the prior probability of $1/2$ to $2/3$. This simple example leads to the following general definition of conditional probability. ◄

For events $A$ and $B$ with $P(A) > 0$, the conditional probability of event $B$, given that event $A$ has occurred, is

$$P(B|A) = \frac{P(B \cap A)}{P(A)}. \tag{4.1}$$

**Example 4.2**

Of all individuals buying a mobile phone, 60% include a 64GB hard disk in their purchase, 40% include a 16 MP camera and 30% include both. If a randomly selected purchase includes a 16 MP camera, what is the probability that a 64GB hard disk is also included?
The conditional probability is given by:

$$P(64\text{GB}|16\text{ MP}) = \frac{P(64\text{GB} \cap 16\text{ MP})}{P(16\text{ MP})} = \frac{0.3}{0.4} = 0.75.$$

◄

## 4.2   Exercises

### 4.1 (Conditional Probability)

1. Event A occurs with probability 0.8. The conditional probability that event B occurs given that A occurs is 0.5. Find the probability that both A and B occur, i.e. find $P(A \cap B)$.
2. An event A occurs with probability 0.5. An event B occurs with probability 0.6. The probability that both A and B occurs is 0.1. Find the conditional probability of A given B.
3. Suppose $P(B|A') = \frac{1}{2}$, $P(A|B') = \frac{3}{5}$, $P(A \cap B) = \frac{1}{8}$. Find $P(A)$ and $P(B)$.
4. Jack is buying two goldfish from a pet shop. The shop's tank contains 7 male and 8 female fish but they all look the same. Find the probability that Jack's fish are both female given they are the same sex.
5. Two events A and B are such that the probability of B occurring given that A has occurred is 4 times the probability of A, and the probability that A occurs given that B has occurred is 9 times the probability that B occurs. If the probability that at least one of the events occurs is 7/48, find the probability of event A occurring.

## 4.3    Multiplication Rule of Conditional Probability

By rearranging the conditional probability definition, we obtain the multiplication rule of conditional probability as follows:

$$P(A \cap B) = P(A)P(B|A).$$

Clearly the roles of $A$ and $B$ could be interchanged leading to:

$$P(A \cap B) = P(B)P(A|B).$$

Hence the multiplication rule of conditional probability for two events is:

$$P(A \cap B) = P(B)P(A|B) = P(A)P(B|A).$$

It is straightforward to show by mathematical induction the following multiplication rule of conditional probability for $k(\geq 2)$ events $A_1, A_2, \ldots, A_k$:

$$P(A_1 \cap A_2 \cap \ldots \cap A_k) = P(A_1)P(A_2|A_1)$$
$$P(A_3|A_1 \cap A_2) \ldots P(A_k|A_1 \cap A_2 \cap \ldots \cap A_{k-1}).$$

**Example 4.3 (Selecting a Committee Continued)**

Return to the committee selection example (Example 3.9), where there are 4 boys (B) and 6 girls (G). We want to select a 2-person committee. Find:

(i)  the probability that both are boys,
(ii)  the probability that one is a boy and the other is a girl.

We have already dealt with this type of urn problems by using the combinatorial method, see Example 3.9 in the previous chapter. Here, the multiplication rule is used instead. Let $B_i$ be the event that the $i$-th person is a boy, and $G_i$ be the event that the $i$-th person is a girl, $i = 1, 2$. Then

$$\text{Probability in (i)} = P(B_1 \cap B_2) = P(B_1)P(B_2|B_1) = \frac{4}{10} \times \frac{3}{9}.$$

Here $P(B_2|B_1)$ has been found by considering the fact that after the first selection of a boy (the event $B_1$), there are 3 boys left among total 9 persons at the time of second selection.

Now the Probability in (ii) is:

$$P(B_1 \cap G_2) + P(G_1 \cap B_2) = P(B_1)P(G_2|B_1) + P(G_1)P(B_2|G_1)$$
$$= \frac{4}{10} \times \frac{6}{9} + \frac{6}{10} \times \frac{4}{9}.$$

Probabilities $P(G_2|B_1)$ and $P(B_2|G_1)$ are found by inspection as previously in the case of $P(B_2|B_1)$. Now we can find the probability that 'both are girls', $P(G_1 \cap G_2)$, in a similar way. ◄

## 4.4   Total Probability Formula

The probability formula in the boy-girl committee Example 4.3 is generalised to what is known as the total probability formula in this section. Before deriving the formula we start with the following example.

**Example 4.4 (Phones)**

Suppose that in our world there are only three phone manufacturing companies: A Pale, B Sung and C Windows, and their market shares are respectively 30, 40 and 30%. Suppose also that respectively 5, 8, and 10% of their phones become faulty within one year. If I buy a phone randomly (ignoring the manufacturer), what is the probability that my phone will develop a fault within one year? After finding the probability, suppose that my phone developed a fault in the first year—what is the probability that it was made by A Pale? ◄

| Company | Market share | Percent defective |
|---------|--------------|-------------------|
| A Pale | 30% | 5% |
| B Sung | 40% | 8% |
| C Windows | 30% | 10% |

To answer this type of questions, we derive two of the most useful results in probability theory: the total probability formula and the Bayes theorem. First, let us derive the total probability formula.

Let $B_1, B_2, \ldots, B_k$ be a set of mutually exclusive, i.e.

$$B_i \cap B_j = \emptyset \text{ for all } 1 \leq i \neq j \leq k,$$

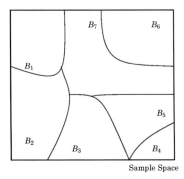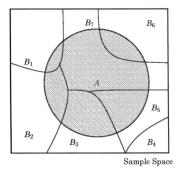

Sample Space                                Sample Space

**Fig. 4.2** The left figure shows the mutually exclusive and exhaustive events $B_1, \ldots, B_7$ (they form a partition of the sample space); the right figure shows another possible event A

and exhaustive events, i.e.:

$$B_1 \cup B_2 \cup \ldots \cup B_k = S.$$

The set of events $B_1, B_2, \ldots, B_k$ is said to form a *partition* of the sample space $S$. It is easy to note that for this partition $\sum_{i=1}^{k} P(B_i) = 1$.

Now any event $A$ can be represented by

$$A = A \cap S = (A \cap B_1) \cup (A \cap B_2) \cup \ldots \cup (A \cap B_k)$$

where $(A \cap B_1), (A \cap B_2), \ldots, (A \cap B_k)$ are mutually exclusive events, see Fig. 4.2. Hence the axiom **A3** of probability gives the *the total probability formula* for the event $PA$:

$$P(A) = P(A \cap B_1) + P(A \cap B_2) + \ldots + P(A \cap B_k)$$
$$= P(B_1)P(A|B_1) + P(B_2)P(A|B_2) + \ldots + P(B_k)P(A|B_k).$$

**Example 4.5 (Phones Continued)**

We can now find the probability of the event, say $A$, that a randomly selected phone develops a fault within one year. Let $B_1, B_2, B_3$ be the events that the phone is manufactured respectively by companies A Pale, B Sung and C Windows. Then we have:

$$P(A) = P(B_1)P(A|B_1) + P(B_2)P(A|B_2) + P(B_3)P(A|B_3)$$
$$= 0.30 \times 0.05 + 0.40 \times 0.08 + 0.30 \times 0.10$$
$$= 0.077.$$

◄

Now suppose that my phone has developed a fault within one year. What is the probability that it was manufactured by A Pale? To answer this we need to introduce the Bayes Theorem.

## 4.5   The Bayes Theorem

Let $A$ be an event, and let $B_1, B_2, \ldots, B_k$ be a set of mutually exclusive and exhaustive events. Then, for $i = 1, \ldots, k,$

$$P(B_i|A) = \frac{P(B_i)P(A|B_i)}{\sum_{j=1}^{k} P(B_j)P(A|B_j)}. \tag{4.2}$$

*Proof* The proof of the Bayes theorem uses the definition of the conditional probability and applications of the multiplication rule of conditional probability and the total probability formula as follows:

$$P(B_i|A) = \frac{P(B_i \cap A)}{P(A)} \qquad \text{[definition]}$$
$$= \frac{P(B_i)P(A|B_i)}{P(A)} \qquad \text{[multiplication rule]}$$
$$= \frac{P(B_i)P(A|B_i)}{\sum_{j=1}^{k} P(B_j)P(A|B_j)}. \quad \text{[total probability formula]}$$

The probability, $P(B_i|A)$ is called the *posterior probability* of $B_i$ and $P(B_i)$ is called the *prior probability*. The Bayes theorem is the rule that converts the prior probability into the posterior probability by using the additional information that some other event, $A$ above, has already occurred.

**Example 4.6 (Phones Continued)**

The probability that my faulty phone was manufactured by A Pale is

$$P(B_1|A) = \frac{P(B_1)P(A|B_1)}{P(A)} = \frac{0.30 \times 0.05}{0.077} = 0.1948.$$

Similarly, the probability that the faulty phone was manufactured by B Sung is 0.4156, and the probability that it was manufactured by C Windows is $1 - 0.1948 - 0.4156 = 0.3896$. ◀

Note that $\sum_{i=1}^{k} P(B_i|A) = 1$. This shows that the total posterior probability is one—parallel to the identity $\sum_{i=1}^{k} P(B_i) = 1$ we have stated before. These posterior probabilities can be used make statistical inference as we will see in Sect. 10.3.

In this section we have learned three important concepts: (i) conditional probability, (ii) total probability formula and the (iii) Bayes theorem. Much of statistical theory depends on these fundamental concepts.

## 4.6    Example: Monty Hall Problem

In this section we discuss the famous Monty Hall problem in probability which originated from a television game show of the same name. In one such game, there are three closed doors behind which there are two goats and one car. The player does not behind which door the car is there. The player chooses one door and then host reveals one of the doors behind which there is a goat. Now the choice is left to the player to either stay with their original choice of the door, swap to the other un-opened door or choose randomly again whether to stay or swap (Fig. 4.3).

In this game the player can win under two mutually exclusive ways if initially they choose:

(i)  the car $(C)$ which has probability $1/3$ and
(ii) a goat$(G)$ which has probability $2/3$.

**Fig. 4.3**  Monty Hall paradox
by Cepheus

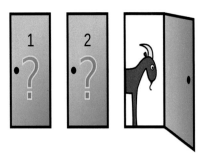

**Table 4.1** Probability calculation for the Monty Hall problem

| Strategy | $P(W|C)$ | $P(W|G)$ | $P(W)$ |
|---|---|---|---|
| Stay | 1 | 0 | $1 \times \frac{1}{3} + 0 \times \frac{2}{3} = \frac{1}{3}$ |
| Swap | 0 | 1 | $0 \times \frac{1}{3} + 1 \times \frac{2}{3} = \frac{2}{3}$ |
| Random | $\frac{1}{2}$ | $\frac{1}{2}$ | $\frac{1}{2} \times \frac{1}{3} + \frac{1}{2} \times \frac{2}{3} = \frac{1}{2}$ |

Hence probability of eventual win (W) is obtained using the following calculation.

$$P(W) = P(W \cap C) + P(W \cap G) \text{ since } C \cup G = S$$
$$= P(W|C)P(C) + P(W|G)P(G)$$
$$= P(W|C)\tfrac{1}{3} + P(W|G)\tfrac{2}{3}$$

The two probabilities $P(W|C)$ and $P(W|G)$ will depend on their strategy after the initial selection. There are three possible strategies: (i) Stay, (ii) Swap and (iii) Randomly choose again. In the Stay strategy (i) the player does not change his choice. In the Swap strategy (ii) they opt for the other unopened door. In the randomly choose again strategy (iii), the player randomly decides to stay or swap with equal probability 0.5. Table 4.1 calculates the probabilities of winning under the three different strategies.

It is clear that the 'Swap' strategy maximises the chance of winning. This example is illustrated using the R function `monty` included in the package `ipsRdbs`. By default this function 'plays', i.e. simulates $N = 1000$ games. The user is given the opportunity to select one of the three strategies from:

1. `stay` which is the default; R command `monty()`
2. `switch`; R command `monty("switch")`
3. `random`; R command `monty("random")`

R prints the results of each of the $N$ games and calculates the proportion of games the user wins by adopting the specific strategy.

## 4.7 Exercises

### 4.2 (Using the Bayes Theorem)

1. A chest has three drawers. The first contains two gold coins, the second contains a gold and silver coin and the third has two silver coins. A drawer is chosen at random and from it a coin is chosen at random. What is the probability that the coin still remaining in the chosen drawer is gold given that the coin chosen is silver?

2. 10% of the boys in a school are left-handed. Of those who are left-handed, 80% are left-footed; of those who are right-handed, 15% are left-footed. If a boy, selected at random, is left-footed, use Bayes Theorem to calculate the probability that he is left-handed.
3. Consider a family with two children. If each child is as likely to be a boy as a girl, what is the probability that both children are boys
   (a) given that the older child is a boy,
   (b) given that at least one of the children is a boy?
4. In a multiple choice examination paper, $n$ answers are given for each question ($n = 4$ in the above two questions). Suppose a candidate knows the answer to a question with probability $p$. If he does not know it he guesses, choosing one of the answers at random. Use the Bayes Theorem to show that the probability that he knew the answer given that his answer is correct, is $\frac{np}{1+(n-1)p}$.
5. A factory buys 10% of its components from supplier A, 30% from supplier B and the rest from supplier C. It is known that 6% of the components it buys from supplier A are faulty. Of the components bought from supplier B, 9% are faulty and of the components bought from supplier C, 3% are faulty. A component is randomly selected from the factory and it was found to be faulty. What is the probability that it was from supplier C?
6. Consider a disease that is thought to occur in 1% of the population. Using a particular blood test a physician observes that out of the patients with disease 98% possess a particular symptom. Also assume that 0.1% of the population without the disease have the same symptom. A randomly chosen person from the population is blood tested and is shown to have the symptom. What is the conditional probability that the person has the disease?
7. Springs produced by a factory are made by two machines of different ages and are either good or defective. Suppose it is known that 5% of the springs made by machine one is defective, that 2% of the springs made by machine two is defective and that 60% of the springs from the factory is made by machine two. What is the probability that a randomly selected spring from this factory is made by machine one given that the selected spring is defective?

## 4.8    Independent Events

The previous section has shown that probability of an event may change if we have additional information. However, in many situations the probabilities may not change. For example, the probability of getting an 'A' in Statistics should not depend on the student's race and sex; the results of two coin tosses should not depend on each other; an expectant mother should not think that she must have a higher probability of having a son given that her previous three children were all girls.

In this section we will learn about the probabilities of independent events. Much of statistical theory relies on the concept of independence.

## 4.8.1  Definition

We have seen examples where prior knowledge that an event $A$ has occurred has changed the probability that event $B$ occurs. There are many situations where this does not happen. The events are then said to be independent. Intuitively, events $A$ and $B$ are independent if the occurrence of one event does not affect the probability of the occurrence of the other event.

This is equivalent to saying that

$$P(B|A) = P(B), \quad \text{where } P(A) > 0, \quad \text{and } P(A|B) = P(A), \quad \text{where } P(B) > 0.$$

These two identities give the following formal definition.

Events $A$ and $B$ are independent if and only if $P(A \cap B) = P(A)P(B)$.

---

**Example 4.7 (Dice Throw)**

Throw a fair dice. Let $A$ be the event that "the result is even" and $B$ be the event that "the result is greater than 3". We want to show that $A$ and $B$ are not independent.

For this, we have $P(A \cap B) = P(\text{either a 4 or 6 thrown}) = 1/3$, but $P(A) = 1/2$ and $P(B) = 1/2$, so that $P(A)P(B) = 1/4 \neq 1/3 = P(A \cap B)$. Therefore $A$ and $B$ are not independent events. ◄

Note that *independence* is not the same as the *mutually exclusive* property. When two events, $A$ and $B$, are mutually exclusive, the probability of their intersection, $A \cap B$, is zero, i.e. $P(A \cap B) = 0$. But if the two events are independent then $P(A \cap B) = P(A) \times P(B)$.

Independence is often assumed on physical grounds, although sometimes incorrectly. There are serious consequences for wrongly assuming independence, e.g. the financial crisis in 2008. However, when the events are independent then the simpler product formula for joint probability is then used instead of the formula involving more complicated conditional probabilities.

---

**Example 4.8**

Two fair dice when shaken together are assumed to behave independently. Hence the probability of two sixes is $1/6 \times 1/6 = 1/36$. ◄

## 4.8.2  Independence of Complementary Events

If $A$ and $B$ are independent, so are $A'$ and $B'$.

**Proof** Given that $P(A \cap B) = P(A)P(B)$, we need to show that $P(A' \cap B') = P(A')P(B')$. This follows from

$$P(A' \cap B') = 1 - P(A \cup B) \quad \text{[using a Venn diagram]}$$
$$= 1 - \{P(A) + P(B) - P(A \cap B)\} \quad \text{[general addition rule]}$$
$$= 1 - \{P(A) + P(B) - P(A)P(B)\} \quad \text{[$A$ and $B$ are independent]}$$
$$= \{1 - P(A)\} - P(B)\{1 - P(A)\}$$
$$= \{1 - P(A)\}\{1 - P(B)\}$$
$$= P(A')P(B'). \qquad\qquad\qquad\qquad\qquad\qquad\qquad\qquad\qquad\qquad \square$$

Similarly, it can be shown that the pairs of events $A'$ and $B$ and $A$ and $B'$ are also independent.

### 4.8.3   Independence of More Than Two Events

The ideas of conditional probability and independence can be extended to *more than two events*.

**Definition** Three events $A$, $B$ and $C$ are defined to be independent if

$$P(A \cap B) = P(A)P(B), \ P(A \cap C) = P(A)P(C), \ P(B \cap C) = P(B)P(C), \tag{4.3}$$

$$P(A \cap B \cap C) = P(A)P(B)P(C) \tag{4.4}$$

Note that (4.3) does NOT imply (4.4), as shown by the next example. Hence, to show the independence of $A$, $B$ and $C$, it is necessary to show that both (4.3) and (4.4) hold.

---

**Example 4.9**

A box contains eight tickets, each labelled with a binary number. Two are labelled with the binary number 111, two are labelled with 100, two with 010 and two with 001. An experiment consists of drawing one ticket at random from the box. Let $A$ be the event "the first digit is 1", $B$ the event "the second digit is 1" and $C$ be the event "the third digit is 1". It is clear that $P(A) = P(B) = P(C) = 4/8 = 1/2$ and $P(A \cap B) = P(A \cap C) = P(B \cap C) = 1/4$, so the events are pairwise independent, i.e. (4.3) holds. However $P(A \cap B \cap C) = 2/8 \neq P(A)P(B)P(C) = 1/8$. So (4.4) does not hold and $A$, $B$ and $C$ are not independent. ◄

### 4.8.4  Bernoulli Trials

The notion of independent events naturally leads to a set of independent trials (or random experiments, e.g. repeated coin tossing). A set of independent trials, where each trial has only two possible outcomes, conveniently called success (S) and failure (F), and the probability of success is the same in each trial are called a set of *Bernoulli trials*. There are lots of fun examples involving Bernoulli trials.

---

**Example 4.10 (Feller's Road Crossing Example)**

The flow of traffic at a certain street crossing is such that the probability of a car passing during any given second is $p$ and cars arrive randomly, i.e. there is no interaction between the passing of cars at different seconds. Treating seconds as indivisible time units, and supposing that a pedestrian can cross the street only if no car is to pass during the next three seconds, find the probability that the pedestrian has to wait for exactly $k = 0, 1, 2, 3, 4$ seconds.
Let $C_i$ denote the event that a car comes in the $i$th second and let $N_i$ denote the event that no car arrives in the $i$th second.

1. Consider $k = 0$. The pedestrian does not have to wait if and only if there are no cars in the next three seconds, i.e. the event $N_1 N_2 N_3$. Now the arrival of the cars in successive seconds are independent and the probability of no car coming in any second is $q = 1 - p$. Hence the answer is $P(N_1 N_2 N_3) = q \cdot q \cdot q = q^3$.
2. Consider $k = 1$. The person has to wait for one second if there is a car in the first second and none in the next three, i.e. the event $C_1 N_2 N_3 N_4$. Hence the probability of that is $pq^3$.
3. Consider $k = 2$. The person has to wait two seconds if and only if there is a car in the 2nd second but none in the next three. It does not matter if there is a car or none in the first second. Hence:

$$P(\text{wait 2 seconds}) = P(C_1 C_2 N_3 N_4 N_5) + P(N_1 C_2 N_3 N_4 N_5)$$
$$= p \cdot p \cdot q^3 + q \cdot p \cdot q^3 = pq^3.$$

4. Consider $k = 3$. The person has to wait for three seconds if and only if a car passes in the 3rd second but none in the next three, $C_3 N_4 N_5 N_6$. Anything can happen in the first two seconds, i.e. $C_1 C_2, C_1 N_2, N_1 C_2, N_1 N_2$—all these four cases are mutually exclusive. Hence,

$$P(\text{wait 3 seconds}) = P(C_1 C_2 C_3 N_4 N_5 N_6) + P(N_1 C_2 C_3 N_4 N_5 N_6)$$
$$+ P(C_1 N_2 C_3 N_4 N_5 N_6) + P(N_1 N_2 C_3 N_4 N_5 N_6)$$
$$= p \cdot p \cdot p \cdot q^3 + p \cdot q \cdot p \cdot q^3 + q \cdot p \cdot p \cdot q^3 + q \cdot q \cdot p \cdot q^3$$
$$= pq^3.$$

5. Consider $k = 4$. This is more complicated because the person has to wait exactly 4 seconds if and only if a car passes in at least one of the first 3 seconds, one passes at the 4th but none pass in the next 3 seconds. The probability that at least one passes in the first three seconds is 1 minus the probability that there is none in the first 3 seconds. This probability is $1 - q^3$. Hence the answer is $(1 - q^3)pq^3$.

◀

We have learned the concept of independent events. It is much easier to calculate probabilities when events are independent. However, there is danger in assuming events to be independent when they are not. For example, there may be serial or spatial dependence! The concept of Bernoulli trials has been introduced.

## 4.9    Fun Examples of Independent Events

This section discusses two substantial examples: one is called system reliability where we have to find the probability of a system, built from several independent components, functioning. For example, we want to find out the probability that a machine/phone or module-based software system will continue to function. In the second substantial example we would like to cleverly estimate probabilities of sensitive events, e.g. do I have HIV/AIDS or did I take any illegal drugs during last summer's music festival?

### 4.9.1    System Reliability

#### 4.9.1.1 Two Components in Series
Suppose each component has a separate operating mechanism. This means that they operate independently.

Let $A_i$ be the event "component $i$ works when required" and let $P(A_i) = p_i$ for $i = 1, 2$. For the system of $A_1$ and $A_2$ in series, the event "the system works" is the event $(A_1 \cap A_2)$. Hence $P(\text{system works}) = P(A_1 \cap A_2) = P(A_1)P(A_2) = p_1 p_2$.

The reliability gets lower when components are included in series. For $n$ components in series, $P(\text{system works}) = p_1 p_2 \cdots p_n$. When $p_i = p$ for all $i$, the reliability of a series of $n$ components is $P(\text{system works}) = p^n$.

#### 4.9.1.2 Two Components in Parallel
For the system of $A_1$ and $A_2$ in parallel, the event "the system works when required" is now given by the event $(A_1 \cup A_2)$. Hence

$$P(\text{system works}) = P(A_1 \cup A_2) = P(A_1) + P(A_2) - P(A_1 \cap A_2) = p_1 + p_2 - p_1 p_2.$$

This is greater than either $p_1$ or $p_2$ so that the inclusion of a (redundant) component in parallel increases the reliability of the system. Another way of arriving at this result uses complementary events:

$$P(\text{system works}) = 1 - P(\text{system fails})$$
$$= 1 - P(A_1' \cap A_2')$$
$$= 1 - P(A_1')P(A_2')$$
$$= 1 - (1 - p_1)(1 - p_2)$$
$$= p_1 + p_2 - p_1 p_2.$$

In general, with $n$ components in parallel, the reliability of the system is

$$P(\text{system works}) = 1 - (1 - p_1)(1 - p_2) \cdots (1 - p_n).$$

If $p_i = p$ for all $i$, we have $P(\text{system works}) = 1 - (1 - p)^n$.

### 4.9.1.3 A General System
The ideas above can be combined to evaluate the reliability of more complex systems.

**Example 4.11 (Switches)**

Six switches make up the circuit shown in the graph in Fig. 4.4.
Each switch has the probability $p_i = P(D_i)$ of closing correctly; the mechanisms are independent; all are operated by the same impulse. Then

$$P(\text{current flows when required}) = p_1 \times [1 - (1 - p_2)(1 - p_2)]$$
◀
$$\times [1 - (1 - p_3)(1 - p_3)(1 - p_3)].$$

## 4.9.2  The Randomised Response Technique

This is an important application of the total probability formula—it is used to try to get honest answers to sensitive questions.

**Fig. 4.4** Illustration of a system with six switches

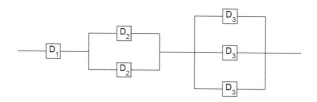

Often we wish to estimate the proportion of people in a population who would not respond 'yes' to some sensitive question such as:

- Have you taken an illegal drug during the last 12 months?
- Have you had an abortion?
- Do you have HIV/AIDs?
- Are you a racist?

It is unlikely that truthful answers will be given in an open questionnaire, even if it is stressed that the responses would be treated with anonymity. Some years ago a randomised response technique was introduced to overcome this difficulty. This is a simple application of conditional probability. It ensures that the interviewee can answer truthfully without the interviewer (or anyone else) knowing the answer to the sensitive question. How can this be possible? We consider two alternative questions, for example:

**Question 1: Was your mother born in January?**
**Question 2: Have you ever taken illegal substances in the last 12 months?**

Question 1 should not be contentious and should not be such that the interviewer could find out the true answer.

The respondent answers only of the two questions. Which question is answered by the respondent is determined by a randomisation device, the result of which is known only to the respondent. The interviewer records only whether the answer given was Yes or No (and he/she does not know which question has been answered). The proportion of Yes answers to the question of interest can be estimated from the total proportion of Yes answers obtained. We carry out this simple experiment:

Toss a coin—do not reveal the result of the coin toss!

If heads—answer Question 1: Was your mother born in January?

If tails—answer Question 2: Have you ever taken illegal substances in the last 12 months?

This survey experiment is carried out with the help of a number of volunteers, $n$ say. Suppose that $r$ of these $n$ volunteers answered yes in the survey, so that an estimate of $P(\text{Yes})$ is $r/n$.

This information can be used to estimate the proportion of Yes answers to the main question of interest, Question 2. Suppose that

- $Q_1$ is the event that 'Q1 was answered'
- $Q_2$ is the event that 'Q2 was answered'

Then, assuming that the coin was unbiased, $P(Q_1) = 0.5$ and $P(Q_2) = 0.5$. Also, assuming that birthdays of mothers are evenly distributed over the months, we have that the probability that the interviewee will answer Yes to Q1 is 1/12. Let $Y$ be the event that a 'Yes' answer is given. Then the total probability formula gives

$$P(Y) = P(Q_1)P(Y|Q_1) + P(Q_2)P(Y|Q_2),$$

which leads to

$$\frac{r}{n} \approx \frac{1}{2} \times \frac{1}{12} + \frac{1}{2} \times P(Y|Q_2).$$

Hence

$$P(Y|Q_2) \approx 2 \cdot \frac{r}{n} - \frac{1}{12}.$$

## 4.10   Exercises

### 4.3 (Independent Events)

1. The probability that Jane can solve a certain problem is 0.4 and that Alice can solve it is 0.3. If they both try independently, what is the probability that it is solved?
2. Recall the definition of the independence of two events A and B.
   (a) Show that if $P(A) = 1$, then $P(A \cap B) = P(B)$.
   (b) Prove that any event $A$ with $P(A) = 0$ or $P(A) = 1$ is independent of every event $B$.
   (c) Prove that if the events $A$ and $B$ are independent, then every pair of events $A'$ and $B$; $A$ and $B'$; $A'$ and $B'$ are independent.
3. A water supply system has three pumps $A$, $B$ and $C$ arranged as below, where the pumps operate independently.

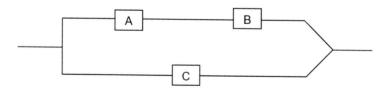

The system functions if either $A$ and $B$ operate or $C$ operates, or all three operate. If $A$ and $B$ have reliability 90% and $C$ has reliability 95%, what is the reliability of the water supply system?
4. (a) A twin-engine aircraft can fly with at least one engine working. If each engine has a probability of failure during flight of $q$ and the engines operate independently, what is the probability of a successful flight?

(b) A four-engine aircraft can fly with at least two of its engines working. If all the engines operate independently and each has a probability of failure of $q$, what is the probability that a flight will be successful?

(c) For what range of values of $q$ would you prefer to fly in a twin-engine aircraft rather than a four-engine aircraft?

(d) Discuss briefly (in no more than two sentences) whether the assumption of independent operation of the engines is reasonable.

5. A system has four components to form two subsystems in parallel, each with two components in series. Sketch the system. If each component has reliability $p$, calculate the system reliability.

# Random Variables and Their Probability Distributions

# 5

**Abstract**

Chapter 5 defines the random variables and their probability distributions. Many properties such as mean, variance, and quantiles of random variables are also defined here. Laws for expectations and variances of linear functions of random variables are also discussed.

The combinatorial probabilities introduced previously are difficult to find and very problem-specific. Instead, this chapter lays the foundation so that we can apply structured methods to calculate many probabilities of interests. The outcomes of random experiments are represented as values of a variable which will be random since the outcomes are random (or impossible to predict with certainty). In so doing, we make our life a lot easier in calculating probabilities in many stylised situations which represent reality. For example, the development of the concepts of random variables introduced in this chapter will allow us to calculate the probability that a computer will make fewer than 10 errors while making $10^{15}$ computations when it has a very tiny chance, $10^{-14}$, of making an erroneous computation, which will be presented in the next chapter. In this chapter we learn the concepts of probability distributions of random variables and their basic properties.

## 5.1 Definition of a Random Variable

A random variable defines a mapping of the sample space consisting of all possible outcomes of a random experiment to the set of real numbers. For example, suppose that a fair coin is tossed. There are two possible equally likely outcomes: head or tail. These two outcomes must be mapped to real numbers. For convenience, we

© The Author(s), under exclusive license to Springer Nature Switzerland AG 2024
S. K. Sahu, *Introduction to Probability, Statistics & R*,
https://doi.org/10.1007/978-3-031-37865-2_5

may define the mapping which assigns the value 1 if head turns up and 0 otherwise. Hence, we have the mapping:

$$\text{Head} \rightarrow 1, \ \text{Tail} \rightarrow 0,$$

which is illustrated in Fig. 5.1. We can conveniently denote the random variable by $X$ which is the number of heads obtained by tossing a single coin. Obviously, in this case all possible values of $X$ are 0 and 1.

Of-course, this is a trivial example. But it is very easy to generalise the concept of random variables. For example, let $X$ denote the number of heads obtained by tossing a fair coin $n$ times. Now the random variable $X$ can take any real positive integer value between 0 and $n$. To give another example, suppose a Professor of Nutrition measures the weight of a randomly selected student as in Example 1.4. The outcome will be a random value which cannot be predicted in advance for sure since we do not know which student will be selected in the first place.

The general concept of a random variable as a one-to-one mapping from the sample space to the real line is illustrated in Fig. 5.2. This figure is a generalisation of the previous Fig. 5.1.

If a random variable has a *finite* or *countably infinite* set of values it is called *discrete*. For example, the number of Apple computer users among 20 randomly selected students, or the number of credit cards a randomly selected person has in their wallet.

When a random variable can take any value on the real line it is called a continuous random variable. For example, the weight of a randomly selected student. A random variable can also take a mixture of discrete and continuous values, e.g. volume of precipitation collected in a day; some days it could be zero, on other days it could be a continuous measurement, e.g. 1.234 millimetre.

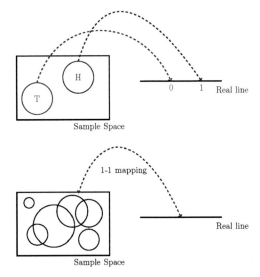

**Fig. 5.1** The random variable, number of heads, takes the values 0 and 1 on the real line

**Fig. 5.2** A random variable is a one-to-one mapping from the sample space (rectangular box on the left) to the real line (on the right). Events are circles on the sample space. The values of the random variables are imagined on the real line

We now introduce two notations: $X$ (or in general the capital letters like $Y$ and $Z$) to denote the random variable, e.g. weight of a randomly selected student, and the corresponding lower case letter $x$ ($y$ and $z$) to denote a particular value, e.g. 77.56432 kilograms. We will follow this convention throughout. For a random variable, say $X$, we will also adopt the notation $P(X \in A)$ to denote that the probability of the event that $X$ belongs to $A$ where now $A$ is a subset of the real line.

## 5.2    Probability Distribution of a Random Variable

The relationship between events in the sample space $S$ and a random variable taking values in the real line ensures the inherited properties that:

1. $P(X \in A) \geq 0$ for any $A$ which is a subset of the real line and
2. $P(-\infty < X < \infty) = 1$.

The inheritance comes directly from the probability axioms **A1** (probabilities are non-negative) and **A2** (total probability is 1) stated in Sect. 3.3.
    We now define the *probability distribution* of a random variable.

A probability distribution distributes the total probability 1 among the possible values of the random variable.

For example, returning to the coin-tossing experiment, if the probability of getting a head with a coin is $p$ (and therefore the probability of getting a tail is $1 - p$), then the probability that $X = 0$ is $1 - p$ and the probability that $X = 1$ is $p$. This gives us the *probability distribution* of $X$, and we say that $X$ has the *probability function* given by:

$$
\begin{array}{ll}
P(X = 0) & = 1 - p \\
P(X = 1) & = p \\
\hline
\text{Total probability} & = 1.
\end{array}
$$

This is an example of the Bernoulli distribution with parameter $p$, perhaps the simplest discrete distribution.

### Example 5.1

Suppose that we consider tossing the coin twice and again defining the random variable $X$ to be the number of heads obtained. The values that $X$ can take are 0, 1 and 2 with probabilities $(1 - p)^2$, $2p(1 - p)$ and $p^2$, respectively. Note that

$X = 1$ corresponds to the two mutually exclusive events $\{TH\}$ and $\{HT\}$, hence the $P(X = 1) = (1 - p)p + p(1 - p)$. Here the distribution is:

| Value($x$) | $P(X = x)$ |
|---|---|
| 0 | $(1 - p)^2$ |
| 1 | $2p(1 - p)$ |
| 2 | $p^2$ |
| Total probability | 1 |

This is a particular case of the Binomial distribution discussed in Sect. 6.2. ◄

In general, for a discrete random variable we define a function $f(x)$ to denote $P(X = x)$ (or $f(y)$ to denote $P(Y = y)$) and call the function $f(x)$ the *probability function (pf)* or *probability mass function (pmf)* of the random variable $X$. Arbitrary functions cannot be a pmf since the total probability must be 1 and all probabilities are non-negative. Hence, for $f(x)$ to be the pmf of a random variable $X$, we require:

1. $f(x) \geq 0$ for all possible values of $x$.
2. $\sum_{\text{all } x} f(x) = 1$.

So for the binomial example, we have the following probability distribution.

| $x$ | $f(x)$ | general  form |
|---|---|---|
| 0 | $(1 - p)^2$ | $^2C_x\, p^x (1 - p)^{2-x}$ |
| 1 | $2p(1 - p)$ | $^2C_x\, p^x (1 - p)^{2-x}$ |
| 2 | $p^2$ | $^2C_x\, p^x (1 - p)^{2-x}$ |
| Total | 1 | 1 |

Note that $f(x) = 0$ for any other value of $x$ and thus $f(x)$ is a discrete function of $x$.

**Example 5.2 (Sum of Scores from Two Throws of a Die)**

Let $X$ denote the sum of the scores from the two throws. By complete enumeration of the 36 possible outcomes we obtain the pmf of $X$: This pmf

| $x$ | 2 | 3 | 4 | 5 | 6 | 7 | 8 | 9 | 10 | 11 | 12 |
|---|---|---|---|---|---|---|---|---|---|---|---|
| $f(x)$ | $\frac{1}{36}$ | $\frac{2}{36}$ | $\frac{3}{36}$ | $\frac{4}{36}$ | $\frac{5}{36}$ | $\frac{6}{36}$ | $\frac{5}{36}$ | $\frac{4}{36}$ | $\frac{3}{36}$ | $\frac{2}{36}$ | $\frac{1}{36}$ |

is drawn to be a graph with spikes at the disllrete values 2 to 12, see Fig. 5.3. ◄

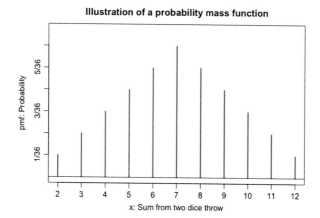

**Fig. 5.3** The pmf of the sum of scores random variable from two throws of a die

## 5.2.1    Continuous Random Variable

In many situations (both theoretical and practical) we often encounter random variables that are inherently continuous because they are measured on a continuum (such as time, length, weight) or can be conveniently well-approximated by considering them as continuous (such as the annual income of adults in a population, closing share prices).

For a continuous random variable, $P(X = x)$ is defined to be zero since we assume that the measurements are continuous and there is zero probability of observing a particular value, e.g. 1.2. The argument goes that a finer measuring instrument will give us an even more precise measurement than 1.2 and so on. Thus for a continuous random variable we adopt the convention that $P(X = x) = 0$ for any particular value $x$ on the real line. But we define probabilities for positive length intervals, e.g. $P(1.2 < X < 1.9)$.

For a continuous random variable $X$ we define its probability by using a continuous function $f(x)$ which we call its *probability density function*, abbreviated as its pdf. With the pdf we define probabilities as integrals (see Fig. 5.4), e.g.

$$P(a \leq X \leq b) = \int_{a}^{b} f(u)\, du,$$

which is naturally interpreted as the area under the curve $f(x)$ inside the interval $[a, b]$. Recall that for a continuous random variable $X$, we do not use $f(x) = P(X = x)$ for any $x$ as by convention we set $P(X = x) = 0$.

Since we are dealing with probabilities which are always between 0 and 1, just any arbitrary function $f(x)$ cannot be a pdf of some random variable. For $f(x)$ to be a pdf, as in the discrete case, we must have:

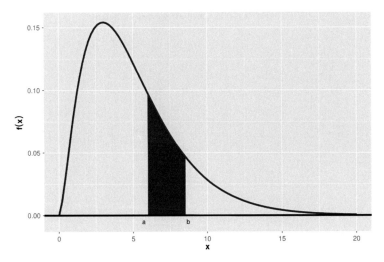

**Fig. 5.4** The shaded area is $P(a \leq X \leq b)$ if the pdf of $X$ is the drawn curve

1. $f(x) \geq 0$ for all possible values of $x$, i.e. $-\infty < x < \infty$.
2. $\int_{-\infty}^{\infty} f(u)du = 1$.

It is very simple to describe the above two requirements: viz. the probabilities are non-negative and the total probability is 1, as stated in the beginning of Sect. 5.2.

## 5.2.2   Cumulative Distribution Function (cdf)

Along with the pdf we also frequently make use of another function which is called the *cumulative distribution function*, abbreviated as the *cdf*. The cdf simply calculates the probability of the random variable up to and including its argument.

### 5.2.2.1 cdf of a Discrete Random Variable
For a discrete random variable, the cdf is the cumulative sum of the pmf $f(u)$ up to (and including) $u = x$. That is, if $X$ is a discrete random variable with pmf $f(x)$:

$$P(X \leq x) \equiv F(x) = \sum_{u \leq x} f(u).$$

**Example 5.3**

Let $X$ be the number of heads in the experiment of tossing two fair coins. Then the probability function is

$$P(X = 0) = 1/4, \quad P(X = 1) = 1/2, \quad P(X = 2) = 1/4.$$

From the definition, the cdf is given by

$$F(x) = \begin{cases} 0 & \text{if } x < 0 \\ 1/4 & \text{if } 0 \leq x < 1 \\ 3/4 & \text{if } 1 \leq x < 2 \\ 1 & \text{if } x \geq 2 \end{cases}.$$

◄

Note that the cdf for a discrete random variable is a step function. The jump-points are the possible values of the random variable, and the height of a jump gives the probability of the random variable taking that value. It is clear that the probability mass function is uniquely determined by the cdf.

**Example 5.4 (Sum of Scores from Two Throws of a Die)**

The cdf for the this example is given by:

| $x$ | 2 | 3 | 4 | 5 | 6 | 7 | 8 | 9 | 10 | 11 | 12 |
|---|---|---|---|---|---|---|---|---|---|---|---|
| $F(x)$ | $\frac{1}{36}$ | $\frac{3}{36}$ | $\frac{6}{36}$ | $\frac{10}{36}$ | $\frac{15}{36}$ | $\frac{21}{36}$ | $\frac{26}{36}$ | $\frac{30}{36}$ | $\frac{33}{36}$ | $\frac{35}{36}$ | $\frac{36}{36}$ |

This cdf is drawn in Fig. 5.5. The cdf is seen to jump at the right end points and is continuous if travelling from the right towards the left. This graph illustrates the general property that cdfs are right continuous. ◄

## 5.2.2.2  cdf of a Continuous Random Variable

For a continuous random variable $X$, the cdf (see Fig. 5.6) is defined as:

$$P(X \leq x) \equiv F(x) = \int_{-\infty}^{x} f(u)\,du.$$

The fundamental theorem of calculus then tells us:

$$f(x) = \frac{dF(x)}{dx}.$$

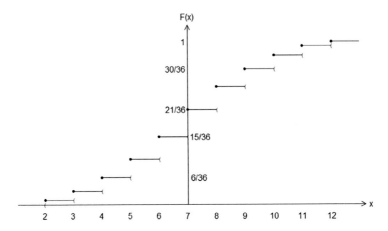

**Fig. 5.5**  An illustration of the cdf of the sum of scores from two throws of a die

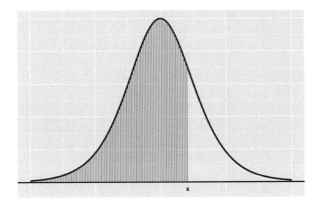

**Fig. 5.6**  The shaded area is $F(a) = P(X \leq a)$ if the pdf of $X$ is the drawn curve

That is, for a continuous random variable the pdf is the derivative of the cdf. Also for any random variable $X$, $P(c \leq X \leq d) = F(d) - F(c)$. Let us consider an example.

**Example 5.5 (Uniform Distribution)**

Suppose,

$$f(x) = \begin{cases} \frac{1}{b-a} & \text{if } a \leq x \leq b \\ 0 & \text{otherwise} \end{cases}.$$

*A random variable having the above density is said to have the Uniform distribution* $U(a, b)$. It has the cdf

$$F(x) = \int_a^x \frac{du}{b-a}$$

$$= \frac{x-a}{b-a}, a \leq x \leq b.$$

A quick check confirms that $F'(x) = f(x)$. If $a = 0, b = 1$ and then $P(0.5 < X < 0.75) = F(0.75) - F(0.5) = 0.25$. ◄

Returning to the general discussion, can any function be taken to be the cdf $P(X \leq x)$ of a random variable, $X$? The general answer is no since a cdf is probability of an event and probabilities must be between 0 and 1. In general, three conditions are required for a function $F(x)$ to be a cdf:

1. $\lim\limits_{x \to -\infty} F(x) = 0$.
2. $\lim\limits_{x \to +\infty} F(x) = 1$.
3. $F(x)$ must be a monotonically non-decreasing function of $x$.

If any of these conditions are violated by the cdf of a random variable, we may end up with negative probability for some values of the random variable. Note that the abbreviation cdf is used for both discrete and continuous random variables, unlike the different abbreviations pmf for discrete distributions and pdf for continuous distributions. Usually the corresponding upper case later is used to denote the cdf, e.g. $F(x)$, $G(y)$ when the corresponding pmf or pdf is written as $f(x)$, $g(y)$.

## 5.3 Exercises

### 5.1 (Random Variables)

1. Suppose that a word is selected at random from the sentence **I DO NOT LIKE STATS ALWAYS**. Let $X$ be the number of letters in the word that is selected. Write down the probability distribution of $X$.
2. An ornithologist carries out a study of the numbers of eggs laid per pair by a species of rare birds in its annual breeding season. He concludes that it may be considered as a discrete random variable $X$ with probability distribution given by:

$$P(X = 0) = 0.2$$

$$P(X = r) = k(4r - r^2) \text{ for } r = 1, 2, 3, 4$$

$$P(X = r) = 0 \qquad \text{otherwise.}$$

Find the value of $k$.

3. A fair die is tossed twice. Let $X$ equal the first score plus the second score. Determine
   (a) the probability function of $X$,
   (b) the cumulative distribution function of $X$ and draw its graph.
4. A coin is tossed three times. If $X$ denotes the number of heads minus the number of tails, find the probability function of $X$ and draw a graph of its cumulative distribution function when
   a. the coin is fair,
   b. the coin is biased so that $P\{H\} = \frac{3}{5}$ and $P\{T\} = \frac{2}{5}$.
5. The continuous random variable $Y$ has cumulative distribution function $F(y)$ given by

$$F(y) = \begin{cases} 0 & y < 1 \\ k\,(y^4 + y^2 - 2) & 1 \le y \le 2 \\ 1 & y > 2. \end{cases}$$

   Find the value of $k$.
6. If $X$ is a random variable with pdf $f(x)$ given by

$$f(x) = \begin{cases} kx^2 & \text{if } 0 \le x \le 1 \\ 0 & \text{otherwise.} \end{cases}$$

   Find the value of $k$.
7. The random variable $X$ has probability density function

$$f(x) = \begin{cases} kx^2(1 - x) & \text{if } 0 \le x \le 1 \\ 0 & \text{otherwise.} \end{cases}$$

   (a) Find the value of $k$.
   (b) Find the probability that $X$ lies in the range $(0, \frac{1}{2})$.
8. A random variable $X$ is said to have a *Cauchy* distribution if its probability density function is given by

$$f(x) = \frac{1}{\pi(1 + x^2)}, \quad -\infty < x < +\infty.$$

   (a) Verify that it is a valid probability density function and sketch its graph.
   (b) Find the cumulative distribution function $F(x)$.
   (c) Find $P(-1 \le X \le 1)$.
9. Let $X$ be a random variable with the following cdf:

$$F(x) = \begin{cases} e^{x-\theta} & \text{for } x \le \theta \\ 1 & \text{for } x > \theta. \end{cases}$$

(a) Find the pdf of $X$.

(b) Find the cdf of $Y = \theta - X$ and then find the pdf of $Y$.

## 5.4 Expectation and Variance of a Random Variable

In this section we learn about the mean or expectation and the variance of a random variable. These two are features or characteristics of a population whose probability distribution is given by a probability mass or density function $f(x)$. This is in contrast to the sample mean and sample variance of the observed sample data $x_1, \ldots, x_n$ we learned in Chap. 1. A simple analogy brings out the difference between sample and population quantities. The expectation and variance that we learn here are like 'climate', what we expect as a long term average and the sample quantities are like 'weather', what we experience in a given time and location. Thus the sample and population characteristics are differently interpreted than the sample characteristics. The mean of a random variable is also called its expectation since it is a value that we can expect.

A random variable having a pmf $f(x)$ or a pdf $f(x)$ is said to have the *expectation* defined by

$$E(X) = \begin{cases} \sum_{\text{all } x} x f(x) & \text{if } X \text{ is discrete,} \\ \int_{-\infty}^{\infty} x f(x) dx & \text{if } X \text{ is continuous.} \end{cases}$$

Note that in the above definition $x$ is the value and $f(x)$ is the probability function and the right hand side is the product of these two being summed over all values of $X$ in the discrete case and integrated in the continuous case. Hence, it is easy to remember that:

the expected value is either the sum or integral of value times probability.

We use the $E(X)$ notation to denote expectation of the random variable $X$. The argument is in upper case since it is the expected value of the random variable which is denoted by an upper case letter. We often use the Greek letter $\mu$ to denote the expectation, $E(X)$.

**Example 5.6 (Discrete)**

Consider the fair-die tossing experiment, with each of the six sides having a probability of 1/6 of landing face up. Let $X$ be the number on the up-face of the die. Then

$$E(X) = \sum_{x=1}^{6} x P(X = x) = \sum_{x=1}^{6} \frac{x}{6} = 3.5.$$

◄

**Example 5.7 (Continuous)**

Consider the uniform distribution $U(a, b)$ which has the pdf $f(x) = \frac{1}{b-a}, a \leq x \leq b$.

$$
\begin{aligned}
E(X) &= \int_{-\infty}^{\infty} x \, f(x) dx \\
&= \int_{a}^{b} \frac{x}{b-a} dx \\
&= \frac{b^2 - a^2}{2(b-a)} = \frac{b+a}{2},
\end{aligned}
$$

the mid-point of the interval $(a, b)$. ◄

The definition of expectation is generalised for random variables which are obtained by transforming the original random variable. For example, $Y = g(X) = X^2$ is a new random variable based on the original random variable $X$. To find the expected value of $Y$, $E(Y)$ we simply use the value times probability rule, i.e. the expected value of $Y$ is either sum or integral of its value, $g(x)$ times probability $f(x)$.

$$
E(Y) = E(g(X)) = 
\begin{cases}
\sum_{\text{all } x} g(x) f(x) & \text{if } X \text{ is discrete,} \\
\int_{-\infty}^{\infty} g(x) f(x) dx & \text{if } X \text{ is continuous.}
\end{cases}
$$

For example, if $X$ is continuous, then $E(X^2) = \int_{-\infty}^{\infty} x^2 f(x) dx$. We now prove one important property of expectation, namely that expectation is a linear operator. Suppose that $a$ and $b$ are known constants and $Y = g(X) = aX + b$. In this case,

$$E(aX + b) = aE(X) + b.$$

***Proof***  The proof of this is given below for the continuous case only. In the discrete case replace the integral $\left(\int\right)$ by the summation $\left(\sum\right)$ sign.

$$
\begin{aligned}
E(Y) &= \int_{-\infty}^{\infty}(ax+b)f(x)dx \\
&= a\int_{-\infty}^{\infty}xf(x)dx + b\int_{-\infty}^{\infty}f(x)dx \\
&= aE(X)+b,
\end{aligned}
$$

since $\int_{-\infty}^{\infty}f(x)dx = 1$, i.e. the fact that the total probability is one.  □

For example, suppose $E(X) = 5$ and $Y = -2X + 549$; then $E(Y) = 539$. This rule provides a method to calculate expectations of linear functions of random variables.

### 5.4.1   Variance of a Random Variable

The variance measures the variability of a random variable and is defined by:

$$
\text{Var}(X) = E(X - \mu)^2 = \begin{cases} \sum\limits_{\text{all } x}(x-\mu)^2 f(x) & \text{if X is discrete,} \\ \int_{-\infty}^{\infty}(x-\mu)^2 f(x)dx & \text{if X is continuous} \end{cases},
$$

where $\mu = E(X)$, and when the sum or integral exists. Note that the sum or integral may not always be assumed to exist. When the variance exists, it is the expectation of $(X - \mu)^2$ where $\mu$ is the mean (or expectation) of $X$. The variance is often denoted by the square of the Greek letter $\sigma$, $\sigma^2$, where the square in $\sigma^2$ emphasises the fact that variance is always non-negative. The square root of the variance is called the *standard deviation* of the random variable and is usually denoted by $\sigma$.

We now derive an easy formula to calculate the variance:

$$
\text{Var}(X) = E(X - \mu)^2 = E\left(X^2\right) - \mu^2.
$$

***Proof***

$$
\begin{aligned}
\text{Var}(X) &= E(X - \mu)^2 \\
&= E(X^2 - 2X\mu + \mu^2) \\
&= E(X^2) - 2\mu E(X) + \mu^2 \\
&= E(X^2) - 2\mu\mu + \mu^2 \\
&= E(X^2) - \mu^2.
\end{aligned}
$$

Thus:

> *the variance of a random variable is the expected value of its square minus the square of its expected value.*

When can the variance be zero? When there is no variation at all in the random variable, i.e. it takes only a single value $\mu$ with probability 1. Hence, there is nothing random about the random variable—we can predict its outcome with certainty.

---

**Example 5.8 (Uniform)**

Consider the uniform $U(a, b)$ random variable $X$.

$$
\begin{aligned}
E(X^2) &= \int_a^b \frac{x^2}{b-a}dx \\
&= \frac{b^3-a^3}{3(b-a)} \\
&= \frac{b^2+ab+a^2}{3},
\end{aligned}
$$

Hence

$$
\begin{aligned}
\text{Var}(X) &= \frac{b^2+ab+a^2}{3} - \left(\frac{b+a}{2}\right)^2 \\
&= \frac{(b-a)^2}{12},
\end{aligned}
$$

after simplification. ◄

We prove the following important property of the variance for linear functions of random variables.

> Suppose $Y = aX + b$;   then $\text{Var}(Y) = a^2\text{Var}(X)$.

The proof is given below for the continuous case only. In the discrete case replace integral ($\int$) by summation ($\sum$).

$$
\begin{aligned}
\text{Var}(Y) &= E(Y - E(Y))^2 \\
&= \int_{-\infty}^{\infty} (ax + b - a\mu - b)^2 f(x)dx \\
&= a^2 \int_{-\infty}^{\infty} (x - \mu)^2 f(x)dx \\
&= a^2\text{Var}(X).
\end{aligned}
$$

This is a very useful result, e.g. suppose $\text{Var}(X) = 25$ and $Y = -X + 5{,}000{,}000$; then $\text{Var}(Y) = \text{Var}(X) = 25$ and the standard deviation, $\sigma = 5$. In words a location shift, $b$, does not change variance but a multiplicative constant, $a$ say, gets squared in variance, $a^2$. Note that the standard deviation of $Y$ is $|a|$ times the standard deviation of $X$. The absolute sign in $|a|$ is required to ensure non-negativity of the measure standard deviation.

## 5.5   Quantile of a Random Variable

Like the sample median and the quartiles, see definitions in Chap. 1, the probability distribution of a random variable, be it either discrete or continuous, also has similar measures which split the distribution in different parts. For example, median, defined as the middle value, splits the distribution in two equal halves, i.e. half of the values are less than the median and the other half of the values are greater than the median. Using the cdf, defined above, we can define the median to be the solution of the equation $F(x) = 0.5$. Similarly, the first quartiles is defined to be the solution of the equation $F(x) = 0.25$.

In general, we define the $p$th quantile of a random variable $X$ having cdf $F(x)$ to be a solution of the equation

$$F(x) = p, \quad \text{for any } 0 < p < 1. \tag{5.1}$$

Thus,

*For a given $0 < p < 1$, the pth quantile (or $100p$ percentile) of the random variable X with cdf F(x) is defined to be the value q for which $F(q) = p$.*

Note that the solution of (5.1) may not always be unique, e.g., in the case of a discrete random variable. This problem is similar to the one we have encountered for the median and quartiles in Chap. 1. However, the solution is unique if $F(x)$ is continuous.

As a way of an example consider the uniform distribution defined by the pdf:

$$f(x) = \frac{1}{b-a}, \quad \text{if } a << x \le b.$$

For this distribution, $F(x) = \frac{x-a}{b-a}$ if $a << x \le b$. Hence the solution to $F(x) = p$ is $x = a + p(b-a)$. By taking $p = \frac{1}{2}$, the median is $\frac{1}{2}(b+a)$, the mid-point of the interval $(a, b)$.

Different terminologies are used to call the quantiles. For example, 50th percentile is the median which is the $p$th quantile for $p = 0.5$. The first quartile is the 25th percentile and so on. We will learn more about the quantiles of many standard distributions in Chaps. 6 and 7.

The 50th percentile is called the median. The 25th and 75th percentiles are called the quartiles.

---

**Example 5.9 (Uniform distribution)**

Consider the uniform distribution $U(a, b)$. Here $F(x) = \frac{x-a}{b-a}$. So for a given $p$, $F(q) = p$ implies $q = a + p(b - a)$.

For the uniform $U(a, b)$ distribution the median is $\frac{b+a}{2}$, and the quartiles are: $\frac{b+3a}{4}$ and $\frac{3b+a}{4}$. ◄

---

## 5.6    Exercises

### 5.2 (Expectation, Variance and Quantiles)

1. Suppose that a word is selected at random from the sentence **I DO NOT LIKE STATS ALWAYS**. Let $X$ be the number of letters in the word that is selected. Write down the probability distribution of $X$. Find the mean and the variance of $X$.

2. The random variable $X$ has probability function

$$p_x = \begin{cases} \frac{1}{14}(1 + x) & \text{if } x = 1, 2, 3, 4 \\ 0 & \text{otherwise.} \end{cases}$$

Find the mean and variance of $X$.

**Hint:** This is a discrete random variable taking only four values. Hence use summation rather than integration to find expectation and variance.

3. Two fair dice are tossed and $X$ equals the larger of the two scores obtained. Find the probability function of $X$ and determine $E(X)$.

4. The random variable $X$ is uniformly distributed on the integers $0, \pm 1, \pm 2, \ldots, \pm n$, i.e.

$$p_x = \begin{cases} \frac{1}{2n+1} & \text{if } x = 0, \pm 1, \pm 2, \ldots, \pm n \\ 0 & \text{otherwise.} \end{cases}$$

Obtain expressions for the mean and variance in terms of $n$. Given that the variance is 10, find $n$.

5. A random variable $X$ has the probability density function $f(x)$ defined as

$$f(x) = \begin{cases} 1 + x & \text{for } -1 < x < 0 \\ 1 - x & \text{for } 0 \le x \le 1 \\ 0 & \text{otherwise.} \end{cases}$$

(a) Sketch the graph of this probability density function.
(b) Determine the probability $P(X \le -1)$.
(c) Calculate the probability $P\left(|X| > \frac{1}{2}\right)$.
(d) Determine the cumulative distribution function $F(x)$ of the random variable $X$.
(e) Calculate the expectation $E(X)$.
(f) Calculate $E(X^2)$ and hence determine the variance of $X$, $\mathrm{Var}(X)$.

6. The probability density function of $X$ has the form

$$f(x) = \begin{cases} a + bx + cx^2 & \text{if } 0 \le x \le 4 \\ 0 & \text{otherwise.} \end{cases}$$

If $E(X) = 2$ and $\mathrm{Var}(X) = \frac{12}{5}$, determine the values of $a$, $b$ and $c$.

7. Suppose that $X$ is a random variable and that $E(X) = \mu$. Show that, for any constant $a$,

$$E[(X - a)^2] = Var(X) + (\mu - a)^2.$$

What value of $a$ will minimise $E[(X - a)^2]$?
**Hint:** Start with the left hand side, write $X - a = X - \mu + \mu - a$ and then expand the square and take expectations.

8. Find the mean (expectation) and median of the following distributions:
(a) $f_X(x) = \binom{4}{x} \left(\frac{1}{4}\right)^x \left(\frac{3}{4}\right)^{4-x}$,   $x \in \{0, 1, 2, 3, 4\}$.
(b) $f_X(x) = 3x^2$,      $x \in (0, 1)$.
(c) $f_X(x) = \frac{1}{\pi(1+x^2)}$,        $x \in \mathcal{R}$.

# Standard Discrete Distributions

# 6

**Abstract**

Chapter 6: This chapter introduces the standard discrete distributions: Bernoulli, binomial, Poisson, geometric, hypergeometric and negative binomial. In each case the basic properties, such as mean and variance are obtained and the R commands to obtain probabilities and cumulative probabilities are illustrated.

This chapter introduces the most widely used standard discrete probability distributions often encountered in practice. These probability distributions are for discrete random variables which count a number of events of interests. Thus these distributions are used as probability models for practical phenomenon in real life world, e.g. the number of road accidents in a city, the number of Covid-19 positive cases for each day during a particular wave and so on.

This chapter presents the Bernoulli, binomial, geometric, hypergeometric, negative binomial and Poisson distributions. For each of these distributions, except for the Bernoulli, the presentations show that the total probability is one and that crucial mathematical identity is exploited to find the expectation (mean) and variance of each particular distribution. In addition, we introduce the R commands to obtain probabilities of interests for each distribution. This makes it un-necessary to consult statistical tables to obtain such probabilities.

## 6.1 Bernoulli Distribution

The Bernoulli distribution arises from the concept of sequence of Bernoulli trials defined in Sect. 4.8.4. Each such trial gives rise to a Bernoulli random variable. A random variable taking only two values, 0 and 1, is said to have the Bernoulli distribution which has the pmf

© The Author(s), under exclusive license to Springer Nature Switzerland AG 2024
S. K. Sahu, *Introduction to Probability, Statistics & R*,
https://doi.org/10.1007/978-3-031-37865-2_6

$$f(x) = \begin{cases} 1 - p, & \text{if } x = 0, \\ p, & \text{if } x = 1, \end{cases}$$

where $0 < p < 1$ is a given parameter. Here we note that the total probability, $P(X = 0) + P(X = 1)$, is $1 \ (= 1 - p + p)$. For $X$,

$$E(X) = 0 \cdot (1 - p) + 1 \cdot p = p,$$
$$E(X^2) = 0^2 \cdot (1 - p) + 1^2 \cdot p = p,$$
$$\text{Var}(X) = E(X^2) - (E(X))^2 = p - p^2 = p(1 - p).$$

Hence note that $\text{Var}(X) < E(X)$.

## 6.2   Binomial Distribution

The binomial distribution arises from a sequence of Bernoulli trials defined in Sect. 4.8.4. Suppose that we have $n$ such trials, where the probability of success in each trial, $P(S)$, is $p$ and the probability of failure is $P(F) = 1 - p$. Let $X$ be the total number of successes in the $n$ trials. Then $X$ is called a binomial random variable with parameters $n$ and $p$, and is denoted by $\text{Bin}(n, p)$.

Below we derive the pmf of the binomial distribution given by:

$$P(X = x) = \binom{n}{x} p^x (1 - p)^{n-x}, \quad x = 0, 1, \ldots, n.$$

**Proof** An outcome of the experiment (of carrying out $n$ Bernoulli trials) is represented by a sequence of $S$'s and $F$'s (such as $SS \ldots FS \ldots SF$) that comprises $x$ successes ($S$) and $(n - x)$ failures ($F$) in any order.
The probability associated with this particular outcome is

$$P(SS \ldots FS \ldots SF) = pp \cdots (1 - p) p \cdots p (1 - p) = p^x (1 - p)^{n-x},$$

since the outcomes of successive Bernoulli trials are independent. For this sequence, $X = x$, but there are many other sequences which will also give $X = x$. In fact, there are $\binom{n}{x}$ such sequences. Hence

$$P(X = x) = \binom{n}{x} p^x (1 - p)^{n-x}, x = 0, 1, \ldots, n.$$

**Remark** A binomial random variable can also be described using the urn model introduced in Sect. 3.5.5 but *with replacement* where items drawn are put back into the urn before the next draw. Suppose we have an urn (population) containing $N$ individuals, a proportion $p$ of which are of type $S$ and a proportion $1 - p$ of type $F$. If we select a sample of $n$ individuals at random **with replacement**, then the number, $X$, of type $S$ individuals in the sample follows the binomial distribution with parameters $n$ and $p$. This is because drawing of a random sample of $n$ individuals can be thought of as $n$ Bernoulli trials each with success probability $p = Np/N$. **How can we guarantee that $\sum_{x=0}^{n} P(X = x) = 1$?** This guarantee is provided by the binomial theorem:

$$(a + b)^n = b^n + \binom{n}{1} ab^{n-1} + \cdots + \binom{n}{x} a^x b^{n-x} + \cdots + a^n,$$

for any $a$, $b$ and positive integer $n$. This theorem can be proved by the method of mathematical induction. In this theorem we choose $a = 1 - p$ and $b = p$ which shows that

$$\sum_{x=0}^{n} \binom{n}{x} p^x (1 - p)^{n-x} = (1 - p + p)^n = 1. \tag{6.1}$$

This may seem to be trivial but it is a powerful identity which essentially states that the total probability under the binomial distribution is 1.

## 6.2.1  Probability Calculation Using R

Probabilities under all the standard distributions have been calculated in R and will be used throughout this book. For the binomial distribution the command `dbinom( x=3, size=5, prob=0.34)` calculates the pmf of $\text{Bin}(n = 5, p = 0.34)$ at $x = 3$. That is, the command `dbinom(x=3, size=5, prob=0.34)` will return the value $P(X = 3) = \binom{5}{3}(0.34)^3(1 - 0.34)^{5-3}$.

The command `pbinom` returns the cdf or the probability up to and including the argument. Thus `pbinom(q=3, size=5, prob=0.34)` will return the value of $P(X \le 3)$ when $X \sim \text{Bin}(n = 5, p = 0.34)$.

---

**Example 6.1**

Suppose that widgets are manufactured in a mass production process with 1% defective. The widgets are packaged in bags of 10 with a money-back guarantee if more than 1 widget per bag is defective. For what proportion of bags would the company have to provide a refund?

Firstly, we want to find the probability that a randomly selected bag has at most 1 defective widget. Note that the number of defective widgets in a bag $X$, $X \sim$ Bin($n = 10$, $p = 0.01$). So, this probability is equal to

$$P(X = 0) + P(X = 1) = (0.99)^{10} + 10(0.01)^1 (0.99)^9 = 0.9957.$$

Hence the probability that a refund is required is $1 - 0.9957 = 0.0043$, i.e. only just over 4 in 1000 bags will incur the refund on average. We can obtain this probability in R using the command `pbinom(q=1, size=10, prob=0.01)`, which returns $0.9957338$. ◄

## 6.2.2  Mean of the Binomial Distribution

Below we show that the mean of the binomial distribution is $np$, i.e., $E(X) = np$ where $X \sim$ Bin($n$, $p$).

***Proof*** By definition

$$E(X) = \sum_{x=0}^{n} x P(X = x) = \sum_{x=0}^{n} x \binom{n}{x} p^x (1 - p)^{n-x}.$$

We evaluate the above sum by using the identity (6.1) as follows. We use the fact that $n! = n(n - 1)!$ for any $n > 0$.

$$
\begin{aligned}
E(X) &= \sum_{x=0}^{n} x \binom{n}{x} p^x (1 - p)^{n-x} \\
&= \sum_{x=1}^{n} x \frac{n!}{x!(n-x)!} p^x (1 - p)^{n-x} \quad \text{[removed the $x = 0$ term]} \\
&= \sum_{x=1}^{n} \frac{n!}{(x-1)!(n-x)!} p^x (1 - p)^{n-x} \\
&= np \sum_{x=1}^{n} \frac{(n-1)!}{(x-1)!(n-1-x+1)!} p^{x-1} (1 - p)^{n-1-x+1} \\
&= np \sum_{y=0}^{n-1} \frac{(n-1)!}{(y)!(n-1-y)!} p^y (1 - p)^{n-1-y} \quad \text{[write $y = x - 1$]} \\
&= np(p + 1 - p)^{n-1} \\
&= np,
\end{aligned}
$$

where we used the total probability identity (6.1) with $n$ replaced by $n - 1$ to conclude that the last sum is equal to 1.                                                      □

## 6.2.3  Variance of the Binomial Distribution

The variance of the binomial distribution is $np(1 - p)$, i.e., Var($X$) $= np(1 - p)$ where $X \sim$ Bin($n$, $p$).

***Proof*** It is difficult to find $E(X^2)$ directly by summation. Instead, we use the factorial structure and mimic the above proof for the mean to find $E[X(X-1)]$. Recall that $k! = k(k-1)(k-2)!$ for any $k > 1$.

$$E[(X(X-1)] = \sum_{x=0}^{n} x(x-1)\binom{n}{x}p^x(1-p)^{n-x}$$

$$= \sum_{x=2}^{n} x(x-1)\frac{n!}{x!(n-x)!}p^x(1-p)^{n-x}$$

$$= \sum_{x=2}^{n} \frac{n!}{(x-2)!(n-x)!}p^x(1-p)^{n-x}$$

$$= n(n-1)p^2 \sum_{x=2}^{n} \frac{(n-2)!}{(x-2)!(n-2-x+2)!}p^{x-2}(1-p)^{n-2-x+2}$$

$$= n(n-1)p^2 \sum_{y=0}^{n-2} \frac{(n-2)!}{(y)!(n-2-y)!}p^y(1-p)^{n-2-y}$$

$$= n(n-1)p^2(p+1-p)^{n-2}.$$

Now, $E(X^2) = E[X(X-1)] + E(X) = n(n-1)p^2 + np$. Hence,

$$\text{Var}(X) = E(X^2) - (E(X))^2 = n(n-1)p^2 + np - (np)^2 = np(1-p).$$

It is illuminating to see these direct proofs. Later on we shall apply statistical theory to prove these results. Notice that the total probability identity (6.1) is used repeatedly to prove the results.

## 6.3    Geometric Distribution

Suppose that we have the same situation as for the binomial distribution but we consider a different random variable $X$, which is defined as the number of failures before the first success. The outcomes for this experiment are:

| Outcome | Value | Probability |
|---------|-------|-------------|
|         | $x$   | $P(X = x)$  |
| $S$     | $X = 0$ | $p$ |
| $FS$    | $X = 1$ | $(1-p)p$ |
| $FFS$   | $X = 2$ | $(1-p)^2 p$ |
| $FFFS$  | $X = 3$ | $(1-p)^3 p$ |
| $\vdots$ | $\vdots$ | $\vdots$ |

In general, we have

$$P(X = x) = (1-p)^x p, \quad x = 0, 1, 2, \ldots$$

This is called the *geometric distribution*, and it has a (countably) infinite domain starting at 0. We write $X \sim \text{Geo}(p)$. We now verify that the above probability defines a valid pmf. We have:

$$
\sum_{x=0}^{\infty} P(X = x) = \sum_{x=0}^{\infty} (1 - p)^x p
$$

$$
= p \, \frac{1}{1 - (1 - p)}
$$

$$
= 1,
$$

where we used the sum of the infinite geometric series

$$
\sum_{y=0}^{\infty} r^y = 1 + r + r^2 + \cdots = \frac{1}{1 - r},
$$

if $|r| < 1$.

A related random variable is $Y =$ total number of trials needed to observe the first success. To find $Y = y$ note that there must be $y - 1$ consecutive failures before the first success at the $y$th trial. Hence, if $X$ counts the number of failures before the first success, then $Y = X + 1$. Thus

$$
P(Y = y) = P(X + 1 = y) = P(X = y - 1) = (1 - p)^{y-1} p
$$

where $y = 1, 2, \ldots$. This probability defines the pmf of the random variable $Y$.

### 6.3.1   Probability Calculation Using R

The R function to calculate $P(X = x)$ is **dgeom**. For example, to calculate $P(X = 1)$ when $p = 0.25$ we can issue the command **dgeom**(x=1, prob=0.25), which outputs $0.1875 (= 0.75 \times 0.25)$. To calculate the cdf $P(X \leq x)$ with $x = 1$ and $p = 0.25$ we can issue the command **pgeom**(x=1, prob=0.25). This calculates $P(X = 0) + P(X = 1)$. The probability $P(X \leq x)$, however, is available in closed form since for any non-negative integer $x$ we have

$$
P(X \leq x) = \sum_{k=0}^{x} P(X = k)
$$

$$
= \sum_{k=0}^{x} p(1 - p)^k
$$

$$
= p \left[ (1 - p)^0 + (1 - p)^1 + \cdots + (1 - p)^x \right]
$$

$$= p \left[ \frac{1 - (1 - p)^{x+1}}{1 - (1 - p)} \right]$$

$$= 1 - (1 - p)^{x+1}.$$

Thus the command `pgeom(q=2, prob=0.25)` returns $0.578125 (= 1 - 0.75^3)$.

## 6.3.2 Negative Binomial Series

To find the mean and variance of the geometric distribution we make use of the negative binomial series which is a generalisation of the above geometric series. For $n > 0$ and $|a| < 1$, the negative binomial series is given by:

$$(1 - a)^{-n} = 1 + na + \frac{1}{2}n(n + 1)a^2 + \frac{1}{6}n(n + 1)(n + 2)a^3$$

$$+ \cdots + \frac{n(n + 1)(n + 2) \cdots (n + k - 1)}{k!} a^k + \cdots$$

Note that the coefficient of $a^k$ in the above expansion is:

$$\frac{n(n + 1)(n + 2) \cdots (n + k - 1)}{k!} = \binom{n + k - 1}{k}.$$

Thus, we have:

$$(1 - a)^{-n} = \sum_{k=0}^{\infty} \binom{n + k - 1}{k} a^k. \tag{6.2}$$

When $n = 2$ the coefficient of $a^k$ in the above expansion is given by:

$$\frac{n(n + 1)(n + 2) \cdots (n + k - 1)}{k!} = \frac{2 \times 3 \times 4 \times \cdots \times (2 + k - 1)}{k!} = k + 1.$$

Thus,

$$(1 - a)^{-2} = 1 + 2a + 3a^2 + 4a^3 + \cdots + (k + 1)a^k + \cdots$$

When $n = 3$ the coefficient of $r^k$ in the above expansion is given by:

$$\frac{n(n+1)(n+2)(n+k-1)}{k!} = \frac{3 \times 4 \times 5 \times \cdots \times (3+k-1)}{k!}$$

$$= \frac{(k+2)!}{2(k!)} = \frac{(k+1)(k+2)}{2}.$$

Thus,

$$(1-a)^{-3} = 1 + 3a + 6a^2 + 10a^3 + \cdots + \frac{(k+1)(k+2)}{2}a^k + \cdots$$

### 6.3.3   Mean of the Geometric Distribution

The mean of the geometric distribution is $\frac{1-p}{p}$, i.e., $E(X) = \frac{1-p}{p}$ where $X \sim$ Geo($p$).

*Proof*

$$
\begin{aligned}
E(X) &= \sum_{x=0}^{\infty} x P(X = x) \\
&= \sum_{x=0}^{\infty} x p (1-p)^x \\
&= p \left[ 0 + (1-p) + 2(1-p)^2 + 3(1-p)^3 + 4(1-p)^4 + \dots \right] \\
&= p(1-p) \left[ 1 + 2(1-p) + 3(1-p)^2 + 4(1-p)^3 + \dots \right] \\
&= p(1-p) [1 - (1-p)]^{-2} \\
&= \frac{1-p}{p}.
\end{aligned}
$$

### 6.3.4   Variance of the Geometric Distribution

The variance of the geometric distribution is $\frac{1-p}{p^2}$, i.e., Var($X$) $= \frac{1-p}{p^2}$ where $X \sim$ Geo($p$).

*Proof*  To obtain the variance we first find $E(X(X-1))$.

$$
\begin{aligned}
E[X(X-1)] &= \sum_{x=0}^{\infty} x(x-1) P(X = x) \\
&= \sum_{x=2}^{\infty} x(x-1) p (1-p)^x \\
&= p(1-p)^2 \sum_{x=2}^{\infty} x(x-1)(1-p)^{x-2} \\
&= p(1-p)^2 \sum_{k=0}^{\infty} (k+2)(k+1)(1-p)^k \\
&= 2p(1-p)^2 \sum_{k=0}^{\infty} \frac{(k+2)(k+1)}{2}(1-p)^k \\
&= 2p(1-p)^2 [1 - (1-p)]^{-3} \\
&= \frac{2(1-p)^2}{p^2}.
\end{aligned}
$$

Hence,

$$\text{Var}(X) = E[X(X-1)] + E(X) - [E(X)]^2$$
$$= \frac{2(1-p)^2}{p^2} + \frac{1-p}{p} - \frac{(1-p)^2}{p^2}$$
$$= \frac{(1-p)}{p^2}.$$

## 6.3.5 Memoryless Property of the Geometric Distribution

The random variable $X$ has an interesting property called the memoryless property, which states that

$$P(X \geq s + k | X \geq k) = P(X \geq s).$$

where $s$ and $k$ are positive integers. This means that the random variable does not remember what has occurred, i.e. the probability of getting an additional $s$ failures given that $k$ failures have already occurred is the same as getting $s$ failures at the start of the sequence. In practice, we say that the geometric distribution 'forgets' its age (denoted by $k$) to determine how long more (denoted by $s$) it will survive!

*Proof* First we obtain the probability that $X \geq k$ for some given natural number $k$:

$$P(X \geq k) = \sum_{x=k}^{\infty} P(X = x)$$

$$= \sum_{x=k}^{\infty} p(1-p)^x$$

$$= p[(1-p)^k + (1-p)^{k+1} + (1-p)^{k+2} + \ldots$$

$$= p(1-p)^k \sum_{y=0}^{\infty} (1-p)^y$$

$$= (1-p)^k. \qquad \square$$

From the definition of conditional probability $P(A|B) = \frac{P(A \cap B)}{P(B)}$, we write,

$$
\begin{aligned}
P(X \geq s + k | X \geq k) &= \frac{P(X \geq s+k, X \geq k)}{P(X \geq k)} \\
&= \frac{P(X \geq s+k)}{P(X \geq k)} \\
&= \frac{(1-p)^{s+k}}{(1-p)^k} \\
&= (1 - p)^s,
\end{aligned}
$$

which does not depend on $k$. Note that the event $X \geq s + k$ and $X \geq k$ implies and is implied by $X \geq s + k$ since $s > 0$.

## 6.4   Hypergeometric Distribution

Suppose we have an urn containing $N$ individuals, a proportion $p$ of which are of type $S$ and a proportion $1 - p$ of type $F$. If we select a sample of $n$ individuals at random **without replacement**, then the number, $X$, of type $S$ individuals in the sample has the hypergeometric distribution:

$$
P(X = x) = \frac{\binom{Np}{x}\binom{N(1-p)}{n-x}}{\binom{N}{n}}, x = 0, 1, \ldots, n,
$$

assuming that $x \leq Np$ and $n - x \leq N(1 - p)$ so that the above combinations are well defined.

To show that the $P(X = x)$ for $x = 0, \ldots, n$ defines a pmf we consider the identity

$$
(1 + a)^{Np}(1 + a)^{N(1-p)} = (1 + a)^N.
$$

Equating the coefficients of $a^n$ on both sides, we get

$$
\sum_{x=0}^{n} \binom{Np}{x}\binom{N(1-p)}{n-x} = \binom{N}{n}.
$$

Hence this proves that $\sum_{x=0}^{n} P(X = x) = 1$.

To find the mean and variance we obtain $E(X)$ and $E[X(X - 1)]$ with the help of the identity just established. Finally, it can be shown that the mean and variance of the hypergeometric distribution are given by

$$
E(X) = np, \text{Var}(X) = npq\frac{N - n}{N - 1}.
$$

Note that hen $N \rightarrow \infty$ the variance converges to the variance of the binomial distribution. Indeed, the hypergeometric distribution is a finite population analogue of the binomial distribution, see the urn model interpretation of the binomial distribution in Sect. 6.2.

## 6.5   Negative Binomial Distribution

The negative binomial distribution is a generalisation of the geometric distribution. We continue to have a sequence of Bernoulli trials and we define the random variable $Y$ to be the total number of trials until the $r$-th success occurs, where $r$ is a given natural number. This is known as the negative binomial distribution with parameters $p$ and $r$. Clearly, if $r = 1$, the negative binomial distribution is just the geometric distribution. We now derive the pmf of the negative binomial distribution.

We first identify the all possible values of $Y$ to be $y = r, r + 1, r + 2, \ldots$ since we need at least $r$ trials to get the $r$th success. Hence there must be $r$ successes and $y - r$ failures in these $y$ trials. But note that the $y$th trial must be a success when the sequence of trials is stopped and in the preceding $y - 1$ trials there must be $r - 1$ successes and $y - r$ failures in any possible order. Hence,

$$P(Y = y) = P(\text{Success in the } r\text{th trial}) \times \binom{y - 1}{r - 1} p^{r-1}(1 - p)^{y-1-(r-1)},$$

applying the binomial probability of having $r - 1$ successes from $n = r - 1$ trials with success probability $p$. Thus, the pmf of $Y$ is given by

$$P(Y = y) = p \times \binom{y - 1}{r - 1} p^{r-1}(1 - p)^{(y-1)-(r-1)}$$

$$= \binom{y - 1}{r - 1} p^r (1 - p)^{y-r}, \quad y = r, r + 1, \ldots$$

This is a particular form of the negative binomial distribution. However, the R software package uses a different form $X = Y - r$ so that $X$ takes the values $0, 1, 2, \ldots$ corresponding to $Y$ taking the values $r, r + 1, r + 2$ and so on. Thus $X$ counts the number of failures before the $r$th success and $Y$ counts the total number of trials to observe the $r$th success.

We find the probability mass function of $X$ from that of $Y$ as follows:

$$P(X = x) = P(Y - r = x)$$

$$= P(Y = r + x)$$

$$= \binom{r + x - 1}{r - 1} p^r (1 - p)^{r+x-r}$$

$$= \binom{r + x - 1}{x} p^r (1 - p)^x, \quad x = 0, 1, \ldots.$$

It is easy to see that

$$\sum_{x=0}^{\infty} P(X = x) = \sum_{x=0}^{\infty} \binom{r + x - 1}{x} p^r (1 - p)^x = 1,$$

using the identity (6.2). We define the *negative binomial distribution* by the random variable $X$ having the above probability mass function. We also define the notation that $X \sim \text{Negbin}(r, p)$ if the random variable $X$ has the above pmf.

## 6.5.1   Probability Calculation Using R

The R functions `dnbinom` and `pnbinom` evaluate the pmf and the cdf of the negative binomial random variable $X$. To obtain $P(X = 2)$ for $r = 4$ and $p = 0.25$ we issue the command, `dnbiom(x=2, size=4, prob=0.25)` which returns 0.022. Thus `size` is the target number of successes, $r$ here.

---

**Example 6.2**

A man plays roulette, betting on red each time. He decides to keep playing until he achieves his second win. The success probability for each game is 18/37 and the results of games are independent. Let $Y$ be the number of games played until he gets his second win. Find $P(Y \geq 4)$ using R. Since R uses the negative binomial distribution counting the additional number of trials, denoted by $X$, after the first $r = 2$ trials, we have

$$P(Y \geq 4) = P(X + 2 \geq 4) = P(X \geq 2) = 1 - P(X \leq 1),$$

which is calculated by the command, `1-pnbinom(q=1, size=2, prob=18/37)` which evaluates to be 0.52. ◄

## 6.5.2   Mean of the Negative Binomial Distribution

To derive the mean we evaluate $E(X)$ by exploiting the identity (6.2).

*Proof*

$$E(X) = \sum_{x=0}^{\infty} x \binom{r+x-1}{x} p^r (1-p)^x$$
$$= p^r \sum_{x=1}^{\infty} x \frac{(r+x-1)!}{x!(r-1)!} 1 - p)^x$$
$$= p^r (1-p) \sum_{x=1}^{\infty} \frac{(r+x-1)!}{(x-1)!(r-1)!} (1-p)^{x-1}$$
$$= p^r r (1-p) \sum_{y=0}^{\infty} \frac{(r+y)!}{y!r!} (1-p)^y$$
$$= p^r r (1-p) \sum_{y=0}^{\infty} \binom{r+y}{y} (1-p)^y$$
$$= p^r r (1-p) \sum_{y=0}^{\infty} \binom{r+1+y-1}{y} (1-p)^y$$
$$= p^r r (1-p) \{1 - (1-p)\}^{-(r+1)}$$
$$= \frac{r(1-p)}{p}.$$

Hence, because of the relationship $Y = X + r$ we have

$$E(Y) = E(X) + r = \frac{r(1-p)}{p} + r = \frac{r}{p}.$$

### 6.5.3  Variance of the Negative Binomial Distribution

To find the variance we need to obtain $E[X(X-1)]$ similarly as in the above derivation for the mean and then we can show that

$$\text{Var}(Y) = \text{Var}(X) = r \frac{1-p}{p^2},$$

see Exercise 6.1 below. This derivation is lengthy. In Sect. 8.7 we will provide an alternative proof for this result.

## 6.6  Poisson Distribution

A random variable $X$ has the Poisson distribution with parameter $\lambda > 0$ if it has the pmf:

$$P(X = x) = e^{-\lambda} \frac{\lambda^x}{x!}, \quad x = 0, 1, 2, \ldots$$

We write $X \sim \text{Poisson}(\lambda)$. It is trivial to show $\sum_{x=0}^{\infty} P(X = x) = 1$, i.e. $\sum_{x=0}^{\infty} e^{-\lambda} \frac{\lambda^x}{x!} = 1$. The identity we need here is simply the expansion of $e^{\lambda}$.

The Poisson distribution can be derived in many ways. It is often used to count rare events in an infinite population as we provide the proof in Sect. 6.6.4. Theoretically, a random variable following the Poisson distribution can take any integer value from 0 to $\infty$. Examples of the Poisson distribution include: the number of breast cancer patients in Southampton; the number of text messages sent (or received) per day by a randomly selected first-year student; the number of credit cards a randomly selected person has in their wallet.

### 6.6.1   Probability Calculation Using R

For the Poisson distribution the command dpois(x=3, lambda=5) calculates the pmf of Poisson($\lambda = 5$) at $x = 3$. That is, the command will return the value $P(X = 3) = e^{-5}\frac{5^3}{3!}$. The command ppois returns the cdf or the probability up to and including the argument. Thus ppois(q=3, lambda=5) will return the value of $P(X \leq 3)$ when $X \sim$ Poisson($\lambda = 5$).

### 6.6.2   Mean of the Poisson Distribution

Let $X \sim$ Poisson($\lambda$). Then $E(X) = \lambda$.

*Proof*

$$E(X) = \sum_{x=0}^{\infty} x P(X = x)$$
$$= \sum_{x=0}^{\infty} x e^{-\lambda}\frac{\lambda^x}{x!}$$
$$= e^{-\lambda} \sum_{x=1}^{\infty} x \frac{\lambda^x}{x!}$$
$$= e^{-\lambda} \sum_{x=1}^{\infty} \frac{\lambda \cdot \lambda^{(x-1)}}{(x-1)!}$$
$$= \lambda e^{-\lambda} \sum_{x=1}^{\infty} \frac{\lambda^{(x-1)}}{(x-1)!}$$
$$= \lambda e^{-\lambda} \sum_{y=0}^{\infty} \frac{\lambda^y}{y!} \quad [y = x - 1]$$
$$= \lambda e^{-\lambda} e^{\lambda} \quad [\text{using the expansion of } e^{\lambda}]$$
$$= \lambda.$$

### 6.6.3   Variance of the Poisson Distribution

Let $X \sim$ Poisson($\lambda$). Then $\text{Var}(X) = \lambda$.

**Proof**

$$E[X(X-1)] = \sum_{x=0}^{\infty} x(x-1)P(X=x)$$

$$= \sum_{x=0}^{\infty} x(x-1)e^{-\lambda}\frac{\lambda^x}{x!}$$

$$= e^{-\lambda} \sum_{x=2}^{\infty} x(x-1)\frac{\lambda^x}{x!}$$

$$= e^{-\lambda} \sum_{x=2}^{\infty} \lambda^2 \frac{\lambda^{x-2}}{(x-2)!}$$

$$= \lambda^2 e^{-\lambda} \sum_{y=0}^{\infty} \frac{\lambda^y}{y!} \quad [y = x-2]$$

$$= \lambda^2 e^{-\lambda} e^{\lambda} = \lambda^2 \quad [\text{using the expansion of } e^{\lambda}]$$

Now, $E(X^2) = E[X(X-1)] + E(X) = \lambda^2 + \lambda$. Hence,

$$\text{Var}(X) = E(X^2) - (E(X))^2 = \lambda^2 + \lambda - \lambda^2 = \lambda.$$

Hence, the mean and variance are the same for the Poisson distribution.

### 6.6.4  Poisson Distribution as a Limit of the Binomial Distribution

The Poisson distribution can be derived from as the limit of the pmf of the binomial distribution with parameters $n$ and $p_n$ where $p_n \rightarrow 0$ as $n \rightarrow \infty$ but the $\lim_{n\to\infty} np_n = \lambda$ where $0 < \lambda < \infty$ is finite number. Note that here we use the notation $p_n$ emphasise the dependence of the success probability $p_n$ on $n$. As $p_n \rightarrow 0$, the underlying event of interest is a rare event, e.g. road accident. We now derive the pmf of the Poisson distribution as the limit of the pmf of the binomial distribution for the rare events.

**Proof** Recall that if $X \sim \text{Bin}(n, p_n)$ then $P(X=x) = \binom{n}{x}p_n^x(1-p_n)^{n-x}$. Now:

$$P(X=x) = \binom{n}{x}p_n^x(1-p_n)^{n-x}$$

$$= \binom{n}{x}\frac{n^n}{n^n}p_n^x(1-p_n)^{n-x}$$

$$= \frac{n(n-1)\cdots(n-x+1)}{n^x x!}(np_n)^x (n(1-p_n))^{n-x}\frac{1}{n^{n-x}}$$

$$= \frac{n}{n}\frac{(n-1)}{n}\cdots\frac{(n-x+1)}{n}\frac{\lambda^x}{x!}\left(1-\frac{\lambda}{n}\right)^{n-x}.$$

$$= \frac{n}{n}\frac{(n-1)}{n}\cdots\frac{(n-x+1)}{n}\frac{\lambda^x}{x!}\left(1-\frac{\lambda}{n}\right)^{n}\left(1-\frac{\lambda}{n}\right)^{-x}.$$

Now it is easy to see that the above tends to

$$e^{-\lambda}\frac{\lambda^x}{x!}$$

as $n \to \infty$ for any fixed value of $x$ in the range 0, 1, 2, .... Note that we have used the exponential limit:

$$e^{-\lambda} = \lim_{n\to\infty} \left(1 - \frac{\lambda}{n}\right)^n,$$

and

$$\lim_{n\to\infty} \left(1 - \frac{\lambda}{n}\right)^{-x} = 1$$

and

$$\lim_{n\to\infty} \frac{n}{n} \frac{(n-1)}{n} \cdots \frac{(n-x+1)}{n} = 1.$$

---

**Example 6.3 (Poisson Approximation of the Binomial Distribution)**

Suppose $X$ follows the binomial distribution with parameters $n = 100$ and $p = 0.02$. Using R calculate $P(X \le 1)$ using binomial distribution and an approximate Poisson distribution.

The Poisson distribution approximation has parameter $\lambda = np = 2$. Hence the two required probabilities are obtained using the R commands:

```
pbinom(q=1, size=100, prob=0.02)
ppois(q=1, lambda=2)
```

These two commands give the results 0.4032717 and 0.4060058, which agree upto two places after the decimal point. ◄

The Poisson distribution can be derived from another consideration when we are waiting for events to occur, e.g. waiting for a bus to arrive or to be served at a supermarket till. The number of occurrences in a given time interval can sometimes be modelled by the Poisson distribution. Here the assumption is that the probability of an event (arrival) is proportional to the length of the waiting time for small time intervals. Such a process is called a Poisson process, and it can be shown that the waiting time between successive events can be modelled by the exponential distribution which is discussed in the next section.

## 6.7    Exercises

### 6.1 (Discrete Distributions)

1. Let $X$ be a Bernoulli random variable with success probability $p$. Find the value of $p$ that maximises the variance of $X$.
2. A factory produces components of which 2% are defective. The components are packed in boxes of 10. A box is selected at random. Find the probability that there is at most 1 defective component in the box.
3. Let $U$ be a discrete random variable with $P(U = u) = p(1 - p)^{u-1}$, for $u = 1, 2, 3, \ldots$ where $0 < p < 1$. Find $P(U \leq u)$.
4. A sequence of independent Bernoulli trials, each with success probability $p$, is undertaken. Find the probability that the second success occurs after exactly $n$ trials (where $n$ is an integer greater than 1).
5. Let $Y$ be a Poisson random variable with $E(Y) = \mu$. Find $E[Y(Y - 1)]$ by assuming $\text{Var}(Y) = \mu$.
6. The number of goals scored by a football team per home match is Poisson distributed with mean 2 throughout a match. For a randomly selected home match find the probability that the team scores exactly one goal in the first half of the match assuming it to be a Poisson random variable.
7. The number of incoming calls at a switchboard in one hour is Poisson distributed with mean $\lambda = 8$. The numbers arriving in non-overlapping time intervals are statistically independent. Find the probability that in 10 non-overlapping one hour periods at least two of the periods have at least 15 calls.
8. During the first UK Covid-19 lockdown Alexa plays a board game with her brother. A player can only start the game if he/she throws a six by tossing a fair 6-faced die; they take turn in throwing the die and the brother is the first player to toss. Assuming independence of the die-throwing results find the probability that Alexa is able to start the game only after her third throw but her brother was lucky enough to start after his second throw.
9. Suppose that the discrete random variable $X$ follows the negative binomial random variable with pmf

$$P(X = x) = \binom{r + x - 1}{x} p^r (1 - p)^x, \quad x = 0, 1, \ldots.$$

   Show that the pmf can be written equivalently as:

$$P(X = x) = \binom{r + x - 1}{x} \left(\frac{r}{\mu + r}\right)^r \left(\frac{\mu}{\mu + r}\right)^x, \quad x = 0, 1, \ldots,$$

   where $\mu = E(X)$.
10. The condition for the Poisson distribution with mean $\mu$ to be a good approximation to the binomial distribution with parameters $n$ and $p$ is:
    (a) $n \to \infty$
    (b) $n \to \infty, \ p \to 0$

(c) $n \to \infty$, $p \to 0$ with $np = \mu$ being a finite number.

(d) $n \to \infty$, $p \to 0$, $\mu \to \infty$

11. The performance of a computer is studied over a period of time, during which it makes $10^{15}$ computations. If the probability that a particular computation is incorrect equals $10^{-14}$, independent of other computations, write down an expression for the probability that fewer than 10 errors occur and calculate an approximation to its value using R.

12. In a certain district of a city, the proportion of people who vote Conservative is 0.3. Let a random sample of 100 voters be selected from the district.

   (a) Calculate the mean, $\mu$ and variance, $\sigma^2$ of the number of people in the sample who vote Conservative.

   (b) If $X$ denotes the number in the sample who vote Conservative and $\mu$ and $\sigma$ are the mean and standard deviation respectively as calculated in (a), use R to evaluate

   i. $P(\mu - \sigma \leq X \leq \mu + \sigma)$,

   ii. $P(\mu - 2\sigma \leq X \leq \mu + 2\sigma)$.

13. Show that $\mathrm{Var}(Y) = r\frac{1-p}{p^2}$ where $Y$ follows the negative binomial distribution with parameter $r$ and $p$.

14. Suppose that over a period of several years the average annual number of deaths from a certain non-contagious disease has been 10. If the number of deaths follows a Poisson model, what is the probability that

   (a) seven people will die from this disease this year,

   (b) 10 or more people will die from this disease this year,

   (c) there will be no deaths from this disease this year?

15. A sequence of Bernoulli trials is undertaken, each with success probability $p$. Let $X$ be the number of trials until the first success.

   (a) Show that $P(X = x) = p(1 - p)^{x-1}$, for $x = 1, 2, 3, \ldots$

   (b) Show that $\sum_{x=1}^{\infty} P(X = x) = 1$.

   (c) Find $P(X > k)$, where $k$ is a positive integer.

   (d) Show that for any two positive integers $s > 0$ and $k > 0$, $P(X > s+k|X > k) = P(X > s)$.

   (e) Show that $E(X) = \frac{1}{p}$. [**Hint:** You may assume the negative binomial series: for $n > 0$ and $|y| < 1$: $(1 - y)^{-n} = 1 + \sum_{k=1}^{\infty} \frac{n(n+1)\cdots(n+k-1)}{k!} y^k$.]

16. Let $X$ be a random variable with pmf:

$$f(x) = C \, \frac{e^{-\lambda}\lambda^x}{x!}, \quad \lambda > 0, \; x = 1, 2, \ldots$$

This is a truncated Poisson distribution.

   (a) Find the value of $C$.

   (b) Find $E(X)$.

# Standard Continuous Distributions

7

**Abstract**

Chapter 7: This chapter introduces standard continuous distributions: exponential, normal, gamma and beta. As in Chap. 6, here we find the means and variances and also discuss the R commands for finding various quantities for each distribution.

This chapter introduces several commonly used continuous probability distributions which are often used probability models in real life practical situations. The presentation style is similar to the previous chapter for discrete distributions but here the summations to obtain probabilities and expectations for discrete random variables are replaced by integration as we have encountered in Chap. 5 for continuous random variables. The distributions we introduce are exponential, normal, gamma and beta distributions.

## 7.1 Exponential Distribution

### 7.1.1 Definition

A continuous random variable $X$ is said to follow the exponential distribution if its pdf is of the form:

$$f(x) = \begin{cases} \beta e^{-\beta x} & \text{if } x > 0 \\ 0 & \text{if } x \leq 0 \end{cases}$$

where $\beta > 0$ is a parameter. We write $X \sim Expo(\beta)$. The distribution only resides in the positive half of the real line, and the tail goes down to zero exponentially as $x \to \infty$. The rate at which that happens is the parameter $\beta$. Hence $\beta$ is known as

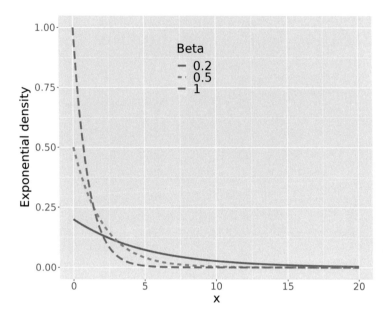

**Fig. 7.1** The pdfs the exponential random variable for $\beta = 0.2, 0.5$ and 1

the *rate parameter*. Figure 7.1 provides a plot of the exponential pdf for three values of the rate parameter $\beta$. The higher the rate parameter, the faster the rate of decline in the density. In Sect. 7.1.4 below we will prove that the mean, $E(X) = 1/\beta$.

It is easy to prove that $\int_0^{\infty} f(x)dx = 1$. This is left as an exercise. The cdf of the exponential distribution is given by:

$$F(x) = P(X \le x) = \int_0^x \beta e^{-\beta u} du = 1 - e^{-\beta x}, \quad x > 0.$$

We have $F(0) = 0$ and $F(x) \to 1$ when $x \to \infty$ and $F(x)$ is non-decreasing in $x$. The cdf can be used to solve many problems. A few examples follow.

### 7.1.2  Using R to Calculate Probabilities

For the exponential distribution the command **dexp**(x=3, rate=1/2) calculates the pdf at $x = 3$ when $\beta = 0.5$. Thus the rate parameter to be supplied is the $\beta$ parameter here. The command **pexp** returns the cdf or the probability up to and including the argument. Thus **pexp**(q=3, rate=1/2) will return the value of $P(X \le 3)$ when $X \sim \text{Expo}(\beta = 0.5)$.

**Example 7.1 (Mobile Phone)**

Suppose that the lifetime of a phone (e.g. the time until the phone does not function even after repairs), denoted by $X$, manufactured by the company A Pale, is exponentially distributed with parameter $\beta = 1/550$. [This implies that $E(X) = 1/\beta = 550$ days—assuming time is measured in days.]

1. Find the probability that a randomly selected phone will still function after two years, i.e. $X > 730$? [Assume there is no leap year in the two years].
2. What are the times by which 25, 50, 75, and 90% of the manufactured phones will have failed?

Here the mean is $1/\beta = 550$. Hence $\beta = 1/550$ is the rate parameter. The solution to the first problem is

$$P(X > 730) = 1 - P(X \leq 730) = 1 - (1 - e^{-730/550}) = e^{-730/550} = 0.2652.$$

The R command to find this is `1-pexp(q=730, rate=1/550)`.
For the second problem we are given the probabilities of failure $(0.25, 0.50$ etc.$)$. We will have to invert the probabilities to find the value of the random variable. In other words, we will have to find a $q$ such that $F(q) = p$, where $p$ is the given probability. For example, what value of $q$ will give us $F(q) = 0.25$, so that 25% of the phones will have failed by time $q$?
Returning to the exponential distribution example, we have $p = F(q) = 1 - e^{-\beta q}$. Find $q$ when $p$ is given.

$$
\begin{aligned}
p &= 1 - e^{-\beta q} \\
\Rightarrow e^{-\beta q} &= 1 - p \\
\Rightarrow -\beta q &= \log(1 - p) \\
\Rightarrow q &= \frac{-\log(1-p)}{\beta} \\
\Rightarrow q &= -550 \times \log(1 - p).
\end{aligned}
$$

Now we have the following Table 7.1:
In  R  you  can  find  these  values  by  `qexp`(p=0.25, rate=1/550), `qexp`(p=0.50, rate=1/550), etc. For fun, you can find `qexp`(p=0.99, rate =1/550) = 6 years and 343 days! The function `qexp`(p, rate) calculates the $100p$ percentile of the exponential distribution with parameter `rate`. ◄

**Table 7.1**  Some quantiles of the exponential distribution with parameter $\beta = 1/550$

| $p$ | $q = -550 \times \log(1 - p)$ |
| --- | --- |
| 0.25 | 158.22 |
| 0.50 | 381.23 |
| 0.75 | 762.46 |
| 0.90 | 1266.422 |

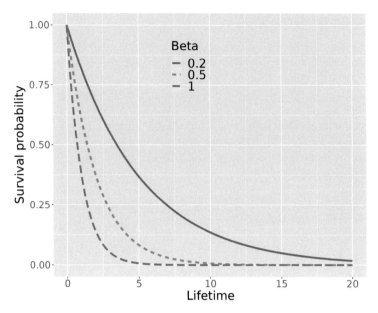

**Fig. 7.2**  $S(t)$ for $\beta = 0.2, 0.5$ and 1

---

**Example 7.2 (Survival Function)**

The exponential distribution is sometimes used to model the survival times in different experiments. For example, an exponential random variable $T$ may be assumed to model the number of days a cancer patient survives after chemotherapy. In such a situation, the function $S(t) = 1 - F(t) = e^{-\beta t}$ is called the survival function. See Fig. 7.2 for an example plot.

Assuming the mean survival time to be 100 days for a fatal late detected cancer, we can expect that half of the patients survive 69.3 days after chemo since `qexp` `(0.50, rate=1/100)` $= 69.3$. ◄

### 7.1.3  Gamma Function

The tasks of finding the mean and variance of the exponential distribution get much easier by using what is called the gamma function defined below. For any positive number $a$,

$$\Gamma(a) = \int_0^\infty x^{a-1} e^{-x} dx \qquad (7.1)$$

is defined to be the gamma function and it has a finite real value. In general, the gamma function cannot be evaluated exactly and must be evaluated by numerical approximation techniques. The R function `gamma` can be used to find an approximate value of the gamma function.

However, there are special cases when the exact value of the gamma function can be evaluated. For example, $\Gamma(1) = \int_0^\infty e^{-x}dx = 1$.

By using integration by parts, it can be easily proved that,

$$\Gamma(a) = (a-1)\Gamma(a-1) \quad \text{if } a > 1, \tag{7.2}$$

which we will use often, here and also in the later sections. Thus we have $\Gamma\left(\frac{3}{2}\right) = \frac{1}{2}\Gamma\left(\frac{1}{2}\right)$ for example.

Result (7.2), together with the fact that $\Gamma(1) = 1$, imply that

$$\Gamma(k) = (k-1)!$$

when $k$ is a positive integer. Thus $\Gamma(2) = \Gamma(1) = 1$.

We will also use another result

$$\Gamma\left(\frac{1}{2}\right) = \sqrt{\pi} \tag{7.3}$$

without a proof for now. The proof of this result will be provided as Example 14.8 in Chap. 14.

### 7.1.4 Mean and Variance of the Exponential Distribution

By definition,

$$
\begin{aligned}
E(X) &= \int_{-\infty}^\infty x f(x)dx \\
&= \int_0^\infty x\beta e^{-\beta x}dx \\
&= \int_0^\infty y e^{-y}\frac{dy}{\beta} \quad [\text{substitute } y = \beta x] \\
&= \frac{1}{\beta}\int_0^\infty y^{2-1}e^{-y}dy \\
&= \frac{1}{\beta}\Gamma(2) \\
&= \frac{1}{\beta} \quad [\text{since } \Gamma(2) = 1! = 1].
\end{aligned}
$$

Now,

$$
\begin{aligned}
E(X^2) &= \int_{-\infty}^{\infty} x^2 f(x)dx \\
&= \int_0^{\infty} x^2 \beta e^{-\beta x} dx \\
&= \beta \int_0^{\infty} \left(\frac{y}{\beta}\right)^2 e^{-y}\frac{dy}{\beta} \quad [\text{substitute } y = \beta x] \\
&= \frac{1}{\beta^2} \int_0^{\infty} y^{3-1} e^{-y} dy \\
&= \frac{1}{\beta^2} \Gamma(3) \\
&= \frac{2}{\beta^2} \quad [\text{since } \Gamma(3) = 2! = 2],
\end{aligned}
$$

and so $\text{Var}(X) = E(X^2) - [E(X)]^2 = 2/\beta^2 - 1/\beta^2 = 1/\beta^2$. Note that for this random variable the mean is equal to the standard deviation.

### 7.1.5  Memoryless Property

Like the geometric distribution, the exponential distribution also has the memoryless property. In simple terms, it means that the probability that the system will survive an additional period $s > 0$ given that it has survived up to time $t$ is the same as the probability that the system survives the period $s$ to begin with. That is, it forgets that it has survived up to a particular time when it is thinking of its future remaining life time.

The proof is exactly as in the case of the geometric distribution, reproduced below. Recall the definition of conditional probability:

$$
P(A|B) = \frac{P(A \cap B)}{P(B)}.
$$

Now the proof,

$$
\begin{aligned}
P(X > s + t | X > t) &= \frac{P(X>s+t, X>t)}{P(X>t)} \\
&= \frac{P(X>s+t)}{P(X>t)} \\
&= \frac{e^{-\beta(s+t)}}{e^{-\beta t}} \\
&= e^{-\beta s} \\
&= P(X > s).
\end{aligned}
$$

Note that the event $X > s + t$ and $X > t$ implies and is implied by $X > s + t$ since $s > 0$.

**Example 7.3**

The time $T$ between any two successive arrivals in a hospital emergency department has probability density function:

$$f(t) = \begin{cases} \lambda e^{-\lambda t} & \text{if } t \geq 0 \\ 0 & \text{otherwise.} \end{cases}$$

Historically, on average the mean of these inter-arrival times is 5 minutes. Calculate (i) $P(0 < T < 5)$, (ii) $P(T < 10 | T > 5)$. An estimate of $E(T)$ is 5. As $E(T) = \frac{1}{\lambda}$ we take $\frac{1}{5}$ as the estimate of $\lambda$.

(i) $P(0 < T < 5) = \int_0^5 \frac{1}{5} e^{-t/5} dt = [-e^{-t/5}]_0^5 = 1 - e^{-1} = 0.63212.$

(ii)

$$P(T < 10 | T > 5) = \frac{P(5 < T < 10)}{P(T > 5)}$$

$$= \frac{\int_5^{10} \frac{1}{5} e^{-t/5} dt}{\int_5^{\infty} \frac{1}{5} e^{-t/5} dt} = \frac{[-e^{-t/5}]_5^{10}}{[-e^{-t/5}]_5^{\infty}}$$

◀ $$= 1 - e^{-1} = 0.63212.$$

## 7.1.6 Exercises

**7.1 (Exponential Distribution)**

1. The random variable $X$ has probability density function

$$f(x) = \begin{cases} \lambda e^{-\lambda x} & \text{if } x \geq 0 \\ 0 & \text{otherwise.} \end{cases}$$

Find expressions for: (i) the mean, (ii) the standard deviation $\sigma$, (iii) the mode, (iv) the median. Show that the interquartile range equals $\sigma \log(3)$.

2. A light bulb has a negative exponential distribution with density function $f(y) = \lambda e^{-\lambda y}$, $y > 0$ for $\lambda > 0$.
   (a) Find the probability that the bulb is still functioning after a time $2/\lambda$.
   (b) A set of 100 independently operating light bulbs as described in this question is put on test. Find the probability that exactly 1 of them is still functioning after a time $1/\lambda$.

3. The continuous random variable $X$ has probability density function $f(x)$ given by

$$f(x) = \begin{cases} 2(x-2), & 2 \le x \le 3 \\ 0 & \text{otherwise} \end{cases}$$

Find the mean, mode and median of the random variable $X$.

4. The time to failure, $T$, measured in months, of a new type of light bulb is thought to have the distribution which has the pdf:

$$f(t) = \begin{cases} kt^2 e^{-\theta t^3} & \text{if } t > 0 \\ 0 & \text{if } t \le 0 \end{cases}$$

where $\theta > 0$.

(a) Show that $k = 3\theta$. [Hint: You may use the substitution $u = \theta t^3$.]
(b) Show that the cumulative distribution function $F(t) = 1 - e^{-\theta t^3}$ for $t > 0$.
(c) Reliability at a value $t > 0$ is defined as $S(t) = 1 - F(t)$. If the reliability at $t = 10$ is 0.9, find $\theta$ and hence the reliability at $t = 20$.
(d) 100 bulbs of this type are put in a new shop. All the bulbs that have failed are replaced at 20 month intervals and none are replaced at other times. If R is the number of bulbs that have to be replaced at the end of the first interval, find the mean and variance of R.

5. A random variable $X$ has the probability density function

$$f(x) = \begin{cases} 4xe^{-2x}, & x > 0, \\ 0 & \text{otherwise.} \end{cases}$$

(a) Show that $\int_{-\infty}^{\infty} f(x)\,dx = 1$.
(b) Using integration by parts show that the cumulative distribution function is:

$$F(x) = \int_{-\infty}^{x} f(u)\,du = 1 - e^{-2x} - 2x\,e^{-2x} \text{ for } x > 0.$$

(c) Show that $E(X) = \mu = 1$.
(d) Obtain the mode of the probability distribution of $X$. [**Hint:** You may take the derivative of $\log(f(x))$ and set it to zero and then check the second derivative condition at the solution.]
(e) Numerically evaluate the value of the cdf, $F(x)$, at the mean $\mu$ and the mode obtained previously.

By giving reasons, write down the correct inequality among: (i) mean < median < mode; (ii) mode < mean < median; (iii) mode < median < mean.

6. A continuous random variable $X$ has probability density function

$$f(x) = \begin{cases} kxe^{-x^2} & \text{if } x > 0 \\ 0 & \text{if } x \le 0 \end{cases}$$

(a) Show that $k = 2$.

(b) Show that the cumulative distribution function is:

$$F(x) = \int_0^x f(u)\,du = 1 - e^{-x^2} \text{ for } x > 0.$$

(c) Find the median of $X$ up to 3 decimal places.

(d) Find $E(X)$ and $\text{Var}(X)$. Is the variance greater than zero?

## 7.2   The Normal Distribution

The normal distribution is the most commonly encountered continuous distribution in statistics and in science in general. Here we learn many properties of this distribution. A random variable $X$ is said to have the normal distribution with parameters $\mu$ and $\sigma^2$ if it has the following pdf:

$$f(x) = \frac{1}{\sqrt{2\pi\sigma^2}} \exp\left\{ -\frac{(x-\mu)^2}{2\sigma^2} \right\}, \quad -\infty < x < \infty \qquad (7.4)$$

where $-\infty < \mu < \infty$ and $\sigma > 0$ are two given constants. In the sub-sections below we prove that

$$\mu = E(X), \quad \text{Var}(X) = \sigma^2.$$

We denote the normal distribution by the notation $N(\mu, \sigma^2)$. Then it is easy to remember the pdf of the normal distribution:

$$f(\text{variable}) = \frac{1}{\sqrt{2\pi\,\text{variance}}} \exp\left\{ -\frac{(\text{variable} - \text{mean})^2}{2\,\text{variance}} \right\}$$

where variable denotes the random variable. The density pdf is much easier to remember and work with when the mean $\mu = 0$ and variance $\sigma^2 = 1$. In this case, we simply write:

$$f(x) = \frac{1}{\sqrt{2\pi}} \exp\left\{-\frac{x^2}{2}\right\} \text{ or } f(\text{variable}) = \frac{1}{\sqrt{2\pi}} \exp\left\{-\frac{\text{variable}^2}{2}\right\}.$$

The above symbol free notation helps to identify the mean and variance of the distribution by looking at the pdf only. For example, if a random variable $U$ has the density

$$f(u) = \sqrt{\frac{\lambda^2}{2\pi}} \exp\left\{-\frac{\lambda^2(u-\gamma)^2}{2}\right\}, \quad -\infty < u < \infty$$

then it can be concluded that $U$ is normally distributed with mean $\gamma$ and variance $1/\lambda^2$. There are further such examples in Sect. 10.3.

### 7.2.1  The Mean and Variance of the Normal Distribution

It is easy to see that $f(x) > 0$ in (7.4) for all $x$. We will now prove that

**R1** $\int_{-\infty}^{\infty} f(x)dx = 1$ or total probability equals 1, so that $f(x)$ defines a valid pdf.
**R2** $E(X) = \mu$, i.e. the mean is $\mu$.
**R3** $\text{Var}(X) = \sigma^2$, i.e. the variance is $\sigma^2$.

The proofs of these results use the properties of odd and even functions in Calculus. Here is a brief review of these functions.

**Even and Odd Functions**
A function $f(x)$ is said to be an *even function* if

$$f(x) = f(-x)$$

for all possible values of $x$. For example, $f(x) = e^{-\frac{x^2}{2}}$ is an even function for real $x$.

A function $f(x)$ is said to be an *odd function* if

$$f(x) = -f(-x)$$

for all possible values of $x$. For example, $f(x) = xe^{-\frac{x^2}{2}}$ is an odd function for real $x$. It can be proved that for any real positive value of $a$,

$$\int_{-a}^{a} f(x)dx = \begin{cases} 2\int_{0}^{a} f(x)dx & \text{if } f(x) \text{ is an even function of } x \\ 0 & \text{if } f(x) \text{ is an odd function of } x. \end{cases}$$

The proof of this is not required. Here is an example for each of even and odd functions. The function $f(x) = x^2$ is an even function. Hence:

$$\int_{-a}^{a} f(x)dx = \int_{-a}^{a} x^2 dx$$
$$= \frac{x^3}{3}\Big|_{-a}^{a}$$
$$= \frac{1}{3}(a^3 + a^3)$$
$$= \frac{2}{3}a^3$$
$$= 2\int_{0}^{a} x^2 dx.$$

Similarly, $f(x) = x$ is an odd function of $x$ and consequently, $\int_{-a}^{a} f(x)dx = 0$ for any $a$.

We now return to the proof of **R1**.

$$\int_{-\infty}^{\infty} f(x)dx = \int_{-\infty}^{\infty} \frac{1}{\sqrt{2\pi\sigma^2}} \exp\left\{-\frac{(x-\mu)^2}{2\sigma^2}\right\} dx$$
$$= \frac{1}{\sqrt{2\pi}} \int_{-\infty}^{\infty} \exp\left\{-\frac{z^2}{2}\right\} dz \quad [\text{substitute } z = \frac{x-\mu}{\sigma} \text{ so that } dx = \sigma dz]$$
$$= \frac{1}{\sqrt{2\pi}} 2\int_{0}^{\infty} \exp\left\{-\frac{z^2}{2}\right\} dz \quad [\text{since the integrand is an even function}]$$
$$= \frac{1}{\sqrt{2\pi}} 2\int_{0}^{\infty} \exp\{-u\} \frac{du}{\sqrt{2u}}$$
$$\quad [\text{substitute } u = \frac{z^2}{2} \text{ so that } z = \sqrt{2u} \text{ and } dz = \frac{du}{\sqrt{2u}}]$$
$$= \frac{1}{2\sqrt{\pi}} 2\int_{0}^{\infty} u^{\frac{1}{2}-1} \exp\{-u\} du \quad [\text{rearrange the terms}]$$
$$= \frac{1}{\sqrt{\pi}} \Gamma\left(\frac{1}{2}\right) \quad [\text{recall the definition of the Gamma function}]$$
$$= \frac{1}{\sqrt{\pi}} \sqrt{\pi} = 1 \quad [\text{as } \Gamma\left(\frac{1}{2}\right) = \sqrt{\pi}].$$

To prove **R2**, i.e. $E(X) = \mu$, we prove the following two results:

$$(i) X \sim N(\mu, \sigma^2) \longleftrightarrow Z \equiv \frac{X - \mu}{\sigma} \sim N(0, 1) \tag{7.5}$$

$$(ii)\ E(Z) = 0. \tag{7.6}$$

Then by the linearity of expectations, i.e. if $X = \mu + \sigma Z$ for constants $\mu$ and $\sigma$ then $E(X) = \mu + \sigma E(Z) = \mu$, the result follows. To prove (7.5), we first calculate the cdf of $Z$, given by:

$$\Phi(z) = P(Z \le z)$$

$$= P\left(\frac{X - \mu}{\sigma} \le z\right)$$

$$= P(X \le \mu + z\sigma)$$

$$= \int_{-\infty}^{\mu + z\sigma} \frac{1}{\sqrt{2\pi\sigma^2}} \exp\left\{-\frac{(x-\mu)^2}{2\sigma^2}\right\} dx$$

$$= \int_{-\infty}^{z} \frac{1}{\sqrt{2\pi}} \exp\left\{-\frac{u^2}{2}\right\} du, \quad [u = (x - \mu)/\sigma] \qquad (7.7)$$

and so the pdf of $Z$ is

$$\frac{d\Phi(z)}{dz} = \frac{1}{\sqrt{2\pi}} \exp\left\{-\frac{z^2}{2}\right\} \quad \text{for } -\infty < z < \infty,$$

by the fundamental theorem of calculus. This proves that $Z \sim N(0, 1)$. The converse is proved just by reversing the steps. Thus we have proved (i) above. We use the $\Phi(\cdot)$ notation to denote the cdf of the standard normal distribution. Now:

$$E(Z) = \int_{-\infty}^{\infty} z f(z) dz$$
$$= \int_{-\infty}^{\infty} z \frac{1}{\sqrt{2\pi}} \exp\left\{-\frac{z^2}{2}\right\} dz$$
$$= \frac{1}{\sqrt{2\pi}} \times 0 = 0,$$

since the integrand $g(z) = z \exp\left\{-\frac{z^2}{2}\right\}$ is an odd function, i.e. $g(z) = -g(-z)$; for an odd function $g(z)$, $\int_{-a}^{a} g(z) dz = 0$ for any $a$. Therefore we have also proved (7.6) and hence **R2**.

To prove **R3**, i.e. $\text{Var}(X) = \sigma^2$, we show that $\text{Var}(Z) = 1$ where $Z = \frac{X-\mu}{\sigma}$ and then claim that $\text{Var}(X) = \sigma^2\text{Var}(Z) = \sigma^2$ from our earlier result. Since $E(Z) = 0$,

$\mathrm{Var}(Z) = E(Z^2)$, which is calculated below:

$$E(Z^2) = \int_{-\infty}^{\infty} z^2 f(z)dz$$

$$= \int_{-\infty}^{\infty} z^2 \frac{1}{\sqrt{2\pi}} \exp\left\{-\frac{z^2}{2}\right\} dz$$

$$= \frac{2}{\sqrt{2\pi}} \int_0^{\infty} z^2 \exp\left\{-\frac{z^2}{2}\right\} dz \quad \text{[since the integrand is an even function]}$$

$$= \frac{2}{\sqrt{2\pi}} \int_0^{\infty} 2u \exp\left\{-u\right\} \frac{du}{\sqrt{2u}}$$

[substituted $u = \frac{z^2}{2}$ so that $z = \sqrt{2u}$ and $dz = \frac{du}{\sqrt{2u}}$]

$$= \frac{4}{2\sqrt{\pi}} \int_0^{\infty} u^{\frac{1}{2}} \exp\left\{-u\right\} du$$

$$= \frac{2}{\sqrt{\pi}} \int_0^{\infty} u^{\frac{3}{2}-1} \exp\left\{-u\right\} du$$

$$= \frac{2}{\sqrt{\pi}} \Gamma\left(\frac{3}{2}\right) \quad \text{[definition of the gamma function]}$$

$$= \frac{2}{\sqrt{\pi}} \left(\frac{3}{2} - 1\right) \Gamma\left(\frac{3}{2} - 1\right) \quad \text{[reduction property of the gamma function]}$$

$$= \frac{2}{\sqrt{\pi}} \frac{1}{2} \sqrt{\pi} \quad \text{[since } \Gamma\left(\frac{1}{2}\right) = \sqrt{\pi}\text{]}$$

$$= 1,$$

as we hoped for. This proves **R3**.

**Linear Transformation of a Normal Random Variable**

Suppose $X \sim N(\mu, \sigma^2)$ and $a$ and $b$ are constants. Then the distribution of $Y = aX + b$ is $N(a\mu + b, a^2\sigma^2)$.

***Proof*** The result that $Y$ has mean $a\mu + b$ and variance $a^2\sigma^2$ can already be claimed from the linearity of the expectations and the variance result for linear functions. What remains to be proved is that the normality of $Y$, i.e. how can we claim that $Y$ will follow the normal distribution too? For this, we note that

$$Y = aX + b = a(\mu + \sigma Z) + b = (a\mu + b) + a\sigma Z$$

since $X = \mu + \sigma Z$. Now we use (7.5) to claim the normality of $Y$.

## 7.2.2   Mode of the Normal Distribution

To find the mode we maximise the pdf $f(x)$ over the values of $x$. In this case we obtain $f'(x)$, solve $f'(x) = 0$ and then check the second derivative condition, $f''(x) < 0$ at the solution. However, often, it is easier to work with $\log(f(x))$ since

$\log(y)$ is a one-to-one function (or bijection) of $y$. Hence, to find the mode, we first solve $\frac{d \log(f(x))}{dx} = 0$. Here,

$$f(x) = \frac{1}{\sqrt{2\pi\sigma^2}} \exp\left\{-\frac{(x-\mu)^2}{2\sigma^2}\right\},$$

and we use the rule:

$$\log\left(\frac{1}{b}e^a\right) = -\log(b) + a,$$

Hence,

$$\log(f(x)) = -\tfrac{1}{2}\log(2\pi\sigma^2) - \frac{(x-\mu)^2}{2\sigma^2}$$
$$\frac{d \log(f(x))}{dx} = -\frac{1}{2\sigma^2}\, 2\,(x-\mu)$$
$$= -\frac{1}{\sigma^2}\,(x-\mu),$$

therefore,

$$\frac{d \log(f(x))}{dx} = 0$$
$$\implies -\frac{1}{\sigma^2}(x-\mu) = 0$$
$$\implies x - \mu = 0$$
$$\implies x = \mu$$

Now we check the second derivative condition:

$$\frac{d \log(f(x))}{dx} = -\frac{1}{\sigma^2}(x-\mu)$$

$$\implies \frac{d^2 \log(f(x))}{dx^2} = -\frac{1}{\sigma^2} < 0 \text{ for all values of } x.$$

Hence we have proved that the mode is at $x = \mu$, same as mean. It is easy to see this from the pdf $f(x)$ as well. Hence the pdf (7.4) $f(x)$ is maximised when $(x-\mu)^2$ is minimised at $x = \mu$.

### 7.2.3   Symmetry of the Normal Distribution

For any $h$, we have,

$$f(\mu+h) = \frac{1}{\sqrt{2\pi\sigma^2}} \exp\left\{-\frac{(\mu+h-\mu)^2}{2\sigma^2}\right\} = \frac{1}{\sqrt{2\pi\sigma^2}} \exp\left\{-\frac{h^2}{2\sigma^2}\right\},$$
$$f(\mu-h) = \frac{1}{\sqrt{2\pi\sigma^2}} \exp\left\{-\frac{(\mu-h-\mu)^2}{2\sigma^2}\right\} = \frac{1}{\sqrt{2\pi\sigma^2}} \exp\left\{-\frac{h^2}{2\sigma^2}\right\}.$$

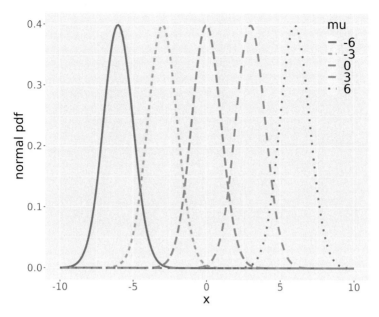

**Fig. 7.3** The pdf of the normal random variables for different values of the mean $\mu$ where $\sigma^2 = 1$

Thus $f(\mu + h) = f(\mu - h)$ for any $h$. Hence the pdf is completely symmetric about $\mu$. Hence, $P(X \leq \mu) = P(X > \mu)$. But since the total probability is 1,

$$P(-\infty < X < \infty) = P(-\infty < X \leq \mu) + P(\mu < X < \infty) = 1,$$

which in turn implies $P(X \leq \mu) = P(X > \mu) = \frac{1}{2}$. This shows that the median of the distribution is $\mu$. Hence, for the normal distribution,

$$\mu = \text{mean} = \text{median} = \text{mode}.$$

Figure 7.3 gives many examples of the pdf of $N(\mu, \sigma^2)$ for different values of $\mu$. Figure 7.4 shows the densities for different values of $\sigma^2$. A shift in the mean value changes the location of the distribution while a shift in the variance changes the height at the peak.

### 7.2.4 Standard Normal Distribution

Now we can claim that the normal pdf (7.4) is symmetric about the mean $\mu$. The spread of the pdf is determined by $\sigma$, the standard deviation of the distribution. When $\mu = 0$ and $\sigma = 1$, the normal distribution $N(0, 1)$ is called the standard normal distribution. The standard normal distribution, often denoted by $Z$, is used

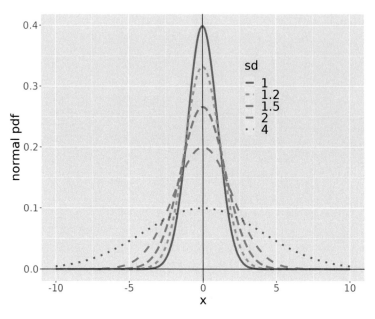

**Fig. 7.4** The pdf of the normal random variables for different values the sd $\sigma$ where $\mu = 0$

to calculate probabilities of interest for any normal distribution because of the following reasons. Suppose $X \sim N(\mu, \sigma^2)$ and we are interested in finding $P(a \leq X \leq b)$ for two constants $a$ and $b$.

$$
\begin{aligned}
P(a \leq X \leq b) &= \int_a^b f(x)dx \\
&= \int_a^b \frac{1}{\sqrt{2\pi\sigma^2}} \exp\left\{-\frac{(x-\mu)^2}{2\sigma^2}\right\} dx \\
&= \frac{1}{\sqrt{2\pi}} \int_{\frac{a-\mu}{\sigma}}^{\frac{b-\mu}{\sigma}} \exp\left\{-\frac{z^2}{2}\right\} dz \ [\text{substituted } z = \frac{x-\mu}{\sigma} \text{ so that } dx = \sigma dz] \\
&= \int_{-\infty}^{\frac{b-\mu}{\sigma}} \frac{1}{\sqrt{2\pi}} \exp\left\{-\frac{z^2}{2}\right\} dz - \int_{-\infty}^{\frac{a-\mu}{\sigma}} \frac{1}{\sqrt{2\pi}} \exp\left\{-\frac{z^2}{2}\right\} dz \\
&= P\left(Z \leq \frac{b-\mu}{\sigma}\right) - P\left(Z \leq \frac{a-\mu}{\sigma}\right) \\
&= \text{cdf of } Z \text{ at } \frac{b-\mu}{\sigma} - \text{cdf of } Z \text{ at } \frac{a-\mu}{\sigma} \\
&= \Phi\left(\frac{b-\mu}{\sigma}\right) - \Phi\left(\frac{a-\mu}{\sigma}\right)
\end{aligned}
$$

where we use the notation $\Phi(z)$ in (7.7) to denote the cdf of $Z$. This result allows us to find the probabilities about a normal random variable $X$ of any mean $\mu$ and variance $\sigma^2$ through the probabilities of the standard normal random variable $Z$. For this reason, only $\Phi(z)$ is tabulated.

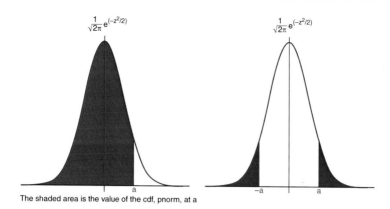

The shaded area is the value of the cdf, pnorm, at a

**Fig. 7.5** The left panel shows the cdf of the standard normal distribution $\Phi(a)$ and the right panel shows the result $\Phi(-a) = 1 - \Phi(a)$

Furthermore, due to the symmetry of the pdf of $Z$, $\Phi(z)$ is tabulated only for positive $z$ values. Suppose $a > 0$, then (see Fig. 7.5)

$$\Phi(-a) = P(Z \leq -a) = P(Z > a)$$
$$= 1 - P(Z \leq a)$$
$$= 1 - \Phi(a).$$

## 7.2.5 Using R for the Normal Distribution

### 7.2.5.1 Probability Calculation Using R

In R , we use the function `pnorm` to calculate the probabilities. The general function is: `pnorm(q, mean = 0, sd = 1, lower.tail = TRUE, log.p = FALSE`
`)`. So, we use the command `pnorm(1)` to calculate $\Phi(1) = P(Z \leq 1)$. We can also use the command `pnorm(15, mean=10, sd=2)` to calculate $P(X \leq 15)$ when $X \sim N(\mu = 10, \sigma^2 = 4)$ directly.

1. $P(-1 < Z < 1) = \Phi(1) - \Phi(-1) = 0.6827$. This means that 68.27% of the probability lies within 1 standard deviation of the mean.
2. $P(-2 < Z < 2) = \Phi(2) - \Phi(-2) = 0.9545$. This means that 95.45% of the probability lies within 2 standard deviations of the mean.
3. $P(-3 < Z < 3) = \Phi(3) - \Phi(-3) = 0.9973$. This means that 99.73% of the probability lies within 3 standard deviations of the mean.

**Example 7.4**

The distribution of the heights of 18-year-old girls may be modelled by the normal distribution with mean 162.5 cm and standard deviation 6 cm. Find the probability that the height of a randomly selected 18-year-old girl is

(i) under 168.5 cm
(ii) over 174.5 cm
(iii) between 168.5 and 174.5 cm.

The answers are:

(i) `pnorm(168.5, mean=162.5, sd=6)` $= 0.841$.
(ii) `1- pnorm(174.5, mean=162.5, sd=6)` $= 0.023$.
(iii) `pnorm(174.5, mean=162.5, sd=6)- pnorm(168.5, mean=162.5,`
◀    `sd=6)` $= 0.136$

---

**Example 7.5**

Historically, the marks in a statistics course follow the normal distribution with mean 58 and standard deviation 32.25.

1. What percentage of students will fail (i.e. score less than 40) in statistics?
   Answer: `pnorm(40, mean=58, sd=32.25)` $= 28.84\%$.
2. What percentage of students will get an A result (score greater than 70)?
   Answer: `1- pnorm(70, mean=58, sd=32.25)` $= 35.49\%$.
3. What is the probability that a randomly selected student will score more than 90? Answer: `1- pnorm(90, mean=58, sd=32.25)` $= 0.1605$.
4. What is the probability that a randomly selected student will score less than
   ◀ 25? Answer: `pnorm(25, mean=58, sd=32.25)` $= 0.1531$.

---

**Example 7.6**

A lecturer set and marked an examination and found that the distribution of marks was $N(42, 14^2)$. The school's policy is to present scaled marks whose distribution is $N(50, 15^2)$. What linear transformation should the lecturer apply to the raw marks to accomplish this and what would the raw mark of 40 be transformed to?
Suppose $X \sim N(\mu_x = 42, \sigma_x^2 = 14^2)$ and $Y \sim N(\mu_y = 50, \sigma_y^2 = 15^2)$. Hence, we should have

$$Z = \frac{X - \mu_x}{\sigma_x} = \frac{Y - \mu_y}{\sigma_y},$$

giving us:

$$Y = \mu_y + \frac{\sigma_y}{\sigma_x}(X - \mu_x) = 50 + \frac{15}{14}(X - 42).$$

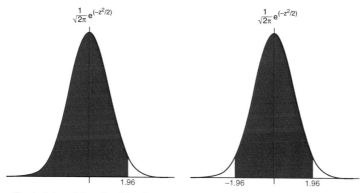

The shaded area is the value of the cdf, pnorm, at a

**Fig. 7.6** Interpreting qnorm in two different ways

Now at raw mark $X = 40$, the transformed mark would be:

$$Y = 50 + \frac{15}{14}(40 - 42) = 47.86.$$

◀

### 7.2.5.2 Quantile Calculation Using R

We are often interested in the quantiles (inverse-cdf of probability, $\Phi^{-1}(\cdot)$ of the normal distribution for various reasons. We find the $p$th quantile by issuing the R command qnorm(p).

1. qnorm(0.95) = $\Phi^{-1}(0.95)$ = 1.645. This means that the 95th percentile of the standard normal distribution is 1.645. This also means that $P(-1.645 < Z < 1.645) = \Phi(1.645) - \Phi(-1.645) = 0.90$.
2. qnorm(0.975) = $\Phi^{-1}(0.975)$ = 1.96. This means that the 97.5th percentile of the standard normal distribution is 1.96. This also means that $P(-1.96 < Z < 1.96) = \Phi(1.96) - \Phi(-1.96) = 0.95$.

Figure 7.6 illustrates two ways to interpret the quantiles of the standard normal distribution obtained by the R function qnorm. We often say that the 95% cut-off point for the standard normal distribution is 1.96.

**Example 7.7**

The random variable $X$ is normally distributed. The mean is twice the standard deviation. It is given that $P(X > 5.2) = 0.9$. Find the standard deviation.

Let $\mu$ and $\sigma$ denote the mean and standard deviation respectively. Here $\mu = 2\sigma$. Now,

$$0.1 = P(X \le 5.2) = P\left(\frac{X-\mu}{\sigma} \le \frac{5.2-\mu}{\sigma}\right)$$

$$= P\left(Z \le \frac{5.2-\mu}{\sigma}\right)$$

$$= \Phi\left(\frac{5.2-\mu}{\sigma}\right).$$

Thus

$$\frac{5.2-2\sigma}{\sigma} = \Phi^{-1}(0.1)$$
$$= qnorm(0.1)$$
$$\frac{5.2}{\sigma} = 2 - 1.282$$

◄                          $\sigma = 7.239$

### Example 7.8

Tyre pressures on a certain type of car independently follow a normal distribution with mean 1.9 bars and standard deviation 0.15 bars. Safety regulations state that the pressures must be between $1.9 - b$ amd $1.9 + b$ bars. It is known that 80% of tyres are within these safety limits. Find the safety limits. ◄

**Solution** Let $X$ denote the tyre pressure of a randomly selected car. It is given that:

$$P(1.9 - b \le X \le 1.9 + b) = P\left(\frac{1.9-b-\mu}{\sigma} \le \frac{X-\mu}{\sigma} \le \frac{1.9+b-\mu}{\sigma}\right)$$

$$= P\left(\frac{1.9-b-1.9}{0.15} \le \frac{X-\mu}{\sigma} \le \frac{1.9+b-1.9}{0.15}\right)$$

$$= P\left(-\frac{b}{0.15} \le \frac{X-\mu}{\sigma} \le \frac{b}{0.15}\right)$$

$$= \Phi\left(\frac{b}{0.15}\right) - \Phi\left(-\frac{b}{0.15}\right)$$

$$= \Phi\left(\frac{b}{0.15}\right) - \left[1 - \Phi\left(\frac{b}{0.15}\right)\right]$$
$$= 2\Phi\left(\frac{b}{0.15}\right) - 1$$

Thus, we have:

$$2\Phi\left(\frac{b}{0.15}\right) - 1 = 0.8$$
$$\Phi\left(\frac{b}{0.15}\right) = 0.9$$
$$\frac{b}{0.15} = 1.282$$
$$b = 0.15(1.282)$$
$$= 0.192$$

Hence the required safety limit is $1.9 \pm 0.192$.

## 7.2.6 Log-Normal Distribution

If $X \sim N(\mu, \sigma^2)$ then the random variable $Y = \exp(X)$ is called a log-normal random variable and its distribution is called a log-normal distribution with parameters $\mu$ and $\sigma^2$. We cannot yet find the pdf of the log-normal random variable $Y$. We will be able to find the pdf after learning further methods in Chap. 14. However, we can find the mean and variance as follows.

The mean of the random variable $Y$ is given by

$$E(Y) = E[\exp(X)]$$

$$= \int_{-\infty}^{\infty} \exp(x) \frac{1}{\sigma\sqrt{2\pi}} \exp\left\{-\frac{(x-\mu)^2}{2\sigma^2}\right\} dx$$

$$= \exp\left\{-\frac{\mu^2 - (\mu + \sigma^2)^2}{2\sigma^2}\right\}$$

$$\int_{-\infty}^{\infty} \frac{1}{\sigma\sqrt{2\pi}} \exp\left\{-\frac{x^2 - 2(\mu + \sigma^2)x + (\mu + \sigma^2)^2}{2\sigma^2}\right\} dx$$

$$= \exp\left\{-\frac{\mu^2 - (\mu + \sigma^2)^2}{2\sigma^2}\right\} \text{ [integrating a } N(\mu + \sigma^2, \sigma^2) \text{ over its domain]}$$

$$= \exp\left\{\mu + \sigma^2/2\right\}$$

where the last integral is the integral of a $N(\mu + \sigma^2, \sigma^2)$ random variable over its domain, which is one.

To find $E(Y^2) = E[\exp(2X)]$ we can proceed similarly and complete the square in the exponent. However, an easier way to find this is to exploit the fact that the random variable $W = 2X$, since it is a linear transformation, follows the normal distribution with mean $2\mu$ and variance $4\sigma^2$. Hence,

$$E[\exp(W)] = \exp\left\{2\mu + 4\sigma^2/2\right\} = \exp\left\{2\mu + 2\sigma^2\right\}.$$

Hence, the variance is given by

$$\mathrm{Var}(Y) = E(Y^2) - (E(Y))^2 = e^{2\mu+2\sigma^2} - e^{2\mu+\sigma^2} = e^{2\mu}\left(e^{2\sigma^2} - e^{\sigma^2}\right).$$

### 7.2.7  Exercises

#### 7.2 (Normal Distribution)

1. A machine dispenses cups of tea. The volumes of tea are Normally distributed with a standard deviation of 1.4 ml. What proportion of cups of tea contain a volume of tea within 2 ml of the mean?
2. A particular variety of sunflower grows to a mean height of 185 cm. The standard deviation is 17 cm. The heights of sunflower plants are Normally distributed. For this variety of sunflower plants, 10% are above a particular height. What is this height?
3. $X$ is a Normally distributed random variable with mean 3 and variance 4. $Z$ is a standard Normal random variable. $P(X > -1)$ is equal to:
   (a)] $P(Z > -1)$   (b) $P(Z < 2)$   (c) $P(Z > 2)$   (d) $P(Z < -1)$
4. The density function of the normally distributed random variable $X$ is $f(x) = \frac{1}{\sqrt{2\pi}}e^{-x^2/2}$, $-\infty < x < \infty$ and the density function of the uniformly distributed random variable $Y$ is $f(y) = 1$, for $0 < y < 1$. It is also given that the R command pnorm(0.5) produces the result 0.6915. The following statements are either true or false.
   (i) The variance of $X$ is 12 times the variance of $Y$.
   (ii) $P\left(X < \frac{1}{2}\right) = 1.383 \times P\left(Y < \frac{1}{2}\right)$. Which combination below correctly describes these two statements
   (a) (i) true and (ii) true
   (b) (i) false and (ii) true
   (c) (i) true and (ii) false
   (d) (i) false and (ii) false
5. The quartiles of a normal distribution are 8 and 14 respectively. Assume that the R command qnorm(0.75) gives the output 0.6745. Find the mean and variance of the normal distribution.
6. If $X$ has the distribution $N(4, 16)$, find
   (a) the number $a$ such that $P(|X - 4| \le a) = 0.95$,
   (b) the median,
   (c) the 90th percentile,
   (d) the interquartile range.
7. Let the random variable $X$ have the pdf

$$f(x) = \frac{C}{\sqrt{2\pi}}e^{-\frac{x^2}{2}}, \quad a < x < \infty$$

where $a$ is a known constant. Let $\phi$ and $\Phi$ denote the pdf and cdf of the standard normal random variable, respectively.

(a) Show that $C = \frac{1}{1-\Phi(a)}$.

(a) Show that $E(X) = \frac{\phi(a)}{1-\Phi(a)}$ where $\phi(a) = \frac{1}{\sqrt{2\pi}}e^{-\frac{a^2}{2}}$. Evaluate this when $a = 0$.

(c) Obtain $F(x)$, the cdf of $X$, in terms of $\Phi$.

(d) What is the value of $x$ when $F(x) = 0.95$ and $a = 0$?

## 7.3 Gamma and Beta Distributions

### 7.3.1 Gamma Distribution

The gamma distribution is a generalisation of the exponential distribution introduced in Sect. 7.1. The gamma distribution also affords a special distribution called the $\chi^2$ (pronounced as chi-squared) distribution, which is often used in statistical inference. The gamma distribution takes only non-negative values and provides a very flexible probability density function given below, which can take any shape between a positively skewed to a symmetric distribution as is illustrated in Fig. 7.7. Further discussions regarding the shape of the gamma densities are provided in Sect. 13.3.

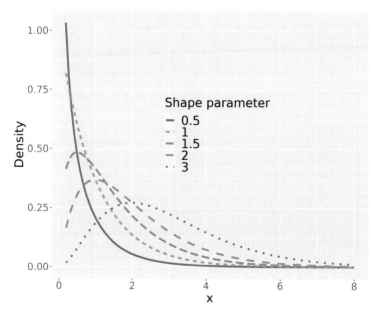

**Fig. 7.7** Illustration of gamma densities for different values of the shape parameter $\alpha$ when $\beta = 1$

Here we do not provide any real life examples of the gamma distribution because in this book we will use this distribution for theoretical purposes, e.g. the use of the $\chi^2$ distribution in deriving theoretical results in Chap. 14. However, the gamma distribution is used in modelling real life data in what is called generalised linear models, which is beyond the scope of this book.

It is said that a random variable $X$ has a *gamma distribution* with parameters $\alpha$ and $\beta$ (both $\alpha, \beta > 0$) if $X$ has a continuous distribution with the pdf:

$$f(x|\alpha, \beta) = \frac{\beta^\alpha}{\Gamma(\alpha)} x^{\alpha-1} e^{-\beta x} \text{ for } x > 0. \tag{7.8}$$

We use the notation $G(\alpha, \beta)$ to denote a gamma random variable having the above pdf.

To show that

$$\int_0^\infty f(x|\alpha, \beta) dx = 1,$$

we use the identity (7.9) given below:

$$\int_0^\infty y^{\alpha-1} e^{-\beta y} dy = \frac{\Gamma(\alpha)}{\beta^\alpha}, \tag{7.9}$$

where $\alpha > 0$ and $\beta > 0$ are parameters. This identity is easily derived from the basic definition of the gamma function (7.1), $\Gamma(\alpha)$, simply by using the substitution, $x = \beta y$.

Thus we have established the fact that the total probability is one for the gamma distribution. Note that when $\alpha = 1$ the gamma distribution reduces to the exponential distribution since the pdf of $G(1, \beta)$ is given by:

$$f(x|1, \beta) = \beta e^{-\beta x} \text{ for } x > 0,$$

which matches with the form in Sect. 7.1.

## 7.3.2   Probability Calculation Using R

The suite of R commands `dgamma`, `pgamma` and `qgamma`, with supplied parameters `shape` ($= \alpha$) and rate ($\beta$), respectively obtain the density, cdf and quantile value. For example, `dgamma`(x=1, shape=3, rate=0.5) yields 0.0379, which is the value of the density (7.8) $\frac{(0.5)^3}{\Gamma(3)}e^{-0.5} = \frac{1}{16}e^{-0.5} = 0.0379$ since $\Gamma(3) = 2$.

### 7.3.3   Mean and Variance of the Gamma Distribution

To find the mean and variance of the gamma distribution we exploit the above identity. We first show that

$$E(X^n) = \frac{1}{\beta^n}\frac{\Gamma(\alpha+n)}{\Gamma(\alpha)}.$$

*Proof*

$$E(X^n) = \int_0^\infty x^n f(x|\alpha,\beta)dx$$
$$= \int_0^\infty x^n \frac{\beta^\alpha}{\Gamma(\alpha)}x^{\alpha-1}e^{-\beta x}dx$$
$$= \frac{\beta^\alpha}{\Gamma(\alpha)}\int_0^\infty x^{\alpha+n-1}e^{-\beta x}dx$$
$$= \frac{\beta^\alpha}{\Gamma(\alpha)}\frac{\Gamma(\alpha+n)}{\beta^{\alpha+n}} \quad \text{[Replace } \alpha \text{ by } \alpha+n \text{ in (7.9)]}$$
$$= \frac{1}{\beta^n}\frac{\Gamma(\alpha+n)}{\Gamma(\alpha)}.$$

Now, taking $n = 1$,

$$E(X) = \frac{1}{\beta}\frac{\Gamma(\alpha+1)}{\Gamma(\alpha)}$$
$$= \frac{1}{\beta}\frac{\alpha\,\Gamma(\alpha)}{\Gamma(\alpha)} \quad \text{using (7.2)}$$
$$= \frac{\alpha}{\beta}.$$

Now, taking $n = 2$,

$$E(X^2) = \frac{1}{\beta^2}\frac{\Gamma(\alpha+2)}{\Gamma(\alpha)}$$
$$= \frac{1}{\beta^2}\frac{(\alpha+1)\,\alpha\,\Gamma(\alpha)}{\Gamma(\alpha)} \quad \text{using (7.2) twice}$$
$$= \frac{(\alpha+1)\,\alpha}{\beta^2}.$$

Hence,

$$\text{Var}(X) = E(X^2) - [E(X)]^2$$
$$= \frac{(\alpha+1)\,\alpha}{\beta^2} - \frac{\alpha^2}{\beta^2}$$
$$= \frac{\alpha}{\beta^2}.$$

---

**Example 7.9 ($\chi^2$ Distribution)**

The $\chi^2$ distribution with $n$ degrees of freedom is obtained if we set $\alpha = \frac{n}{2}$ and $\beta = \frac{1}{2}$. Hence the pdf of the random variable $X$, which follows the $\chi^2$ with $n$ degrees of freedom is given by

$$f(x) = \frac{1}{2^{n/2}\Gamma(n/2)} x^{\frac{n}{2}-1} e^{-\frac{1}{2}x}, \quad x > 0.$$

Mean of this distribution (often denoted by $\chi_n^2$) is $n$ and variance is $2n$. For large $n$ it can be shown that $\chi_n^2$ approaches the normal distribution. ◄

### 7.3.4  Beta Distribution

A random variable $X$ has the *beta* distribution with parameters $\alpha$ and $\beta$ (both positive) if it has the pdf:

$$f(x|\alpha, \beta) = \frac{1}{B(\alpha, \beta)} x^{\alpha-1}(1-x)^{\beta-1} \text{ for } 0 < x < 1, \qquad (7.10)$$

where

$$B(\alpha, \beta) = \int_0^1 x^{\alpha-1}(1-x)^{\beta-1} dx, \qquad (7.11)$$

is the *beta function* which takes a finite value which can be evaluated numerically, for example using the R function **beta**. We denote this distribution by using the Beta($\alpha, \beta$) notation. Figure 7.8 illustrates the beta distribution when $\alpha = \beta$, in which case the distribution is symmetric about 0.5. Figure 7.9 illustrates the beta distribution in the particular asymmetric case when $\beta = 2\alpha$.

The following result holds for the beta function.

$$B(\alpha, \beta) = \frac{\Gamma(\alpha)\Gamma(\beta)}{\Gamma(\alpha + \beta)}.$$

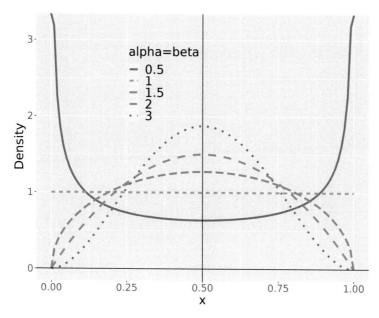

**Fig. 7.8** Illustrations of pdfs of the beta distribution in the symmetric case for $\alpha = \beta$

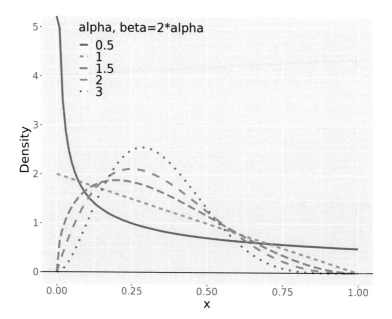

**Fig. 7.9** Illustrations of pdfs of un-symmetric beta distribution for $\beta = 2 \times \alpha$

The proof of this result is given in Example 14.9.

---

**Example 7.10 (Beta Function Evaluation)**

Find (i) $B(1, 1)$, (ii) $B(3, 2)$ and (iii) $B\left(\frac{3}{2}, \frac{3}{2}\right)$.

(i)

$$B(1, 1) = \frac{\Gamma(1)\Gamma(1)}{\Gamma(2)} = 1.$$

(ii)

$$B(3, 2) = \frac{\Gamma(3)\Gamma(2)}{\Gamma(5)}$$
$$= \frac{2!1!}{4!}$$
$$= \frac{1}{12}$$

(iii)

$$B\left(\frac{3}{2}, \frac{3}{2}\right) = \frac{\left[\Gamma\left(\frac{3}{2}\right)\right]^2}{\Gamma(3)}$$
$$= \frac{\left[\left(\frac{3}{2}-1\right)\Gamma\left(\frac{1}{2}\right)\right]^2}{\Gamma(3)}$$
$$= \frac{\left(\frac{1}{2}\sqrt{\pi}\right)^2}{2!}$$
$$= \frac{1}{8}\pi.$$

◄

To find the mean and variance of this distribution, we first show:

$$E(X^n) = \frac{\Gamma(\alpha + n)\,\Gamma(\alpha + \beta)}{\Gamma(\alpha)\,\Gamma(\alpha + \beta + n)}$$

for any positive integer $n$.

***Proof***

$$E(X^n) = \int_0^1 x^n \frac{1}{B(\alpha, \beta)} x^{\alpha-1}(1 - x)^{\beta-1} dx$$
$$= \frac{1}{B(\alpha, \beta)} \int_0^1 x^{\alpha+n-1}(1 - x)^{\beta-1} dx$$
$$= \frac{B(\alpha+n, \beta)}{B(\alpha, \beta)}$$
$$= \frac{\Gamma(\alpha+n)\,\Gamma(\alpha+\beta)}{\Gamma(\alpha)\,\Gamma(\alpha+\beta+n)}.$$

By taking $n = 1$, we obtain:

$$E(X) = \frac{\Gamma(\alpha+1)\,\Gamma(\alpha+\beta)}{\Gamma(\alpha)\,\Gamma(\alpha+\beta+1)}$$
$$= \frac{\alpha\,\Gamma(\alpha+1)\,\Gamma(\alpha+\beta)}{\Gamma(\alpha)\,(\alpha+\beta)\Gamma(\alpha+\beta)}$$
$$= \frac{\alpha}{\alpha+\beta}.$$

By taking $n = 2$ we can find $E(X^2)$ and then the variance is obtained as:

$$\mathrm{Var}(X) = \frac{\alpha\beta}{(\alpha+\beta)^2(\alpha+\beta+1)}.$$

### 7.3.5 Exercises

**7.3 (Gamma and Beta Distributions)**

1. A random variable $X$ has the pdf

$$f(x) = \begin{cases} kx^{\alpha-1}e^{-\beta x} & \text{if } x > 0 \\ 0 & \text{otherwise,} \end{cases}$$

for known positive values of $\alpha$ and $\beta$. Show that $k = \frac{\beta^{\alpha}}{\Gamma(\alpha)}$, where $\Gamma(\alpha)$ is the gamma function defined by:

$$\Gamma(\alpha) = \int_0^{\infty} y^{\alpha-1}e^{-y}\,dy$$

for $\alpha > 0$. **Hint:** In the integral first substitute $y = \beta x$ and then use the definition of the gamma function.

2. A random variable $X$ has the pdf

$$f(x) = \begin{cases} \frac{\lambda^r x^{r-1}e^{-\lambda x}}{(r-1)!} & \text{if } x > 0 \\ 0 & \text{otherwise,} \end{cases}$$

for known positive integer values of $r$ and $\lambda > 0$. Show that $E(X) = \frac{r}{\lambda}$ and $\mathrm{Var}(X) = \frac{r}{\lambda^2}$.

3. In a survey of telephone usage, it was found that the length of conversation $T$ has the pdf

$$f(t) = \frac{1}{2}\lambda^3 t^2 e^{-\lambda t}, \quad t \geq 0$$

for $\lambda > 0$. Show that $E(T) = \frac{3}{\lambda}$ and $\text{Var}(T) = \frac{3}{\lambda^2}$. Suppose $\lambda = 1$, using R find (i) $P(T < 3)$, (ii) $P(T > 10)$.

4. Suppose that a random variable $X$ has the pdf

$$\frac{1}{\sqrt{2\pi}} x^{-\frac{1}{2}} e^{-\frac{x}{2}}, 0 < x < \infty.$$

By comparing this pdf with the familiar ones give the names and the parameters of two distributions that it matches.

5. Suppose that a random variable $X$ has the pdf given by

$$f(x) = Cx^{-\frac{1}{2}}(1-x)^2, \quad 0 < x < 1$$

for a positive constant $C$. Find $C$ by recognising the distribution of $X$.

# Joint Distributions and the CLT

# 8

**Abstract**

Chapter 8: This chapter introduces the joint probability distribution for multiple random variables. It also discusses conditional and marginal distributions, conditional expectations, covariance and correlation. Finally it introduces the central limit theorem for the sum of independent random variables.

Often we need to study more than one random variable, e.g. height and weight, simultaneously, so that we can exploit the relationship between them to make inferences about their properties. Multiple random variables are studied in this chapter through their joint probability distributions.

## 8.1 Joint Distribution of Discrete Random Variables

If $X$ and $Y$ are discrete, the quantity $f(x, y) = P(X = x \cap Y = y)$ is called the *joint probability mass function* (joint pmf) of $X$ and $Y$. The notation $P(X = x \cap Y = y)$ is used as the probability $P(A \cap B)$ of the intersection of the two events $A$ and $B$. To be a joint pmf, $f(x, y)$ needs to satisfy two conditions below (following the axioms **A1** and **A2** of probability):

$$(i) \quad f(x, y) \geq 0$$

for all $x$ and $y$ and

$$(ii) \quad \sum_{\text{All } x} \sum_{\text{All } y} f(x, y) = 1.$$

The marginal probability mass functions (marginal pmf's) of $X$ and $Y$ are respectively

$$f_X(x) = \sum_{\text{All } y} f(x, y), \quad f_Y(y) = \sum_{\text{All } x} f(x, y).$$

We can use the identity in (ii) above to prove that $f_X(x)$ and $f_Y(y)$ are also probability mass functions.

---

**Example 8.1**

Suppose that two fair die are tossed independently one after the other. Let

$$X = \begin{cases} -1 & \text{if the result from dice 1 is larger} \\ 0 & \text{if the results are equal} \\ 1 & \text{if the result from dice 1 is smaller.} \end{cases}$$

Let $Y = |\text{difference between the results from the two die}|$. There are 36 possible outcomes. Each of them gives a pair of values of $X$ and $Y$. Here $Y$ can take any of the values 0, 1, 2, 3, 4, 5. Below we construct the joint probability table for $X$ and $Y$.

| Results | x | y | Results | x | y | Results | x | y |
|---------|---|---|---------|---|---|---------|---|---|
| 1  1 | 0 | 0 | 3  1 | −1 | 2 | 5  1 | −1 | 4 |
| 1  2 | 1 | 1 | 3  2 | −1 | 1 | 5  2 | −1 | 3 |
| 1  3 | 1 | 2 | 3  3 | 0 | 0 | 5  3 | −1 | 2 |
| 1  4 | 1 | 3 | 3  4 | 1 | 1 | 5  4 | −1 | 1 |
| 1  5 | 1 | 4 | 3  5 | 1 | 2 | 5  5 | 0 | 0 |
| 1  6 | 1 | 5 | 3  6 | 1 | 3 | 5  6 | 1 | 1 |
| 2  1 | −1 | 1 | 4  1 | −1 | 3 | 6  1 | −1 | 5 |
| 2  2 | 0 | 0 | 4  2 | −1 | 2 | 6  2 | −1 | 4 |
| 2  3 | 1 | 1 | 4  3 | −1 | 1 | 6  3 | −1 | 3 |
| 2  4 | 1 | 2 | 4  4 | 0 | 0 | 6  4 | −1 | 2 |
| 2  5 | 1 | 3 | 4  5 | 1 | 1 | 6  5 | −1 | 1 |
| 2  6 | 1 | 4 | 4  6 | 1 | 2 | 6  6 | 0 | 0 |

Each pair of results above (and hence pair of values of $X$ and $Y$) has the same probability $1/36$. Hence the joint probability table is given in Table 8.1
◄

The marginal probability distributions are just the row totals or column totals depending on whether you want the marginal distribution of $X$ or $Y$. For example, the marginal distribution of $X$ is given in Table 8.2.

**Table 8.1** Joint probability distribution of $X$ and $Y$

|   |   | $y$ | | | | | | |
|---|---|---|---|---|---|---|---|---|
|   |   | 0 | 1 | 2 | 3 | 4 | 5 | Total |
| $x$ | $-1$ | 0 | $\frac{5}{36}$ | $\frac{4}{36}$ | $\frac{3}{36}$ | $\frac{2}{36}$ | $\frac{1}{36}$ | $\frac{15}{36}$ |
|   | 0 | $\frac{6}{36}$ | 0 | 0 | 0 | 0 | 0 | $\frac{6}{36}$ |
|   | 1 | 0 | $\frac{5}{36}$ | $\frac{4}{36}$ | $\frac{3}{36}$ | $\frac{2}{36}$ | $\frac{1}{36}$ | $\frac{15}{36}$ |
|   | Total | $\frac{6}{36}$ | $\frac{10}{36}$ | $\frac{8}{36}$ | $\frac{6}{36}$ | $\frac{4}{36}$ | $\frac{2}{36}$ | 1 |

**Table 8.2** Marginal probability distribution of $X$

| $x$ | $P(X = x)$ |
|---|---|
| $-1$ | $\frac{15}{36}$ |
| 0 | $\frac{6}{36}$ |
| 1 | $\frac{15}{36}$ |
| Total | 1 |

Similarly the marginal probability distribution of $Y$ is given by:

| $y$ | 0 | 1 | 2 | 3 | 4 | 5 | Total |
|---|---|---|---|---|---|---|---|
| $P(Y = y)$ | $\frac{6}{36}$ | $\frac{10}{36}$ | $\frac{8}{36}$ | $\frac{6}{36}$ | $\frac{4}{36}$ | $\frac{2}{36}$ | 1 |

We can now obtain the mean and variance of $Y$.

## 8.2 Continuous Bivariate Distributions

If $X$ and $Y$ are continuous, a non-negative real-valued function $f(x, y)$ is called the *joint probability density function* (joint pdf) of $X$ and $Y$ if

$$\int_{-\infty}^{\infty} \int_{-\infty}^{\infty} f(x, y) dx dy = 1.$$

The marginal pdf's of $X$ and $Y$ are respectively

$$f_X(x) = \int_{-\infty}^{\infty} f(x, y) dy, \quad f_Y(y) = \int_{-\infty}^{\infty} f(x, y) dx.$$

**Example 8.2 (Continuous Bivariate Distribution)**

Define a joint pdf by

$$f(x, y) = \begin{cases} 6xy^2 & \text{if } 0 < x < 1 \text{ and } 0 < y < 1 \\ 0 & \text{otherwise.} \end{cases}$$

◄

How can we show that the above is a pdf? It is non-negative for all $x$ and $y$ values. But does it integrate to 1? We are going to use the following rule.

Suppose that a real-valued function $f(x, y)$ is continuous in a region $D$ where $a < x < b$ and $c < y < d$, then

$$\int \int_D f(x, y)dxdy = \int_c^d dy \int_a^b f(x, y)dx.$$

Here $a$ and $b$ may depend upon $y$ but $c$ and $d$ should be free of $x$ and $y$. When we evaluate the inner integral $\int_a^b f(x, y)dx$, we treat $y$ as constant.

**Notes**  To evaluate a bivariate integral over a region $A$ we:

- draw a picture of $A$ whenever possible.
- rewrite the region $A$ as an intersection of two one-dimensional intervals. The first interval is obtained by treating one variable as constant.
- perform two one-dimensional integrals.

**Example 8.3 (Continued)**

$$\int_0^1 \int_0^1 f(x, y)dxdy = \int_0^1 \int_0^1 6xy^2 \, dxdy$$
$$= 6\int_0^1 y^2 dy \int_0^1 x \, dx$$
$$= 3\int_0^1 y^2 dy \quad [\text{as } \int_0^1 x \, dx = \tfrac{1}{2}]$$
$$= 1. \quad [\text{as } \int_0^1 y^2 \, dy = \tfrac{1}{3}]$$

Now we can find the marginal pdf's as well.

$$f_X(x) = 2x, 0 < x < 1 \text{ and } f_Y(y) = 3y^2, 0 < y < 1.$$

◄

**Example 8.4 (Continued)**

For the bivariate continuous distribution in Example 8.2 find $P(X + Y \geq 1)$. Let $A = \{(x, y) : x + y \geq 1, \ 0 < x < 1; \ 0 < y < 1\}$. We can rewrite $A$ as: $A = \{(x, y) : 0 < y < 1; \ 1 - y \leq x < 1\}$.

$$P(X + Y \geq 1) = \int \int_A f(x, y) dx dy$$
$$= 6 \int_0^1 y^2 \int_{1-y}^1 x \, dx$$
$$= 3 \int_0^1 y^2 \, dy\{1 - (1 - y)^2\}$$
$$= 3 \int_0^1 y^2 (2y - y^2) \, dy$$
$$= \frac{9}{10}.$$

◄

More of this type of examples will be provided in Sect. 15.1, which assumes the reader's familiarity with multivariate calculus.

## 8.3 Covariance and Correlation

We first define the expectation of a real-valued scalar function $g(X, Y)$ of $X$ and $Y$:

$$E[g(X, Y)] = \begin{cases} \sum_x \sum_y g(x, y) f(x, y) & \text{if } X \text{ and } Y \text{ are discrete} \\ \int_{-\infty}^{\infty} \int_{-\infty}^{\infty} g(x, y) f(x, y) dx dy & \text{if } X \text{ and } Y \text{ are continuous.} \end{cases}$$

**Example 8.5 (Example 8.1 Continued)**

Let $g(x, y) = xy$.

$$E(XY) = (-1)(0)0 + (-1)(1)\frac{5}{36} + \cdots + (1)(5)\frac{1}{36} = 0.$$

◄

For $g(x, y) = x$, $E[g(X, Y)]$ will be the same thing as $E(X) = \sum_x x f_X(x)$. Suppose that two random variables $X$ and $Y$ have joint pmf or pdf $f(x, y)$ and let $E(X) = \mu_x$ and $E(Y) = \mu_y$. The covariance between $X$ and $Y$ is defined by

$$\text{Cov}(X, Y) = E\left[(X - \mu_x)(Y - \mu_y)\right] = E(XY) - \mu_x \mu_y.$$

Let $\sigma_x^2 = \text{Var}(X) = E(X^2) - \mu_x^2$ and $\sigma_y^2 = \text{Var}(Y) = E(Y^2) - \mu_y^2$. The correlation coefficient between $X$ and $Y$ is defined by:

$$\text{Corr}(X, Y) = \frac{\text{Cov}(X, Y)}{\sqrt{\text{Var}(X)\,\text{Var}(Y)}} = \frac{E(XY) - \mu_x \mu_y}{\sigma_x \, \sigma_y}. \tag{8.1}$$

Note that $\text{Corr}(X, Y)$, often denoted by $\rho$ is free of the measuring units of $X$ and $Y$ as the units cancel in the ratio. Below in Theorem 8.1 we prove that for any two random variables, $-1 \le \text{Corr}(X, Y) \le 1$. The correlation $\rho$ is a measure of linear dependency between two random variables $X$ and $Y$, where $|\rho| \approx 1$ indicates a strong linear relationship and $|\rho| = 0$ indicates a weak linear relationship. This will be discussed further in Chap. 17.

Suppose $X$ and $Y$ are a pair of continuous random variables with joint pdf $f(x, y)$. Then we have the following results for the random variable $Z = aX + bY$ where $a$ and $b$ are constants.

$$E(aX + bY) = aE(X) + bE(Y)$$

$$\text{Var}(aX + bY) = a^2\text{Var}(X) + b^2\text{Var}(Y) + 2ab\text{Cov}(X, Y) \tag{8.2}$$

The first result confirms that expectation of a linear function of two random variables is the linear function of the expectations. The second result shows that to find the variance of a linear function of two random variables one needs to square the linear function and in that expansion one replaces the square terms of the random variables by the variances and the cross-product term by the covariance between the two random variables. The proof is given below.

**Proof** Note that $\int_{-\infty}^{\infty} f(x, y)dy = f_X(x)$, which is the marginal pdf of $X$. and $E(X) = \int_{-\infty}^{\infty} x f_X(x)dx$. Similarly, $\int_{-\infty}^{\infty} f(x, y)dx = f_Y(y)$, which is the marginal pdf of $Y$. and $E(Y) = \int_{-\infty}^{\infty} y f_Y(y)dy$. Now,

$$
\begin{aligned}
E(aX + bY) &= \int_{-\infty}^{\infty}\int_{-\infty}^{\infty} (ax + by)f(x, y)dxdy \\
&= \int_{-\infty}^{\infty}\int_{-\infty}^{\infty} axf(x, y)dxdy + \int_{-\infty}^{\infty}\int_{-\infty}^{\infty} byf(x, y)dxdy \\
&= a\int_{-\infty}^{\infty} xdx \int_{-\infty}^{\infty} f(x, y)dy + b\int_{-\infty}^{\infty} ydy \int_{-\infty}^{\infty} f(x, y)dx \\
&= a\int_{-\infty}^{\infty} x f_X(x)dx + b\int_{-\infty}^{\infty} y f_Y(y)dy \\
&= aE(X) + bE(Y)
\end{aligned}
$$

$$\text{Var}(aX + bY) = E\left[aX + bY - E(aX + bY)\right]^2$$
$$= E\left[aX - aE(X) + bY - bE(Y)\right]^2$$
$$= E\left[a^2\{X - E(X)\}^2 + b^2\{Y - E(Y)\}^2\right]$$
$$+ 2abE\{(X - E(X))(Y - E(Y))\}$$
$$= a^2\text{Var}(X) + b^2\text{Var}(Y) + 2ab\text{Cov}(X, Y).$$

**Theorem 8.1**   *If X and Y are any two random variables,*

(a) $-1 \le \rho_{XY} \le 1$.
(b) $|\rho_{XY}| = 1$ *if and only if there exist numbers* $a \ne 0$ *and b such that* $P(Y = aX + b) = 1$. *If* $\rho_{XY} = 1$ *then* $a > 0$ *and if* $\rho_{XY} = -1$ *then* $a < 0$.

**Proof**   Substitute $U = X - \mu_x$ and $V = Y - \mu_y$ in the Cauchy-Schwarz inequality proved as Theorem 8.2 below. Then

$$[\text{Cov}(X, Y)]^2 \le \sigma_x^2 \sigma_y^2,$$

which proves the main inequality in part **a**.

For part **b** note that equality ensues in the Cauchy-Schwarz if and only if $aU + bV$ is a constant for a suitable pair of values of $a$ and $b$. But $aU + bV = c$ implies a linear relationship between $X$ and $Y$ as well. That relationship is expressed as $P(Y = aX + b) = 1$ in the statement of the theorem.                                              □

**Theorem 8.2 (Cauchy-Schwarz Inequality)**   *For any random variables U and V,*

$$[E(UV)]^2 \le E(U^2)\, E(V^2).$$

**Proof**   It is trivial if either $E(U^2)$ or $E(V^2)$ is zero. Then $U$ or $V$ must be zero with probability 1 since zero expectation of a non-negative random variable (either $U^2$ or $V^2$) implies that the random variable must take the value zero with probability 1. In this case the left hand side, $[E(UV)]^2$ will be zero as well.

Let us assume that both $E(U^2)$ and $E(V^2)$ are finite and positive. The random variables $(aU + bV)^2$ and $(aU - bV)^2$ are both non-negative. Hence their expectations must be non-negative as well. Now using (8.2) we have the following inequalities for any numbers $a$ and $b$

$$0 \le E\left[(aU + bV)^2\right] = a^2\, E(U^2) + b^2\, E(V^2) + 2ab\, E(UV)$$

and

$$0 \leq E\left[(aU - bV)^2\right] = a^2\,E(U^2) + b^2\,E(V^2) - 2ab\,E(UV).$$

Let $a = [E(V^2)]^{1/2}$ and $b = [E(U^2)]^{1/2}$. Then the above two inequalities yield respectively,

$$E(UV) \geq -\left[E(U^2)\,E(V^2)\right]^{1/2}$$

and

$$E(UV) \leq \left[E(U^2)\,E(V^2)\right]^{1/2}$$

These two inequalities are combined to write:

$$-\left[E(U^2)\,E(V^2)\right]^{1/2} \leq E(UV) \leq \left[E(U^2)\,E(V^2)\right]^{1/2},$$

which is re-written as,

$$[E(UV)]^2 \leq E(U^2)\,E(V^2),$$

which is the Cauchy-Schwarz inequality.

Note that this proof is similar to the proof for the discrete version of the Cauchy-Schwarz inequality in Exercise 1.5.

## 8.4   Independence

Independence is an important concept. Recall that we say two events $A$ and $B$ are independent if $P(A \cap B) = P(A) \times P(B)$. We use the same idea here. Two random variables $X$ and $Y$ having the joint pdf or pmf $f(x, y)$ are said to be independent if and only if

$$f(x, y) = f_X(x) \times f_Y(y) \text{ for } \textbf{ALL } x \text{ and } y.$$

It is important that the above factorisation holds for all values of the random variables $X$ and $Y$.

### Example 8.6 (Discrete Case)

$X$ and $Y$ are independent if *each* cell probability, $f(x, y)$, is the product of the corresponding row and column totals. In our very first dice example (Example 8.1) $X$ and $Y$ are not independent. Verify that in the following example $X$ and $Y$ are independent. We need to check all 9 cells. ◄

| | | $y$ | | | |
|---|---|---|---|---|---|
| | | 1 | 2 | 3 | Total |
| $x$ | 0 | $\frac{1}{6}$ | $\frac{1}{12}$ | $\frac{1}{12}$ | $\frac{1}{3}$ |
| | 1 | $\frac{1}{4}$ | $\frac{1}{8}$ | $\frac{1}{8}$ | $\frac{1}{2}$ |
| | 2 | $\frac{1}{12}$ | $\frac{1}{24}$ | $\frac{1}{24}$ | $\frac{1}{6}$ |
| | Total | $\frac{1}{2}$ | $\frac{1}{4}$ | $\frac{1}{4}$ | 1 |

### Example 8.7

Let $f(x, y) = 6xy^2$, $0 < x < 1$, $0 < y < 1$. Here we can easily check that $X$ and $Y$ are independent. ◄

### Example 8.8

Let $f(x, y) = 2x$, $0 \leq x \leq 1$, $0 \leq y \leq 1$. Here also we can easily check that $X$ and $Y$ are independent. ◄

### Example 8.9 (Deceptive)

The joint pdf may look like something you can factorise. But $X$ and $Y$ may not be independent because they may be related in the domain. The random variables $X$ and $Y$ are not independent in the following two examples.

1. $f(x, y) = \frac{21}{4}x^2y$, $x^2 \leq y \leq 1$.
2. $f(x, y) = e^{-y}$, $0 < x < y < \infty$.

◄

### Consequences of Independence
- Suppose that $X$ and $Y$ are independent random variables. Then

$$P(X \in A, Y \in B) = P(X \in A) \times P(Y \in B)$$

for any events $A$ and $B$. That is, the joint probability can be obtained as the product of the marginal probabilities. We will use this result in the next section.

For example, suppose Jack and Jess are two randomly selected students. Let $X$ denote the height of Jack and $Y$ denote the height of Jess. Then we have,

$$P(X < 182 \text{ and } Y > 165) = P(X < 182) \times P(Y > 165).$$

Obviously this has to be true for any numbers other than the example numbers 182 and 165, and for any inequalities.

• Further, let $g(x)$ be a function of $x$ only and $h(y)$ be a function of $y$ only. Then, if $X$ and $Y$ are independent, it is easy to prove that

$$E[g(X)h(Y)] = E[g(X)] \times E[h(Y)].$$

As a special case, let $g(x) = x$ and $h(y) = y$. Then we have

$$E(XY) = E(X) \times E(Y).$$

Consequently, for independent random variables $X$ and $Y$, $\text{Cov}(X, Y) = 0$ and $\text{Corr}(X, Y) = 0$. But the converse is not true in general. That is, merely having $\text{Corr}(X, Y) = 0$ does not imply that $X$ and $Y$ are independent random variables. Thus for two independent random variables $X$ and $Y$,

$$\text{Var}(aX + bY) = a^2 \text{Var}(X) + b^2 \text{Var}(Y),$$

where $a$ and $b$ are two constants.

## 8.5   Conditional Distribution

Conditional distributions play many important roles in statistics. This section introduces the main important concepts regarding these. Conditional distributions are analogously defined as the conditional probabilities, see Eq. 4.1 in Chap. 4, namely

$$P(A|B) = \frac{P(A \cap B)}{P(B)} \quad \text{if} \quad P(B) > 0.$$

Suppose that $X$ and $Y$ are random variables with joint pmf or pdf $f(x, y)$ and the marginal pmf or pdf $f_X(x)$ and $f_Y(y)$. If the random variables are discrete then we can interpret $f(x, y) = P(X = x \cap Y = y)$ and $f_X(x) = P(X = x)$.

For any $x$ such that $f_X(x) > 0$, the *conditional pmf or pdf of $Y$ given that $X = x$* is defined by

$$f(y|x) = \frac{f(x, y)}{f_X(x)}. \tag{8.3}$$

The conditional probability function $f(y|x)$ is easily interpreted when $X$ and $Y$ are both discrete random variables. In this case $f(x, y) = P(X = x \cap Y = y)$ and $f_X(x) = P(X = x)$. Hence $f(y|x) = \frac{f(x,y)}{f_X(x)} = P(Y = y|X = x)$. This interpretation is also carried over to the continuous case although in that case we do not interpret $f_X(x)$ as $P(X = x)$ for any value $x$.

The conditional probability function $f(y|x)$ defines a valid pmf or pdf, which is guaranteed by the total probability identity

$$\sum_{\text{All } x} \sum_{\text{All } y} f(x, y) = 1$$

in the discrete case and

$$\int_{-\infty}^{\infty} \int_{-\infty}^{\infty} f(x, y)dxdy = 1$$

in the continuous case. Hence, we are able to define all the properties of the random variable $Y|X = x$, such as the conditional expectation, $E(Y|X = x)$, and variance $\text{Var}(Y|X = x)$, and also the quantiles of the distribution of $Y|X = x$.

## 8.5.1 Conditional Expectation

We define conditional mean or expectation using the familiar *sum or integral value times probability* way, i.e.

$$E(Y|X = x) = \begin{cases} \sum_{\text{all } y} y\, f(y|x) & \text{if } Y \text{ is discrete} \\ \int_{-\infty}^{\infty} y\, f(y|x)dy & \text{if } Y \text{ is continuous.} \end{cases}$$

The conditional expectation $E(Y|X = x)$ is very important in statistics. It provides the best guess at $Y$ based on the knowledge $X = x$. This is often called regression as discussed in Chap. 17.

### Example 8.10

We continue with Example 8.1 and find the conditional distribution of $Y|X = 1$. The joint probability distribution has been provided as Table 8.1.

| | | $y$ | | | | | | |
|---|---|---|---|---|---|---|---|---|
| | | 0 | 1 | 2 | 3 | 4 | 5 | Total |
| $x$ | $-1$ | 0 | $\frac{5}{36}$ | $\frac{4}{36}$ | $\frac{3}{36}$ | $\frac{2}{36}$ | $\frac{1}{36}$ | $\frac{15}{36}$ |
| | 0 | $\frac{6}{36}$ | 0 | 0 | 0 | 0 | 0 | $\frac{6}{36}$ |
| | 1 | 0 | $\frac{5}{36}$ | $\frac{4}{36}$ | $\frac{3}{36}$ | $\frac{2}{36}$ | $\frac{1}{36}$ | $\frac{15}{36}$ |
| | Total | $\frac{6}{36}$ | $\frac{10}{36}$ | $\frac{8}{36}$ | $\frac{6}{36}$ | $\frac{4}{36}$ | $\frac{2}{36}$ | 1 |

Given $X = 1$ the third row of the joint probability table is the relevant row.

| | | $y$ | | | | | | |
|---|---|---|---|---|---|---|---|---|
| | | 0 | 1 | 2 | 3 | 4 | 5 | Total |
| $X = 1$ | | 0 | $\frac{5}{36}$ | $\frac{4}{36}$ | $\frac{3}{36}$ | $\frac{2}{36}$ | $\frac{1}{36}$ | $\frac{15}{36}$ |

Now $P(X = 1) = f_X(1) = \frac{15}{36}$. Hence: $f(y|X = 1) = \frac{f(x,y)}{\frac{15}{36}}$. Therefore, $f(y|X = 1)$ is given by:

| $y$ | 1 | 2 | 3 | 4 | 5 | Total |
|---|---|---|---|---|---|---|
| $f(y\|X = 1)$ | $\frac{5}{15}$ | $\frac{4}{15}$ | $\frac{3}{15}$ | $\frac{2}{15}$ | $\frac{1}{15}$ | 1 |

Now we can find:

$$E(Y|X = 1) = 1 \times \frac{5}{15} + 2 \times \frac{4}{15} + 3 \times \frac{3}{15} + 4 \times \frac{2}{15} + 5 \times \frac{1}{15} = \frac{35}{15} = \frac{7}{3}.$$

Note that here $E(Y) = \frac{35}{18}$ which does not equal $E(Y|X = 1)$. We can similarly find $E(Y^2|X = 1)$ and $\text{Var}(Y|X = 1)$ as follows.

$$E(Y^2|X = 1) = 1^2 \times \frac{5}{15} + 2^2 \times \frac{4}{15} + 3^2 \times \frac{3}{15} + 4^2 \times \frac{2}{15} + 5^2 \times \frac{1}{15} = \frac{105}{15} = 7.$$

Now

$$\text{Var}(Y|X = 1) = 7 - \left(\frac{7}{3}\right)^2 = \frac{14}{9}.$$

◀

## 8.5.2   Conditional Distribution Under Independence

If $X$ and $Y$ are independent, then

$$f(y|X = x) = \frac{f(x, y)}{f_X(x)} = \frac{f_X(x) f_Y(y)}{f_X(x)} f_Y(y) \text{ for all } x.$$

Thus, if $X$ and $Y$ are independent, then conditional distribution of $Y$ given $X = x$ is same as its marginal distribution. This implies that the additional knowledge that $X = x$ has no effect on the probability distribution of $Y$. Thus we can evaluate probabilities for the random variable $Y$ without needing to know a particular value of the random variable $X$.

## 8.5.3   Some Further Remarks on Conditional Distribution

We can define the other type of conditional distributions as well. If $f_Y(y) > 0$ then the *conditional pmf or pdf of X given that $Y = y$* is defined by

$$f(x|y) = \frac{f(x, y)}{f_Y(y)}.$$

Here also, if $X$ and $Y$ are independent, then $f(x|Y = y) = f_X(x)$ for all $y$.
Regardless of independence, we can write:

$$f(x, y) = f(y|X = x) f_X(x) = f(x|Y = y) f_Y(y).$$

This shows that joint distribution of $X$ and $Y$ can be specified by a conditional and marginal distribution. This result had a parallel in probability theory, namely, the multiplication rule of conditional probability:

$$P(A \cap B) = P(B|A)P(A) = P(A|B)P(B).$$

This is referred to as hierarchical specification of a joint density $f(x, y)$.

## 8.6   Exercises

### 8.1 (Joint Distribution)

1. Two random variables $X$ and $Y$ have the following joint probability table where $k$ is a constant. Find $k$.

|   | Y |   |   |
|---|---|---|---|
|   | 1 | 2 | 3 |
| 1 | 0 | k | k |
| X 2 | k | 0 | k |
| 3 | k | k | 0 |

2. Two random variables $X$ and $Y$ have the following joint probability table. Find $\text{Cov}(X, Y)$.

|   |   | y |   |
|---|---|---|---|
|   | −1 | 0 | 1 |
| 1 | 0.2 | 0.1 | 0 |
| x 2 | 0.2 | 0.2 | 0.1 |
| 3 | 0.1 | 0 | 0.1 |

3. Two random variables $X$ and $Y$ have the joint probability table as in the previous question. Find $\text{Var}(X + Y)$.
4. $X$ and $Y$ are binary random variables satisfying:

$$P(X = 0, Y = 0) = P(X = 1, Y = 1) = 0.3$$

$$P(X = 0, Y = 1) = P(X = 1, Y = 0) = 0.2.$$

Find the covariance of $X$ and $Y$ upto two decimal places.
5. The random variables $X$, $Y$ have the following joint probability table.

|   | Y |   |   |
|---|---|---|---|
|   | 1 | 2 | 3 |
| 1 | $\frac{1}{16}$ | $\frac{1}{8}$ | $\frac{1}{16}$ |
| X 2 | $\frac{1}{8}$ | $\frac{1}{8}$ | $\frac{1}{8}$ |
| 3 | $\frac{1}{16}$ | $\frac{1}{4}$ | $\frac{1}{16}$ |

(a) Find the marginal probability functions of $X$ and $Y$.
(b) Show that $P(X = 1, Y = 2) = P(X = 1)P(Y = 2)$.
   Can it be concluded that $X$ and $Y$ are independent?
6. Two random variables $X$ and $Y$ each take only the values 0 and 1, and their joint probability table is as follows.

|   | | Y | |
|---|---|---|---|
|   |   | 0 | 1 |
| X | 0 | $a$ | $b$ |
|   | 1 | $c$ | $d$ |

(a) Show that $\text{Cov}(X, Y) = ad - bc$.

(b) Use the definition of independence to verify that $ad = bc$ when $X$ and $Y$ are independent.

(c) Show that $X$ and $Y$ are independent when $ad = bc$.

7. In a gathering there are 3 first year students, 4 second year students and 5 third year students. Two students are selected at random to form a two member committee. Let $X$ and $Y$ be the number of first year and second year students respectively in the committee.

|   |   | $y$ | | | |
|---|---|---|---|---|---|
|   |   | 0 | 1 | 2 | Total |
| $x$ | 0 | $\frac{10}{66}$ | $\frac{20}{66}$ | – | – |
|   | 1 | $\frac{15}{66}$ | – | – | – |
|   | 2 | – | – | – | $\frac{3}{66}$ |
|   | Total | – | – | – | 1 |

(a) Complete the table. Are $X$ and $Y$ independent?

(b) Find $P(X + Y = 2)$ and $P(Y = 2 | X = 0)$.

8. The joint probability distribution of a pair of random variables $X$ and $Y$ is given by the following table.

|   |   | $y$ | | | |
|---|---|---|---|---|---|
|   |   | 1 | 2 | 3 | Total |
| $x$ | 1 | $\frac{1}{27}$ | $\frac{3}{27}$ | – | $\frac{5}{27}$ |
|   | 2 | $\frac{4}{27}$ | – | $\frac{6}{27}$ | – |
|   | 3 | $\frac{4}{27}$ | $\frac{4}{27}$ | – | $\frac{12}{27}$ |
|   | Total | – | $\frac{7}{27}$ | $\frac{11}{27}$ | 1 |

(a) Complete the table. Are $X$ and $Y$ independent?

(b) Find the conditional distribution of $Y$ given $X = 1$ and $E(Y | X = 1)$.

(c) Calculate $\text{Cov}(X, Y) = E(XY) - E(X) E(Y)$.

## 8.7    Properties of Sum of Independent Random Variables

In this section we consider sums of independent random variables, which arise
frequently in both theory and practice of statistics. For example, the mark achieved
in an examination is the sum of the marks for several questions. Here we will also
use this theory to reproduce some of the results we obtained before, e.g. finding the
mean and variance of the binomial and negative binomial distributions.

The independent random variables are often obtained by drawing random
samples from a population. For example, when a fair coin is tossed $n$ times we may
use the notation $X_i$ to denote the random variable taking the value 1 if head appears
in the $i$th trial and 0 otherwise for $i = 1, \ldots, n$. This produces a random sample
$X_1, \ldots, X_n$ from the Bernoulli distribution with probability of success equal to 0.5
since the coin has been assumed to be fair.

In general, suppose we have obtained a random sample from a distribution
with pmf or pdf $f(x)$, so that $X$ can either be a discrete or a continuous random
variable. Let $X_1, \ldots, X_n$ denote the random sample of size $n$ where $n$ is a positive
integer. Note that the upper case letters are used here since each member of the
random sample is a random variable. A particular realisation will be denoted
by the corresponding lower case letters. For example, we let $x_1, \ldots, x_n$ denote
observations of the random variables $X_1, \ldots, X_n$.

For a random sample $X_1, \ldots, X_n$ the joint probability distribution is the product
of the individual probability distributions, i.e.

$$f(x_1, \ldots, x_n) = f(x_1)f(x_2) \cdots f(x_n).$$

This result has been discussed previously for two random variables $X$ and $Y$ in
Sect. 8.4. The above is a generalisation for any $n > 1$. A consequence of this is the
result that:

$$P(X_1 \in A_1, X_2 \in A_2, \ldots, X_n \in A_n)$$

$$= P(X_1 \in A_1) \times P(X_2 \in A_2) \times \cdots P(X_n \in A_n)$$

for any set of events, $A_1, A_2, \ldots A_n$. That is, the joint probability can be obtained
as the product of individual probabilities. An example of this for $n = 2$ was given
in Sect. 8.4.

**Example 8.11 (Distribution of the Sum of Independent Binomial Random Variables)**

Suppose $X \sim \text{Bin}(m, p)$ and $Y \sim \text{Bin}(n, p)$ independently. Note that $p$ is
the same in both distributions. Using the above fact that joint probability is the
multiplication of individual probabilities, we can conclude that $Z = X + Y$
has the binomial distribution. A property such as this is called the reproductive
property of random variables. It is intuitively clear that this should happen
since $X$ comes from $m$ Bernoulli trials and $Y$ comes from $n$ Bernoulli trials

independently, so $Z$ comes from $m + n$ Bernoulli trials with common success probability $p$. We can prove the result mathematically as well, by finding the probability mass function of $Z = X + Y$ directly and observing that it is of the appropriate form. First, note that

$$P(Z = z) = P(X = x, Y = y)$$

subject to the constraint that $x + y = z, 0 \leq x \leq m, 0 \leq y \leq n$. Thus,

$$
\begin{aligned}
P(Z = z) &= \sum_{x+y=z} P(X = x, Y = y) \\
&= \sum_{x+y=z} \binom{m}{x} p^x (1-p)^{m-x} \binom{n}{y} p^y (1-p)^{n-y} \\
&= \sum_{x+y=z} \binom{m}{x}\binom{n}{y} p^z (1-p)^{m+n-z} \\
&= p^z (1-p)^{m+n-z} \sum_{x+y=z} \binom{m}{x}\binom{n}{y} \\
&= \binom{m+n}{z} p^z (1-p)^{m+n-z}.
\end{aligned}
$$

Here we have used the result that

$$\sum_{x+y=z} \binom{m}{x}\binom{n}{y} = \binom{m+n}{z}$$

which follows from the total probability identity of the hypergeometric distribution discussed in Sect. 6.4. Thus, we have proved that the sum of independent binomial random variables with common success probability is binomial as well. ◄

The reproductive property of random variables seen in the Example 8.11 also holds for other distributions such as the Poisson, normal and gamma distributions. We will prove these results as exercises and also later having learned further techniques in Sect. 13.2.2.

It is often a much more challenging task to obtain the distribution of sum of independent random variables having arbitrary probability distributions. We consider this problem in the next section where we use an approximation for the distribution when the number of independent random variables, $n$, tends to be large. In this section, we aim to derive the mean and variance of the distribution of the sum of independent random variables. In fact, in general we consider the new random variable

$$Y = a_1 X_1 + a_2 X_2 + \cdots + a_n X_n$$

where $a_1, a_2, \ldots, a_n$ are constants. Note that we can simply choose $a_i = 1$ to obtain $Y$ as the sum of $n$ independent random variables $X_1, \ldots, X_n$. Suppose that $E(X_i) = \mu_i$ and $\text{Var}(X_i) = \sigma_i^2$ for all $i = 1, \ldots, n$. Then we have the results:

1. $E(Y) = a_1 \mu_1 + a_2 \mu_2 + \cdots + a_n \mu_n$.

2. $\text{Var}(Y) = a_1^2\sigma_1^2 + a_2^2\sigma_2^2 + \cdots + a_n^2\sigma_n^2$, if $X_1, \ldots, X_n$ are independent random variables.

These are generalisations of the two results (8.2) from Sect. 8.3 in the case of more than two *independent* random variables.

**Proof** The proofs are sketched for the continuous random variables only where we use integration to define expectation. The proofs for the discrete case can be obtained by replacing integration $\int$ with summation $\sum$. To write the expression for $E(Y)$ we will use the 'integral of value times probability' definition we noted in Sect. 5.4. Here the value of the random variable $Y$ is $a_1x_1 + a_2x_2 + \cdots + a_nx_n$ and the pdf is the joint pdf $f(x_1, \ldots, x_n)$. The integral in the definition of expectation will involve the multiple integral integrating over $x_1, x_2, \ldots, x_n$. However, it is not necessary to evaluate multiple integrals since $Y$ involves sum of individual random variables and, for each $i = 1, \ldots, n$

$$\underbrace{\int_{-\infty}^{\infty} \cdots \int_{-\infty}^{\infty}}_{n \text{ times}} a_i x_i f(x_1, \ldots, x_n)dx_1 \cdots dx_n$$

$$= \int_{-\infty}^{\infty} a_i x_i dx_i \underbrace{\int_{-\infty}^{\infty} \cdots \int_{-\infty}^{\infty}}_{n\text{-}1 \text{ times}} f(x_1, \ldots, x_n)dx_1 dx_{(-i)} \cdots dx_n,$$

where the notation $dx_{(-i)}$ for a particular value of $i$ is used to denote the omission of the term $dx_i$. Now,

$$\underbrace{\int_{-\infty}^{\infty} \cdots \int_{-\infty}^{\infty}}_{n\text{-}1 \text{ times}} f(x_1, \ldots, x_n)dx_1 dx_{(-i)} \cdots dx_n = f(x_i),$$

since the $n - 1$ variables $x_1, \ldots, x_{-(i)}, \ldots, x_n$ are all integrated over their whole domain. Hence,

$$\underbrace{\int_{-\infty}^{\infty} \cdots \int_{-\infty}^{\infty}}_{n \text{ times}} a_i x_i f(x_1, \ldots, x_n)dx_1 \cdots dx_n$$

$$= \int_{-\infty}^{\infty} a_i x_i f(x_i)$$
$$= a_i E(X_i) = a_i \mu_i.$$

Hence,

$$E(Y) = \underbrace{\int_{-\infty}^{\infty} \cdots \int_{-\infty}^{\infty}}_{n \text{ times}} (a_1 x_1 + a_2 x_2 + \cdots + a_n x_n) f(x_1, x_2, \ldots, x_n) dx_1 \cdots dx_n$$

$$= \int_{-\infty}^{\infty} a_1 x_1 f(x_1) + \int_{-\infty}^{\infty} a_2 x_2 f(x_2) dx_2 + \cdots + \int_{-\infty}^{\infty} a_n x_n f(x_n) dx_n$$

$$= a_1 \mu_1 + a_2 \mu_2 + \cdots + a_n \mu_n.$$

To prove the result for the variance of the sum $Y$, we need to assume that the random variables $X_1, X_2, \ldots, X_n$ are independent. We have:

$$\begin{aligned}
\text{Var}(Y) &= E\left[Y - E(Y)\right]^2 \\
&= E\left[\sum_{i=1}^{n} a_i X_i - \sum_{i=1}^{n} a_i \mu_i\right]^2 \\
&= E\left[\sum_{i=1}^{n} a_i (X_i - \mu_i)\right]^2 \\
&= E\left[\sum_{i=1}^{n} a_i^2 (X_i - \mu_i)^2 + \sum_{i=1}^{n} \sum_{j \neq i} a_i a_j (X_i - \mu_i)(X_j - \mu_j)\right] \\
&= \sum_{i=1}^{n} a_i^2 E(X_i - \mu_i)^2 + \sum_{i=1}^{n} \sum_{j \neq i} a_i a_j E\left[(X_i - \mu_i)(X_j - \mu_j)\right] \\
&= \sum_{i=1}^{n} a_i^2 \text{Var}(X_i)
\end{aligned}$$

since, for $i \neq j$

$$E\left[(X_i - \mu_i)(X_j - \mu_j)\right] = \text{Cov}(X_i, X_j) = 0$$

because $X_i$ and $X_j$ are independent random variables, see Sect. 8.4.                □

Note that in the above proof we have used the identity:

$$(b_1 + b_2 + \cdots + b_n)^2 = \sum_{i=1}^{n} b_i^2 + \sum_{j \neq i} b_i b_j,$$

which can be proved by the method of induction.
If $a_i = 1$ for all $i = 1, \ldots, n$, the two results above imply that:

1. $E(X_1 + \cdots + X_n) = \mu_1 + \mu_2 + \cdots + \mu_n.$
2. $\text{Var}(X_1 + \cdots + X_n) = \sigma_1^2 + \sigma_2^2 + \cdots + \sigma_n^2.$

The expectation of the sum of random variables is the sum of the expectations of the individual random variables

the variance of the sum of *independent* random variables is the sum of the variances of the individual random variables.

Notice that we do not require the assumption of independence of the random variables for the first result on expectation but this assumption is required for the second result on variance.

**Example 8.12 (Mean and Variance of Binomial Distribution)**

Suppose $Y \sim \text{Bin}(n, p)$. Then we can write:

$$Y = X_1 + X_2 + \ldots + X_n$$

where each $X_i$ is an independent Bernoulli trial with success probability $p$. In Sect. 6.1 we have shown that, $E(X_i) = p$ and $\text{Var}(X_i) = p(1 - p)$ by direct calculation. Now the above two results imply that:

$$E(Y) = E\left(\sum_{i=1}^{n} X_i\right) = p + p + \ldots + p = np.$$

$$\text{Var}(Y) = \text{Var}(X_1) + \cdots + \text{Var}(X_n) = p(1 - p) + \ldots + p(1 - p) = np(1 - p).$$

Thus we avoided the complicated sums used to derive $E(X)$ and $\text{Var}(X)$ in Sect. 6.2. ◄

**Example 8.13 (Mean and Variance of the Negative Binomial Distribution)**

From Sect. 6.5 consider the form of the negative binomial random variable $Y$ which counts the total number of Bernoulli trials, with success probability $p$, required to get the $r$th success for $r > 0$. Let $Y_i$ be the number of trials needed after the $(i - 1)$-th success to obtain the $i$-th success for $i = 1, \ldots, r$. It is easy to see that each $Y_i$ is a geometric random variable and $Y = Y_1 + \cdots + Y_r$. We know from Sect. 6.3 that

$$E(Y_i) = \frac{1}{p} \quad \text{and Var}(Y_i) = \frac{(1 - p)}{p^2}.$$

Hence,

$$E(Y) = E(Y_1) + \cdots + E(Y_r) = \frac{1}{p} + \cdots + \frac{1}{p} = \frac{r}{p}.$$

and

$$\text{Var}(Y) = \text{Var}(Y_1) + \cdots + \text{Var}(Y_r) = \frac{(1 - p)}{p^2} + \cdots + \frac{(1 - p)}{p^2} = \frac{r(1 - p)}{p^2}.$$

◄

**Example 8.14 (Sum of Independent Normal Random Variables)**

Suppose that $X_i \sim N(\mu_i, \sigma_i^2)$, $i = 1, 2, \ldots, n$ are independent random variables. Let $a_1, a_2, \ldots, a_m$ be constants and suppose that

$$Y = a_1 X_1 + \cdots + a_n X_n.$$

Then we can prove that:

$$Y \sim N \left( \sum_{i=1}^{n} a_i \mu_i, \sum_{i=1}^{n} a_i^2 \sigma_i^2 \right).$$

It is clear that $E(Y) = \sum_{i=1}^{n} a_i \mu_i$ and $\text{Var}(Y) = \sum_{i=1}^{n} a_i^2 \sigma_i^2$. But that $Y$ has the normal distribution cannot yet be proved with the theory we know so far. The proof of this will be provided in Sect. 13.2.2 in Chap. 13.

As a consequence of the stated result we can easily see the following. Suppose $X_1$ and $X_2$ are independent $N(\mu, \sigma^2)$ random variables. Then $2X_1 \sim N(2\mu, 4\sigma^2)$, $X_1 + X_2 \sim N(2\mu, 2\sigma^2)$, and $X_1 - X_2 \sim N(0, 2\sigma^2)$. Note that $2X_1$ and $X_1 + X_2$ have different distributions.

Suppose that $X_i \sim N(\mu, \sigma^2)$, $i = 1, \ldots, n$ are independent. Then

$$X_1 + \cdots + X_n \sim N(n\mu, n\sigma^2),$$

and consequently,

$$\bar{X}_n = \frac{1}{n}(X_1 + \cdots + X_n) \sim N \left( \mu, \frac{\sigma^2}{n} \right).$$

◀

In this section we learned that the sample sum or the mean are random variables in their own right. Also, we have obtained the distribution of the sample sum for the binomial random variable. We have stated two important results regarding the mean and variance of the sample sum. Moreover, we have stated without proof that the distribution of the sample mean is normal if the samples are from the normal distribution itself. This is also an example of the reproductive property of the distributions. These results help us to introduce the central limit theorem in the next section.

## 8.8 The Central Limit Theorem

The sum (and average) of independent random variables show a remarkable behaviour in practice which is captured by the Central Limit Theorem (CLT). These independent random variables do not even have to be continuous, all we require is that they are independent and each of them has a finite mean and a finite variance.

There are many different versions of the CLT in statistics. Below we state a version of the CLT for independently distributed random variables having finite variance.

### 8.8.1 Statement of the Central Limit Theorem (CLT)

Let $X_1, \ldots, X_n$ be independent random variables with $E(X_i) = \mu_i$ and finite $\mathrm{Var}(X_i) = \sigma_i^2$. Define $Y_n = \sum_{i=1}^{n} X_i$ where we have included the subscript $n$ to emphasise the fact that $Y_n$ is the sum of $n$ independent random variables. Then, for a sufficiently large $n$, *the central limit theorem states that* $Y_n$ is approximately normally distributed with

$$E(Y_n) = \sum_{i=1}^{n} \mu_i, \quad \mathrm{Var}(Y_n) = \sum_{i=1}^{n} \sigma_i^2.$$

This also implies that $\bar{X}_n = \frac{1}{n} Y_n$ also follows the normal distribution approximately, as the sample size $n \to \infty$. In particular, if $\mu_i = \mu$ and $\sigma_i^2 = \sigma^2$, i.e. all means are equal and all variances are equal, then the CLT states that, as $n \to \infty$,

$$\bar{X}_n \sim N\left(\mu, \frac{\sigma^2}{n}\right).$$

Equivalently,

$$\frac{\sqrt{n}(\bar{X}_n - \mu)}{\sigma} \sim N(0, 1)$$

as $n \to \infty$. The notion of convergence is explained by the convergence of distribution of $\bar{X}_n$ to that of the normal distribution with the appropriate mean and variance. It means that the cdf of the random variable $\sqrt{n}\frac{(\bar{X}_n - \mu)}{\sigma}$, converges to the cdf of the standard normal random variable, $\Phi(\cdot)$. In other words,

$$\lim_{n \to \infty} P\left(\sqrt{n}\frac{(\bar{X}_n - \mu)}{\sigma} \le z\right) = \Phi(z), \quad -\infty < z < \infty.$$

So for "large samples", we can use $N(0, 1)$ as an approximation to the probability distribution of $\sqrt{n}(\bar{X}_n - \mu)/\sigma$. This result is 'exact', i.e. no approximation is required, if the distribution of the $X_i$'s are normal in the first place—as noted previously in Sect. 8.7.

How large does $n$ have to be before this approximation becomes usable? There is no definitive answer to this, as it depends on how "close to normal" the distribution of $X$ is. However, it is often a pretty good approximation for sample sizes as small as 20, or even smaller. It also depends on the skewness of the distribution of $X$; if the $X$-variables are highly skewed, then $n$ will usually need to be larger than for corresponding symmetric $X$-variables for the approximation to be good. We illustrate this numerically using R.

## 8.8.2 Application of CLT to Binomial Distribution

We know that a binomial random variable $Y_n$ with parameters $n$ and $p$ is the number of successes in a set of $n$ independent Bernoulli trials, each with success probability $p$. We have also learnt that

$$Y_n = X_1 + X_2 + \cdots + X_n,$$

where $X_1, \ldots, X_n$ are independent Bernoulli random variables with success probability $p$. It follows from the CLT that, for a sufficiently large $n$, $Y_n$ is approximately normally distributed with expectation $E(Y_n) = np$ and variance $\text{Var}(Y_n) = np(1 - p)$ (Figs. 8.1 and 8.2).

Hence, for given integers $y_1$ and $y_2$ between 0 and $n$ and a suitably large $n$, we have

$$P(y_1 \leq Y_n \leq y_2) = P\left\{ \frac{y_1 - np}{\sqrt{np(1 - p)}} \leq \frac{Y_n - np}{\sqrt{np(1 - p)}} \leq \frac{y_2 - np}{\sqrt{np(1 - p)}} \right\}$$

$$\approx P\left\{ \frac{y_1 - np}{\sqrt{np(1 - p)}} \leq Z \leq \frac{y_2 - np}{\sqrt{np(1 - p)}} \right\},$$

where $Z \sim N(0, 1)$.

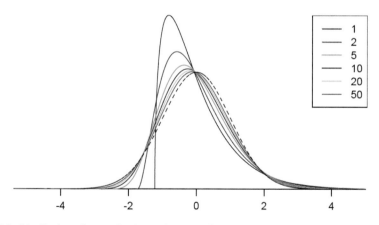

**Fig. 8.1** Distribution of normalised sample means for samples of different sizes. Initially very skew (original distribution, $n = 1$) becoming rapidly closer to standard normal (dashed line) with increasing $n$. This plot can be reproduced using the code provided in the helpfile ? see_the_clt_for_uniform in the R package ipsRdbs

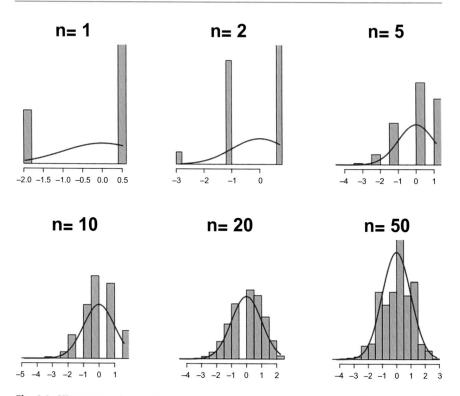

**Fig. 8.2** Histograms of normalised sample means for Bernoulli ($p = 0.8$) samples of different sizes. The pdf of the limiting standard normal distribution has been superimposed in all the plots. This plot can be reproduced using the code provided in the helpfile ? `see_the_clt_for_Bernoulli` in the R package `ipsRdbs`

We should take account of the fact that the binomial random variable $Y_n$ is integer-valued, and so $P(y_1 \leq Y_n \leq y_2) = P(y_1 - f_1 \leq Y_n \leq y_2 + f_2)$ for any two fractions $0 < f_1, f_2 < 1$. This is called *continuity correction* and we take $f_1 = f_2 = 0.5$ in practice.

$$P(y_1 \leq Y_n \leq y_2) = P(y_1 - 0.5 \leq Y_n \leq y_2 + 0.5)$$

$$= P\left\{ \frac{y_1 - 0.5 - np}{\sqrt{np(1-p)}} \leq \frac{Y_n - np}{\sqrt{np(1-p)}} \leq \frac{y_2 + 0.5 - np}{\sqrt{np(1-p)}} \right\}$$

$$\approx P\left\{ \frac{y_1 - 0.5 - np}{\sqrt{np(1-p)}} \leq Z \leq \frac{y_2 + 0.5 - np}{\sqrt{np(1-p)}} \right\}.$$

What do we mean by a suitably large $n$? A commonly-used guideline is that the approximation is adequate if $np \geq 5$ and $n(1-p) \geq 5$.

**Example 8.15**

A producer of natural yogurt believed that the market share of their brand was 10%. To investigate this, a survey of 2500 yogurt consumers was carried out. It was observed that only 205 of the people surveyed expressed a preference for their brand. Should the producer be concerned that they might be losing market share?

Assume that the conjecture about market share is true. Then the number of people $Y$ who prefer this product follows a binomial distribution with $p = 0.1$ and $n = 2500$. So the mean is $np = 250$, the variance is $np(1 - p) = 225$, and the standard deviation is 15. The exact probability of observing ($Y \leq 205$) is given by the sum of the binomial probabilities up to and including 205, which is difficult to compute, although the R command `pbinom(205, size=2500, prob =0.1)` finds it to be 0.001174. However, this can be approximated by using the CLT:

$$P(Y_n \leq 205) = P(Y_n \leq 205.5)$$

$$= P\left\{ \frac{Y_n - np}{\sqrt{np(1-p)}} \leq \frac{205.5 - np}{\sqrt{np(1-p)}} \right\}$$

$$\approx P\left\{ Z \leq \frac{205.5 - np}{\sqrt{np(1-p)}} \right\}$$

$$= P\left\{ Z \leq \frac{205.5 - 250}{15} \right\}$$

$$= \Phi(-2.967) = 0.001505$$

This approximation is accurate upto the 4th place after the decimal point, and hence is very accurate. This probability is so small that it casts doubt on the validity of the assumption that the market share is 10%. ◀

In this section we have learned about the central limit theorem. This basically states that the probability distribution (also called the sampling distribution) of the sample sum (and also the mean) is an approximate normal distribution regardless of the probability distribution of the original random variables, provided that those random variables have finite means and variances.

## 8.9 Exercises

### 8.2 (Central Limit Theorem)

1. In typesetting a book of 50 pages typesetting errors occur. Suppose that the average number of typesetting errors per page is 0.1. What is the probability (2 digits accuracy) that the entire book is typeset without any error? To answer this

question suppose that $X_1, \ldots, X_{50}$ are independently distributed Poisson(0.1) random variables where $X_i$ is number of typeset errors on page $i$. Make use of the fact that if $X_1, \ldots, X_n$ are independently distributed Poisson($\lambda$) random variables then $\sum_{i=1}^{n} X_i \sim$ Poisson($n\lambda$).

2. The fractional parts of 100 numbers are distributed uniformly between 0 and 1. The numbers are first rounded to the nearest integer and then added. Using the CLT, find an approximation for the probability that the error in the sum due to rounding lies between –0.5 and 0.5. **Hint:** First argue that the individual errors are uniformly distributed and find the mean and variance of the uniform distribution.

3. A man buys a new dice and throws it 600 times. He does not yet know if the dice is fair!
   (a) Show that the probability that he obtains between 90 and 100 'sixes' if the dice is fair is 0.3968 approximately.
   (b) From the first part we see that $P(90 \leq X \leq 100) \approx 0.3968$ where $X$ denotes the number of 'sixes' from 600 throws. If the dice is fair we can expect about 100 'sixes' from 600 throws. Between what two limits symmetrically placed about 100 would the number sixes obtained lie with probability 0.95 if the dice is fair? That is, find $N$ such that $P(100 - N \leq X \leq 100 + N) = 0.95$. You may use the continuity correction of $\frac{1}{2}$ on each of the limits inside the probability statement.
   (c) What might he conclude if he obtained 120 sixes?

4. A revision guide contains 100 multiple questions. There are 5 answers to choose from for each question. Use an approximating distribution to find a number, $x$, such that a student who guesses on every question has a smaller than 1% chance of getting, $x$, or more, questions right.

5. $X$ is a Binomial random variable with $n = 40$ and $p = 0.01$. A reasonable approximation to the distribution of X is:
   (a) Normal with mean 0.4 and variance 0.396,
   (b) Poisson with mean 0.4,
   (c) Geometric with mean 0.4,
   (d) Normal with mean 0.4 and variance 0.099.

6. There are six hundred students in the business school of a university who all are studying the course *Statistics 101* and *Statistics Made Simple* is the prescribed textbook for this course. Suppose that the probability is 0.05 that any randomly chosen student will need to borrow a copy of the text book from the university library. How many copies of the text book *Statistics Made Simple* should be kept in the library so that the probability may be greater than 0.90 that none of the students needing a copy from the library has to come back disappointed? Assume that qnorm(0.90) = 1.282.

# Part III

# Introduction to Statistical Inference

This is Part III of this book where we introduce the basic ideas of statistical inference.

# Introduction to Statistical Inference

9

**Abstract**

Chapter 9: This chapter introduces the basic concepts of statistical inference and statistical modelling. It distinguishes between population distributions and sample statistics (quantities). The concepts of estimators and their sampling (probability) distributions are also introduced. The properties of bias and mean square errors of estimators and defined.

In the last three chapters we learned the probability distributions of common standard random variables that are used in probability calculations. These calculations are based on our assumption of a probability distribution with known parameter values. Statistical inference is the process by which we try to learn about those probability distributions using only random observations. Hence, if our aim is to learn about some typical characteristics of the population of students in a university, we simply randomly select a number of students, observe their characteristics and then try to generalise, as has been discussed in Sect. 1.1 to define statistics. To give a specific example, suppose we are interested in learning what proportion of students are of Indian origin. We may then select a number of students at random and observe the sample proportion of Indian origin students. We will then claim that the sample proportion is really our guess for the population proportion. But obviously we may be making grave errors since we are inferring about some unknown based on only a tiny fraction of total information. Statistical inference methods formalise these aspects. We aim to learn some of these methods in this chapter and also in few subsequent chapters. However, before discussing more on statistical inference we first introduce some preliminary concepts and methods for drawing random samples from a population with a finite number of individuals.

## 9.1    Drawing Random Samples

The primary purpose of this section is to consider issues around the sampling of data in scientific experiments. For example, suppose we are interested in estimating the unknown percentage of ethnic minority students in a university. Assuming that we have access to a list of all students and their university contact information. How can we sample effectively to estimate the percentage, which is a population characteristic?

To sample effectively we must aim to give some positive probability to selection of the population of all students so that the sample is a true representative of the population. This is called *probability sampling*. A probability sampling scheme is called *simple random sampling*, SRS, if it gives the same probability of selection for each member of the population. There are two kinds of SRS depending on whether we select the individuals one by one with or without returning the sampled individuals to the population. When we return the selected individuals immediately back to the population, we perform *simple random sampling with replacement* or SRSWR, and if we do not return the sampled individuals back, we perform *simple random sampling without replacement* or SRSWOR. The UK National Lottery draw of six numbers each week is an example of SRSWOR.

Suppose there are $N$ individuals in the population of interest and we are drawing a sample of $n$ individuals. In SRSWR, the same unit of the population may occur more than once in the sample; there are $N^n$ possible samples (using multiplication rules of counting), and each of these samples has equal chance of $1/N^n$ to materialise. In the case of SRSWOR, no member of the population can occur more than once in the sample. There are $\binom{N}{n}$ possible samples and each has equal probability of inclusion $1/\binom{N}{n}$.

### How Can We Draw Random Samples?

A quick method is drawing numbers out of hat. But this is cumbersome and manual. In practice, we use a suitable random number series to draw samples at random. A random sampling number series is an arrangement, which may be looked upon either as linear or rectangular, in which each place has been filled in with one of the digits $0, 1, \ldots, 9$. The digit occupying any place is selected at random from these 10 digits and independently of the digits occurring in other positions. Different random number series are available in books and computers. In R we can easily use the `sample` command to draw random samples either using SRSWR or SRSWOR. For example, suppose the problem is to select 50 students out of the 200 in a specific lecture theatre. In this experiment, we may number the students 1 to 200 in any way possible, e.g. alphabetically by surname. We can then issue the command `sample(200, size=50)` for SRSWOR and `sample(200, size=50, replace=T)` for SRSWR.

There are a huge number of considerations and concepts to design good sample surveys avoiding bias from many possible sources. For example, there may be response bias where we only select the samples where the values of a response variable are higher, e.g., air pollution, observational bias, biases from non-response, interviewer bias, bias due to defective sampling technique, bias due to substitution, bias due to faulty differentiation of sampling units and so on. However, discussion of such topics is beyond the scope of this book.

## 9.2 Foundations of Statistical Inference

Statistical analysis (or inference) involves drawing conclusions, and making predictions and decisions, using the evidence provided to us by observed data. To do this we use probability distributions, often called *statistical models*, to describe the process by which the observed data were generated. For example, we may suppose that the true proportion of Indian origin students is $p$, $0 < p < 1$, and if we have selected $n$ students at random, that each of those students gives rise to a Bernoulli distribution which takes the value 1 if the student is of Indian origin and 0 otherwise. The success probability of the Bernoulli distribution will be the unknown $p$. The underlying statistical model is then the Bernoulli distribution.

To illustrate with another example, suppose we have observed fast food waiting times in the morning and afternoon. If we assume time, measured in number of whole seconds, to be discrete, then a suitable model for the random variable $X =$ "the number of seconds waited" would be the Poisson distribution. However, if we treat time as continuous then the random variable $X =$ "the waiting time" could be modelled as a normal random variable. In general, it is clear that:

- The form of the assumed model helps us to understand the real-world process by which the data were generated.
- If the model explains the observed data well, then it should also inform us about future (or unobserved) data, and hence help us to make predictions (and decisions contingent on unobserved data).
- The use of statistical models, together with a carefully constructed methodology for their analysis, also allows us to quantify the uncertainty associated with any conclusions, predictions or decisions we make.

We will continue to use the notation $x_1, x_2, \ldots, x_n$ to denote $n$ observations of the random variables $X_1, X_2, \ldots, X_n$ (corresponding capital letters). For the fast food waiting time example, we have $n = 20$, $x_1 = 38$, $x_2 = 100, \ldots, x_{20} = 70$, and $X_i$ is the waiting time for the $i$th person in the sample.

### 9.2.1 Statistical Models

Suppose we denote the complete data by the vector $x = (x_1, x_2, \ldots, x_n)$ and use $X = (X_1, X_2, \ldots, X_n)$ for the corresponding random variables. A statistical model specifies a probability distribution for the random variables $X$ corresponding to the data observations $x$. Providing a specification for the distribution of $n$ jointly varying random variables can be a daunting task, see Chap. 8. However, this task is made much easier if we can make some simplifying assumptions, such as

1. $X_1, X_2, \ldots, X_n$ are *independent* random variables,
2. $X_1, X_2, \ldots, X_n$ have the same probability distribution (so $x_1, x_2, \ldots, x_n$ are observations of a single random variable $X$).

Assumption 1 depends on the sampling mechanism and is very common in practice. If we are to make this assumption for the university student sampling experiment, we need to select randomly among all possible students. We should not get the sample from an event in the Indian or Chinese Student Association as that will give us a biased result. The assumption will be violated when samples are correlated either in time or in space, e.g. the daily air pollution level in a city for the last year. In this book we will only consider data sets where Assumption 1 is valid. Assumption 2 is not always appropriate, but is often reasonable when we are modelling a single variable. In the fast food waiting time example, if we assume that there are no differences between the AM and PM waiting times, then we can say that $X_1, \ldots, X_{20}$ are *independent and identically distributed* (or i.i.d. for short).

## 9.2.2  A Fully Specified Model

Sometimes a model completely specifies the probability distribution of $X = (X_1, X_2, \ldots, X_n)$. For example, if we assume that the waiting time $X \sim N(\mu, \sigma^2)$ where $\mu = 60$, and $\sigma^2 = 64$, then this is a fully specified model. In this case, there is no need to collect any data as there is no need to make any inference about any unknown quantities, although we may use the data to judge the plausibility of the model.

However, a fully specified model would be appropriate when for example, there is some external (to the data) theory as to why the model (in particular the values of $\mu$ and $\sigma^2$) was appropriate. Fully specified models such as this are uncommon as we rarely have external theory which allows us to specify a model so precisely.

## 9.2.3  A Parametric Statistical Model

A parametric statistical model specifies a probability distribution for a random sample apart from the value of a number of parameters in that distribution. This could be confusing in the first instance—a parametric model does not specify parameters! Here the word parametric signifies the fact that the probability distribution is completely specified by a few parameters in the first place. For example, the Poisson distribution is parameterised by the parameter $\lambda$ which happens to be the mean of the distribution; the normal distribution is parameterised by two parameters, the mean $\mu$ and the variance $\sigma^2$.

When a parametric statistical model is assumed with some unknown parameters, statistical inference methods use data to *estimate* the unknown parameters, e.g. $\lambda$, $\mu$, $\sigma^2$. Estimation will be discussed in more detail in the following sections.

## 9.2.4   A Nonparametric Statistical Model

Sometimes it is not appropriate, or we want to avoid, making a precise specification for the distribution which generated $X_1, X_2, \ldots, X_n$. For example, when the data histogram does not show a bell-shaped distribution, it would be wrong to assume a normal distribution for the data. In such a case, although we can attempt to use some other non-bell-shaped parametric model, we can decide altogether to abandon parametric models. We may then still assume that $X_1, X_2, \ldots, X_n$ are i.i.d. random variables, but from a nonparametric statistical model which cannot be written down, having a probability function which only depends on a *finite* number of parameters. Such analysis approaches are also called distribution-free methods.

---

**Example 9.1 (Return to the Computer Failure Example)**

Let $X$ denote the count of computer failures per week. We want to estimate how often will the computer system fail at least once per week in the next year? The answer is $52 \times (1 - P(X = 0))$. But how could we estimate $P(X = 0)$? Consider two approaches.

1. **Nonparametric.** Estimate $P(X = 0)$ by the relative frequency of number of zeros in the above sample, which is 12 out of 104. Thus our estimate of $P(X = 0)$ is 12/104. Hence, our estimate of the number of weeks when there will be at least one computer failure is $52 \times (1 - 12/104) = 46$.
2. **Parametric.** Suppose we assume that $X$ follows the Poisson distribution with parameter $\lambda$. Then the answer to the above question is

$$52 \times (1 - P(X = 0)) = 52 \times \left(1 - e^{-\lambda} \tfrac{\lambda^0}{0!}\right)$$
$$= 52 \times \left(1 - e^{-\lambda}\right)$$

which involves the unknown parameter $\lambda$. For the Poisson distribution we know that $E(X) = \lambda$. Hence we could use the *sample mean* $\bar{X}$ to estimate $E(X) = \lambda$. Thus our estimate $\hat{\lambda} = \bar{x} = 3.75$. This type of estimator is called a *moment estimator* as discussed in Sect. 10.1 in Chap. 10. Now our answer is $52 \times \left(1 - e^{-3.75}\right) = $ `52 * (1- exp(-3.75))` $= 50.78 \approx 51$, which is very different compared to our answer of 46 from the nonparametric approach. Which of the two estimates, 46 and 51 should we use?

To answer the above question we examine the quality of the fit of the Poisson model. Figure 9.1 plots the observed and Poisson model fitted frequencies in a bar plot. Clearly, the Poisson distribution based fitted values in red differ a

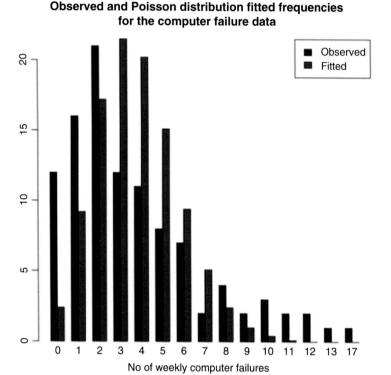

**Fig. 9.1** Plotted values are observed frequencies and fitted frequencies $= n \times e^{-\hat{\lambda}} \frac{\hat{\lambda}^x}{x!}$. This plot can be reproduced by using the code provided in the help file **help** (cfail)

lot from the observed frequencies. In this case the Poisson distribution fails to provide a good fit and hence the non-parametric method should be preferred. Note that in the non-parametric case the observed frequencies are also taken to be fitted frequencies.

◄

### 9.2.5   Modelling Cycle

In general, if a model fails then we re-fit the data by searching for a better model. For example, we may investigate if a negative-binomial model may be more suitable instead of the Poisson distribution. In general statistical modelling we follow the algorithm below.

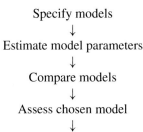

Specify models
↓
Estimate model parameters
↓
Compare models
↓
Assess chosen model
↓
If model is adequate: Base predictions/decisions on chosen model
↓
Else, find alternative model

---

**Example 9.2 (Return to the Bomb Hits Example)**

Recall that we have data on the number of flying-bomb hits in the south of London during World War II. The observed frequency distribution is:

| $x$ | 0 | 1 | 2 | 3 | 4 | 5 | Total |
|-----|-----|-----|----|----|---|---|-------|
| Frequency | 229 | 211 | 93 | 35 | 7 | 1 | 576 |

The table implies that 229 areas had no hits, 211 areas had exactly one hit and so on.

Here the random variable, $X$, is the number of bomb hits in a given area. Let $x_i$ denote the observed number of hits in area $i$ for $i = 1, \ldots, n = 576$. Hence, $x_1, \ldots, x_n$ are random observation of $X_1, \ldots, X_n$. A parametric model assumes that $X_1, \ldots, X_n \sim$ i.i.d Poisson($\lambda$).

The Poisson population mean $\lambda$ is estimated by the sample mean $\bar{x}$. Here

$$\bar{x} = \frac{0 \times 229 + 1 \times 211 + 2 \times 93 + 3 \times 35 + 4 \times 7 + 5 \times 1}{576} = \frac{535}{576} = 0.9288.$$

Thus we obtain the table of observed and fitted frequencies as presented in Table 9.1, and plotted in Fig. 9.2.

The Poisson distribution is a good fit for the data. Hence, model (Poisson distribution) based inference can now proceed. In general, often model-based analysis is preferred because it is more precise and accurate, and we can find estimates of uncertainty in such analysis based on the structure of the model. We will illustrate this with more examples later.

◄

**Table 9.1** Observed and fitted frequencies for the bombhits data example

| $x$ | Freq | Rel. freq $\frac{Freq}{576}$ | Poisson prob. $e^{\hat{\lambda}}\frac{(\hat{\lambda})^x}{x!}$ | Fitted freq. $576 \times e^{\hat{\lambda}}\frac{(\hat{\lambda})^x}{x!}$ |
|---|---|---|---|---|
| 0 | 229 | 0.3976 | 0.3950 | 227.5 |
| 1 | 211 | 0.3663 | 0.3669 | 211.3 |
| 2 | 93 | 0.1614 | 0.1704 | 98.1 |
| 3 | 35 | 0.0608 | 0.0528 | 30.5 |
| 4 | 7 | 0.0122 | 0.0123 | 7.2 |
| 5 | 1 | 0.0017 | 0.0024 | 1.4 |
| Total | 576 | 1 | 1 | 576 |

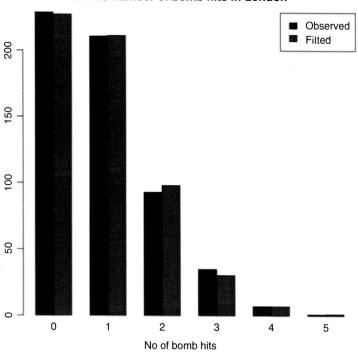

## Observed and Poisson distribution fitted frequencies for the number of bomb hits in London

**Fig. 9.2** Plotted values are observed frequencies and fitted frequencies $= n \times e^{-\hat{\lambda}}\frac{\hat{\lambda}^x}{x!}$. This plot can be reproduced by using the code provided in the help file **help**(bombhits)

## 9.3 Estimation

Once we have collected data and proposed a statistical model for our data, the initial statistical analysis usually involves *estimation*.

- For a parametric model, we need to estimate the unknown (unspecified) parameter $\lambda$. For example, if our model for the computer failure data is that they are i.i.d. Poisson, we need to estimate the mean ($\lambda$) of the Poisson distribution.
- For a nonparametric model, we may want to estimate the properties of the data-generating distribution. For example, if our model for the computer failure data is that they are i.i.d., following the distribution of an unspecified common random variable $X$, then we may want to estimate $\mu = E(X)$ or $\sigma^2 = \mathrm{Var}(X)$.

In the following discussion, we use the generic notation $\theta$ to denote the *estimand* (what we want to estimate or the parameter). For example, $\theta$ is the parameter $\lambda$ in the first example, and $\theta$ may be either $\mu$ or $\sigma^2$ or both in the second example.

### 9.3.1 Population and Sample

Recall that a statistical model specifies a probability distribution for the random variables $X$ corresponding to the data observations $x$.

- The observations $x = (x_1, \ldots, x_n)$ are called the *sample*, and quantities derived from the sample are sample quantities. For example, as in Chap. 1, we call

$$\bar{x} = \frac{1}{n} \sum_{i=1}^{n} x_i$$

the *sample mean*.
- The probability distribution for $X$ specified in our model represents all possible observations which might have been observed in our sample, and is therefore sometimes referred to as the *population*. Quantities derived from this distribution are population quantities.

  For example, if our model is that $X_1, \ldots, X_n$ are i.i.d., following the common distribution of a random variable $X$, then we call $E(X)$ the *population mean*.

### 9.3.2 Statistic, Estimator and Sampling Distribution

A *statistic* $T(x)$ is any function of the observed data $x_1, \ldots, x_n$ alone (and therefore does not depend on any parameters or other unknowns).

An *estimate* of $\theta$ is any statistic which is used to estimate $\theta$ under a particular statistical model. We will use $\tilde{\theta}(x)$ (sometimes shortened to $\tilde{\theta}$) to denote an estimate of $\theta$.

An estimate $\tilde{\theta}(x)$ is an observation of a corresponding random variable $\tilde{\theta}(X)$ which is called an *estimator*. Thus an estimate is a particular observed value, e.g. 1.2, but an estimator is a random variable which can take values which are called estimates.

An estimate is a particular numerical value, e.g. $\bar{x}$; an estimator is a random variable, e.g. $\bar{X}$.

The probability distribution of any estimator $\tilde{\theta}(X)$ is called its *sampling distribution*. The estimate $\tilde{\theta}(x)$ is an observed value (a number), and is a single observation from the sampling distribution of $\tilde{\theta}(X)$.

### Example 9.3 (Uniform Distribution)

Suppose that we have a random sample $X_1, \ldots, X_n$ from the uniform distribution on the interval $[0, \theta]$ where $\theta > 0$ is unknown. Suppose that $n = 5$ and we have the sample observations $x_1 = 2.3, x_2 = 3.6, x_3 = 20.2, x_4 = 0.9, x_5 = 17.2$. Our objective is to estimate $\theta$. How can we proceed?

Here the pdf is $f(x) = \frac{1}{\theta}$ for $0 \le x \le \theta$ and 0 otherwise. Hence $E(X) = \int_0^\theta \frac{1}{\theta} x\, dx = \frac{\theta}{2}$. There are many possible estimators for $\theta$, e.g. $\hat{\theta}_1(X) = 2\bar{X}$, which is motivated by the method of moments (detailed in Chap. 10) because $\theta = 2E(X)$. A second estimator is $\hat{\theta}_2(X) = \max\{X_1, X_2, \ldots, X_n\}$, which is intuitive since $\theta$ must be greater than or equal to all observed values and thus the maximum of the sample value will be closest to $\theta$. This is also the maximum likelihood estimate of $\theta$.

How could we choose between the two estimators $\hat{\theta}_1$ and $\hat{\theta}_2$? This is where we need to learn the sampling distribution of an estimator to determine which estimator will be unbiased, i.e. correct on average, and which will have minimum variability. We will formally define these in a minute, but first let us derive the sampling distribution, i.e. the pdf, of $\hat{\theta}_2$. Note that $\hat{\theta}_2$ is a random variable since the sample $X_1, \ldots, X_n$ is random. We will first find its cdf and then differentiate the cdf to get the pdf. For ease of notation, suppose $Y$ is taken to be the sample maximum, i.e.,

$$ Y = \hat{\theta}_2(X) = \max\{X_1, X_2, \ldots, X_n\}. $$

For any $0 < y < \theta$, the cdf of $Y$, $F(y) = P(Y \le y)$ is found below using several arguments. Key among those is the fact that the event $\max\{X_1, X_2, \ldots, X_n\} \le y$ is equivalent to the event that each one of $X_1 \le y, X_2 \le y, \ldots, X_n$ is less than

$y$. To fix ideas, set a value of $y = 2.7$, say and see where should the numbers $X_1, X_2, \ldots, X_n$ lie so that $\max\{X_1, X_2, \ldots, X_n\} \leq 2.7$. Now,

$$
\begin{aligned}
P(Y \leq y) &= P(\max\{X_1, X_2, \ldots, X_n\} \leq y) \\
&= P(X_1 \leq y, X_2 \leq y, \ldots, X_n \leq y)) \\
&\quad [\max \leq y \text{ if and only if each } \leq y] \\
&= P(X_1 \leq y)P(X_2 \leq y) \cdots P(X_n \leq y) \\
&\quad [\text{since the } X\text{'s are independent}] \\
&= \frac{y}{\theta}\frac{y}{\theta} \cdots \frac{y}{\theta} \\
&= \left(\frac{y}{\theta}\right)^n .
\end{aligned}
$$

Now the pdf of $Y$ is $f(y) = \frac{dF(y)}{dy} = n\frac{y^{n-1}}{\theta^n}$ for $0 \leq y \leq \theta$. For an assumed value of $\theta$ we can plot this as a function of $y$ to examine the pdf. Now easy integration yields $E(\hat{\theta}_2) = E(Y) = \frac{n}{n+1}\theta$ and $\text{Var}(\hat{\theta}_2) = \frac{n\theta^2}{(n+2)(n+1)^2}$. ◄

### 9.3.3 Bias and Mean Square Error

In the uniform distribution example we saw that the estimator $\hat{\theta}_2 = Y = \max\{X_1, X_2, \ldots, X_n\}$ is a random variable and its pdf is given by $f(y) = n\frac{y^{n-1}}{\theta^n}$ for $0 \leq y \leq \theta$. This probability distribution is called the sampling distribution of $\hat{\theta}_2$. For this we have seen that $E(\hat{\theta}_2) = \frac{n}{n+1}\theta$.

In general, we define the *bias* of an estimator $\tilde{\theta}(X)$ of $\theta$ to be

$$
\text{bias}(\tilde{\theta}) = E(\tilde{\theta}) - \theta.
$$

An estimator $\tilde{\theta}(X)$ is said to be *unbiased* if

$$
\text{bias}(\tilde{\theta}) = 0, \quad \text{i.e. if} \quad E(\tilde{\theta}) = \theta.
$$

So an estimator is unbiased if the expectation of its sampling distribution is equal to the quantity we are trying to estimate. Unbiased means "getting it right on average", i.e. under repeated sampling (relative frequency interpretation of probability).

Thus for the uniform distribution example, $\hat{\theta}_2$ is a biased estimator of $\theta$ and

$$
\text{bias}(\hat{\theta}_2) = E(\hat{\theta}_2) - \theta = \frac{n}{n+1}\theta - \theta = -\frac{1}{n+1}\theta,
$$

which goes to zero as $n \to \infty$. However, $\hat{\theta}_1 = 2\bar{X}$ is unbiased since $E(\hat{\theta}_1) = 2E(\bar{X}) = 2\frac{\theta}{2} = \theta$.

Unbiased estimators are "correct on average", but that does not mean that they are guaranteed to provide estimates which are close to the estimand $\theta$. A better measure of the quality of an estimator than bias is the *mean squared error* (or m.s.e.), defined as

$$\text{m.s.e.}(\tilde{\theta}) = E\left[(\tilde{\theta} - \theta)^2\right].$$

Therefore, if $\tilde{\theta}$ is unbiased for $\theta$, i.e. if $E(\tilde{\theta}) = \theta$, then m.s.e.$(\tilde{\theta}) = \text{Var}(\tilde{\theta})$. In general, we have the following result:

**Theorem 9.1 (Bias Variance Identity)**

$$m.s.e.(\tilde{\theta}) = Var(\tilde{\theta}) + bias(\tilde{\theta})^2.$$

A proof of the above result is similar to a proof we used to show that the sample mean minimises the sum of squares of the deviations in Sect. 1.3. Here we add and subtract $E\left(\tilde{\theta}\right)$ inside the square in the definition of the m.s.e. and proceed by expanding it.

$$
\begin{aligned}
\text{m.s.e.}(\tilde{\theta}) &= E\left[(\tilde{\theta} - \theta)^2\right] \\
&= E\left[\left(\tilde{\theta} - E\left(\tilde{\theta}\right) + E\left(\tilde{\theta}\right) - \theta\right)^2\right] \\
&= E\left[\left(\tilde{\theta} - E\left(\tilde{\theta}\right)\right)^2 + \left(E\left(\tilde{\theta}\right) - \theta\right)^2 + 2\left(\tilde{\theta} - E\left(\tilde{\theta}\right)\right)\left(E\left(\tilde{\theta}\right) - \theta\right)\right] \\
&= E\left[\tilde{\theta} - E\left(\tilde{\theta}\right)\right]^2 + E\left[E\left(\tilde{\theta}\right) - \theta\right]^2 \\
&\quad + 2E\left[\left(\tilde{\theta} - E\left(\tilde{\theta}\right)\right)\left(E\left(\tilde{\theta}\right) - \theta\right)\right] \\
&= \text{Var}\left(\tilde{\theta}\right) + \left[E\left(\tilde{\theta}\right) - \theta\right]^2 + 2\left(E\left(\tilde{\theta}\right) - \theta\right)E\left[\left(\tilde{\theta} - E\left(\tilde{\theta}\right)\right)\right] \\
&= \text{Var}\left(\tilde{\theta}\right) + \text{bias}(\tilde{\theta})^2 + 2\left(E\left(\tilde{\theta}\right) - \theta\right)\left[E\left(\tilde{\theta}\right) - E\left(\tilde{\theta}\right)\right] \\
&= \text{Var}\left(\tilde{\theta}\right) + \text{bias}(\tilde{\theta})^2.
\end{aligned}
$$

Hence, the mean squared error incorporates both the bias and the variability (sampling variance) of $\tilde{\theta}$. We are then faced with the bias-variance trade-off when selecting an optimal estimator. We may allow the estimator to have a little bit of bias if we can ensure that the variance of the biased estimator will be much smaller than that of any unbiased estimator.

---

**Example 9.4 (Uniform Distribution)**

Continuing with the uniform distribution $U[0, \theta]$ example, we have seen that $\hat{\theta}_1 = 2\bar{X}$ is unbiased for $\theta$ but $\text{bias}(\hat{\theta}_2) = -\frac{1}{n+1}\theta$. How do these estimators compare with respect to the m.s.e.? Since $\hat{\theta}_1$ is unbiased, its m.s.e. is its variance. In the next section, we will prove that for random sampling from any population

$$\text{Var}(\bar{X}) = \frac{\text{Var}(X)}{n},$$

where $\text{Var}(X)$ is the variance of the population sampled from. Returning to our example, we know that if $X \sim U[0, \theta]$ then $\text{Var}(X) = \frac{\theta^2}{12}$. Therefore we have:

$$\text{m.s.e.}(\hat{\theta}_1) = \text{Var}\left(\hat{\theta}_1\right) = \text{Var}\left(2\bar{X}\right) = 4\text{Var}\left(\bar{X}\right) = 4\frac{\theta^2}{12n} = \frac{\theta^2}{3n}.$$

Now, for $\hat{\theta}_2$ we know that:

1. $\text{Var}(\hat{\theta}_2) = \frac{n\theta^2}{(n+2)(n+1)^2}$;
2. $\text{bias}(\hat{\theta}_2) = -\frac{1}{n+1}\theta$.

Now

$$\begin{aligned}
\text{m.s.e.}(\hat{\theta}_2) &= \text{Var}\left(\hat{\theta}_2\right) + \text{bias}(\hat{\theta}_2)^2 \\
&= \frac{n\theta^2}{(n+2)(n+1)^2} + \frac{\theta^2}{(n+1)^2} \\
&= \frac{\theta^2}{(n+1)^2}\left(\frac{n}{n+2} + 1\right) \\
&= \frac{\theta^2}{(n+1)^2}\frac{2n+2}{n+2}.
\end{aligned}$$

Clearly, the m.s.e. of $\hat{\theta}_2$ is an order of magnitude (of order $n^2$ rather than $n$) smaller than the m.s.e. of $\hat{\theta}_1$, providing justification for the preference of $\hat{\theta}_2 = \max\{X_1, X_2, \ldots, X_n\}$ as an estimator of $\theta$. ◄

### 9.3.4 Section Summary

In this section we have learned the basics of estimation. We have learned that estimates are particular values and estimators have probability distributions. We have also learned the concepts of the bias and variance of an estimator. We have proved a key fact that the mean squared error of an estimator is composed of two pieces, namely bias and variance. Sometimes there may be bias-variance trade-off where a little bias can lead to much lower variance. We have illustrated this with an example.

## 9.4    Estimation of Mean, Variance and Standard Error

Often, one of the main tasks of a statistician is to estimate a population average or mean. However the estimates, using whatever procedure, will not be usable or scientifically meaningful if we do not know their associated uncertainties. For example, a statement such as: "the Arctic ocean will be completely ice-free in the summer in the next few decades" provides little information as it does not communicate the extent or the nature of the uncertainty in it. Perhaps a more precise statement could be: "the Arctic ocean will be completely ice-free in the summer some time in the next 20–30 years". This last statement not only gives a numerical value for the number of years for complete ice-melt in the summer, but also acknowledges the uncertainty of $\pm 5$ years in the estimate. A statistician's main job is to estimate such uncertainties. In this section, we will get started with estimating uncertainties when we estimate a population mean. We will introduce the standard error of an estimator.

### 9.4.1    Estimation of a Population Mean

Suppose that $X_1, X_2, \ldots, X_n$ is a random sample from any probability distribution $f(x)$, which may be discrete or continuous. Suppose that we want to estimate the unknown population mean $E(X) = \mu$ and variance, $\mathrm{Var}(X) = \sigma^2$. In order to do this, it is not necessary to make any assumptions about $f(x)$, so this may be thought of as *non-parametric* inference. Suppose

$$\bar{X} = \frac{1}{n} \sum_{i=1}^{n} X_i$$

is the sample mean and

$$S^2 = \frac{1}{n-1} \sum_{i=1}^{n} (X_i - \bar{X})^2$$

is the sample variance. Note that the sample variance is defined with the divisor $n - 1$. We have the following results:

**R1** $E(\bar{X}) = \mu$. Thus, the sample mean is an unbiased estimator of the population mean $E(X) = \mu$.

**R2** $\mathrm{Var}(\bar{X}) = \sigma^2/n$. Thus, the variance of the sample mean is the variance of the population divided by the sample size $n$.

**R3** $E(S^2) = \sigma^2$. Thus the sample variance (with the divisor $n - 1$) is an unbiased estimator of the population variance $\sigma^2$.

We prove **R1** as follows.

$$E[\bar{X}] = \frac{1}{n} \sum_{i=1}^{n} E(X_i) = \frac{1}{n} \sum_{i=1}^{n} E(X) = E(X),$$

so $\bar{X}$ is an unbiased estimator of $E(X)$.

We prove **R2** using the result that for independent random variables the variance of the sum is the sum of the variances from Sect. 8.7. Thus,

$$\text{Var}[\bar{X}] = \frac{1}{n^2} \sum_{i=1}^{n} \text{Var}(X_i) = \frac{1}{n^2} \sum_{i=1}^{n} \text{Var}(X) = \frac{n}{n^2} \text{Var}(X) = \frac{\sigma^2}{n},$$

so the m.s.e. of $\bar{X}$ is $\text{Var}(X)/n$. This proves the following assertion we made earlier:

Variance of the sample mean = Variance of the population divided by the sample size.

We now want to prove **R3**, i.e. show that the sample variance, $S^2$, with divisor $n-1$ is an unbiased estimator of the population variance $\sigma^2$, i.e. $E(S^2) = \sigma^2$. We have

$$S^2 = \frac{1}{n-1} \sum_{i=1}^{n} (X_i - \bar{X})^2 = \frac{1}{n-1} \left[ \sum_{i=1}^{n} X_i^2 - n\bar{X}^2 \right].$$

To evaluate the expectation of the above, we need $E(X_i^2)$ and $E(\bar{X}^2)$. In general, we know for any random variable,

$$\text{Var}(Y) = E(Y^2) - (E(Y))^2 \Rightarrow E(Y^2) = \text{Var}(Y) + (E(Y))^2.$$

Thus, we have

$$E(X_i^2) = \text{Var}(X_i) + (E(X_i))^2 = \sigma^2 + \mu^2,$$

and

$$E(\bar{X}^2) = \text{Var}(\bar{X}) + (E(\bar{X}))^2 = \sigma^2/n + \mu^2,$$

from **R1** and **R2**. Now

$$
\begin{aligned}
E(S^2) &= E\left\{\frac{1}{n-1}\left[\sum_{i=1}^{n} X_i^2 - n\bar{X}^2\right]\right\} \\
&= \frac{1}{n-1}\left[\sum_{i=1}^{n} E(X_i^2) - nE(\bar{X}^2)\right] \\
&= \frac{1}{n-1}\left[\sum_{i=1}^{n}(\sigma^2 + \mu^2) - n(\sigma^2/n + \mu^2)\right] \\
&= \frac{1}{n-1}\left[n\sigma^2 + n\mu^2 - \sigma^2 - n\mu^2)\right] \\
&= \sigma^2 \equiv \mathrm{Var}(X).
\end{aligned}
$$

In words, this proves that

> The sample variance is an unbiased estimator of the population variance.

### 9.4.2 Standard Deviation and Standard Error

It follows that, for an unbiased (or close to unbiased) estimator $\tilde{\theta}$,

$$
\mathrm{m.s.e.}(\tilde{\theta}) = \mathrm{Var}(\tilde{\theta})
$$

and therefore the sampling variance of the estimator is an important summary of its quality.

We usually prefer to focus on the standard deviation of the sampling distribution of $\tilde{\theta}$,

$$
\mathrm{s.d.}(\tilde{\theta}) = \sqrt{\mathrm{Var}(\tilde{\theta})}.
$$

In practice we will not know s.d.$(\tilde{\theta})$, as it will typically depend on unknown features of the distribution of $X_1, \ldots, X_n$. However, we may be able to estimate $s.d.(\tilde{\theta})$ using the observed sample $x_1, \ldots, x_n$. We define the *standard error*, s.e.$(\tilde{\theta})$, of an estimator $\tilde{\theta}$ to be *an estimate of the standard deviation of its sampling distribution, s.d.$(\tilde{\theta})$.*

> Standard error of an estimator is an **estimate** of the standard deviation of its sampling distribution

We proved that

$$
\mathrm{Var}[\bar{X}] = \frac{\sigma^2}{n} \quad \Rightarrow \quad \mathrm{s.d.}(\bar{X}) = \frac{\sigma}{\sqrt{n}}.
$$

As $\sigma$ is unknown, we cannot calculate this standard deviation. However, we know that $E(S^2) = \sigma^2$, i.e. that the sample variance is an unbiased estimator of the population variance. Hence $S^2/n$ is an unbiased estimator for $\text{Var}(\bar{X})$. Therefore we obtain the *standard error of the mean*, s.e.$(\bar{X})$, by plugging in the estimate

$$s = \left( \frac{1}{n-1} \sum_{i=1}^{n} (x_i - \bar{x})^2 \right)^{1/2}$$

of $\sigma$ into s.d.$(\bar{X})$ to obtain

$$\text{s.e.}(\bar{X}) = \frac{s}{\sqrt{n}}.$$

Therefore, for the computer failure data, our estimate, $\bar{x} = 3.75$, for the population mean is associated with a standard error

$$\text{s.e.}(\bar{X}) = \frac{3.381}{\sqrt{104}} = 0.332.$$

Note that this is 'a' standard error, so other standard errors may be available. Indeed, for parametric inference, where we make assumptions about $f(x)$, alternative standard errors are available. For example, $X_1, \ldots, X_n$ are i.i.d. Poisson($\lambda$) random variables. $E(X) = \lambda$, so $\bar{X}$ is an unbiased estimator of $\lambda$. $\text{Var}(X) = \lambda$, so another s.e.$(\bar{X}) = \sqrt{\hat{\lambda}/n} = \sqrt{\bar{x}/n}$. In the computer failure data example, this is $\sqrt{\frac{3.75}{104}} = 0.19$.

### 9.4.3   Section Summary

In this section we have defined the standard error of an estimator. This is very important in practice, as the standard error tells us how precise our estimate is through how concentrated the sampling distribution of the estimator is. For example, in the age guessing example in R lab session 3, a standard error of 15 years indicates hugely inaccurate guesses. We have learned three key results: the sample mean is an unbiased estimate of the population mean; the variance of the sample mean is the population variance divided by the sample size; and the sample variance with divisor $n - 1$ is an unbiased estimator of the population variance.

## 9.5    Exercises

### 9.1 (Estimation Concepts)

1. A random sample of 10 boys and 10 girls from a large sixth form college were weighed with the following results.

| Boy's weight (kg) | 77 | 67 | 65 | 60 | 71 | 62 | 67 | 58 | 65 | 81 |
|---|---|---|---|---|---|---|---|---|---|---|
| Girl's weight (kg) | 42 | 57 | 46 | 49 | 64 | 61 | 52 | 50 | 44 | 59 |

Find
(a) unbiased estimates of $\mu_b$ and $\sigma_b^2$, the mean and variance of the weights of the boys;
(b) unbiased estimates of $\mu_g$ and $\sigma_g^2$, the mean and variance of the weights of the girls;
(c) an unbiased estimate of $\mu_b - \mu_g$.
Assuming that $\sigma_b^2 = \sigma_g^2 = \sigma^2$, calculate an unbiased estimate of $\sigma^2$ using both sets of weights.

2. The time that a customer has to wait for service in a restaurant has the probability density function

$$f(x) = \begin{cases} \frac{3\theta^3}{(x+\theta)^4} & \text{if } x \geq 0 \\ 0 & \text{otherwise,} \end{cases}$$

where $\theta$ is an unknown positive constant. Let $X_1, X_2, \ldots, X_n$ denote a random sample from this distribution. Show that

$$\hat{\theta} = \frac{2}{n} \sum_{i=1}^{n} X_i$$

is an unbiased estimator for $\theta$. Find the standard error of $\hat{\theta}$.

3. Show that if $X_1, X_2, \ldots X_n$ is a random sample from a distribution having mean $\mu$ and variance $\sigma^2$, then

$$Y = \sum_{i=1}^{n} a_i X_i = a_1 X_1 + a_2 X_2 + \cdots + a_n X_n$$

is an unbiased estimator for $\mu$ provided $\sum_{i=1}^{n} a_i = 1$, and find an expression for $Var(Y)$.

In the case $n = 4$, determine which of the following estimators are unbiased for $\mu$:

(a) $Y_1 = (X_1 + X_2 + X_3 + X_4)/4$,
(b) $Y_2 = X_1 - X_2 + X_3 - X_4$,
(c) $Y_3 = (2X_1 + 3X_4)/5$,
(d) $Y_4 = (X_1 + X_3)/2$.

**Hint:** Find the expectation of each $Y_i$ and see which one equals $\mu$. Which is the most efficient of those estimators which are unbiased?

**Hint: The most efficient estimator is the one with the least variance.**

# Methods of Point Estimation

# 10

**Abstract**

Chapter 10: This chapter discusses three important methods for point estimation: method of moments, maximum likelihood and Bayesian methods.

Estimation of unknown population characteristics or parameters is one of the primary tasks in statistics. Random samples are used to perform such estimation as suggested by the main idea of statistics, viz. collect samples and generalise to make inference regarding population proportions. Estimation, however, can be performed in many different formal and informal ways. Chief among such informal ways is the crude method of estimating the population mean by the sample mean as we have discussed in Sect. 9.4. This is an example of 'the method of moments' generalised below in Sect. 10.1. However, such a crude method may not always yield very good estimators as we have seen in the uniform distribution example in Sect. 9.3.3. Hence we need to develop better methods of estimation.

This chapter outlines three most popular methods of estimation. The first of these is the method of moments mentioned in the above paragraph. The second method, detailed in Sect. 10.2, is the maximum likelihood method which maximises what is called the likelihood function to obtain optimal parameter estimates. The third method in Sect. 10.3 is the Bayesian method of estimation which not only uses information from the sample data but also exploits information from past similar experiments which in turn allows for incorporation of subjective beliefs, based on subjective probability mentioned in Sect. 3.1, of the experimenter. In addition to these three, there is an important practical method known as least squares estimation method for linear statistical models. That method is introduced in Sect. 17.3.1.

© The Author(s), under exclusive license to Springer Nature Switzerland AG 2024
S. K. Sahu, *Introduction to Probability, Statistics & R*,
https://doi.org/10.1007/978-3-031-37865-2_10

## 10.1    Method of Moments

The method of moments is a straightforward and general method for deriving estimators for any parameter or function of parameters. Suppose that $X_1, \ldots, X_n$ are i.i.d. random variables with pdf (or pmf) $f_X(x; \boldsymbol{\theta})$, which is completely specified except for the values of $K$ unknown parameters $\boldsymbol{\theta} = (\theta_1, \ldots, \theta_K)$. The observations in the sample are denoted by $x_1, \ldots, x_n$ which are $n$ given numbers. Note that we now write the $\boldsymbol{\theta}$ vector in bold face as the dimension $K$ could be greater than one when there are multiple parameters. For example, if $X \sim N(\mu, \sigma^2)$ then $\boldsymbol{\theta} = (\mu, \sigma^2)$ and $K = 2$.

The term 'moment' in the 'method of moments' refers to the expected values of the $k$th power of the random variable $X$ for $k = 1, \ldots, K$. For example, the population moments are $E(X), E(X^2), E(X^3)$ and so on. These population moments are functions of the parameters $\boldsymbol{\theta}$. For example, if $X \sim N(\mu, \sigma^2)$ then $E(X) = \mu$ and $E(X^2) = \mu^2 + \sigma^2$ because of the definition $\mathrm{Var}(X) \equiv \sigma^2 = E(X^2) - \mu^2$. Moments are discussed in more detail in Chap. 13.

The sample moments are sample analogues of the population moments. For example,

$$\overline{X^k} \equiv \frac{1}{n} \sum_{i=1}^{n} X_i^k, k = 1, 2, \ldots, K$$

is the $k$th sample moment from a random sample of size $n$. Thus $\bar{X}$, the sample mean, is the first sample moment.

The most crucial fact in the 'method of moments' is the observation that, the $k$th sample moment, $\overline{X^k} \equiv \frac{1}{n} \sum_{i=1}^{n} X_i^k$, is an unbiased estimator of the $k$th population moment, $E(X^k)$ for any $k$. This is because,

$$E\left(\overline{X^k}\right) = \frac{1}{n} \sum_{i=1}^{n} E\left(X_i^k\right) = E\left(X^k\right)$$

for any $k$, since $X_1, \ldots, X_n$ are i.i.d. random variables. Therefore, $\bar{X}, \overline{X^2}, \ldots, \overline{X^K}$ are unbiased estimators of $E(X), E(X^2), \ldots, E(X^K)$ respectively.

The method of moments estimation for estimating $K$ parameters in $\boldsymbol{\theta}$ involves deriving expressions for the first $K$ population moments of $X$ as functions of $\boldsymbol{\theta}$:

$$
\begin{aligned}
E(X) &= g_1(\theta_1, \ldots, \theta_K) \\
E(X^2) &= g_2(\theta_1, \ldots, \theta_K) \\
&\vdots \\
E(X^K) &= g_K(\theta_1, \ldots, \theta_K),
\end{aligned}
$$

and then solving the $K$ simultaneous equations

$$\overline{X} = g_1(\theta_1, \ldots, \theta_K)$$
$$\overline{X^2} = g_2(\theta_1, \ldots, \theta_K)$$
$$\vdots$$
$$\overline{X^K} = g_K(\theta_1, \ldots, \theta_K)$$

for the $K$ unknown parameters $\theta_1, \ldots, \theta_K$. The solutions, which we denote by $\hat{\theta}_1, \ldots, \hat{\theta}_K$, are called the *method of moments* estimators of $\boldsymbol{\theta}$ (assuming that a solution to the equations exists). The most straightforward examples occur when $K = 1$, and $f_X$ depends on only a single unknown parameter, $\theta$. Then

$$E(X) = g(\theta) \quad \Rightarrow \quad \hat{\theta} = g^{-1}(\overline{X}).$$

In this case the 'estimate', observed value of the estimator, is simply given by $\hat{\theta} = g^{-1}(\bar{x})$.

---

**Example 10.1 (Method of Moments)**

1. $X_1, \ldots, X_n$ are i.i.d. Exponential($\beta$) random variables

$$E(X) = 1/\beta \quad \Rightarrow \quad \hat{\beta} = 1/\overline{X}.$$

2. $X_1, \ldots, X_n$ are i.i.d. Geometric($p$) random variables.

$$E(X) = (1 - p)/p \quad \Rightarrow \quad \hat{p} = 1/(\overline{X} + 1).$$

3. $X_1, \ldots, X_n$ are i.i.d. N($\mu, \sigma^2$) random variables.

$$
\begin{aligned}
E(X) &= \mu \\
E(X^2) &= \sigma^2 + \mu^2
\end{aligned}
\quad \Rightarrow \quad
\begin{aligned}
\hat{\mu} &= \overline{X} \\
\hat{\sigma}^2 &= \overline{X^2} - \overline{X}^2.
\end{aligned}
$$

4. $X_1, \ldots, X_n$ are i.i.d. Gamma($\alpha, \beta$) random variables.

$$
\begin{aligned}
E(X) &= \alpha/\beta \\
E(X^2) &= \alpha(\alpha + 1)/\beta^2
\end{aligned}
\quad \Rightarrow \quad
\begin{aligned}
\hat{\alpha} &= \overline{X}^2/(\overline{X^2} - \overline{X}^2) \\
\hat{\beta} &= \overline{X}/(\overline{X^2} - \overline{X}^2).
\end{aligned}
$$

◀

Here are a few remarks that describe various characteristics of the method of moments estimator.

- Although, unbiased estimators for $E(X), E(X^2), \ldots$ are used in the derivation, the method of moments estimators $\hat{\theta}_1, \hat{\theta}_2 \ldots$ are not generally unbiased. For example, if $X_1, \ldots, X_n$ are i.i.d. $N(\mu, \sigma^2)$ random variables, the method of moments estimator for $\sigma^2$ is

$$\hat{\sigma}^2 = \overline{X^2} - \overline{X}^2 = \frac{1}{n}\sum_{i=1}^{n} X_i^2 - \bar{X}^2 = \frac{n-1}{n}S^2,$$

see Sect. 1.3.2. We know that $\sigma^2 = \text{Var}(X)$ and $S^2$ is always unbiased for $\text{Var}(X)$ from **R3** in Sect. 9.4, so $\text{bias}(\hat{\sigma}^2) = -\sigma^2/n$.

- Method of moments estimators are coherent, in the sense that, if $\hat{\theta}$ is the method of moments estimator of $\theta$, then $g(\hat{\theta})$ is the method of moments estimator of $g(\theta)$. For example, if $X_1, \ldots, X_n$ are i.i.d. Exponential($\beta$) random variables, the method of moments estimator for $\text{Var}(X) = 1/\beta^2$ is

$$\widehat{\text{Var}}(X) = 1/\hat{\beta}^2 = \overline{X}^2.$$

- Often, it is not possible to calculate exact standard errors for method of moments estimators. For example, if $X_1, \ldots, X_n$ are i.i.d. exponential($\beta$) random variables, and $\hat{\beta} = 1/\overline{X}$, then an expression for $\text{Var}(\hat{\beta})$ is not available in closed form, although it is possible to find an approximation.

Thus the method of moments suffers from various shortcomings and is not the most widely used in estimation. Hence, we turn to maximum likelihood estimation in the next section.

## 10.2  Maximum Likelihood Estimation

Perhaps the most important concept for parametric statistical inference, particularly for complex statistical models is *likelihood*.

Recall that for parametric statistical inference, we assume that the observed data $x_1, \ldots, x_n$ are observations of random variables $X_1, \ldots, X_n$ with joint density $f_{\mathbf{X}}(\mathbf{x}; \boldsymbol{\theta})$, a function known except for the values of parameters $\theta_1, \theta_2, \ldots, \theta_K$ (denoted $\boldsymbol{\theta}$).

Until now, we have thought of the joint density $f_{\mathbf{X}}(\mathbf{x}; \boldsymbol{\theta})$ as a function of $\mathbf{x}$ for fixed $\boldsymbol{\theta}$, which describes the relative probabilities of different possible (sets of) $\mathbf{x}$, given a particular set of parameters $\boldsymbol{\theta}$. However, in statistical inference, we have observed $x_1, \ldots, x_n$ (values of $X_1, \ldots, X_n$). Knowledge of the probability of alternative possible realisations of $X_1, \ldots, X_n$ is largely irrelevant since we must estimate $\boldsymbol{\theta}$ using the observed values $\mathbf{x}$.

Our only link between the observed data $x_1, \ldots, x_n$ and $\boldsymbol{\theta}$ is through the function $f_{\mathbf{X}}(\mathbf{x}; \boldsymbol{\theta})$. Therefore, it seems sensible that parametric statistical inference should be based on this function. We can think of $f_{\mathbf{X}}(\mathbf{x}; \boldsymbol{\theta})$ as a function of $\boldsymbol{\theta}$ for fixed $\mathbf{x}$, which describes the relative *likelihoods* of different possible (sets of) $\boldsymbol{\theta}$, given observed data $x_1, \ldots, x_n$. When $f_{\mathbf{X}}(\mathbf{x}; \boldsymbol{\theta})$ is considered as a function of $\boldsymbol{\theta}$ for fixed (observed) $x_1, \ldots, x_n$, we call it the *likelihood function*. The below example clearly illustrates the difference between the concepts of the likelihood function and the joint probability function both denoted by $f_{\mathbf{X}}(\mathbf{x}; \boldsymbol{\theta})$.

---

**Example 10.2 (Lion's Appetite Taken from Liero and Zwanzig [12])**

Suppose that the appetite of a lion, see Fig. 10.1, has three different stages:

$$\{\text{hungry, moderate, lethargic}\} = \{\theta_1, \theta_2, \theta_3\}.$$

These are all possible values of the single parameter $\theta$. Let the random variable $X$ be the number of people the lion has eaten in a particular night. Suppose that the probability $P(X = x) = f_\theta(x)$ is given by the following table: In this table three different probability distributions are there for three different possible value of $\theta$. Thus each row of the table provides a probability distribution. The *likelihood function*, however, for a given observed value of $x$ is given as the column for that particular value of $x$. For example, if $x = 2$ then the likelihood function takes the values 0.05, 0.8 and 0.05 corresponding to $\theta = \theta_1, \theta_2$ and $\theta_3$ respectively. Hence when $x = 2$ the likelihood function peaks at $\theta = \theta_2$=moderate. ◄

The likelihood function is of central importance in parametric statistical inference (Table 10.1). It provides a means for comparing different possible values of $\boldsymbol{\theta}$, based on the probabilities (densities) that they assign to the observed data $x_1, \ldots, x_n$. Here are some important remarks describing various aspects of the likelihood function.

- Often the likelihood function is written as $L(\boldsymbol{\theta})$ or $L(\boldsymbol{\theta}; \mathbf{x})$. We shall continue to use $f_{\mathbf{X}}(\mathbf{x}; \boldsymbol{\theta})$. Wherever parametric statistical inference is our concern, we treat $f_{\mathbf{X}}(\mathbf{x}; \boldsymbol{\theta})$ as a function of $\boldsymbol{\theta}$ and call it the likelihood.

**Fig. 10.1** A lion cub.
Author: flowcomm. Source:
https://commons.wikimedia.
org/ License: CC BY-SA 2.0

**Table 10.1** Probability distribution of number of people eaten overnight by a lion

| x | 0 | 1 | 2 | 3 | 4 | Total |
|---|---|---|---|---|---|-------|
| $\theta_1$ | 0 | 0.05 | 0.05 | 0.8 | 0.1 | 1 |
| $\theta_2$ | 0.05 | 0.05 | 0.8 | 0.1 | 0 | 1 |
| $\theta_3$ | 0.9 | 0.05 | 0.05 | 0 | 0 | 1 |

- Frequently it is more convenient to consider the log-likelihood function $\log f_{\mathbf{X}}(\mathbf{x}; \boldsymbol{\theta})$ because the log-likelihood is an additive function for independent samples $\mathbf{x}$ as the examples below will show. The log-likelihood function is often denoted by $l(\boldsymbol{\theta})$ or $l(\boldsymbol{\theta}; \mathbf{x})$.
- Nothing in the definition of the likelihood requires $x_1, \ldots, x_n$ to be observations of independent or identically distributed random variables, although we shall frequently make these assumptions, in which case

$$f_{\mathbf{X}}(\mathbf{x}; \boldsymbol{\theta}) = \prod_{i=1}^{n} f_X(x_i; \boldsymbol{\theta}),$$

where $f_X$ denotes the common pdf of $X_1, \ldots, X_n$.
- Any multiplicative factors which depend on $\mathbf{x}$ alone (and not on $\boldsymbol{\theta}$) can be ignored when writing down the likelihood function. Such factors give no information about the relative likelihoods of different possible values of $\boldsymbol{\theta}$.
- The dependence between $\boldsymbol{\theta}$ and $\mathbf{X}$ should be completely described by the likelihood function. Therefore, if the sample space for $\mathbf{X}$ depends on $\boldsymbol{\theta}$, this should be reflected by the likelihood function, see the uniform distribution example below.

## 10.2.1  Examples of the Log-Likelihood Function

**Example 10.3 (Bernoulli Distribution)**

Suppose that $x_1, \ldots, x_n$ are observations of $X_1, \ldots, X_n$, i.i.d. Bernoulli($p$) random variables. Here $\boldsymbol{\theta} = (p)$ and

$$f_{\mathbf{X}}(\mathbf{x}; p) = \prod_{i=1}^{n} p^{x_i}(1 - p)^{1-x_i} = p^{\sum_{i=1}^{n} x_i}(1 - p)^{n-\sum_{i=1}^{n} x_i}.$$

Hence the log-likelihood function is:

$$\log f_{\mathbf{X}}(\mathbf{x}; p) = n\bar{x}\log(p) + n(1 - \bar{x})\log(1 - p).$$

This function is illustrated in Fig. 10.2 for three different values of $\bar{x}$, viz. 0.4, 0.6 and 0.8, when $n = 25$ ◄

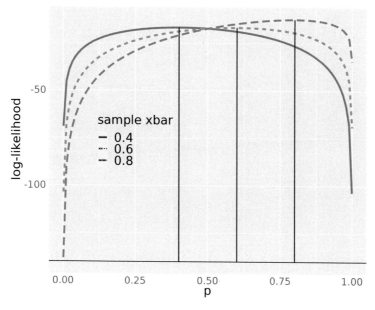

**Fig. 10.2** Illustration of the log-likelihood function (upto a constant of proportionality) of $p$ for three different values of $\bar{x}$. The three black vertical lines are at the three corresponding $\bar{x}$ values

---

### Example 10.4 (Poisson Distribution)

Suppose that $x_1, \ldots, x_n$ are observations of $X_1, \ldots, X_n$, i.i.d. Poisson($\lambda$) random variables. Here $\boldsymbol{\theta} = (\lambda)$ and

$$f_{\mathbf{X}}(\mathbf{x}; \lambda) = \prod_{i=1}^{n} \frac{\exp(-\lambda)\lambda^{x_i}}{x_i!} = \frac{\exp(-n\lambda)\lambda^{\sum_{i=1}^{n} x_i}}{\prod x_i!} \propto \exp(-n\lambda)\lambda^{\sum_{i=1}^{n} x_i}$$

Hence the log-likelihood function is:

$$\log f_{\mathbf{X}}(\mathbf{x}; \lambda) = -n\lambda + n\bar{x} \log \lambda - \sum_{i=1}^{n} \log(x_i!).$$

This function is illustrated in Fig. 10.3 for three different values of $\bar{x}$, viz. 1.25, 2.5 and 5, when $n = 10$ ◄

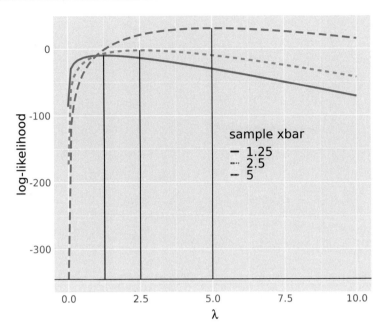

**Fig. 10.3** Illustration of the log-likelihood function (upto a constant of proportionality) of $\lambda$ for three different values of $\bar{x}$. The three vertical lines are at the three corresponding $\bar{x}$ values

---

**Example 10.5 (Normal Distribution)**

Suppose that $x_1, \ldots, x_n$ are observations of $X_1, \ldots, X_n$, i.i.d. $N(\mu, \sigma^2)$ random variables. Here $\boldsymbol{\theta} = (\mu, \sigma^2)$ and

$$
\begin{aligned}
f_{\mathbf{X}}(\mathbf{x}; \mu, \sigma^2) &= \prod_{i=1}^{n} \frac{1}{\sqrt{2\pi\sigma^2}} \exp\left(-\frac{1}{2\sigma^2}(x_i - \mu)^2\right) \\
&= (2\pi\sigma^2)^{-\frac{n}{2}} \exp\left(-\frac{1}{2\sigma^2}\textstyle\sum_{i=1}^{n}(x_i - \mu)^2\right) \\
&\propto (\sigma^2)^{-\frac{n}{2}} \exp\left(-\frac{1}{2\sigma^2}\textstyle\sum_{i=1}^{n}(x_i - \mu)^2\right).
\end{aligned}
$$

Hence the log-likelihood function is:

$$
\log f_{\mathbf{X}}(\mathbf{x}; \mu, \sigma^2) = -\frac{n}{2}\log(2\pi) - \frac{n}{2}\log(\sigma^2) - \frac{1}{2\sigma^2}\textstyle\sum_{i=1}^{n}(x_i - \mu)^2.
$$

Note that this function depends on two parameters $\mu$ and $\sigma^2$. ◀

### Example 10.6 (Gamma Distribution)

Suppose that $x_1, \ldots, x_n$ are observations of $X_1, \ldots, X_n$, i.i.d. gamma$(\alpha, \beta)$ random variables. Here $\theta = (\alpha, \beta)$ and

$$
\begin{aligned}
f_{\mathbf{X}}(\mathbf{x}; \alpha, \beta) &= \prod_{i=1}^{n} \frac{\beta^{\alpha}}{\Gamma(\alpha)} x_i^{\alpha-1} \exp(-\beta x_i) \\
&= \frac{\beta^{n\alpha}}{\Gamma(\alpha)^n} \prod_{i=1}^{n} x_i^{\alpha-1} \exp\left(-\beta \sum_{i=1}^{n} x_i\right) \\
&\propto \frac{\beta^{n\alpha}}{\Gamma(\alpha)^n} \prod_{i=1}^{n} x_i^{\alpha-1} \exp\left(-\beta \sum_{i=1}^{n} x_i\right).
\end{aligned}
$$

Hence the log-likelihood function is:

$$
\log f_{\mathbf{X}}(\mathbf{x}; \alpha, \beta) = n\alpha \log \beta - n \log \Gamma(\alpha) + (\alpha - 1)\sum \log x_i - \beta \sum_{i=1}^{n} x_i.
$$

◄

### Example 10.7 (Uniform Distribution)

Suppose that $x_1, \ldots, x_n$ are observations of $X_1, \ldots, X_n$, i.i.d. uniform$(\alpha, \beta)$ random variables. Here $\theta = (\alpha, \beta)$ and

$$
\begin{aligned}
f_X(x; \alpha, \beta) &= \frac{1}{\beta - \alpha}, \qquad x \in (\alpha, \beta) \\
&= \frac{1}{\beta - \alpha} I[x \in (\alpha, \beta)], \qquad x \in \mathcal{R}
\end{aligned}
$$

where $I[A] = 1$ if $A$ is true and 0 otherwise. Therefore,

$$
\begin{aligned}
f_{\mathbf{X}}(\mathbf{x}; \alpha, \beta) &= \prod_{i=1}^{n} \frac{1}{\beta - \alpha} I[x_i \in (\alpha, \beta)] \\
&= \frac{1}{(\beta - \alpha)^n} I[x_1 \in (\alpha, \beta)] \cdots I[x_n \in (\alpha, \beta)] \\
&= \frac{1}{(\beta - \alpha)^n} I[\alpha < \min(x_1, \ldots, x_n)] I[\beta > \max(x_1, \ldots, x_n)]
\end{aligned}
$$

Suppose that $\alpha$ is assumed to be zero. Then $\theta = (\beta)$ and

$$
f_{\mathbf{X}}(\mathbf{x}; \beta) = \frac{1}{\beta^n} I[\beta > \max(x_1, \ldots, x_n)].
$$

An illustration of this likelihood function is provided in Fig. 10.4 below. ◄

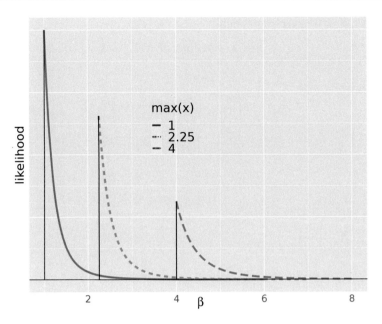

**Fig. 10.4** Illustration of the likelihood function (upto a constant of proportionality) for three different values of $\max(x_1, \ldots, x_n)$

## 10.2.2 Maximum Likelihood Estimates

Consider the value of $\theta$, which maximises the likelihood function. This is the 'most likely' value of $\theta$, the one which makes the observed data 'most probable'. When we are searching for an estimate of $\theta$, this would seem to be a good candidate.

We call the value of $\theta$ which maximises the likelihood $f_{\mathbf{X}}(\mathbf{x}; \theta)$ the *maximum likelihood estimate* (m.l.e.) of $\theta$, denoted by $\hat{\theta}(\mathbf{x})$. $\hat{\theta}$ depends on $\mathbf{x}$, as different observed data samples lead to different likelihood functions. The corresponding function of $\mathbf{X}$, $\hat{\theta}(\mathbf{X})$ is called the *maximum likelihood estimator* (also m.l.e.). We will use $\hat{\theta}$ as shorthand for both $\hat{\theta}(\mathbf{x})$ and $\hat{\theta}(\mathbf{X})$.

Note that as $\theta = (\theta_1, \ldots, \theta_p)$, the m.l.e. for any component of $\theta$ is given by the corresponding component of $\hat{\theta} = (\hat{\theta}_1, \ldots, \hat{\theta}_p)$. Similarly, the m.l.e. for any function of parameters $g(\theta)$ is given by $g(\hat{\theta})$, so maximum likelihood estimation is coherent. This follows by noting that the maximum possible $f_{\mathbf{X}}(\mathbf{x}; \theta)$ is given by $f_{\mathbf{X}}(\mathbf{x}; \hat{\theta})$, which corresponds to $g(\theta) = g(\hat{\theta})$. If there exists a value of $g(\theta)$ which can result in a larger $f_{\mathbf{X}}(\mathbf{x}; \theta)$, then $\hat{\theta}$ cannot be the m.l.e.

As log is a strictly increasing function, the value of $\theta$ which maximises $f_{\mathbf{X}}(\mathbf{x}; \theta)$ also maximises $\log f_{\mathbf{X}}(\mathbf{x}; \theta)$. It is almost always easier to maximise $\log f_{\mathbf{X}}(\mathbf{x}; \theta)$. This is achieved in the usual way; finding a stationary point by differentiating $\log f_{\mathbf{X}}(\mathbf{x}; \theta)$ with respect to $\theta_1, \ldots, \theta_p$ and solving the resulting $p$ simultaneous equations. It should also be checked that the stationary point is a maximum.

### 10.2.3  Examples of m.l.e

**Example 10.8 (Bernoulli Distribution)**

Suppose that $x_1, \ldots, x_n$ are observations of $X_1, \ldots, X_n$, i.i.d. Bernoulli$(p)$ random variables. Here $\boldsymbol{\theta} = (p)$ and

$$\log f_{\mathbf{X}}(\mathbf{x}; p) = n\overline{x} \log p + n(1 - \overline{x}) \log(1 - p)$$
$$\frac{\partial}{\partial p} \log f_{\mathbf{X}}(\mathbf{x}; p) = \frac{n\overline{x}}{p} - \frac{n(1 - \overline{x})}{1 - p}$$
$$\Rightarrow \quad 0 = \frac{n\overline{x}}{\hat{p}} - \frac{n(1 - \overline{x})}{1 - \hat{p}}$$
$$\Rightarrow \quad \hat{p} = \overline{x}.$$

Note that:

1. $\frac{\partial^2}{\partial p^2} \log f_{\mathbf{X}}(\mathbf{x}; p) = -n\overline{x}/p^2 - n(1 - \overline{x})/(1 - p)^2 < 0$ everywhere, so the stationary point is clearly a maximum.
2. $\hat{p}$ is an unbiased estimator of $p$.

◄

**Example 10.9 (Poisson Distribution)**

$x_1, \ldots, x_n$ are observations of $X_1, \ldots, X_n$, i.i.d. Poisson$(\lambda)$ random variables. Here $\boldsymbol{\theta} = (\lambda)$ and

$$\log f_{\mathbf{X}}(\mathbf{x}; \lambda) = -n\lambda + n\overline{x} \log \lambda - n\sum_{i=1}^{n} \log(x_i!)$$
$$\frac{\partial}{\partial \lambda} \log f_{\mathbf{X}}(\mathbf{x}; \lambda) = -n + \frac{n\overline{x}}{\lambda}$$
$$\Rightarrow \quad 0 = -n + \frac{n\overline{x}}{\hat{\lambda}}$$
$$\Rightarrow \quad \hat{\lambda} = \overline{x}.$$

1. $\frac{\partial^2}{\partial \lambda^2} \log f_{\mathbf{X}}(\mathbf{x}; \lambda) = -n\overline{x}/\lambda^2 < 0$ everywhere, so the stationary point is clearly a maximum.
2. $\hat{\lambda}$ is an unbiased estimator of $\lambda$.

◄

**Example 10.10 (Normal Distribution)**

$x_1, \ldots, x_n$ are observations of $X_1, \ldots, X_n$, i.i.d. $N(\mu, \sigma^2)$ random variables. Here $\boldsymbol{\theta} = (\mu, \sigma^2)$ and

$$\log f_{\mathbf{X}}(\mathbf{x}; \mu, \sigma^2) = -\frac{n}{2}\log(2\pi) - \frac{n}{2}\log(\sigma^2) - \frac{1}{2\sigma^2}\sum_{i=1}^{n}(x_i - \mu)^2$$

$$\frac{\partial}{\partial \mu}\log f_{\mathbf{X}}(\mathbf{x}; \mu, \sigma^2) = \frac{1}{\sigma^2}\sum_{i=1}^{n}(x_i - \mu) = \frac{n(\overline{x} - \mu)}{\sigma^2}$$

$$\Rightarrow \quad 0 = \frac{n(\overline{x} - \hat{\mu})}{\hat{\sigma}^2} \quad (1)$$

$$\frac{\partial}{\partial \sigma^2}\log f_{\mathbf{X}}(\mathbf{x}; \mu, \sigma^2) = -\frac{n}{2\sigma^2} + \frac{1}{2(\sigma^2)^2}\sum_{i=1}^{n}(x_i - \mu)^2$$

$$\Rightarrow \quad 0 = -\frac{n}{2\hat{\sigma}^2} + \frac{1}{2(\hat{\sigma}^2)^2}\sum_{i=1}^{n}(x_i - \hat{\mu})^2 \quad (2)$$

Solving (1) and (2), we obtain

$$\hat{\mu} = \overline{x}$$

$$\hat{\sigma}^2 = \frac{1}{n}\sum_{i=1}^{n}(x_i - \hat{\mu})^2 = \frac{1}{n}\sum_{i=1}^{n}(x_i - \overline{x})^2.$$

Strictly, to show that this stationary point is a maximum, we need to show that the Hessian matrix (the matrix of second derivatives with elements $[\mathbf{H}(\boldsymbol{\theta})]_{ij} = \frac{\partial^2}{\partial \theta_i \partial \theta_j}\log f_{\mathbf{X}}(\mathbf{x}; \boldsymbol{\theta})$) is negative definite at $\boldsymbol{\theta} = \hat{\boldsymbol{\theta}}$, that is $\mathbf{a}^T\mathbf{H}(\hat{\boldsymbol{\theta}})\mathbf{a} < 0$ for every $\mathbf{a} \neq \mathbf{0}$. Here we have

$$\mathbf{H}(\hat{\mu}, \hat{\sigma}^2) = \begin{pmatrix} -\frac{n}{\hat{\sigma}^2} & 0 \\ 0 & -\frac{n}{2(\hat{\sigma}^2)^2} \end{pmatrix}$$

which is clearly negative definite.

Note that $\hat{\mu}$ is an unbiased estimator of $\mu$, but $\text{bias}(\hat{\sigma}^2) = -\sigma^2/n$. ◀

**Example 10.11 (Uniform Distribution)**

$x_1, \ldots, x_n$ are observations of $X_1, \ldots, X_n$, i.i.d. uniform$[0, \beta]$ random variables. Here the sample space is $[0, \beta]$, $\boldsymbol{\theta} = (\beta)$ and

$$f_{\mathbf{X}}(\mathbf{x}; \beta) = \frac{1}{\beta^n}I[\beta \geq \max(x_1, \ldots, x_n)].$$

Clearly $\hat{\beta} = \max(x_1, \ldots, x_n)$.

Now, let $Y = \max\{X_1, \ldots, X_n\}$, so $Y = \hat{\beta}$. Then

$$
\begin{aligned}
F_Y(y) = P(Y \le y) &= P(\max(X_1, \ldots, X_n) \le y) \\
&= P(X_1 \le y, \ldots, X_n \le y) \\
&= \prod_{i=1}^{n} P(X_i < y) \\
&= \left(\frac{y}{\beta}\right)^n \qquad 0 < y \le \beta
\end{aligned}
$$

Therefore

$$
f_Y(y) = \frac{n}{\beta}\left(\frac{y}{\beta}\right)^{n-1} \qquad 0 \le y \le \beta,
$$

and

$$
\begin{aligned}
E(\hat{\beta}) = E(Y) &= \frac{n}{n+1}\beta \\
\Rightarrow \quad \mathrm{bias}(\hat{\beta}) &= -\beta/(n+1).
\end{aligned}
$$

◀

## 10.3  Bayesian Estimation and Inference

Bayesian theory (named after the Rev. Thomas Bayes, an amateur eighteenth century English mathematician), provides an approach to statistical inference which is different in spirit from the familiar classical approach. We do not think of Bayesian statistics as a separate area within Statistics. Any statistical problem (Survival analysis; Multivariate analysis; Generalised linear models *etc.*) can be approached in a Bayesian way.

The basic philosophy underlying Bayesian inference is that **the only sensible measure of uncertainty is probability**. Data are still assumed to come from one of a parameterised family of distributions. However, whereas classical statistics considers the parameters to be *fixed but unknown*, the Bayesian approach treats them as random variables in their own right. Prior beliefs about $\theta$ are represented by the **prior**, $\pi(\theta)$, a probability density (or mass) function. The **posterior** density (mass function), $\pi(\theta|x_1, \ldots, x_n)$ represents our *modified* belief about $\theta$ in the light of the observed data. We will do this in quite detail. Let us start from the basics.

### 10.3.1  Prior and Posterior Distributions

We start by recalling the Bayes Theorem as stated in Sect. 4.5. Let $B_1, B_2, \ldots, B_k$ be a set of mutually exclusive and exhaustive events. For any new event $A$,

$$
P(B_i|A) = \frac{P(B_i \cap A)}{P(A)} = \frac{P(A|B_i)P(B_i)}{\sum_{i=1}^{k} P(A|B_i)P(B_i)}. \tag{10.1}
$$

**Example 10.12**

We can understand the theorem using a simple example. Consider a disease that is thought to occur in 1% of the population. Using a particular blood test a physician observes that out of the patients with disease 98% possess a particular symptom. Also assume that 0.1% of the population without the disease have the same symptom. A randomly chosen person from the population is blood tested and is shown to have the symptom. What is the conditional probability that the person has the disease?

**Solution** Here $k = 2$ and let $B_1$ be the event that a randomly chosen person has the disease and $B_2$ is the complement of $B_1$. Let $A$ be the event that a randomly chosen person has the symptom. The problem is to determine $Pr\{B_1|A\}$.

We have $Pr(B_1) = 0.01$ since 1% of the population has the disease, and $Pr(A|B_1) = 0.98$. Also $Pr(B_2) = 0.99$ and $Pr(A|B_2) = 0.001$. Now

$$Pr(\text{disease} \mid \text{symptom}) = Pr(B_1|A) = \frac{Pr(A|B_1)\,Pr(B_1)}{Pr(A|B_1)\,Pr(B_1)+Pr(A|B_2)\,Pr(B_2)}$$
$$= \frac{0.98 \times 0.01}{0.98 \times 0.01 + 0.001 \times 0.99}$$
$$= 0.9082.$$

This probability will be much smaller if $Pr(A|B_2)$ is much higher. For example, with $Pr(A|B_2) = 0.1$ we have $Pr(B_1|A) = 0.09$. ◄

**Notation** We are used to the notation that $X$ is the random variable and $x$ is its value. Now we will relax that little bit for the random variable $\theta$ only. We will use $\theta$ to denote the random variable and $\Theta$ to denote the parameter space. Often, however, $\Theta$ will be generically taken as $(-\infty, \infty)$.

We first state the following version of the Bayes theorem for random variables. Suppose that two random variables $X$ and $\theta$ are given with pdfs $f(x|\theta)$ and $\pi(\theta)$. The the conditional pdf of $\theta$ given $X = x$, also called the posterior pdf, is given by:

$$\pi(\theta|x) = \frac{f(x|\theta)\pi(\theta)}{\int_{-\infty}^{\infty} f(x|\theta)\pi(\theta)d\theta}. \tag{10.2}$$

The above theorem allows us to work in the Bayesian inference framework where we assume:

- $x$ to be the data.
- $\theta$ to be the unknown parameters.
- $f(x|\theta)$ the likelihood of data given unknown parameters $\theta$.
- $\pi(\theta) = $ Prior distribution or prior belief for the unknown parameter $\theta$.

Suppose that we observe $n$ independent sample values, $x_1, x_2, \ldots, x_n$. In this case, we replace $x$ by $x_1, x_2, \ldots, x_n$ and $f(x|\theta)$ by $f(x_1, \ldots, x_n|\theta)$ in the above Bayes Theorem. The modified Bayes theorem is written as:

$$\pi(\theta|x_1, \ldots, x_n) = \frac{f(x_1, \ldots, x_n|\theta)\pi(\theta)}{\int_{-\infty}^{\infty} f(x_1, \ldots, x_n \mid \theta)\pi(\theta)d\theta}.$$

This distribution is called the **posterior distribution**. Bayesian inference proceeds from this distribution. In practice, the denominator of the above equation needn't usually be calculated, and Bayes' rule is often just written,

$$\pi(\theta|x_1, \ldots, x_n) \propto f(x_1, \ldots, x_n \mid \theta)\pi(\theta).$$

Hence we always know the posterior distribution up-to the constant of proportionality, which is often called the normalising constant. Often we will be able to identify the posterior distribution of $\theta$ just by looking at the numerator. By Bayes Theorem we "update" $\pi(\theta)$ to $\pi(\theta|\mathbf{x})$.

### 10.3.1.1  Bayesian Learning

Suppose that we observe $x_1$ and $x_2$ independently one after another. Then,

$$\begin{aligned} \pi(\theta|x_1) &\propto f(x_1|\theta)\pi(\theta) \\ \text{and } \pi(\theta|x_1, x_2) &\propto f(x_2|\theta)f(x_1|\theta)\pi(\theta) \\ &\propto f(x_2|\theta)\pi(\theta|x_1) \end{aligned}$$

Thus the Bayes theorem shows how the knowledge about the state of nature represented by $\theta$ is continually modified as new data becomes available.

---

**Example 10.13 (Binomial Distribution)**

Suppose $X \sim \text{binomial}(n, \theta)$ where $n$ is known and we assume the Beta$(\alpha, \beta)$ prior distribution for $\theta$. See Sect. 7.3.4 for the definition of the beta distribution. Here the likelihood function is given by:

$$L(\theta; x) = f(x|\theta) = \binom{n}{x}\theta^x(1-\theta)^{n-x}.$$

The prior distribution is given by:

$$\pi(\theta) = \frac{1}{B(\alpha, \beta)}\theta^{\alpha-1}(1-\theta)^{\beta-1},$$

see the pdf of the beta distribution in (7.10). Hence

$$\pi(\theta|x) \propto \theta^{x+\alpha-1}(1-\theta)^{n-x+\beta-1}.$$

Note that we have only written down the terms involving $\theta$ from the likelihood × the prior. We do not need to write the other terms which do not involve $\theta$, like the $\binom{n}{x}$ or the constant $\frac{1}{B(\alpha,\beta)}$ because these cancel in the ratio. Now the posterior distribution is recognised to be a beta distribution with parameters $x + \alpha$ and $n - x + \beta$. Hence:

$$\pi(\theta|x) = \frac{1}{B(x+\alpha, n-x+\beta)}\theta^{x+\alpha-1}(1-\theta)^{n-x+\beta-1}.$$

To illustrate suppose that we are interested in the proportion of female births, $\theta$. Fictitiously suppose that we observe $x = 2419$ female births out of $n = 4934$ total number of births in a town over a period of time. These numbers are fictitious but the proportion mimics real data observed for Paris sometime during 1700s.

A reasonable prior here is the uniform distribution on $(0, 1)$ which is beta distribution with parameters $\alpha = \beta = 1$. Although this prior distribution is workable, it implies a horizontal straight line density, drawn in purple colour, lying at the bottom of the vertical axis in Fig. 10.5. To illustrate with another prior distribution we choose the Beta($\alpha = 51, \beta = 51$) prior distribution which is a symmetric prior distribution about the prior mean 0.5. The parameters $\alpha = 51$ and $\beta = 51$ may be interpreted as the prior derived from 50 female and 50 male births, emphasising prior neutrality.

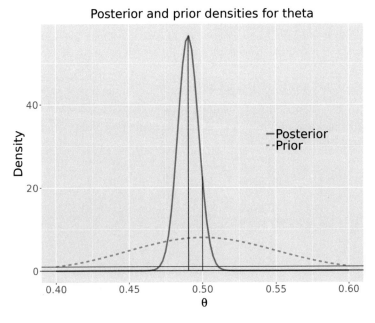

**Fig. 10.5**  Prior and posterior densities for $\theta$ in the binomial example

The posterior distribution is the beta distribution with parameters $2470(= 2419 + 51)$ and $2566(= 2515 + 51)$. Using the R function `dbeta` we draw the prior and the posterior distributions in Fig. 10.5. The Figure also shows the location of the mean of the posterior distribution at $0.49 \approx \frac{2470}{2419 + 2565 + 102}$ of $\theta$ with a black vertical line. A blue coloured vertical line marks the posterior spot at 0.5 (the prior mean and the median). Notice that $P(\theta \geq 0.5|x)$ is calculated to be 0.088 from `1-pbeta(0.5, 2470, 2566)` in R. This value is much lower than the 0.5, `1-pbeta(0.5, 51, 51)`, value for the prior distribution. ◄

### Example 10.14 (Poisson Distribution)

Return to the London bomb hits Example 1.3 and define the random variable, $X$, to be the number of bomb hits in a given area. The frequency distribution given there summarises a sample of size $n = 576$ and in Sect. 9.2 we have obtained $\bar{x} = 0.9288$.

Assuming the random sample $X_1, \ldots, X_n$ to an i.i.d. sample from the Poisson distribution with unknown parameter $\lambda$ we obtain the likelihood function:

$$
\begin{aligned}
L(\lambda; x_1, \ldots, x_n) &= f(x_1, \ldots, x_n|\lambda) \\
&= f(x_1|\lambda) f(x_2|\lambda) \cdots f(x_n|\lambda) \\
&= e^{-\lambda} \frac{\lambda^{x_1}}{x_1!} e^{-\lambda} \frac{\lambda^{x_2}}{x_2!} \cdots e^{-\lambda} \frac{\lambda^{x_n}}{x_n!} \\
&= e^{-n\lambda} \frac{\lambda^{x_1 + x_2 + \cdots + x_n}}{x_1! x_2! \cdots x_n!}.
\end{aligned}
$$

The term $x_1! x_2! \cdots x_n!$ in the denominator does not depend on the unknown parameter $\theta$. Hence we write,

$$
L(\lambda; x_1, \ldots, x_n) \propto e^{-n\lambda} \lambda^{x_1 + x_2 + \cdots + x_n},
$$

for $\lambda > 0$. Suppose that we assume a gamma prior distribution $G(\alpha, \beta)$ for some known values of $\alpha$ and $\beta$. Hence

$$
\pi(\lambda|\alpha, \beta) = \frac{\beta^\alpha}{\Gamma(\alpha)} \lambda^{\alpha - 1} e^{-\beta\lambda}
$$

according to the gamma density (7.8). Now the posterior distribution, $\pi(\lambda|x_1, \ldots, x_n)$ is proportional to the likelihood times the prior. Hence,

$$
\begin{aligned}
\pi(\lambda|x_1, \ldots, x_n, \alpha, \beta) &\propto e^{-n\lambda} \lambda^{x_1 + x_2 + \cdots + x_n} \lambda^{\alpha - 1} e^{-\beta\lambda} \\
&\propto e^{-(n+\beta)\lambda} \lambda^{n\bar{x} + \alpha - 1}.
\end{aligned}
$$

Now compare the last expression with the gamma density (7.8) to conclude that the posterior distribution of $\lambda$ is a gamma distribution with parameters: $n\bar{x} + \alpha$ and $n + \beta$.

Suppose that we take $\alpha = 1$ and $\beta = 1$. In this case both the prior mean and variance of $\lambda$ are 1. The posterior distribution of $\lambda$ in this case is given by $G(536,$

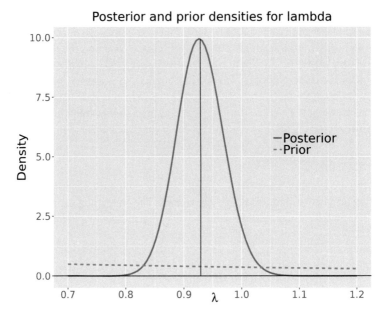

**Fig. 10.6**   Prior and posterior densities for λ in the bomb hits example

577). The posterior mean and variance are obtained as the mean and variance of this gamma distribution, which are 0.9289 and 0.0016 respectively. The prior and the posterior distributions are shown in Fig. 10.6. ◄

### Example 10.15 (Exponential Distribution)

As a motivating example consider the wealth of billionaires in Example 1.6. A histogram of the wealth data confirms that the exponential distribution will be an appropriate candidate model. In general, let $X_1, \ldots, X_n$ denote a random sample from the distribution with pdf

$$f(x|\theta) = \theta e^{-\theta x}, \quad x > 0, \theta > 0.$$

Suppose that the prior distribution for $\theta$ is also assumed to be exponential with a known parameter value $\beta$, say. Hence,

$$\pi(\theta) = \beta e^{-\beta \theta}, \quad \theta > 0.$$

Then the likelihood function is:

$$
\begin{aligned}
L(\theta; x_1, x_2, \ldots, x_n) &= f(x_1, x_2, \ldots, x_n | \theta) \\
&= f(x_1 | \theta) f(x_2 | \theta) \cdots f(x_n | \theta) \\
&= \theta e^{-\theta x_1} \theta e^{-\theta x_n} \cdots, \theta e^{-\theta x_n} \\
&= \theta^n e^{-\theta \sum_{i=1}^{n} x_i}.
\end{aligned}
$$

Therefore, the posterior distribution, which is proportional to the Likelihood × Prior is:

$$
\begin{aligned}
\pi(\theta | x_1, \ldots, x_n) &\propto \theta^n e^{-\theta \sum_{i=1}^{n} x_i} \beta e^{-\beta \theta} \\
&= \theta^n e^{-\theta (\beta + \sum_{i=1}^{n} x_i)},
\end{aligned}
$$

for $\theta > 0$. This is recognised as the pdf of a gamma random variable with parameters $n+1$ and $\beta + \sum_{i=1}^{n} x_i$. For the billionaire's data example the posterior distribution is gamma with parameters 226 and 506 when we assume $\beta = 1$. This posterior distribution can be drawn as well. ◄

## Example 10.16 (Normal Distribution)

Suppose $X_1, \ldots, X_n \sim N(\theta, \sigma^2)$ independently, where $\sigma^2$ is known. Let $\pi(\theta) \sim N(\mu, \tau^2)$ for known values of $\mu$ and $\tau^2$. The likelihood function of $\theta$ is:

$$
\begin{aligned}
L(\theta; x_1, \ldots, x_n) &= f(x_1, x_2, \ldots, x_n | \theta) \\
&= \prod_{i=1}^{n} \frac{1}{\sqrt{2\pi\sigma^2}} e^{-\frac{1}{2} \frac{(x_i - \theta)^2}{\sigma^2}} \\
&= \left( \frac{1}{2\pi\sigma^2} \right)^{n/2} e^{-\frac{1}{2} \sum_{i=1}^{n} \frac{(x_i - \theta)^2}{\sigma^2}}.
\end{aligned}
$$

Note that here $\sigma^2$ is assumed to be known so that the likelihood is only a function of $\theta$. The density of the prior distribution is:

$$
\pi(\theta) = \frac{1}{\sqrt{2\pi\tau^2}} e^{-\frac{1}{2} \frac{(\theta - \mu)^2}{\tau^2}}.
$$

The posterior distribution is proportional to the likelihood × prior. Hence, in the posterior distribution, we keep the terms involving $\theta$ only.

$$
\begin{aligned}
\pi(\theta | x_1, \ldots, x_n) &\propto \left( \frac{1}{2\pi\sigma^2} \right)^{n/2} e^{-\frac{1}{2} \sum_{i=1}^{n} \frac{(x_i - \theta)^2}{\sigma^2}} \times \frac{1}{\sqrt{2\pi\tau^2}} e^{-\frac{1}{2} \frac{(\theta - \mu)^2}{\tau^2}} \\
&\propto e^{-\frac{1}{2} \left[ \sum_{i=1}^{n} \frac{(x_i - \theta)^2}{\sigma^2} + \frac{(\theta - \mu)^2}{\tau^2} \right]} \\
&= e^{-\frac{1}{2} M},
\end{aligned}
$$

where $M$ is the expression inside the square bracket. Notice that $M$ is a quadratic in $\theta$. Hence we can complete a square in $\theta$, as we did in the log-normal distribution Sect. 7.2.6 although for a different purpose. Now:

$$
\begin{aligned}
M &= \sum_{i=1}^{n} \frac{(x_i - \theta)^2}{\sigma^2} + \frac{(\theta - \mu)^2}{\tau^2} \\
&= \frac{\sum_{i=1}^{n} x_i^2 - 2\theta \sum_{i=1}^{n} x_i + n\theta^2}{\sigma^2} + \frac{\theta^2 - 2\theta\mu + \mu^2}{\tau^2} \\
&= \theta^2 \left( \frac{n}{\sigma^2} + \frac{1}{\tau^2} \right) - 2\theta \left( \frac{n\bar{x}}{\sigma^2} + \frac{\mu}{\tau^2} \right) + \sum_{i=1}^{n} \frac{x_i^2}{\sigma^2} + \frac{\mu^2}{\tau^2} \\
&= \theta^2 a - 2\theta b + c
\end{aligned}
$$

where

$$
a = \frac{n}{\sigma^2} + \frac{1}{\tau^2}, b = \frac{n\bar{x}}{\sigma^2} + \frac{\mu}{\tau^2}, c = \frac{\sum_{i=1}^{n} x_i^2}{\sigma^2} + \frac{\mu^2}{\tau^2}.
$$

Note that none of $a, b$ and $c$ involves $\theta$. These are defined just for writing convenience. Now

$$
\begin{aligned}
M &= a \left( \theta^2 - 2\theta \frac{b}{a} \right) + c \\
&= a \left( \theta^2 - 2\theta \frac{b}{a} + \frac{b^2}{a^2} - \frac{b^2}{a^2} \right) + c \\
&= a \left( \theta - \frac{b}{a} \right)^2 - \frac{b^2}{a} + c
\end{aligned}
$$

Note again that none of $a, b$ and $c$ involves $\theta$, hence in $M$ the first term only involves $\theta$ and the last two can be ignored in following proportionality:

$$
\pi(\theta | x_1, \ldots, x_n) \propto e^{-\frac{1}{2} a (\theta - \frac{b}{a})^2}
$$

which is easily recognised to be the pdf of a normal distribution with mean $\frac{b}{a}$ and variance $\frac{1}{a}$. More explicitly

$$
\theta | \mathbf{x} \sim N \left( \frac{\frac{n\bar{x}}{\sigma^2} + \frac{\mu}{\tau^2}}{\frac{n}{\sigma^2} + \frac{1}{\tau^2}}, \frac{1}{\frac{n}{\sigma^2} + \frac{1}{\tau^2}} \right).
$$

variance $\frac{1}{a}$. For some specific values of $\mu$, $\tau^2$, $n$ and $\bar{x}$, the prior and posterior densities may be seen as illustrated in Fig. 10.7. ◄

## 10.3.2  Bayes Estimators

Given $\pi(\theta | x_1, \ldots, x_n)$, we require a mechanism to choose a reasonable estimator $\hat{\theta}$. Suppose the true parameter is $\theta_0$ which is unknown. Let $a$ be our guess for it. In real life we may not have $a = \theta_0$. Then it is sensible to measure the penalty we have to pay for guessing incorrectly. The penalty may be measured by $(a - \theta_0)^2$ or $|a - \theta_0|$ or some other function. We should choose that value of $a$ which minimises the

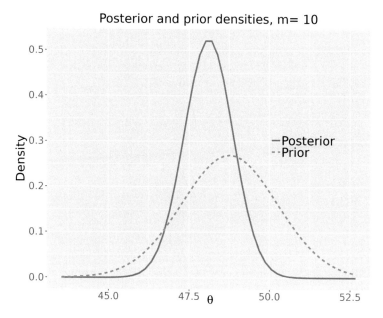

**Fig. 10.7**  Prior and posterior distributions

expected loss $E[L(a, \theta)]$, sometimes called the **risk**, where the expectation is taken with respect to the posterior distribution $\pi(\theta|x_1, \ldots, x_n)$ of $\theta$. Note that $a$ should not be a function of $\theta$, rather it should be a function of $x_1, \ldots, x_n$, the random sample. The minimiser, $\hat{\theta}$ say, is called the Bayes estimator of $\theta$.

Note that the calculations in the following sub-sections are similar to what we have encountered in Sect. 1.3.1 for sample data. But here the calculations are using the probability distribution $\pi(\theta|x_1, \ldots, x_n)$ - not just a sample of observations $\theta_1, \theta_2, \ldots$.

### 10.3.3  Squared Error Loss Function

We consider the loss function:

$$L(a, \theta) = (a - \theta)^2,$$

which gives equal amount of losses for both under and over estimation. For example, if the true value of $\theta$ is 5, then the loss is same for guesses $a_1 = \theta + \delta$ and $a_2 = \theta - \delta$ for any value of $\delta > 0$. It is possible to adopt other, e.g., asymmetric loss functions as well.

Note that $\theta$ is unknown and it is now assumed to be a random variable, which has a probability distribution, called the posterior distribution, $\pi(\theta|x_1, \ldots, x_n)$ in the light of the observed data $x_1, \ldots, x_n$. Hence the loss function will vary as $\theta$ varies according to $\pi(\theta|x_1, \ldots, x_n)$. Hence, it is sensible to look at the expected value of the loss function, called the risk for any estimator (guess) $a$. This expected value of $L(a, \theta)$ is with respect to the uncertainty distribution of $\theta$, the posterior distribution. This makes sense since intuitively we should incur more losses for those value of *theta* which have higher probabilities as measured by $\pi(\theta|x_1, \ldots, x_n)$. Let

$$b = E_{\pi(\theta|x_1, x_2, \ldots, x_n)}(\theta) = \int \theta \pi(\theta|x_1, x_2, \ldots, x_n)d\theta.$$

$$\begin{aligned} E[L(a, \theta)] &= \int L(a, \theta)\pi(\theta|x_1, \ldots, x_n)d\theta \\ &= \int (a - b + b - \theta)^2 \pi(\theta|x_1, \ldots, x_n)d\theta \\ &= (a - b)^2 + \int (b - \theta)^2 \pi(\theta|x_1, \ldots, x_n)d\theta \\ &\geq \int (b - \theta)^2 \pi(\theta|x_1, \ldots, x_n)d\theta, \end{aligned}$$

for any value of $a$. When will the above inequality be an equality? The obvious answer is when $a = b$. Note that $b$ is the posterior mean of $\theta$. Hence we say that **the Bayes estimator under squared error loss is the posterior mean.**

### 10.3.4  Absolute Error Loss Function

What happens when we assume the absolute error loss, $L(a, \theta) = |a - \theta|$. Then it can be shown that $E[|a - \theta|]$ is minimised by taking $a$ to be the median of the posterior distribution of $\theta$ essentially by using similar arguments as in the proof of the fact that the "sample median minimises the sum of the absolute deviations" in Sect. 1.3.1. Hence for the squared error loss function the Bayes estimator is the posterior median. Recall from Sect. 5.5 that the median of a random variable $Y$ with pdf $g(y)$ is defined as the value $\mu$ which solves:

$$\int_{-\infty}^{\mu} g(y)dy = \tfrac{1}{2}.$$

This is hard to find except for symmetric distributions. For example, for the normal example the Bayes estimator of $\theta$ under the absolute error loss is still the posterior mean because it is also the posterior median. For the other examples, we need a computer to find the posterior medians.

### 10.3.5  Step Function Loss

We consider the loss function:

$$\begin{aligned} L(a, \theta) &= 0 \text{ if } |a - \theta| \leq \delta \\ &= 1 \text{ if } |a - \theta| > \delta \end{aligned}$$

where $\delta$ is a given small positive number. Now let us find the expected loss, i.e. the risk. Note that expectation is to be taken under the posterior distribution.

$$
\begin{aligned}
E[L(a,\theta)] &= \int_\Theta I(|a-\theta|>\delta)\pi(\theta|\mathbf{x})d\theta \\
&= \int_\Theta (1-I(|a-\theta|\le\delta))\,\pi(\theta|\mathbf{x})d\theta \\
&= 1 - \int_{a-\delta}^{a+\delta}\pi(\theta|\mathbf{x})d\theta \\
&\approx 1 - 2\delta\pi(a|\mathbf{x})
\end{aligned}
$$

where $I(\cdot)$ is the indicator function. In order to minimise the risk we need to maximise $\pi(a|\mathbf{x})$ with respect to $a$ and the Bayes estimator is the maximiser.

Therefore, the Bayes estimator is that value of $\theta$ which maximises the posterior, i.e. the modal value. This estimator is called the maximum a-posteriori (MAP) estimator.

---

**Example 10.17 (Binomial)**

Consider Example 10.13. The Bayes estimator under squared error loss is

$$
\hat\theta = \frac{x+\alpha}{x+\alpha+n-x+\beta} = \frac{x+\alpha}{n+\alpha+\beta}.
$$

◄

---

**Example 10.18 (Exponential)**

Consider Example 10.15. The Bayes estimator is

$$
\hat\theta = \frac{n+1}{\mu+\sum_{i=1}^n x_i}.
$$

Note that $\mu$ is a given constant. ◄

---

**Example 10.19 (Normal)**

Consider Example 10.16. The Bayes estimator under all three loss functions is

$$
\hat\theta = \frac{n\bar{x}/\sigma^2 + \mu/\tau^2}{n/\sigma^2 + 1/\tau^2}
$$

◄

Example 10.20

Let $X_1, \ldots, X_n \sim Poisson(\theta)$ independently. Also suppose the prior is $\pi(\theta) = e^{-\theta}, \theta > 0$. Then the likelihood function is:

$$f(x_1, \ldots, x_n | \theta) = \prod_{i=1}^{n} \frac{1}{x_i!} e^{-\theta} \theta^{x_i} = e^{-n\theta} \theta^{\sum_{i=1}^{n} x_i} \frac{1}{x_1! x_2! \cdots x_n!}$$

Now we obtain the posterior distribution of $\theta$ as the Likelihood times the prior. We only collect the terms involving $\theta$ only.

$$\pi(\theta | x_1, \ldots, x_n) \propto e^{-n\theta} \theta^{\sum_{i=1}^{n} x_i} e^{-\theta}$$
$$\propto e^{-(n+1)\theta} \theta^{\sum_{i=1}^{n} x_i},$$

which is easily seen to be the pdf of $\mathrm{Gamma}\left(1 + \sum_{i=1}^{n} x_i, \frac{1}{n+1}\right)$. Hence, the Bayes estimator of $\theta$ under squared error loss is:

$$\hat{\theta} = \text{posterior mean} = \frac{1 + \sum_{i=1}^{n} x_i}{1 + n}.$$

◀

## 10.3.6 Credible Regions

In this section we aim to obtain suitable credible intervals, or sets in general, which contain the unknown parameter with high probability. Hence, we aim to choose a set $A$ such that

$$P(\theta \in A | \mathbf{x}) = 1 - \alpha.$$

Such a set $A$ is called $100(1 - \alpha)\%$ credible region for $\theta$.

The set $A$ is called a *Highest Posterior Density* (HPD) credible region if $\pi(\theta | \mathbf{x}) \geq \pi(\psi | \mathbf{x})$ for all $\theta \in A$ and $\psi \notin A$. Hence a HPD credible region contains the highest probability region under the posterior distribution.

Example 10.21 (Normal–Normal)

Return to Example 10.16. We wish to find a 95% HPD credible region for $\theta$. It is given by

$$\frac{n\bar{x}/\sigma^2 + \mu/\tau^2}{n/\sigma^2 + 1/\tau^2} \pm 1.96 \sqrt{\frac{1}{n/\sigma^2 + 1/\tau^2}}.$$

Put $\sigma^2 = 1, \mu = 0, n = 12, \bar{x} = 5, \tau^2 = 10$ to construct a numerical example. ◀

## 10.4  Exercises

### 10.1 (Estimation Methods)

1. A random variable $X$ is said to have a beta($\alpha, \beta$) distribution if its probability density function takes the form

$$f_X(x) = \frac{\Gamma(\alpha + \beta)}{\Gamma(\alpha)\Gamma(\beta)} x^{\alpha-1}(1 - x)^{\beta-1}, \qquad 0 < x < 1$$

For a beta-distributed random variable

$$E(X) = \frac{\alpha}{\alpha + \beta} \quad \text{and} \quad E(X^2) = \frac{\alpha(\alpha + 1)}{(\alpha + \beta)(\alpha + \beta + 1)}.$$

If $x_1, \ldots, x_n$ are observations of i.i.d. beta($\alpha, \beta$) random variables, $X_1, \ldots, X_n$, derive the method of moments estimators of $\alpha$ and $\beta$.

2. Suppose that $x_1, \ldots, x_n$ are observations of random variables $X_1, \ldots, X_n$ which are i.i.d. with pdf (or pmf) $f_X(x; \theta)$ for a scalar parameter $\theta$. Derive the maximum likelihood estimate of $\theta$ when
   (a) $f_X(x; \theta) = \theta \exp(-\theta x)$, $\quad x \in \mathcal{R}_+$ (exponential distribution).
   (b) $f_X(x; \theta) = \theta x^{\theta-1}$, $\quad x \in (0, 1)$.
   (c) $f_X(x; \theta) = \theta(1 - \theta)^{x-1}$, $\quad x \in \{1, 2, 3, \ldots\}$ (geometric distribution).

3. Suppose that $x_1, \ldots, x_n$ are observations of discrete random variables $X_1, \ldots, X_n$ which are i.i.d. with probability function

$$f_X(x; \alpha, \theta) = (1 - \theta)\theta^{(x-\alpha)}, \qquad x \in \{\alpha, \alpha + 1, \alpha + 2, \ldots\}$$

where $\alpha \in \mathcal{Z}$ and $\theta \in (0, 1)$. Derive the maximum likelihood estimators of $\alpha$ and $\theta$.

4. A random variable $X$ has a Rayleigh distribution if its probability density function takes the form

$$f_X(x) = \frac{x}{\theta} \exp\left(\frac{-x^2}{2\theta}\right), \qquad x > 0$$

where $\theta$ is a positive parameter. It can be shown that $E(X) = (\theta\pi/2)^{1/2}$, $E(X^2) = 2\theta$ and $E(X^4) = 8\theta^2$. (Here $\pi$ is the ratio of the circumference of a circle to its diameter and not an unknown parameter.) Let $X_1, \ldots, X_n$ be independent and identically distributed copies of $X$.
   (a) Find the method of moments estimator for $\theta$.
   (b) Find the maximum likelihood estimator for $\theta$.
   (c) Find the bias of the method of moments estimator and show that the maximum likelihood estimator is unbiased for $\theta$.
   (d) Write down the maximum likelihood estimator for $E(X)$.

5. Suppose we have two independent observations $X_1$ and $X_2$ from a discrete distribution with probability mass function of the form

$$f(x) = \left(\frac{\theta}{2}\right)^{|x|} (1 - \theta)^{1-|x|}, \qquad x = -1, 0, 1; \; 0 \le \theta \le 1.$$

Show that the maximum likelihood estimator of $\theta$ is $\hat{\theta} = (|X_1| + |X_2|)/2$
List all possible outcomes for the pair $(X_1, X_2)$ and give the probability of each outcome. Hence or otherwise write down the sampling distribution of the maximum likelihood estimator $\hat{\theta}$. Find the bias and variance of $\hat{\theta}$.
Show that the estimator $T = I(X_1 = 1) + I(X_2 = 1)$, where $I(\cdot)$ is the indicator function, is unbiased for $\theta$ and then show that its variance is larger than that of $\hat{\theta}$. Finally, show that $T$ can take values outside the parameter space with positive probability.

6. Suppose that the number of defects on a roll of magnetic recording tape has a Poisson distribution for which the mean $\theta$ is unknown and that the prior distribution of $\theta$ is a gamma distribution with parameters $\alpha = 3$ and $\beta = 1$. When five rolls of this tape are selected at random and inspected, the number of defects found on the rolls are 2, 2, 6, 0, and 3. If the squared error loss function is used what is the Bayes estimate of $\theta$?

7. Suppose that $X_1, \ldots, X_n$ is a random sample from the distribution with pdf

$$f(x|\theta) = \begin{cases} \theta x^{\theta-1} & \text{if } 0 < x < 1, \\ 0 & \text{otherwise.} \end{cases}$$

Suppose also that the value of the parameter $\theta$ is unknown ($\theta > 0$) and that the prior distribution of $\theta$ is a gamma distribution with parameters $\alpha$ and $\beta$ ($\alpha > 0$ and $\beta > 0$). Determine the posterior distribution of $\theta$ and hence obtain the Bayes estimator of $\theta$ under a squared error loss function.

8. Suppose that $X_1, \ldots, X_n$ is a random sample of size $n$ from a normal distribution with unknown mean $\theta$ and variance 1. Suppose also that the prior distribution of $\theta$ is also a normal distribution with mean 0 and given variance $\tau^2$. Derive the posterior distribution of $\theta$ given $X_1, \ldots, X_n$. From the posterior distribution write the Bayes estimator under (i) squared error loss and (ii) absolute error loss, clearly stating the standard results used here.

# Interval Estimation

# 11

**Abstract**

Chapter 11 discusses techniques such as the pivoting method for interval estimation. The central limit theorem is used to derive confidence intervals for the mean parameters of binomial, Poisson and normal distributions. For the normal distribution we also discuss the exact confidence interval using the t-distribution without actually deriving the sampling distribution of the t-statistic.

In any estimation problem it is very hard to guess the exact true value, but it is often much better to provide an interval where the true value is very likely to fall. For example, think of guessing the age of a stranger. It may be easier to guess whether a person is in his teens, 20's, 30's etc. than to say the exact age, i.e. 45.

An estimate $\hat{\theta}$ of a parameter $\theta$ is sometimes referred to as a *point estimate*. The usefulness of a point estimate is enhanced if some kind of measure of its precision can also be provided. Usually, for an unbiased estimator, this will be a standard error, an estimate of the standard deviation of the associated estimator, as we have discussed previously. An alternative summary of the information provided by the observed data about the location of a parameter $\theta$ and the associated precision is an *interval estimate* or *confidence interval*. This chapter is devoted to finding confidence intervals for the mean parameters of the three most commonly used distributions: normal, Bernoulli and Poisson using different techniques.

## 11.1   Pivoting Method

The method of pivoting is a widely used approach to obtain a confidence interval for a parameter starting with a suitable probability identity. This section details the method for a generic parameter.

Suppose that $x_1, \ldots, x_n$ are observations of random variables $X_1, \ldots, X_n$ whose joint pdf is specified apart from a single parameter $\theta$. To construct a confidence interval for $\theta$, we need to find a random variable $T(\mathbf{X}, \theta)$ whose distribution does not depend on $\theta$ and is therefore known. *This random variable $T(\mathbf{X}, \theta)$ is called a pivot for $\theta$.* Hence we can find numbers $h_1$ and $h_2$ such that

$$P(h_1 \leq T(\mathbf{X}, \theta) \leq h_2) = 1 - \alpha, \qquad (11.1)$$

where $1 - \alpha$ is any specified probability. If (11.1) can be 'inverted' (or manipulated), we can write it as

$$P[g_1(\mathbf{X}) \leq \theta \leq g_2(\mathbf{X})] = 1 - \alpha, \qquad (11.2)$$

where $g_1(\mathbf{X})$ and $g_2(\mathbf{X})$ are functions of the random sample $X_1, \ldots, X_n$ but are free of the parameter of interest $\theta$. Hence, the identity (11.2) implies that with probability $1 - \alpha$, the parameter $\theta$ will lie between the random variables $g_1(\mathbf{X})$ and $g_2(\mathbf{X})$. Alternatively, the random interval $[g_1(\mathbf{X}), g_2(\mathbf{X})]$ includes $\theta$ with probability $1 - \alpha$. Now, when we observe $x_1, \ldots, x_n$, we observe a single observation of the random interval $[g_1(\mathbf{X}), g_2(\mathbf{X})]$, which can be evaluated as $[g_1(\mathbf{x}), g_2(\mathbf{x})]$. We do not know if $\theta$ lies inside or outside of this interval, but we do know that if we observed repeated samples, then $100(1 - \alpha)\%$ of the resulting intervals would contain $\theta$. Hence, if $1 - \alpha$ is high, we can be reasonably confident that our observed interval contains $\theta$. We call the observed interval $[g_1(\mathbf{x}), g_2(\mathbf{x})]$ a $100(1 - \alpha)\%$ confidence interval for $\theta$. It is common to present intervals with high confidence levels, usually 90, 95 or 99%, so that $\alpha = 0.1, 0.05$ or $0.01$ respectively.

**Example 11.1 (Confidence Interval for a Normal Mean)**

Let $X_1, \ldots, X_n$ be i.i.d. $N(\mu, \sigma^2)$ random variables. We stated in Sect. 8.7

$$\bar{X} \sim N(\mu, \sigma^2/n) \quad \Rightarrow \quad \sqrt{n}\frac{(\bar{X} - \mu)}{\sigma} \sim N(0, 1).$$

Assume that we know the value of $\sigma$ so that $\sqrt{n}(\bar{X} - \mu)/\sigma$ is a pivot for $\mu$. Then we can use the cdf of the standard normal distribution, $\Phi(\cdot)$, to find values $h_1$ and $h_2$ such that

$$P\left(h_1 \leq \sqrt{n}\frac{(\bar{X} - \mu)}{\sigma} \leq h_2\right) = 1 - \alpha$$

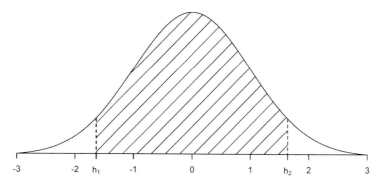

**Fig. 11.1** Illustration of left and right cut off points for $100(1 - \alpha)\%$ confidence intervals

for a chosen value of $1 - \alpha$ which is called the *confidence level*. So $h_1$ and $h_2$ are chosen so that the shaded area in Fig. 11.1 is equal to the confidence level $1 - \alpha$. It is common practice to make the interval symmetric, so that the two unshaded areas are equal (to $\alpha/2$), in which case

$$- h_1 = h_2 \equiv h \quad \text{and} \quad \Phi(h) = 1 - \frac{\alpha}{2}.$$

The most common choice of confidence level is $1 - \alpha = 0.95$, in which case $h = 1.96 = \texttt{qnorm(0.975)}$. We may also occasionally see 90% ($h = 1.645 = \texttt{qnorm(0.95)}$) or 99% ($h = 2.58 = \texttt{qnorm(0.995)}$) intervals. We discussed these values in Sect. 7.2.4. We generally use the 95% intervals for a reasonably high level of confidence without making the interval unnecessarily wide. Therefore, we have

$$P\left(-1.96 \leq \sqrt{n}\frac{(\bar{X} - \mu)}{\sigma} \leq 1.96\right) = 0.95$$

$$\Rightarrow P\left(\bar{X} - 1.96\frac{\sigma}{\sqrt{n}} \leq \mu \leq \bar{X} + 1.96\frac{\sigma}{\sqrt{n}}\right) = 0.95.$$

Hence, $\bar{X} - 1.96\frac{\sigma}{\sqrt{n}}$ and $\bar{X} + 1.96\frac{\sigma}{\sqrt{n}}$ are the endpoints of a random interval which includes $\mu$ with probability 0.95. The observed value of this interval, $\left(\bar{x} \pm 1.96\frac{\sigma}{\sqrt{n}}\right)$, is called a *95% confidence interval* for $\mu$. ◄

**Example 11.2**

For the fast food waiting time data, we have $n = 20$ data points combined from the morning and afternoon data sets. We have $\bar{x} = 67.85$ and $n = 20$. Hence, under the normal model assuming (just for the sake of illustration) $\sigma = 18$, a 95% confidence interval for $\mu$ is

$$67.85 - 1.96(18/\sqrt{20}) \le \mu \le 67.85 + 1.96(18/\sqrt{20})$$
$$\Rightarrow \quad 59.96 \le \mu \le 75.74$$

◀

The R command to obtain the above confidence interval is `mean(a)+ c(-1, 1)*qnorm(0.975)*18/sqrt(20)`, assuming a is the vector containing 20 waiting times. If $\sigma$ is unknown, we need to seek alternative methods for finding the confidence intervals.

## 11.2   Interpreting Confidence Intervals

1. Notice that $\bar{x}$ is an unbiased estimate of $\mu$, $\frac{\sigma}{\sqrt{n}}$ is the standard error of the estimate (when $\sigma$ is assumed to be known) and 1.96 (in general $h$ in the above discussion) is a critical value from the associated known sampling distribution. The formula $\left(\bar{x} \pm 1.96 \, \frac{\sigma}{\sqrt{n}}\right)$ for the confidence interval is then generalised as:

$$\text{Estimate} \pm \text{Critical value} \times \text{Standard error,}$$

where the estimate is $\bar{x}$, the critical value is 1.96 and the standard error is $\frac{\sigma}{\sqrt{n}}$. This is an easy to remember formula that we use often for obtaining a confidence interval for a mean parameter, see e.g. Chaps. 17 and 18.

2. Confidence intervals are frequently used, but also frequently misinterpreted. A $100(1-\alpha)\%$ confidence interval for $\theta$ is a single observation of a random interval which, under repeated sampling, would include $\theta$ $100(1 - \alpha)\%$ of the time.
   The following example from the National Lottery in the UK clarifies the interpretation. We collected six chosen lottery numbers (sampled at random from 1 to 49) for 20 weeks and then constructed 95% confidence intervals for the population mean $\mu$, which is 25 since we can think of the six lottery numbers as a random sample of size six from the discrete uniform distribution, which takes values 1 to 49 each with equal probability $1/49$. Figure 11.2 plots the intervals along with the observed sample means, shown as points. It can be seen that exactly one out of 20 (5%) of the intervals do not contain the true population

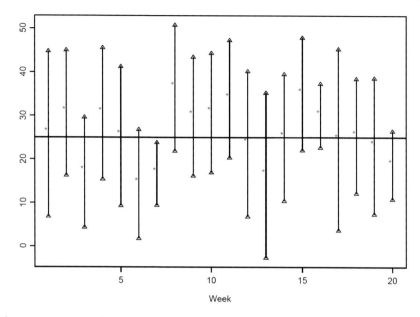

**Fig. 11.2** Confidence intervals for the known population mean $\mu = 25$ obtained by using 6 lottery numbers in each of the 20 weeks

mean 25. Although this is a coincidence, it explains the main point that if we construct the random intervals with $100(1 - \alpha)\%$ confidence levels again and again for hypothetical repetition of the data, on average $100(1 - \alpha)\%$ of them will contain the true parameter.

3. A confidence interval is not a probability interval. We should avoid making statements like $P(1.3 < \theta < 2.2) = 0.95$. In the classical approach to statistics we can only make probability statements about random variables, and $\theta$ is assumed to be a constant.

4. If a confidence interval is interpreted as a probability interval, this may lead to problems. For example, suppose that $X_1$ and $X_2$ are i.i.d. $U[\theta - \frac{1}{2}, \theta + \frac{1}{2}]$ random variables. Then $P[\min(X_1, X_2) < \theta < \max(X_1, X_2)] = \frac{1}{2}$ so $[\min(x_1, x_2), \max(x_1, x_2)]$ is a 50% confidence interval for $\theta$, where $x_1$ and $x_2$ are the observed values of $X_1$ and $X_2$. Now suppose that $x_1 = 0.3$ and $x_2 = 0.9$. What is $P(0.3 < \theta < 0.9)$? Here we do not have a probability distribution for $\theta$ as $\theta$ is a fixed constant and not a random variable. Hence we cannot calculate the probability, although we can determine the value of the probability to be either 0 or 1, if $\theta$ is known. For example, if $\theta = 0.5$ then $P(0.3 < \theta < 0.9) = 1$ and if $\theta = 0.2$ then $P(0.3 < \theta < 0.9) = 0$.

In this section we have learned to obtain confidence intervals by using an appropriate statistic in the pivoting technique. The main task is then to invert the inequality so that the unknown parameter is in the middle by itself and the two end points are functions of the sample observations. The most difficult task is to correctly interpret confidence intervals, which are not probability intervals but have long-run properties. That is, the interval will contain the true parameter with the stipulated confidence level only under infinitely repeated sampling.

## 11.3 Confidence Intervals Using the CLT

Confidence intervals are generally difficult to find. The difficulty lies in finding a pivot, i.e. a statistic $T(\mathbf{X}, \theta)$ in (11.1) whose probability distribution is a known distribution. There is also the requirement that the pivot identity can be inverted to put the unknown $\theta$ in the middle of the inequality inside the probability statement as stated in (11.2). One solution to this problem is to use the powerful Central Limit Theorem, stated in Sect. 8.8 to claim normality, and then basically follow the above normal example 11.1 for known variance.

### 11.3.1 Confidence Intervals for $\mu$ Using the CLT

The CLT allows us to assume the large sample approximation

$$\sqrt{n}\frac{(\bar{X} - \mu)}{\sigma} \overset{approx}{\sim} N(0, 1) \text{ as } n \to \infty.$$

So a general confidence interval for $\mu$ can be constructed, just as before in Sect. 11.1. Thus a 95% confidence interval (CI) for $\mu$ is given by $\bar{x} \pm 1.96\frac{\sigma}{\sqrt{n}}$. But note that $\sigma$ is unknown so this CI cannot be used unless we can estimate $\sigma$, i.e. replace the unknown standard deviation of $\bar{X}$ by its estimated standard error. In this case, we get the CI in the familiar form: Estimate $\pm$ Critical value $\times$ Standard error.

Suppose that we do not assume any distribution for the sampled random variable $X$ but assume only that $X_1, \ldots, X_n$ are i.i.d, following the distribution of $X$ where $E(X) = \mu$ and $\text{Var}(X) = \sigma^2$. We know that the standard error of $\bar{X}$ is $s/\sqrt{n}$ where $s$ is the sample standard deviation with divisor $n - 1$. Then the following provides a 95% CI for $\mu$:

$$\bar{x} \pm 1.96\frac{s}{\sqrt{n}}.$$

If we can assume a probability distribution for $X$, i.e. a parametric model for $X$, then we can do slightly better in estimating the standard error of $\bar{X}$ and as a result we can improve upon the previously obtained 95% CI, see the examples below.

**Example 11.3 (Computer Failure)**

For the computer failure data, $\bar{x} = 3.75$, $s = 3.381$ and $n = 104$. Under the model that the data are observations of i.i.d. random variables with population mean $\mu$ (but no other assumptions about the underlying distribution), we compute a 95% confidence interval for $\mu$ to be

$$\left(3.75 - 1.96\frac{3.381}{\sqrt{104}}, 3.75 + 1.96\frac{3.381}{\sqrt{104}}\right) = (3.10, 4.40).$$

The R command `?cfail` shows this code line. ◄

**Example 11.4 (Poisson)**

If $X_1, \ldots, X_n$ are modelled as i.i.d. Poisson($\lambda$) random variables, then $\mu = \lambda$ and $\sigma^2 = \lambda$. We know Var($\bar{X}$) $= \sigma^2/n = \lambda/n$. Hence a standard error is $\sqrt{\hat{\lambda}/n} = \sqrt{\bar{x}/n}$ since $\hat{\lambda} = \bar{X}$ is an unbiased estimator of $\lambda$. Thus a 95% CI for $\mu = \lambda$ is given by

$$\bar{x} \pm 1.96\sqrt{\frac{\bar{x}}{n}}.$$

For the computer failure data, $\bar{x} = 3.75$, $s = 3.381$ and $n = 104$. Under the model that the data are observations of i.i.d. random variables following a Poisson distribution with population mean $\lambda$, we compute a 95% confidence interval for $\lambda$ as

$$\bar{x} \pm 1.96\sqrt{\frac{\bar{x}}{n}} = 3.75 \pm 1.96\sqrt{3.75/104} = (3.38, 4.12).$$

We see that this interval is narrower ($0.74 = 4.12 - 3.38$) than the earlier interval ($3.10, 4.40$), which has a length of 1.3. We may prefer a narrower confidence intervals as those facilitate more accurate inference regarding the unknown parameter. However, we need to make sure that the model assumption is justified for the data. However, in this example from the discussion on Fig. 9.1 we know that the Poisson distribution is not a good fit to the data. Hence, it is better to use the wider confidence interval instead of the narrower Poisson distribution based interval. ◄

Example 11.5 (Bernoulli)

If $X_1, \ldots, X_n$ are modelled as i.i.d. Bernoulli($p$) random variables, then $\mu = p$ and $\sigma^2 = p(1-p)$. We know $\text{Var}(\bar{X}) = \sigma^2/n = p(1-p)/n$. Hence a standard error is $\sqrt{\hat{p}(1-\hat{p})/n} = \sqrt{\bar{x}(1-\bar{x})/n}$, since $\hat{p} = \bar{X}$ is an unbiased estimator of $p$. Thus a 95% CI for $\mu = p$ is given by

$$\bar{x} \pm 1.96\sqrt{\frac{\bar{x}(1-\bar{x})}{n}}.$$

For the example, suppose $\bar{x} = 0.2$ and $n = 10$. Then we obtain the 95% CI as

$$0.2 \pm 1.96\sqrt{(0.2 \times 0.8)/10} = (-0.048, 0.448).$$

This confidence interval allows for negative values of the probability parameter $p$. Hence, a quick fix is to quote the confidence interval as $(0, 0.448)$. However, that is arbitrary, hence unsatisfactory, and we need to look for other alternatives which may work better even for a small value of $n$, 10 here. ◄

## 11.3.2 Confidence Interval for a Bernoulli $p$ by Quadratic Inversion

It turns out that for the Bernoulli and Poisson distributions we can find alternative confidence intervals *without using the approximation for standard error* but still using the CLT. This is more complicated and requires us to solve a quadratic equation. We consider the two distributions separately.

We start with the CLT and obtain the following statement:

$$P\left(-1.96 \le \sqrt{n}\frac{(\bar{X}-p)}{\sqrt{p(1-p)}} \le 1.96\right) = 0.95$$
$$\Leftrightarrow \quad P\left(-1.96\sqrt{p(1-p)} \le \sqrt{n}(\bar{X}-p) \le 1.96\sqrt{p(1-p)}\right) = 0.95$$
$$\Leftrightarrow \quad P\left(-1.96\sqrt{p(1-p)/n} \le (\bar{X}-p) \le 1.96\sqrt{p(1-p)/n}\right) = 0.95$$
$$\Leftrightarrow \quad P\left(p - 1.96\sqrt{p(1-p)/n} \le \bar{X} \le p + 1.96\sqrt{p(1-p)/n}\right) = 0.95$$
$$\Leftrightarrow \quad P\left(L(p) \le \bar{X} \le R(p)\right) = 0.95,$$

where $L(p) = p - h\sqrt{p(1-p)/n}$, $R(p) = p + h\sqrt{p(1-p)/n}$, $h = 1.96$. Now, consider the inverse mappings $L^{-1}(x)$ and $R^{-1}(x)$ so that:

$$P\left[L(p) \le \bar{X} \le R(p)\right] = 0.95$$
$$\Leftrightarrow P\left[R^{-1}(\bar{X}) \le p \le L^{-1}(\bar{X})\right] = 0.95$$

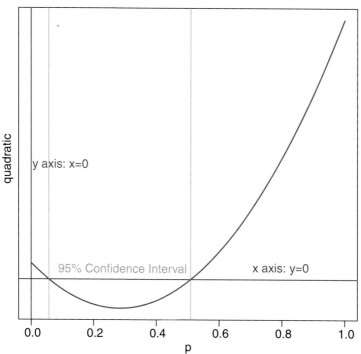

**Fig. 11.3**   Plot showing a quadratic function of $p$ for finding a confidence interval

which now defines our confidence interval $(R^{-1}(\bar{X}), L^{-1}(\bar{X}))$ for $p$. We can obtain $R^{-1}(\bar{x})$ and $L^{-1}(\bar{x})$ by solving the equations $R(p) = \bar{x}$ and $L(p) = \bar{x}$ for $p$, treating $n$ and $\bar{x}$ as known quantities (see Fig. 11.3). Thus we have,

$$R(p) = \bar{x}, \; L(p) = \bar{x}$$
$$\Leftrightarrow \quad (\bar{x} - p)^2 = h^2 p(1-p)/n, \qquad \text{where } h = 1.96$$
$$\Leftrightarrow \quad p^2(1 + h^2/n) - p(2\bar{x} + h^2/n) + \bar{x}^2 = 0.$$

The endpoints of the confidence interval are the roots of the quadratic. Hence, the endpoints of the 95% confidence interval for $p$ are:

$$\frac{\left(2\bar{x} + \frac{h^2}{n}\right) \pm \left[\left(2\bar{x} + \frac{h^2}{n}\right)^2 - 4\bar{x}^2\left(1 + \frac{h^2}{n}\right)\right]^{1/2}}{2\left(1 + \frac{h^2}{n}\right)}$$

$$= \frac{\left(\bar{x} + \frac{h^2}{2n}\right) \pm \left[\left(\bar{x} + \frac{h^2}{2n}\right)^2 - \bar{x}^2\left(1 + \frac{h^2}{n}\right)\right]^{1/2}}{\left(1 + \frac{h^2}{n}\right)}$$

$$= \frac{\bar{x} + \frac{h^2}{2n} \pm \frac{h}{\sqrt{n}}\left[\frac{h^2}{4n} + \bar{x}(1 - \bar{x})\right]^{1/2}}{\left(1 + \frac{h^2}{n}\right)}.$$

This is sometimes called the *Wilson Score Interval*. Code can be written in R to calculate this interval for given $n$, $\bar{x}$ and confidence level $\alpha$, which determines the value of $h$. Returning to the previous example, $n = 10$ and $\bar{x} = 0.2$, the 95% CI obtained from this method is (0.057, 0.510) compared to the previous illegitimate one $(-0.048, 0.448)$. In fact you can see that the intervals obtained by quadratic inversion are more symmetric and narrower as $n$ increases, and are also more symmetric for $\bar{x}$ closer to 0.5. See Table 11.1.

For smaller $n$ and $\bar{x}$ closer to 0 (or 1), the approximation required for the plug-in estimate of the standard error is insufficiently reliable. However, for larger $n$ it is adequate.

**Table 11.1** End points of 95% Confidence intervals for $p$ using quadratic inversion and plug-in estimation

| | | Quadratic inversion | | Plug-in s.e. estimation | |
|---|---|---|---|---|---|
| $n$ | $\bar{x}$ | Lower end | Upper end | Lower end | Upper end |
| 10 | 0.2 | 0.057 | 0.510 | −0.048 | 0.448 |
| 10 | 0.5 | 0.237 | 0.763 | 0.190 | 0.810 |
| 20 | 0.1 | 0.028 | 0.301 | −0.031 | 0.231 |
| 20 | 0.2 | 0.081 | 0.416 | 0.025 | 0.375 |
| 20 | 0.5 | 0.299 | 0.701 | 0.281 | 0.719 |
| 50 | 0.1 | 0.043 | 0.214 | 0.017 | 0.183 |
| 50 | 0.2 | 0.112 | 0.330 | 0.089 | 0.311 |
| 50 | 0.5 | 0.366 | 0.634 | 0.361 | 0.639 |

### 11.3.3 Confidence Interval for a Poisson λ by Quadratic Inversion

Here we proceed as in the Bernoulli case and using the CLT claim that a 95% CI for λ is given by:

$$P\left(-1.96 \leq \sqrt{n}\frac{(\bar{X} - \lambda)}{\sqrt{\lambda}} \leq 1.96\right) = 0.95 \quad \Rightarrow \quad P\left(n\frac{(\bar{X} - \lambda)^2}{\lambda} \leq 1.96^2\right) = 0.95.$$

Now the confidence interval for λ is found by solving the (quadratic) equality for λ by treating $n, \bar{x}$ and $h$ to be known:

$$n\frac{(\bar{x} - \lambda)^2}{\lambda} = h^2, \quad \text{where } h = 1.96$$

$$\Rightarrow \bar{x}^2 - 2\lambda\bar{x} + \lambda^2 = h^2\lambda/n$$

$$\Rightarrow \lambda^2 - \lambda(2\bar{x} + h^2/n) + \bar{x}^2 = 0.$$

Hence, the endpoints of the 95% confidence interval for λ are:

$$\frac{\left(2\bar{x} + \frac{h^2}{n}\right) \pm \left[\left(2\bar{x} + \frac{h^2}{n}\right)^2 - 4\bar{x}^2\right]^{1/2}}{2} = \bar{x} + \frac{h^2}{2n} \pm \frac{h}{n^{1/2}}\left[\frac{h^2}{4n} + \bar{x}\right]^{1/2}.$$

#### Example 11.6

For the computer failure data, $\bar{x} = 3.75$ and $n = 104$. For a 95% confidence interval (CI), $h = 1.96$. Hence, we calculate the above CI using the R commands:

```
x <- cfail
n <- length(x)
h <- qnorm(0.975)
mean(x) + (h*h)/(2*n) + c(-1, 1) * h/sqrt(n) * sqrt(h*h/(4*n)
  + mean(x))
```

The result is (3.40, 4.14), which compares well with the earlier interval (3.38, 4.12). ◄

### 11.3.4 Summary

In this section we learned how to find confidence intervals using the CLT, when the sample size is large. We have seen that we can make more accurate inferences if we can assume a model, e.g. the Poisson model for the computer failure data. However, we have also encountered problems when applying the method for a small sample size. In such cases we should use alternative methods for calculating confidence intervals. For example, we learned a technique of finding confidence intervals which does not require us to approximately estimate the standard errors for Bernoulli and Poisson distributions. In the next section, we will learn how to find an exact confidence interval for the normal mean $\mu$ using the t-distribution.

## 11.4    Exact Confidence Interval for the Normal Mean

Recall that we can obtain better quality inferences if we can justify a precise model for the data. This saying is analogous to the claim that a person can better predict and infer in a situation when there are established rules and regulations, i.e. the analogue of a statistical model. In this section, we will discuss a procedure for finding confidence intervals based on the statistical modelling assumption that the data are from a normal distribution *but with unknown variance*. This assumption will enable us to find an exact confidence interval for the mean rather than an approximate one using the central limit theorem.

For normal models we do not have to rely on large sample approximations, because it turns out that the distribution of

$$T = \frac{\sqrt{n}(\bar{X} - \mu)}{S},$$

where $S^2$ is the sample variance with divisor $n - 1$, is standard (easily calculated) and thus the statistic $T = T(\mathbf{X}, \mu)$ can be an exact pivot for any sample size $n > 1$. The point about easy calculation is that for any given $1 - \alpha$, e.g. $1 - \alpha = 0.95$, we can calculate the critical value $h$ such that $P(-h < T < h) = 1 - \alpha$. Note also that the pivot $T$ does not involve the other unknown parameter of the normal model, namely the variance $\sigma^2$. If indeed, we can find $h$ for any given $1 - \alpha$, we can proceed as follows to find the exact CI for $\mu$:

$$P(-h \le T \le h) = 1 - \alpha$$

$$\text{i.e. } P\left(-h \le \sqrt{n}\frac{(\bar{X} - \mu)}{S} \le h\right) = 0.95$$

$$\Rightarrow P\left(\bar{X} - h\frac{S}{\sqrt{n}} \le \mu \le \bar{X} + h\frac{S}{\sqrt{n}}\right) = 0.95$$

The observed value of this interval, $(\bar{x} \pm h\frac{s}{\sqrt{n}})$, is the 95% confidence interval for $\mu$. Remarkably, this also of the general form, Estimate $\pm$ Critical value $\times$ Standard error, where the Critical value is $h$ and the standard error of the sample mean is $\frac{s}{\sqrt{n}}$. Now, how do we find the critical value $h$ for a given $1 - \alpha$? We need to introduce the $t$-distribution.

Let $X_1, \ldots, X_n$ be i.i.d $N(\mu, \sigma^2)$ random variables. Define $\bar{X} = \frac{1}{n} \sum_{i=1}^{n} X_i$ and

$$S^2 = \frac{1}{n-1} \left( \sum_{i=1}^{n} X_i^2 - n\bar{X}^2 \right) = \frac{1}{n-1} \sum_{i=1}^{n} (X_i - \bar{X})^2.$$

Then, it will be proved in Sect. 14.6 that

$$\sqrt{n} \frac{(\bar{X} - \mu)}{S} \sim t_{n-1},$$

where $t_{n-1}$ denotes the standard $t$ *distribution* with $n - 1$ *degrees of freedom*. The standard $t$ distribution is a family of distributions which depend on one parameter called the degrees-of-freedom (df) which is $n - 1$ here. The concept of degrees of freedom is that it is usually the number of independent random samples, $n$ here, minus the number of linear parameters estimated, 1 here for $\mu$. Hence the df is $n - 1$.

The probability density function of the $t_k$ distribution, derived in Sect. 14.5, is similar to a standard normal, in that it is symmetric around zero and 'bell-shaped', but the t-distribution is more *heavy-tailed*, giving greater probability to observations further away from zero. Figure 11.4 illustrates the $t_k$ density function for $k = 1, 2, 5, 10$ together with the standard normal pdf (solid line).

The values of $h$ for a given $1 - \alpha$ have been tabulated using the standard $t$-distribution and can be obtained using the R command qt (abbreviation for quantile

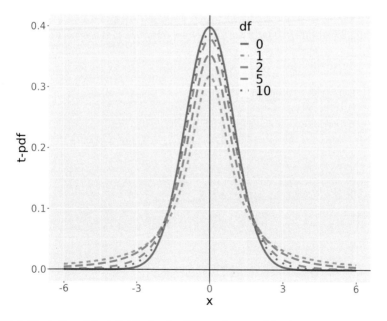

**Fig. 11.4**  Illustration of the t-distribution for different degrees of freedom

of $t$). For example, if we want to find $h$ for $1 - \alpha = 0.95$ and $n = 20$ then we issue the command: `qt(0.975, df=19)` $= 2.093$. Note that it should be 0.975 so that we are splitting 0.05 probability between the two tails equally and the df should be $n - 1 = 19$. Indeed, using the above command repeatedly, we obtain the following critical values for the 95% interval for different values of the sample size $n$.

| $n$ | 2 | 5 | 10 | 15 | 20 | 30 | 50 | 100 | $\infty$ |
|---|---|---|---|---|---|---|---|---|---|
| $h$ | 12.71 | 2.78 | 2.26 | 2.14 | 2.09 | 2.05 | 2.01 | 1.98 | 1.96 |

Note that the critical value approaches 1.96 (which is the critical value for the normal distribution) as $n \to \infty$, since the $t$-distribution itself approaches the normal distribution for large values of its df parameter.

> If we can justify that the underlying distribution is normal then we can use the $t$-distribution-based confidence interval.

**Example 11.7 (Fast Food Waiting Time Revisited)**

We would like to find a confidence interval for the true mean waiting time. If $X$ denotes the waiting time in seconds, we have $n = 20$, $\bar{x} = 67.85$, $s = 18.36$. Hence, recalling that the critical value $h = 2.093$, from the command `qt(0.975 , df=19)`, see Fig. 11.5, a 95% confidence interval for $\mu$ is

$$67.85 - 2.093 \times 18.36/\sqrt{20} \le \mu \le 67.85 + 2.093 \times 18.36/\sqrt{20}$$

$$\Rightarrow \quad 59.26 \le \mu \le 76.44.$$

In R we issue the commands:

```
a <- c(ffood$AM, ffood$PM)
mean(a) + c(-1, 1) * qt(0.975, df=19) * sqrt(var(a))/sqrt(20)
```

The above commands give the result (59.25, 76.45). If we want a 90% confidence interval then we issue the command:

```
mean(a) + c(-1, 1) * qt(0.95, df=19) * sqrt(var(a))/sqrt(20)
```

which gives the result (60.75, 74.95).
If we want a 99% confidence interval then we issue the command:

```
mean(a) + c(-1, 1) * qt(0.995, df=19) * sqrt(var(a))/sqrt(20)
```

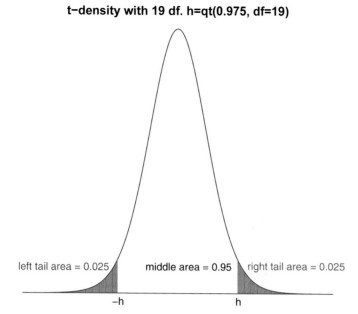

**t−density with 19 df. h=qt(0.975, df=19)**

left tail area = 0.025          middle area = 0.95          right tail area = 0.025

−h                    h

**Fig. 11.5** Illustration of a t-distribution based critical value

which gives (56.10, 79.60). We can see clearly that the interval is getting wider as the level of confidence is getting higher. ◄

**Example 11.8 (Weight Gain Revisited)**

We would like to find a confidence interval for the true average weight gain (final weight—initial weight). Here $n = 68, \bar{x} = 0.8672$ and $s = 0.9653$. Hence, a 95% confidence interval for $\mu$ is

$$0.8672 - 1.996 \times 0.9653/\sqrt{68} \leq \mu \leq 0.8672 + 1.996 \times 0.9653/\sqrt{68}$$

$$\Rightarrow \quad 0.6335 \leq \mu \leq 1.1008$$

[In R, we obtain the critical value 1.996 by `qt(0.975, df=67)` or `-qt(0.025, df=67)`]

To obtain a 95% confidence interval for the mean weight gain, the R command is: `mean(x)+ c(-1, 1)*qt(0.975, df=67)*sqrt(var(x)/68)` if the vector x contains the 68 weight gain differences. We may obtain this by issuing the commands:

```
x <- wgain$final - wgain$initial
mean(x) + c(-1, 1) * qt(0.975, df=67) * sqrt(var(x)/68)
```

Note that the interval here does not include the value 0, so it is very likely that the weight gain is significantly positive, which we will justify using what is called testing of hypothesis, in Chap. 12. ◄

In this section we have learned how to find an exact confidence interval for a population mean based on the assumption that the population is normal. The confidence interval is based on the $t$-distribution which is a very important distribution in statistics. The $t$-distribution converges to the normal distribution when its only parameter, called the degrees of freedom, becomes very large. If the assumption of the normal distribution for the data can be justified, then the method of inference based on the $t$-distribution is best when the variance parameter, sometimes called the nuisance parameter, is unknown.

## 11.5 Exercises

### 11.1 (Interval Estimation)

1. In an experiment, 100 observations were taken from a normal distribution with variance 16. The experimenter quoted [1.545, 2.861] as the confidence interval for $\mu$. What level of confidence was used?
   **Hint: The variance is known and the distribution is normal. Hence you will have to use the normal distribution to obtain confidence intervals.**
2. The heights of $n$ randomly selected seven-year-old children were measured. The sample mean and standard deviation were found to be 121 cm and 5 cm respectively. Assuming that height is normally distributed, calculate the following confidence intervals for the mean height of seven-year-old children:
   (a) 90% with $n = 16$,
   (b) 99% with $n = 16$,
   (c) 95% with $n = 16, 25, 100, 225, 400$.
   **Hint: You will have to use the $t$-distribution based intervals.**
3. At the end of a severe winter a certain insurance company found that of 972 policy holders living in a large city who had insured their homes with the company, 357 had suffered more than $500-worth of snow and frost damage. Calculate an approximate 95% confidence interval for the proportion of all homeowners in the city who suffered more than $500-worth of damage. State any assumptions that you make.
4. A random variable is known to be normally distributed, but its mean $\mu$ and variance $\sigma^2$ are unknown. A 95% confidence interval for $\mu$ based on 9 observations was found to be [22.4, 25.6]. Calculate unbiased estimates of $\mu$ and $\sigma^2$.
5. The wavelength of radiation from a certain source is 1.372 microns. The following 10 independent measurements of the wavelength were obtained using a measuring device:

   1.359, 1.368, 1.360, 1.374, 1.375, 1.372, 1.362, 1.372, 1.363, 1.371.

Assuming that the measurements are normally distributed, calculate 95% confidence limits for the mean error in measurements obtained with this device and comment on your result.

6. In five independent attempts, a girl completed a Rubik's cube in 135.4, 152.1, 146.7, 143.5 and 146.0 seconds. In five further attempts, made two weeks later, she completed the cube in 133.1, 126.9, 129.0, 139.6 and 144.0 seconds. Find a 90% confidence interval for the change in the mean time taken to complete the cube. State your assumptions.

7. In an experiment to study the effect of a certain concentration of insulin on blood glucose levels in rats, each member of a random sample of 10 rats was treated with insulin. The blood glucose level of each rat was measured both before and after treatment. The results, in suitable units, were as follows.

| Rat | 1 | 2 | 3 | 4 | 5 | 6 | 7 | 8 | 9 | 10 |
|---|---|---|---|---|---|---|---|---|---|---|
| Level before | 2.30 | 2.01 | 1.92 | 1.89 | 2.15 | 1.93 | 2.32 | 1.98 | 2.21 | 1.78 |
| Level after | 1.98 | 1.85 | 2.10 | 1.78 | 1.93 | 1.93 | 1.85 | 1.67 | 1.72 | 1.90 |

Let $\mu_1$ and $\mu_2$ denote respectively the mean blood glucose levels of a randomly selected rat before and after treatment with insulin. By considering the differences of the measurements on each rat and assuming that they are normally distributed, find a 95% confidence interval for $\mu_1 - \mu_2$.

8. The heights (in metres) of 10 fifteen-year-old boys were as follows:

1.59, 1.67, 1.55, 1.63, 1.69, 1.58, 1.66, 1.62, 1.64, 1.61.

Assuming that heights are normally distributed, find a 99% confidence interval for the mean height of fifteen-year-old boys.

If you were told that the true mean height of boys of this age was 1.67m, what would you conclude?

# Hypothesis Testing

12

**Abstract**

Chapter 12 discusses testing of statistical hypotheses called null and alternative hypothesis. Definintions of many related keywords, e.g. critical region, types of errors while testing statistical hypothesis, power function, sensitivity and specificity are provided. These are illustrated with the t-test for testing hypothesis regarding the mean of one ir two normal distributions. This chapter ends with a discussion on designs of experiments for estimation and testing purposes.

The manager of a new fast food chain claims that the average waiting time to be served in their restaurant is less than a minute. The marketing department of a mobile phone company claims that their phones never break down in the first 3 years of their lifetime. A professor of nutrition claims that students gain significant weight in the first year of their life in college away form home. How can we verify these claims? We will learn the procedures of hypothesis testing for such problems.

In statistical inference, we use observations $x_1, \ldots, x_n$ of univariate random variables $X_1, \ldots, X_n$ in order to draw inferences about the probability distribution $f(x)$ of the underlying random variable $X$. So far, we have mainly been concerned with estimating features (usually unknown parameters) of $f(x)$. It is often of interest to compare alternative specifications for $f(x)$. If we have a set of competing probability models which might have generated the observed data, we may want to determine which of the models is most appropriate. A proposed (hypothesised) model for $X_1, \ldots, X_n$ is then referred to as a *hypothesis*, and pairs of models are compared using hypothesis tests.

For example, we may have two competing alternatives, $f^{(0)}(x)$ (model $H_0$) and $f^{(1)}(x)$ (model $H_1$) for $f(x)$, both of which completely specify the joint distribution of the sample $X_1, \ldots, X_n$. Completely specified statistical models are called *simple* hypotheses. Usually, $H_0$ and $H_1$ both take the same parametric form $f(x, \theta)$, but with different values $\theta^{(0)}$ and $\theta^{(1)}$ of $\theta$. Thus the joint distribution of the sample given by $f(\mathbf{X})$ is completely specified apart from the values of the unknown parameter $\theta$ and $\theta^{(0)} \neq \theta^{(1)}$ are specified alternative values.

More generally, competing hypotheses often do not completely specify the joint distribution of $X_1, \ldots, X_n$. For example, a hypothesis may state that $X_1, \ldots, X_n$ is a random sample from the probability distribution $f(x; \theta)$ where $\theta < 0$. This is not a completely specified hypothesis, since it is not possible to calculate probabilities such as $P(X_1 < 2)$ when the hypothesis is true, as we do not know the exact value of $\theta$. Such an hypothesis is called a *composite* hypothesis.

Examples of hypotheses:

$X_1, \ldots, X_n \sim N(\mu, \sigma^2)$ with $\mu = 0, \sigma^2 = 2$, which is a simple hypothesis.

$X_1, \ldots, X_n \sim N(\mu, \sigma^2)$ with $\mu = 0, \sigma^2 \in \mathcal{R}_+$, which is a composite hypothesis.

$X_1, \ldots, X_n \sim N(\mu, \sigma^2)$ with $\mu \neq 0, \sigma^2 \in \mathcal{R}_+$, which is a composite hypothesis.

$X_1, \ldots, X_n \sim \text{Bernoulli}(p)$ with $p = \frac{1}{2}$, which is a simple hypothesis.

$X_1, \ldots, X_n \sim \text{Bernoulli}(p)$ with $p \neq \frac{1}{2}$, which is a composite hypothesis.

$X_1, \ldots, X_n \sim \text{Bernoulli}(p)$ with $p > \frac{1}{2}$, which is a composite hypothesis.

$X_1, \ldots, X_n \sim \text{Poisson}(\lambda)$ with $\lambda = 1$, which is a simple hypothesis.

$X_1, \ldots, X_n \sim \text{Poisson}(\theta)$ with $\theta > 1$, which is a composite hypothesis.

## 12.1  Testing Procedure

A hypothesis test provides a mechanism for comparing two competing statistical models, $H_0$ and $H_1$. A hypothesis test does not treat the two hypotheses (models) symmetrically. One hypothesis, $H_0$, is given special status, and referred to as the *null hypothesis*. The null hypothesis is the reference model, and is assumed to be appropriate unless the observed data strongly indicate that $H_0$ is inappropriate, and that $H_1$ (the *alternative* hypothesis) should be preferred.

Hence, the fact that a hypothesis test does not reject $H_0$ should not be taken as evidence that $H_0$ is true and $H_1$ is not, or that $H_0$ is better-supported by the data than $H_1$, merely that the data does not provide significant evidence to reject $H_0$ in favour of $H_1$.

A hypothesis test is defined by its *critical region* or *rejection region*, which we shall denote by $C$. $C$ is a subset of $\mathcal{R}^n$ (see Fig. 12.1), and is the set of possible observed values of $\mathbf{X}$ which, if observed, would lead to rejection of $H_0$ in favour of $H_1$, *i.e.*

If $\mathbf{x} \in C$ $H_0$ is rejected in favour of $H_1$
If $\mathbf{x} \notin C$ $H_0$ is not rejected.

As $\mathbf{X}$ is a random variable, there remains the possibility that a hypothesis test will give an erroneous result. We define two types of error:

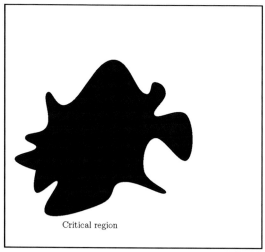

Critical region

Sample Space

**Fig. 12.1** Illustration of a critical rejection (rejection) region $C$ inside the sample space $S$

**Table 12.1** Possible decisions and errors in hypothesis testing

| | $H_0$ true | $H_0$ false |
|---|---|---|
| Reject $H_0$ | Type I error | Correct decision |
| Do not reject $H_0$ | Correct decision | Type II error |

Type I error: $H_0$ is rejected when it is true
Type II error: $H_0$ is not rejected when it is false

The above table (Table 12.1) helps to understand further. When $H_0$ and $H_1$ are simple hypotheses, we can define

$$\alpha = P(\text{Type I error}) = P(\mathbf{X} \in C) \quad \text{if } H_0 \text{ is true}$$
$$\beta = P(\text{Type II error}) = P(\mathbf{X} \notin C) \quad \text{if } H_1 \text{ is true}$$

**Example 12.1 (Uniform)**

Suppose that we have **one** observation from the uniform distribution on the range $(0, \theta)$. In this case, $f(x) = 1/\theta$ if $0 < x < \theta$ and $P(X \leq x) = \frac{x}{\theta}$ for $0 < x < \theta$. We want to test $H_0 : \theta = 1$ against the alternative $H_1 : \theta = 2$. Suppose we decide arbitrarily that we will reject $H_0$ if $X > 0.75$. Then

$$\alpha = P(\text{Type I error}) = P(X > 0.75) \quad \text{if } H_0 \text{ is true}$$
$$\beta = P(\text{Type II error}) = P(X \leq 0.75) \quad \text{if } H_1 \text{ is true}$$

which will imply:

$$\alpha = P(X > 0.75|\theta = 1) = 1 - 0.75 = \frac{1}{4},$$

$$\beta = P(X < 0.75|\theta = 2) = 0.75/2 = \frac{3}{8}.$$

Here the notation | means "given that". ◀

**Example 12.2 (Poisson)**

The daily demand for a product has a Poisson distribution with mean $\lambda$, the demands on different days being statistically independent. It is desired to test the hypotheses $H_0 : \lambda = 0.7$, $H_1 : \lambda = 0.3$. The null hypothesis is to be accepted if in 20 days the number of days with no demand is less than 15. Calculate the Type I and Type II error probabilities.
Let $p$ denote the probability that the demand on a given day is zero.
Then

$$p = e^{-\lambda} = \begin{cases} e^{-0.7} & \text{under } H_0 \\ e^{-0.3} & \text{under } H_1. \end{cases}$$

If $X$ denotes the number of days out of 20 with zero demand, it follows that

$$X \sim B(20, e^{-0.7}) \text{ under } H_0,$$

$$X \sim B(20, e^{-0.3}) \text{ under } H_1.$$

Thus

$$\alpha = P(\text{Reject } H_0|H_0 \text{ true})$$

$$= P(X \geq 15|X \sim B(20, e^{-0.7}))$$

$$= 1 - P(X \leq 14|X \sim B(20, 0.4966))$$

$$= 1 - 0.98028$$

$$= 0.01923 \, (1 - \text{pbinom}(14, \text{size} = 20, \text{prob} = 0.4966) \text{ in R}).$$

Furthermore

$$\beta = P(\text{Accept } H_0|H_1 \text{ true})$$

$$= P(X \leq 14|X \sim B(20, e^{-0.3}))$$

$$= P(X \leq 14|X \sim B(20, 0.7408))$$

$$= 0.42023 \, (\text{pbinom}(14, \text{size} = 20, \text{prob} = 0.7408) \text{ in R}).$$

◀

Sometimes $\alpha$ is called the *size* (or *significance level*) of the test and $\omega \equiv 1 - \beta$ is called the *power* of the test. Thus, power is defined as the probability of the correct decision to reject the null hypothesis when the alternative is true, see Sect. 12.3 for further discussion regarding this.

Ideally, we would like to avoid error during a hypothesis testing procedure so we would like to make both $\alpha$ and $\beta$ as small as possible. In other words, a good test will have small size, but large power. However, it is not possible to make $\alpha$ and $\beta$ both arbitrarily small. For example if $C = \emptyset$ then $\alpha = 0$, but $\beta = 1$. On the other hand if $C = \mathbf{S} = \mathcal{R}^n$ then $\beta = 0$, but $\alpha = 1$.

The general hypothesis testing procedure is to fix $\alpha$ to be some small value (often 0.05), so that the probability of a Type I error is limited. In doing this, we are giving $H_0$ precedence over $H_1$, and acknowledging that Type I error is potentially more serious than Type II error. (Note that for discrete random variables, it may be difficult to find $C$ so that the test has exactly the required size). Given our specified $\alpha$, we try to choose a test, defined by its rejection region $C$, to make $\beta$ as small as possible, *i.e.* we try to find the most powerful test of a specified size. Where $H_0$ and $H_1$ are *simple* hypotheses this can be achieved easily.

Note that tests are usually based on a one-dimensional *test statistic* $T(\mathbf{X})$ whose sample space is some subset of $\mathcal{R}$. The rejection region is then a set of possible values for $T(\mathbf{X})$, so we also think of $C$ as a subset of $\mathcal{R}$. In order to be able to ensure the test has size $\alpha$, the distribution of the test statistic under $H_0$ should be known.

## 12.2 The Test Statistic

We perform a hypothesis test by computing a *test statistic*, $T(X)$. A test statistic must (obviously) be a statistic (i.e. a function of $X$ and other known quantities only). Furthermore, the random variable $T(X)$ must have a distribution which is known under the null hypothesis. The easiest way to construct a test statistic is to obtain a pivot for $\theta$. If $T(X, \theta)$ is a pivot for $\theta$ then its sampling distribution is known and, therefore, under the null hypothesis ($\theta = \theta_0$) the sampling distribution of $T(X, \theta_0)$ is known. Hence $T(x, \theta_0)$ is a test statistic, as it depends on observed data $x$ and the hypothesised value $\theta_0$ only. We then assess the plausibility of $H_0$ by evaluating whether $T(x, \theta_0)$ seems like a *reasonable observation from its (known) distribution*. This is all rather abstract. How does it work in a concrete example?

### Example 12.3 (Testing a Normal Mean $\mu$)

Suppose that we observe data $x_1, \ldots, x_n$ which are modelled as observations of i.i.d. $N(\mu, \sigma^2)$ random variables $X_1, \ldots, X_n$, and we want to test the null hypothesis

$$H_0 : \mu = \mu_0$$

against the alternative hypothesis

$$H_1 : \mu \neq \mu_0,$$

where $\sigma^2$ is unknown.
We recall that

$$\sqrt{n}\frac{(\bar{X} - \mu)}{S} \sim t_{n-1}$$

and therefore, when $H_0$ is true, often written as under $H_0$,

$$\sqrt{n}\frac{(\bar{X} - \mu_0)}{S} \sim t_{n-1}$$

so $\sqrt{n}(\bar{X} - \mu_0)/s$ is a test statistic for this test. The sampling distribution of the test statistic when the null hypothesis is true is called the null distribution of the test statistic. In this example, the null distribution is the t-distribution with $n - 1$ degrees of freedom.

This test is called a *t-test*. We reject the null hypothesis $H_0$ in favour of the alternative $H_1$ if the observed test statistic *seems unlikely to have been generated by the null distribution.* ◄

---

**Example 12.4 (Weight Gain Data)**

For the weight gain data, if $x$ denotes the differences in weight gain, we have $\bar{x} = 0.8672$, $s = 0.9653$ and $n = 68$. Hence our test statistic for the null hypothesis $H_0 : \mu = \mu_0 = 0$ is

$$\sqrt{n}\frac{(\bar{x} - \mu_0)}{s} = 7.41.$$

The observed value of 7.41 does not seem reasonable from the graph in Fig. 12.2. The graph has plotted the density of the *t*-distribution with 67 degrees of freedom, and a vertical line is drawn at the observed value of 7.41. So there may be evidence here to reject $H_0 : \mu = 0$. ◄

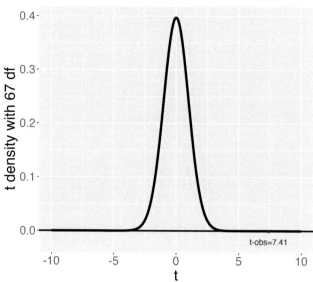

Fig. 12.2 The null distribution of the $t$-statistic and the observed value for the weight gain data example

---

**Example 12.5 (Fast Food Waiting Time Revisited)**

Suppose the manager of the fast food outlet claims that the average waiting time is only 60 seconds. So, we want to test $H_0 : \mu = 60$. We have $n = 20, \bar{x} = 67.85, s = 18.36$. Hence our test statistic for the null hypothesis $H_0 : \mu = \mu_0 = 60$ is

$$\sqrt{n}\frac{(\bar{x} - \mu_0)}{s} = \sqrt{20}\frac{(67.85 - 60)}{18.36} = 1.91.$$

The observed value of 1.91 may or may not be reasonable from the graph in Fig. 12.3. The graph has plotted the density of the t-distribution with 19 degrees of freedom and a vertical line is drawn at the observed value of 1.91. This value is a bit out in the tail but we are not sure, unlike in the previous weight gain example. So how can we decide whether to reject the null hypothesis? ◄

## 12.2.1 The Significance Level

In the weight gain example, it seems clear that there is no evidence to reject $H_0$, but how extreme (far from the mean of the null distribution) should the test statistic be

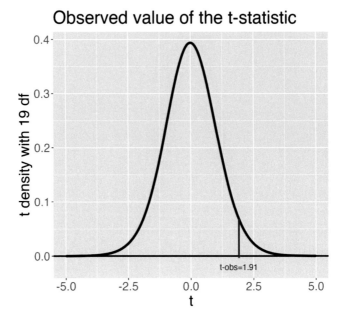

**Fig. 12.3** The null distribution of the *t*-statistic and the observed value for the fast food data example

in order for $H_0$ to be rejected? The *significance level* of the test, $\alpha$, is the probability that we will erroneously reject $H_0$ (called *Type I error* as discussed before). Clearly we would like $\alpha$ to be small, but making it too small risks failing to reject $H_0$ even when it provides a poor model for the observed data (*Type II error*). Conventionally, $\alpha$ is usually set to a value of 0.05, or 5%. Therefore we reject $H_0$ when the test statistic lies in a *rejection region* which has probability $\alpha = 0.05$ under the null distribution.

## 12.2.2 Rejection Region for the t-Test

For the t-test, the null distribution is $t_{n-1}$ where $n$ is the sample size, so the rejection region for the test corresponds to a region of total probability $\alpha = 0.05$ comprising the 'most extreme' values in the direction of the alternative hypothesis. If the alternative hypothesis is two-sided, e.g. $H_1 : \mu \neq \mu_0$, then this is obtained as illustrated in Fig. 12.4, where the two shaded regions both have area (probability) $\alpha/2 = 0.025$.

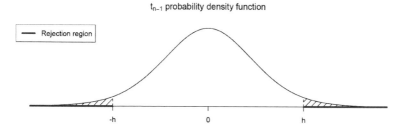

**Fig. 12.4** Illustration of a rejection region for a $t$-test

**Table 12.2** Critical values of the $t$-distribution for various degrees of freedom for a right tail test at 5% level of significance

| $n$ | 2 | 5 | 10 | 15 | 20 | 30 | 50 | 100 | $\infty$ |
|---|---|---|---|---|---|---|---|---|---|
| $h$ | 6.31 | 2.13 | 1.83 | 1.76 | 1.73 | 1.70 | 1.68 | 1.66 | 1.64 |

The value of $h$ depends on the sample size $n$ and can be found by issuing the `qt` command. Here are a few examples obtained from `qt(0.975, df=c(1, 4, 9, 14, 19, 29, 49, 99))`:

$$
\begin{array}{c|ccccccccc}
n & 2 & 5 & 10 & 15 & 20 & 30 & 50 & 100 & \infty \\
h & 12.71 & 2.78 & 2.26 & 2.14 & 2.09 & 2.05 & 2.01 & 1.98 & 1.96
\end{array}
$$

Note that we need to put $n - 1$ in the df argument of `qt` and the last value for $n = \infty$ is obtained from the normal distribution.

However, if the alternative hypothesis is one-sided, e.g. $H_1 : \mu > \mu_0$, then the critical region will only be in the right tail. Consequently, we need to leave an area $\alpha$ on the right and as a result the critical values will be from a command such as: `qt(0.95, df=c(1, 4, 9, 14, 19, 29, 49, 99))`, which produces the results in Table 12.2.

### 12.2.3 t-Test Summary

Suppose that we observe data $x_1, \ldots, x_n$ which are modelled as observations of i.i.d. $N(\mu, \sigma^2)$ random variables $X_1, \ldots, X_n$ and we want to test the null hypothesis $H_0 : \mu = \mu_0$ against the alternative hypothesis $H_1 : \mu \neq \mu_0$:

1. Compute the test statistic

$$
t = \sqrt{n}\frac{(\bar{x} - \mu_0)}{s}.
$$

2. For chosen significance level $\alpha$ (usually 0.05) calculate the rejection region for $t$, which is of the form $|t| > h$ where $-h$ is the $\alpha/2$ percentile of the null distribution, $t_{n-1}$.
3. If the computed $t$ lies in the rejection region, i.e. $|t| > h$, we report that $H_0$ is rejected in favour of $H_1$ at the chosen level of significance. If $t$ does not lie in the rejection region, we report that $H_0$ is not rejected. [Never refer to 'accepting' a hypothesis.]

**Example 12.6 (Fast Food Waiting Time)**

We would like to test $H_0 : \mu = 60$ against the alternative $H_1 : \mu > 60$, as this alternative will refute the claim of the store manager that customers only wait for a maximum of 1 minute. We calculated the observed value to be 1.91. This is a one-sided test and for a 5% level of significance, the critical value $h$ will come from `qt(0.95, df=19)`=1.73. Thus the observed value is higher than the critical value so we will reject the null hypothesis, disputing the manager's claim regarding a minute wait. ◀

**Example 12.7 (Weight Gain)**

For the weight gain example $\bar{x} = 0.8671$, $s = 0.9653$, $n = 68$. Then, we would be interested in testing $H_0 : \mu = 0$ against the alternative hypothesis $H_1 : \mu \neq 0$ in the model that the data are observations of i.i.d. $N(\mu, \sigma^2)$ random variables.

- We obtain the test statistic

$$t = \sqrt{n}\frac{(\bar{x} - \mu_0)}{s} = \sqrt{68}\frac{(0.8671 - 0)}{0.9653} = 7.41.$$

- Under $H_0$ this is an observation from a $t_{67}$ distribution. For significance level $\alpha = 0.05$ the rejection region is $|t| > 1.996$.
- Our computed test statistic lies in the rejection region, i.e. $|t| > 1.996$, so $H_0$ is rejected in favour of $H_1$ at the 5% level of significance.

In R we can perform the test as follows:

```
x <- wgain$final - wgain$initial
t.test(x)
```

This gives the results: t = 7.4074, and df = 67. ◀

## 12.2.4 p-Values

The result of a test is most commonly summarised by rejection or non-rejection of $H_0$ at the stated level of significance. An alternative, which we may see in practice, is the computation of a *p-value*. This is the probability that the reference distribution would have generated the actual observed value of the statistic *or something more extreme*. A small p-value is evidence against the null hypothesis, as it indicates that the observed data were unlikely to have been generated by the reference distribution. In many examples a threshold of 0.05 is used, below which the null hypothesis is rejected as being insufficiently well-supported by the observed data. Hence for the t-test with a two-sided alternative, the p-value is given by:

$$p = P(|T| > |t_{\text{obs}}|) = 2P(T > |t_{\text{obs}}|),$$

where $T$ has a $t_{n-1}$ distribution and $t_{\text{obs}}$ is the observed sample value.

However, if the alternative is one-sided and to the right then the p-value is given by:

$$p = P(T > t_{\text{obs}}),$$

where $T$ has a $t_{n-1}$ distribution and $t_{\text{obs}}$ is the observed sample value.

A small p-value corresponds to an observation of $T$ that is improbable (since it is far out in the low probability tail area) under $H_0$ and hence provides evidence against $H_0$. The p-value should *not* be misinterpreted as the probability that $H_0$ is true. $H_0$ is not a random event (under our models) and so cannot be assigned a probability. The null hypothesis is rejected at significance level $\alpha$ if the p-value for the test is less than $\alpha$.

Reject $H_0$ if p-value $< \alpha$.

The p-value, denoted by $p$, is a way of trying to present "evidence" regarding $H_0$. We sometimes need to make a decision to accept or reject $H$. This is more complicated theoretically (requiring specification of costs which change for each problem) but in practice can be done by setting agreed cut off levels for $p$. Conventionally, in scientific problems we adopt the following guides (Table 12.3):

Note that we reject $H_0$ when the evidence against is strong, in the sense that the observed data are unlikely to have been generated by a model consistent with $H_0$. When $H_0$ has not been rejected, that is exactly what has happened—it has not been rejected (and neither has it been 'accepted').

**Table 12.3** Guides for judging strength of evidence based on p-values

| | |
|---|---|
| $0.1 < p$ | No evidence against $H_0$, |
| $0.05 < p \leq 0.1$ | Weak evidence against $H_0$, |
| $0.01 < p \leq 0.05$ | Moderate evidence against $H_0$ (reject at the 5% level), |
| $p \leq 0.01$ | Strong evidence against $H_0$ (reject at the 1% level). |

---

**Example 12.8 (p-Value Examples)**

In the fast food example, a test of $H_0 : \mu = 60$ against $H_1 : \mu > 60$ resulted in a test statistic $t = 1.91$ with 19 df. Then the p-value is given by:

$$
\begin{aligned}
p &= P(T > 1.91) \\
&= 1 - P(T \le 1.91) \\
&= 1 - \mathrm{pt}(1.91, \mathrm{df} = 19) \\
&= 0.036.
\end{aligned}
$$

This is the area of the shaded region in Fig. 12.5. In R it is: `1 - pt(1.91, df =19)`. The p-value 0.036 indicates some evidence against the manager's claim at the 5% level of significance but not the 1% level of significance.

When the alternative hypothesis is two-sided the p-value has to be calculated from $P(|T| > t_{\mathrm{obs}})$, where $t_{\mathrm{obs}}$ is the observed value and $T$ follows the $t$-distribution with $n - 1$ df. For the weight gain example, because the alternative is two-sided, the p-value is given by:

$$
p = P(|T| > 7.41) = 2.78 \times 10^{-10} \approx 0.0, \quad \text{when } T \sim t_{67}.
$$

This very small p-value for the second example indicates very strong evidence against the null hypothesis of no weight gain in the first year of university. ◄

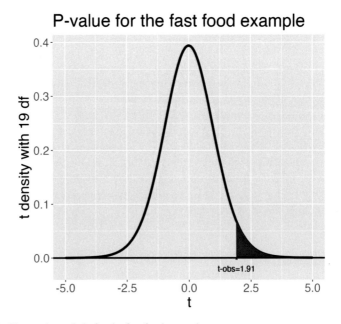

**Fig. 12.5** Observed t-statistic for the fast food example

## 12.3   Power Function, Sensitivity and Specificity

Power has been defined in Sect. 12.1 as the probability of the correct decision to reject the null hypothesis when the alternative is true. This concept of power is often used to solve various statistical problems, e.g sample size determination in scientific experiments where the underlying distribution  is normal. To illustrate suppose that we have random samples $X_1, \ldots, X_n$ from the normal distribution $N(\theta, 1)$. It is desired to test $H_0 : \theta = \theta_0 = 0$ against $H_1 : \theta > \theta_0$. Using all the samples, we have that $\bar{X} \sim N\left(\theta, \frac{1}{n}\right)$, see Example 8.14. The statistic $Z = \sqrt{n}(\bar{X} - \theta_0) \sim N(0, 1)$ when $H_0$ is true and this is often used in testing $H_0$ against $H_1$. A sensible decision rule is to reject $H_0$ if $Z > 1.645$ at 5% level of significance. Notice that $P(Z > 1.645) = 0.05$ when $Z \sim N(0, 1)$. For this test, the power is a function of $\theta$ and it is given by:

$$
\begin{aligned}
\omega(\theta) &= P\,(\text{Reject } H_0 | H_0 \text{ is false}) \\
&= P\left(\sqrt{n}(\bar{X} - \theta_0) > 1.645 | H_1 \text{ is true}\right) \\
&= P\left(\sqrt{n}(\bar{X} - \theta + \theta - \theta_0) > 1.645 | H_1 \text{ is true}\right) \\
&= P\left(\sqrt{n}(\bar{X} - \theta) > 1.645 - \sqrt{n}(\theta - \theta_0) | H_1 \text{ is true}\right) \\
&= P\left(Z > 1.645 - \sqrt{n}(\theta - \theta_0) | Z \sim N(0, 1)\right) \\
&= 1 - \Phi\left(1.645 - \sqrt{n}(\theta - \theta_0)\right),
\end{aligned}
$$

where $\Phi(\cdot)$ is the cdf of the standard normal distribution, see Sect. 7.2. Clearly, $\omega(\theta)$ is a function of $\theta$, the value of the mean under the alternative hypothesis and $n$ the sample size. Because $\Phi(\cdot)$ is a cdf, it is plain to see that $\omega(\theta)$ is an increasing function of the difference $\theta - \theta_0$ and also an increasing function of $n$, see Fig. 12.6. Hence, the power will be higher for higher values of $\theta > \theta_0$. Moreover, for a fixed value of $\theta$, the power will increase as $n$ increases. The sample size determination problem alluded to earler is to determine $n$ for a given high value of power, say 80%, for a given value of the mean difference $\theta - \theta_0$.

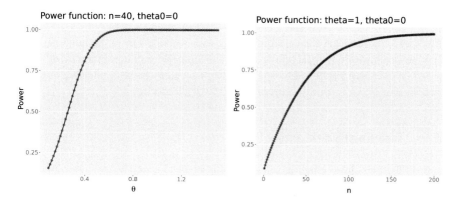

**Fig. 12.6**  Illustrations of the power function for varying values of the mean $\theta$ under the alternative hypothesis (left) and for varying sample size $n$ (right)

The associated concepts of sensitivity and specificity are used in testing for the presence of a disease (or condition) given the presence of a symptoms, see e.g. Sect. 4.1. Sensitivity is defined the probability of the correct decision which rejects the null hypothesis when it is false and specificity is the probability of not rejecting the null hypothesis when it is indeed true. Sometimes, sensitivity and specificity are also defined as the true positive and true negative rate respectively. Here the word 'positive' implies that the test produces a significant result, i.e. the test yields a small p-value. Similarly 'negative' means that the test does not produce a significant result. Sensitivity and specificity, being the probabilities of correct decisions, must be very high for a good testing procedure. The decisions and errors shown in Table 12.1 help us determine sensitivity and specificity in practical problems.

## 12.4  Equivalence of Testing and Interval Estimation

Note that the 95% confidence interval for $\mu$ in the weight gain example has previously been calculated to be (0.6335, 1.1008) in Sect. 11.4. This interval does not include the hypothesised value 0 of $\mu$. Hence we can conclude that the hypothesis test at the 5% level of significance will reject the null hypothesis $H_0 : \mu_0 = 0$.

In general, we can make the following argument. Let $C$ be the critical critical region of the test. Suppose that we do not reject the test, i.e.,

$$|T_{obs}| < h$$

$$\begin{aligned}
\leftrightarrow \quad & -h < & T_{obs} & < h \\
\leftrightarrow \quad & -h < & \frac{\sqrt{n}(\bar{x}-\mu_0)}{s} & < h \\
\leftrightarrow \quad & -hs < & \sqrt{n}(\bar{x}-\mu_0) & < hs \\
\leftrightarrow \quad & -h\frac{s}{\sqrt{n}} < & (\bar{x}-\mu_0) & < h\frac{s}{\sqrt{n}} \\
\leftrightarrow \quad -\bar{x} - h\frac{s}{\sqrt{n}} < & & -\mu_0 & < -\bar{x} + h\frac{s}{\sqrt{n}} \\
\leftrightarrow \quad \bar{x} - h\frac{s}{\sqrt{n}} < & & \mu_0 & < \bar{x} + h\frac{s}{\sqrt{n}}.
\end{aligned}$$

This shows that the acceptance region of the test is equivalent to the hypothesised value $\mu_0$ being within the end points of the $100(1-\alpha)\%$ confidence interval defined by the critical value $h$. For this reason we often just calculate the confidence interval and take the reject/do not reject decision merely by inspection.

## 12.5  Two Sample t-Tests

Suppose that we observe two samples of data, $x_1, \ldots, x_n$ and $y_1, \ldots, y_m$, and that we propose to model them as observations of

$$X_1, \ldots, X_n \overset{i.i.d.}{\sim} N(\mu_X, \sigma_X^2)$$

and

$$Y_1, \ldots, Y_m \overset{i.i.d.}{\sim} N(\mu_Y, \sigma_Y^2)$$

respectively, where it is also assumed that the $X$ and $Y$ variables are independent of each other. Suppose that we want to test the hypothesis that the distributions of $X$ and $Y$ are identical, that is

$$H_0 : \mu_X = \mu_Y, \quad \sigma_X = \sigma_Y = \sigma$$

against the alternative hypothesis

$$H_1 : \mu_X \neq \mu_Y.$$

A generalisation of this test for more than two samples is presented in Chap. 19.

**Two Sample t-Test Statistic** In Sect. 8.7 we stated that

$$\bar{X} \sim N(\mu_X, \sigma_X^2/n) \quad \text{and} \quad \bar{Y} \sim N(\mu_Y, \sigma_Y^2/m)$$

and therefore, using the theory in Sect. 13.2, it can be shown that

$$\bar{X} - \bar{Y} \sim N\left(\mu_X - \mu_Y, \frac{\sigma_X^2}{n} + \frac{\sigma_Y^2}{m}\right).$$

Hence, under $H_0$,

$$\bar{X} - \bar{Y} \sim N\left(0, \sigma^2\left[\frac{1}{n} + \frac{1}{m}\right]\right) \quad \Rightarrow \quad \sqrt{\frac{nm}{n+m}} \frac{(\bar{X} - \bar{Y})}{\sigma} \sim N(0, 1).$$

The involvement of the (unknown) $\sigma$ above means that this is not a pivotal test statistic. It can be proved that if $\sigma$ is replaced by its unbiased estimator $S$, which here is the *two-sample estimator of the common standard deviation*, given by

$$S^2 = \frac{\sum_{i=1}^{n}(X_i - \bar{X})^2 + \sum_{i=1}^{m}(Y_i - \bar{Y})^2}{n+m-2},$$

then

$$\sqrt{\frac{nm}{n+m}} \frac{(\bar{X} - \bar{Y})}{S} \sim t_{n+m-2}.$$

Hence

$$t = \sqrt{\frac{nm}{n+m}} \frac{(\bar{x} - \bar{y})}{s}$$

is a test statistic for this test. The rejection region is $|t| > h$ where $-h$ is the $\alpha/2$ (usually 0.025) percentile of $t_{n+m-2}$.

**Confidence Interval for $\mu_X - \mu_Y$**
From the hypothesis testing, a $100(1 - \alpha)\%$ confidence interval is given by

$$\bar{x} - \bar{y} \pm h\sqrt{\frac{n+m}{nm}} s,$$

where $-h$ is the $\alpha/2$ (usually 0.025) percentile of $t_{n+m-2}$.

---

**Example 12.9 (Fast Food Waiting Time as a Two Sample t-Test)**

In this example, we would like to know if there are significant differences between the AM and PM waiting times. We test this by performing the test of hypotheses

$$H_0 : \mu_x = \mu_y$$

against

$$H_1 : \mu_x \neq \mu_y,$$

where $\mu_x$ is the mean of the morning waiting times and $\mu_y$ is the mean for the afternoon waiting times.
Here the 10 morning waiting times $(x)$ are: 38, 100, 64, 43, 63, 59, 107, 52, 86, 77 and the 10 afternoon waiting times $(y)$ are: 45, 62, 52, 72, 81, 88, 64, 75, 59, 70. Here $n = m = 10$, $\bar{x} = 68.9$, $\bar{y} = 66.8$, $s_x^2 = 538.22$ and $s_y^2 = 171.29$. From this we calculate,

$$s^2 = \frac{(n-1)s_x^2 + (m-1)s_y^2}{n+m-2} = 354.8,$$

$$t_{\text{obs}} = \sqrt{\frac{nm}{n+m}} \frac{(\bar{x} - \bar{y})}{s} = 0.25.$$

This is not significant as the critical value $h = $ qt$(0.975, 18) = 2.10$ is larger in absolute value than 0.25. This can be achieved by calling the R function t.test with two arguments containing the sample values:

```
t.test(ffood$AM, ffood$PM)
```

The above command automatically calculates the test statistic as 0.249 and a p-value of 0.8067. It also obtains the 95% CI given by $(-15.94, 20.14)$ for the mean difference between the morning and afternoon waiting times. ◄

## 12.6   Paired t-Test

Sometimes the assumption that the $X$ and $Y$ variables are independent of each other is unlikely to be valid, due to the design of the study. The most common example of this is where $n = m$ and data are *paired*. For example, a measurement has been made on patients before treatment $(X)$ and then again on the same set of patients after treatment $(Y)$. Recall the weight gain example is exactly of this type. In such examples, we proceed by computing data on the differences

$$z_i = x_i - y_i, \quad i = 1, \ldots, n$$

and modelling these differences as observations of i.i.d. $N(\mu_z, \sigma_z^2)$ variables $Z_1, \ldots, Z_n$. Then, a test of the hypothesis $\mu_x = \mu_y$ is achieved by testing $H_0 : \mu_z = 0$, which is just a standard (one sample) t-test, as described previously.

### Example 12.10 (Paired t-Test)

Water-quality researchers wish to measure the biomass to chlorophyll ratio for phytoplankton (in milligrams per litre of water). There are two possible tests, the second is less expensive than the first. To see whether the two tests give the same results, ten water samples were taken and each was measured both ways. The results are as follows:

$$\begin{array}{l|l} Test1(x) & 45.9 \ 57.6 \ 54.9 \ 38.7 \ 35.7 \ 39.2 \ 45.9 \ 43.2 \ 45.4 \ 54.8 \\ Test2(y) & 48.2 \ 64.2 \ 56.8 \ 47.2 \ 43.7 \ 45.7 \ 53.0 \ 52.0 \ 45.1 \ 57.5 \end{array}$$

To test the null-hypothesis

$$H_0 : \mu_z = 0 \ \text{against} \ H_1 : \mu_z \neq 0$$

we use the test statistic $t = \sqrt{n}\frac{\bar{z}}{s_z}$, where $s_z^2 = \frac{1}{n-1}\sum_{i=1}^{n}(z_i - \bar{z})^2$.

**Confidence interval for $\mu_z$**
From the hypothesis testing, a $100(1 - \alpha)\%$ confidence interval is given by $\bar{z} \pm h\frac{s_z}{\sqrt{n}}$, where $h$ is the critical value of the $t$ distribution with $n - 1$ degrees of freedom. In R we perform the test as follows:

```
x <- c(45.9, 57.6, 54.9, 38.7, 35.7, 39.2, 45.9, 43.2, 45.4,
   54.8)
y <- c(48.2, 64.2, 56.8, 47.2, 43.7, 45.7, 53.0, 52.0, 45.1,
   57.5)
t.test(x, y, paired=T)
```

These commands yield the test statistic $t_{\text{obs}} = -5.0778$ with a df of 9 and a p-value = 0.0006649. Thus we reject the null hypothesis. The associated 95% CI is $(-7.53, -2.89)$, printed by R.

**Interpretation** The values of the second test are significantly higher than the ones of the first test, and so the second less expensive test cannot be considered as a replacement for the first.

We have introduced the two-sample t-test for testing whether the distributions of two independent samples of data are identical. We have also learned about the paired t-test, which we use in circumstances where the assumption of independence is not valid.

## 12.7   Design of Experiments

In a medical experiment, suppose the aim is to prove that the new drug is better than the existing drug for treating a mental health disorder. How should we select patients to allocate the drugs, often called treatments in statistical jargon? Obviously it would be wrong to administer the new drug to the male individuals and the old to the females as any difference between the effects of the new and existing drugs will be completely mixed-up with the difference between the effect of the sexes. Hence, effective sampling techniques and optimal design of the data collection experiments are necessary to make valid statistical inferences. This section will discuss several methods for statistical data collection.

An *experiment* is a means of getting an answer to the question that the experimenter has in mind. This may be to decide which of the several pain-relieving tablets that are available over the counter is the most effective or equally effective. An experiment may be conducted to study and compare the British and Chinese methods of teaching mathematics in schools. In planning an experiment, we clearly state our objectives and formulate the hypotheses we want to test. We now give some key definitions.

*Treatment* The different procedures under comparison in an experiment are the different treatments. For example, in a chemical engineering experiment different factors such as Temperature (T), Concentration (C) and Catalyst (K) may affect the yield value from the experiment.

*Experimental Unit*  An experimental unit is the material to which the treatment is applied and on which the variable under study is measured. In a human experiment in which the treatment affects the individual, the individual will be the experimental unit.

*Design of Experiments*  is a systematic and rigorous approach to problem-solving in many disciplines such as engineering, medicine etc. that applies principles and techniques at the data collection stage, so as to ensure valid inferences for the hypotheses of interest.

## 12.7.1  Three Principles of Experimental Design

1. *Randomisation.* This is necessary to draw valid conclusions and minimise bias. In an experiment to compare two pain-relief tablets we should allocate the tablets randomly among participants—not one tablet to the boys and the other to the girls.
2. *Replication.* A treatment is repeated a number of times in order to obtain a more reliable estimate than a single observation. In an experiment to compare two diets for children, we can plan the experiment so that no particular diet is favoured in the experiment, i.e. each diet is applied approximately equally among all types of children (boys, girls, their ethnicity etc.).
   The most effective way to increase the precision of an experiment is to increase the number of replications. Remember, $\text{Var}(\bar{X}) = \sigma^2/n$, which says that the standard deviation decreases proportional to the square root of the number of replications. However, replication beyond a limit may be impractical due to cost and other considerations.
3. *Local control.* In the simplest case of local control, the experimental units are divided into homogeneous groups or blocks. The variation among these blocks is eliminated from the error and thereby efficiency is increased. These considerations lead to the topic of construction of block designs, where random allocation of treatments to the experimental units may be restricted in different ways in order to control experimental error. Another means of controlling error is through the use of confounded designs where the number of treatment combinations is very large, e.g. in factorial experiments.

**Factorial Experiment**  A thorough discussion of construction of block designs and factorial experiments is beyond the scope of this textbook. In the remainder of this section, we simply discuss an example of a factorial experiment and how to estimate different effects.

### Example 12.11 (A Three Factor Experiment)

Chemical engineers wanted to investigate the yield, the value of the outcome of the experiment which is often called the *response*, from a chemical process. They identified three factors that might affect the yield: Temperature (T), Concentration (C) and catalyst (K), with levels reported in Table 12.4.

To investigate how factors jointly influence the response, they should be investigated in an experiment in which they are all varied. Even when there are no factors that interact, a factorial experiment gives greater accuracy. Hence they are widely used in science, agriculture and industry. Here we will consider factorial experiments in which each factor is used at only two levels. This is a very common form of experiment, especially when many factors are to be investigated. We will *code* the levels of each factor as 0 (low) and 1 (high). Each of the 8 combinations of the factor levels were used in the experiment. Thus the treatments in standard order were:

$$000, 001, 010, 011, 100, 101, 110, 111.$$

Each treatment was used in the manufacture of one batch of the chemical and the yield (amount in grams of chemical produced) was recorded. Before the experiment was run, a decision had to be made on the order in which the treatments would be run. To avoid any unknown feature that changes with time being confounded with the effects of interest, a random ordering was used; see Table 12.5. The response data are also shown in the table. ◀

**Table 12.4** Chemical yield data

| Temperature ($T$) | 160C | 180C |
|---|---|---|
| Concentration ($C$) | 20% | 40% |
| Catalyst ($K$) | A | B |

**Table 12.5** Chemical yield data and design

| Standard order | Randomised order | $T$ | $C$ | $K$ | Yield | Code |
|---|---|---|---|---|---|---|
| 1 | 6 | 0 | 0 | 0 | 60 | 000 |
| 2 | 2 | 0 | 0 | 1 | 52 | 001 |
| 3 | 5 | 0 | 1 | 0 | 54 | 010 |
| 4 | 8 | 0 | 1 | 1 | 45 | 011 |
| 5 | 7 | 1 | 0 | 0 | 72 | 100 |
| 6 | 4 | 1 | 0 | 1 | 83 | 101 |
| 7 | 3 | 1 | 1 | 0 | 68 | 110 |
| 8 | 1 | 1 | 1 | 1 | 80 | 111 |

## 12.7.2 Questions of Interest

1. How much is the response changed when the level of one factor is changed from high to low?
2. Does this change in response depend on the level of another factor?

For simplicity, we first consider the case of two factors only and call them factors $A$ and $B$, each having two levels, 'low' (0) and 'high' (1). The four treatments in the experiment are then 00, 01, 10, 11, and suppose that we have just one response measured for each treatment combination. We denote the four response values by $yield_{00}$, $yield_{01}$, $yield_{10}$ and $yield_{11}$.

## 12.7.3 Main Effects

For this particular experiment, we can answer the first question by measuring the difference between the average yields at the two levels of $A$:
The average yield at the high level of A is $\frac{1}{2}(yield_{11} + yield_{10})$.
The average yield at the low level of A is $\frac{1}{2}(yield_{01} + yield_{00})$.
These are represented by the open stars in the Fig. 12.7. The main effect of $A$ is defined as the difference between these two averages, that is

$$A = \frac{1}{2}(yield_{11} + yield_{10}) - \frac{1}{2}(yield_{01} + yield_{00})$$

$$= \frac{1}{2}(yield_{11} + yield_{10} - yield_{01} - yield_{00}),$$

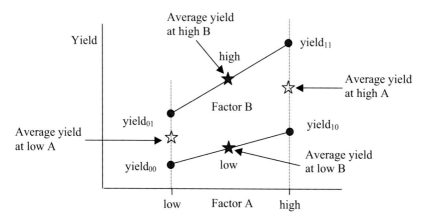

**Fig. 12.7**  Figure showing factorial effects

which is represented by the difference between the two open stars in Fig. 12.7. Notice that $A$ is used to denote the main effect of a factor as well as its name. This is a common practice. This quantity measures how much the response changes when factor $A$ is changed from its low to its high level, averaged over the levels of factor $B$.

Similarly, the main effect of $B$ is given by

$$B = \frac{1}{2}(yield_{11} + yield_{01}) - \frac{1}{2}(yield_{10} + yield_{00})$$

$$= \frac{1}{2}(yield_{11} - yield_{10} + yield_{01} - yield_{00}),$$

which is represented by the difference between the two black stars in Fig. 12.7.

We now consider question 2, that is, whether this change in response is consistent across the two levels of factor $A$.

### 12.7.4  Interaction Between Factors $A$ and $B$

**Case 1: No Interaction**

When the effect of factor $B$ at a given level of $A$ (difference between the two black stars) is the same, regardless of the level of $A$, the two factors $A$ and $B$ are said not to interact with each other. The response lines are parallel in Fig. 12.8.

**Case 1: Non-zero Interaction**

When the effect of factor $B$ (difference between the two black stars) is different from the corresponding differences for different levels of $A$, the two factors $A$ and $B$ are said to interact with each other. The response lines are not parallel in Fig. 12.9.

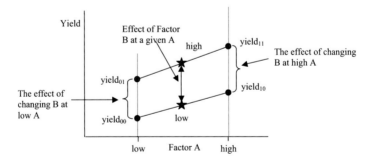

**Fig. 12.8**  Plot showing no interaction effects

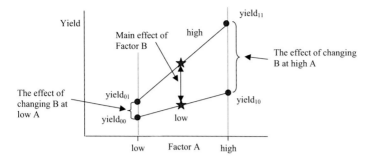

**Fig. 12.9**  Plot showing interaction effects

## 12.7.5  Computation of Interaction Effect

We define the interaction between factors $A$ and $B$ as one half of the differences
between

- the effect of changing $B$ at the high level of $A$, $(yield_{11} - yield_{10})$, and
- the effect of changing $B$ at the low level of $A$, $(yield_{01} - yield_{00})$, that is

$$AB = \frac{1}{2}(yield_{11} - yield_{10} - yield_{01} + yield_{00}).$$

If the lines are parallel then this interaction, $AB$, will be small. Also, note that
if we interchange the roles of $A$ and $B$ in the above expression we obtain the same
formula. Hence the interaction effect of A and B is the same as the interaction effect
of B and A. When there is a large interaction between two factors, the two main
effects cannot be interpreted separately and they must be studied jointly. The main
effects and interaction effects are known collectively as the *factorial effects*.

## 12.7.6  Summary

In this section we have discussed techniques of random sampling and the main
ideas of design of experiments. A three factor experiment example has demonstrated
estimation of the main effects of the factors and their interactions. Further analysis
using R can be performed by following the `yield` example online. Estimation and
testing of interaction effects will be further considered in Chap. 19.

## 12.8   Exercises

### 12.1 (Hypothesis Testing)

1. The random variable $X$ has a normal distribution with standard deviation 3.5 but
   unknown mean $\mu$. The hypothesis $H_0 : \mu = 3$ is to be tested against $H_1 : \mu = 4$
   by taking a random sample of size 50 from the distribution of $X$ and rejecting
   $H_0$ if the sample mean exceeds 3.4. Calculate the Type I and Type II error
   probabilities of this procedure.
2. The daily demand for a product has a Poisson distribution with mean $\lambda$, the
   demands on different days being statistically independent. It is desired to test the
   hypotheses $H_0 : \lambda = 0.7$, $H_1 : \lambda = 0.3$. The null hypothesis is to be accepted
   if in 20 days the number of days with no demand is less than 15. Calculate the
   Type I and Type II error probabilities.
3. A wholesale greengrocer decides to buy a field of winter cabbages if he can
   convince himself that their mean weight exceeds 1.2 kg. Accordingly he cuts
   12 cabbages at random and weighs them with the following results (in kg):

   $$1.26, 1.19, 1.17, 1.24, 1.23, 1.25, 1.20, 1.18, 1.23, 1.21, 1.19, 1.17.$$

   Should the greengrocer buy the cabbages? Use a 10% level of significance.
   **Hint: Let $\mu$ denote the true mean of the cabbages. Test the hypotheses $H_0$ :
   $\mu = 1.2$; $H_1 : \mu > 1.2$.**
4. A market gardener, wishing to compare the effectiveness of two fertilisers, used
   one fertiliser throughout the growing season on half of his plants and used the
   second one on the other half. The yields of the surviving plants are shown below,
   measured in kilograms.

| Fertiliser A | 6.72, | 9.17, | 7.43, | 6.53, | 10.20, | 6.88, | 6.28, | 6.73, |
|---|---|---|---|---|---|---|---|---|
|  | 7.38, | 6.95, | 7.03, | 8.05, | 7.57, | 6.90, | 7.63. |  |
| Fertiliser B | 7.73, | 10.34, | 8.59, | 5.71, | 5.42, | 7.94, | 8.29, |  |
|  | 7.63, | 9.41, | 9.35, | 9.62, | 8.94. |  |  |  |

   Stating all assumptions made, test at a 10% significance level the hypothesis
   that the fertilisers are equally effective against the alternative that they are not.
5. Eight young English county cricket batsmen were awarded scholarships which
   enabled them to spend the winter in Australia playing club cricket. Their first-
   class batting averages in the preceding and following seasons were as follows.

| Batsman | 1 | 2 | 3 | 4 | 5 | 6 | 7 | 8 |
|---|---|---|---|---|---|---|---|---|
| Average before | 29.43 | 21.21 | 31.23 | 36.27 | 22.28 | 30.06 | 27.60 | 43.19 |
| Average after | 31.26 | 24.95 | 29.74 | 33.43 | 28.50 | 30.35 | 29.16 | 47.24 |

Is there a significant improvement in their batting averages between seasons? Could any change be attributed to the winter practice?

6. In an experiment to study the effect of a certain concentration of insulin on blood glucose levels in rats, each member of a random sample of 10 rats was treated with insulin. The blood glucose level of each rat was measured both before and after treatment. The results, in suitable units, were as follows.

| Rat | 1 | 2 | 3 | 4 | 5 | 6 | 7 | 8 | 9 | 10 |
|---|---|---|---|---|---|---|---|---|---|---|
| Level before | 2.30 | 2.01 | 1.92 | 1.89 | 2.15 | 1.93 | 2.32 | 1.98 | 2.21 | 1.78 |
| Level after | 1.98 | 1.85 | 2.10 | 1.78 | 1.93 | 1.93 | 1.85 | 1.67 | 1.72 | 1.90 |

Find a 95% confidence interval for the mean difference stating all the assumptions.

7. A sociologist wishes to estimate the proportion of wives in a city who are happy with their marriage. To overcome the difficulty that a wife, if asked directly, may say that her marriage is happy even when it is not, the following procedure is adopted. Each member of a random sample of 500 married women is asked to toss a fair coin but not to disclose the result. If the coin lands 'heads', the question to be answered is 'Does your family own a car?'. If it lands 'tails', the question to be answered is 'Is your marriage a happy one?'. In either case the response is to be either 'Yes' or 'No'. The respondent knows that the sociologist has no means of knowing which question has been answered. Suppose that of the 500 responses, 350 are 'Yes' and 150 are 'No'. Assuming that every response is truthful and given that 75% of families own a car, estimate the proportion of women who are happy with their marriage and obtain an approximate 90% confidence interval for this proportion. Based on this confidence interval should we reject the null hypothesis that 80% of women are happy with their marriage?

8. The time, $X$, (in minutes) that a customer has to wait for service in a restaurant has the probability density function

$$f(x) = \begin{cases} \frac{3\theta^3}{(x+\theta)^4} & \text{if } x \geq 0 \\ 0 & \text{otherwise,} \end{cases}$$

where $\theta$ is an unknown positive constant. Let $X_1, X_2, \ldots, X_n$ denote a random sample from this distribution.

(a) Show that

$$\hat{\theta} = \frac{2}{n} \sum_{i=1}^{n} X_i$$

is an unbiased estimator for $\theta$.

(b)  Show that $\text{Var}(X) = \frac{3\theta^2}{4}$.

(c)  Find the standard error of $\hat{\theta}$.

(d)  State the central limit theorem (CLT) for $\hat{\theta}$.

(e)  By using the CLT derive an approximate $100(1 - \alpha)\%$ confidence interval for $\theta$ for a given value of $0 < \alpha < 1$.

(f)  Suppose that the manager of the restaurant observes $\bar{x} = 11.25$ from 48 randomly chosen customers over a certain period. Calculate the approximate 95% confidence interval for $\theta$. Based on the calculated confidence interval, state if the decision is to reject the null hypotheses: (i) $H_0 : \theta = 15$ and (ii) $H_0 : \theta = 20$ each at 5% level of significance.

9.  Suppose that $x_1, \ldots, x_n$ are measurements of the electrical resistance of particular components and are modelled as observations of i.i.d. random variables $X_1, \ldots, X_n$ following the distribution of the random variable $X$. Suppose that $E(X) = \mu$ and $\text{Var}(X) = \sigma^2$, where $\sigma^2 < \infty$, but that the distribution of $X$ is not a normal distribution. Define $\bar{X} = \frac{1}{n} \sum_{i=1}^{n} X_i$ and $S^2 = \frac{1}{n-1} \sum_{i=1}^{n} (X_i - \bar{X})^2$.

(a)  Prove that $E(\bar{X}) = \mu$.

(b)  State the central limit theorem (CLT) for $\bar{X}$ assuming a known value of $\sigma^2$. Re-write the statement of the CLT by replacing $\sigma^2$ by its unbiased estimator, $S^2$.

   Suppose that $n = 100$, $\bar{x} = 9$, and $s = 3.5$. Assume that: qnorm(0.95) = 1.645 and qnorm(0.975) = 1.96.

(c)  Calculate a 95% confidence interval for $\mu$.

(d)  A manufacturer of the electrical components claims that the mean electrical resistance of the components produced in her factory is 10. Use the above data to test the hypothesis that $\mu = 10$. Do the data support the manufacturer's claim?

(e)  Suppose that an alternative model is proposed where $X$ has the distribution with $\mu = \theta$ and variance $\sigma^2 = \theta^2/2$.

   i.  Derive an expression for a 95% confidence interval for $\theta \ (= \mu)$ in terms of $n$ and $\bar{X}$ by using the first version of the CLT stated in part (b) of this question.

   ii.  Obtain the numerical values of the end points of the confidence interval.

   iii.  Based on the confidence interval obtained here, is it possible to support the manufacturer's claim of $\mu = 10$?

10.  Suppose that $x_1, \ldots, x_n$ are measurements of agricultural yields of a particular variety of crops using a fertiliser type A. These are modelled as observations of i.i.d. random variables $X_1, \ldots, X_n$ following the distribution of the random variable $X$. Suppose that $E(X) = \mu_x$ and $\text{Var}(X) = \sigma_x^2$, where $\sigma_x^2 < \infty$. Define

$$\bar{x} = \frac{1}{n} \sum_{i=1}^{n} x_i, \text{ and } s_x^2 = \frac{1}{n-1} \sum_{i=1}^{n} (x_i - \bar{x})^2.$$

Suppose that $y_1, \ldots, y_m$ are measurements of yields of the same variety of crops but using a different fertiliser type B. These are modelled as observations of i.i.d. random variables $Y_1, \ldots, Y_m$ following the distribution of the random variable $Y$. Suppose that $E(Y) = \mu_y$ and $\text{Var}(Y) = \sigma_y^2$, where $\sigma_y^2 < \infty$. Define

$$\bar{y} = \frac{1}{m} \sum_{j=1}^{m} y_j, \text{ and } s_y^2 = \frac{1}{m-1} \sum_{j=1}^{m} (y_j - \bar{y})^2.$$

Suppose that

$$n = 25, \ \bar{x} = 16.6, \ s_x^2 = 47.6$$
$$m = 15, \ \bar{y} = 9.5, \ s_y^2 = 16.6$$

Assume that: qt(0.95, df=24) = 1.71 and qt(0.975, df=24) = 2.06, qt(0.975, df=38)=2.02 and t(0.95, df=38)=1.69.

(a) Stating any assumptions you make regarding the distributions of $X$, calculate a 95% confidence interval for $\mu_x$ based on the $t$-distribution. Please use one place after the decimal point.

(b) Stating any assumptions regarding the distributions of $X$ and $Y$, calculate a 95% confidence interval for the difference $\mu_x - \mu_y$ based on the $t$-distribution. Please use one place after the decimal point.

(c) Using the confidence interval, test the hypothesis that $H_0 : \mu_x - \mu_y = 0$ against $H_0 : \mu_x - \mu_y \neq 0$ at 5% level of significance.

(d) Write down the expression for the test statistic that can be used to perform a testing of the above hypotheses and state its null distribution. Calculate the value of the test statistic and state its degrees of freedom. Please use two digits after the decimal point. Write down and sketch the rejection region using a diagram.

Write down the R command that can be used to calculate the P-value of the test. Obtain this p-value upto 4 places after the decimal point.

# Part IV

# Advanced Distribution Theory and Probability

This is Part II of this book where we assume a higher level of mathematical background in multivariate calculus and some familiarity with basic linear algebra. This part develops statistical distribution theory so that we are able to learn more mathematically advanced and sound statistical methods. One of the main objectives of this part is to prove the many identities and fact that we have avoided so far. For example, in Sect. 8.14 we were unable to prove that the sum of two independent normally distributed random variables is also normally distributed. There are many more, e.g. $\Gamma(1/2) = \sqrt{\pi}$ and the fact that $T = \frac{\sqrt{n}(\bar{X}-\mu)}{S}$, follows the the $t$-distribution with $n - 1$ degrees of freedom in Sect. 11.4.

# Generating Functions

# 13

**Abstract**

Chapter 13 starts the Part III of this book on advanced distribution theory and probability. It discusses the moment generating function, cumulant generating function and probability generating function for discrete random variables. The uniqueness theorem for the moment generating function is also stated here to facilitate many proofs in statistical distribution theory.

This chapter introduces three different types of generating functions: moment generating function (mgf), probability generating function (pgf) and cumulant generating function (cgf). These are expectations of certain functions of random variables. Using the mgfs we will not only be able to evaluate the moments, defined as the expectations $E(X^k)$ for integer values of $k$, easily avoiding the required complicated sums or integrals, but also able to determine the probability distributions of sample quantities of interests. This is a great result in theoretical statistics. The greatness comes from the following fact. The mgfs are mere expectations of a function of the random variables, but by examining these expectations we are able to determine entire probability distribution. For example, knowing $E(X) = 5$ and $\text{Var}(X) = 1$ we cannot conclude that $X \sim N(5, 1)$. However, mgf defined by $E\left(e^{tX}\right)$ for $t$ being in an interval containing zero we are able to determine the probability distribution of the random variable $X$. This is especially useful when we arrive at a statistic, e.g. the sample sum, whose probability distribution, usually called the sampling distribution, is unknown. The pgf performs similar tasks in generating the probability functions, but only for the discrete random variables. The cgf helps us further in studying the probability distributions more deeply.

© The Author(s), under exclusive license to Springer Nature Switzerland AG 2024
S. K. Sahu, *Introduction to Probability, Statistics & R*,
https://doi.org/10.1007/978-3-031-37865-2_13

## 13.1   Moments, Skewnees and Kurtosis

Moments are generalisations of means and variances. These are needed to study many properties of the probability distributions.

**Definition 13.1**  For each integer $n$, the $n$th *moment* of $X$ is (sometimes called $n$th-order moment)

$$\mu'_n = E(X^n).$$

Note that prime notation here is not used to denote differentiation. The $n$th *central moment* of $X$ is

$$\mu_n = E\left[(X - \mu)^n\right],$$

where $\mu = E(X)$.                                                                                  $\diamond$

---

**Example 13.1**

Note that

$$\mu'_1 = E(X) = \mu, \quad \text{and} \quad \mu_2 = E\left[(X - \mu)^2\right] = \text{Var}(X).$$

Hence the first moment is simply the mean and the second central moment is the variance. The first central moment is $\mu_1 = E(X - \mu) = 0$. ◄

Moments are useful for investigating the shape of a probability distribution. For this we use the moments to define the following measures of skewness and kurtosis of a probability distribution:

$$\beta_1 = \frac{\mu_3}{\mu_2^{\frac{3}{2}}}, \quad \beta_2 = \frac{\mu_4}{\mu_2^2}. \tag{13.1}$$

The first measure $\beta_1$ gives a measure of skewness and the second measure $\beta_2$ gives a measure of kurtosis or peakedness. For symmetric distributions $\beta_1 = 0$. Positive values of $\beta_1$ imply positively skewed distributions, while negative values correspond to negatively skewed distributions. The measure of kurtosis, $\beta_2$, is shown to be three for the normal distribution in Example 13.6 below. The level of peakedness of a distribution can be judged relative to this value of three for the normal distribution. See the gamma densities in Fig. 7.7 which are always positively skewed with more peakedness than the normal densities in Fig. 7.3 (proved below in Sect. 13.3).

## 13.2 Moment Generating Function

It is quite cumbersome, and usually difficult, to calculate the moments one by one for the purposes of exploring probability distributions. The moment generating functions, defined below, alleviate this task quite considerably.

**Definition 13.2** The function

$$M(t) = E\left(e^{tX}\right)$$

is called the *moment generating function* (mgf) of the random variable $X$ provided that it exists for all $t$ in some neighbourhood of 0. Note that $M(0) = 1$.  ◇

---

**Example 13.2 (Exponential)**

Suppose that the random variable $X$ has the exponential distribution with parameter $\beta$. Hence the pdf of $X$ is $f(x) = \beta e^{-\beta x}$ for $0 < x < \infty$. The mgf is:

$$\begin{aligned}
M(t) = E\left(e^{tX}\right) &= \int_0^\infty e^{tx} \beta e^{-\beta x} dx \\
&= \beta \int_0^\infty e^{-(\beta-t)x} dx \\
&= \frac{\beta}{\beta-t} \quad \text{if } t < \beta.
\end{aligned}$$

We needed the condition $t < \beta$ to ensure existence of the integral. The integral exists if the co-efficient of $-x$ in the exponent of $e$ is greater than 0, i.e., $\beta - t > 0$. ◀

---

**Example 13.3 (Gamma)**

The mgf of the gamma distribution with parameters $\alpha$ and $\beta$ is:

$$\begin{aligned}
M(t) = E(e^{tX}) &= \int_0^\infty e^{tx} \frac{\beta^\alpha}{\Gamma(\alpha)} x^{\alpha-1} e^{-\beta x} dx \\
&= \frac{\beta^\alpha}{\Gamma(\alpha)} \int_0^\infty x^{\alpha-1} e^{-x(\beta-t)} dx \\
&= \frac{\beta^\alpha}{\Gamma(\alpha)} \frac{\Gamma(\alpha)}{(\beta-t)^\alpha}, \quad \text{for } t < \beta, \text{ [see identity (7.9)]} \\
&= \left(\frac{\beta}{\beta-t}\right)^\alpha.
\end{aligned}$$

The mgf of the $\chi^2$ distribution, which is gamma with parameters $\alpha = \frac{n}{2}$ and $\beta = \frac{1}{2}$, is given by

$$M(t) = \left(\frac{1}{1-2t}\right)^{n/2}, t < \frac{1}{2}.$$

◄

**Example 13.4 (Normal)**

Suppose that $X$ is the standard normal random variable. The mgf of $X$ is:

$$
\begin{aligned}
M(t) = E(e^{tX}) &= \int_{-\infty}^{\infty} e^{tx} \frac{1}{\sqrt{2\pi}} e^{-\frac{1}{2}x^2} dx \\
&= \frac{1}{\sqrt{2\pi}} \int_{-\infty}^{\infty} e^{-\frac{1}{2}(x^2 - 2tx)} dx \\
&= \frac{1}{\sqrt{2\pi}} \int_{-\infty}^{\infty} e^{-\frac{1}{2}(x^2 - 2tx + t^2 - t^2)} dx \quad \text{[Add and subtract } t^2 \text{]} \\
&= \frac{1}{\sqrt{2\pi}} \int_{-\infty}^{\infty} e^{-\frac{1}{2}(x-t)^2 + \frac{1}{2}t^2} dx \\
&= e^{\frac{1}{2}t^2} \int_{-\infty}^{\infty} \frac{1}{\sqrt{2\pi}} e^{-\frac{1}{2}(x-t)^2} dx \\
&= e^{\frac{1}{2}t^2}. \qquad \text{[Since the last integral is 1.]}
\end{aligned}
$$

◄

The following theorem obtains a general result to find the moments by differentiating the mgf of a random variable.

**Theorem 13.1** *If X has mgf M(t), then*

$$\mu_n' = E(X^n) = \frac{d^n}{dt^n} M(t) \Big|_{t=0},$$

*for any positive integer n.*

***Proof*** Assume that we can differentiate under the integral sign.

$$
\begin{aligned}
\frac{d}{dt} M(t) &= \frac{d}{dt} \int_{-\infty}^{\infty} e^{tx} f(x) dx \\
&= \int_{-\infty}^{\infty} \left(\frac{d}{dt} e^{tx}\right) f(x) dx \\
&= \int_{-\infty}^{\infty} (x e^{tx}) f(x) dx \\
&= E\left(X e^{tX}\right)
\end{aligned}
$$

Hence

$$\frac{d}{dt}M(t)\Big|_{t=0} = E(Xe^{tX})\Big|_{t=0} = E(X).$$

By successive differentiation we get the general result.                                    □

### Example 13.5 (Exponential)

Suppose that $X$ follows the exponential distribution, which has the mgf $M(t) = \frac{\beta}{\beta-t}$ if $t < \beta$. Therefore,

$$M'(t) = \frac{\beta}{(\beta-t)^2} \text{ and } M''(t) = \frac{2\beta}{(\beta-t)^3}.$$

Hence $E(X) = M'(0) = \frac{1}{\beta}$ and $E(X^2) = M''(0) = \frac{2}{\beta^2}$. Now $\text{Var}(X) = E(X^2) - [E(X)]^2 = \frac{2}{\beta^2}$. Thus the mgf provides an easy method to calculate the mean and variance avoiding the required integrations as we have done in Sect. 7.1.4 for the exponential distribution. ◄

### Example 13.6 (Normal)

Suppose that $X$ is the standard normal random variable. The mgf of $X$ is $M(t) = e^{\frac{1}{2}t^2}$. Now,

$$e^{\frac{1}{2}t^2} = 1 + \left(\frac{t^2}{2}\right) + \frac{1}{2!}\left(\frac{t^2}{2}\right)^2 + \frac{1}{3!}\left(\frac{t^2}{2}\right)^3 + \cdots \tag{13.2}$$

Hence

$$E(X) = \frac{d}{dt}e^{\frac{1}{2}t^2}\Big|_{t=0} = 0, \quad E(X^2) = \frac{d^2}{dt^2}e^{\frac{1}{2}t^2}\Big|_{t=0} = 1,$$

$$E(X^3) = \frac{d^3}{dt^3}e^{\frac{1}{2}t^2}\Big|_{t=0} = 0, \quad E(X^4) = \frac{d^4}{dt^4}e^{\frac{1}{2}t^2}\Big|_{t=0} = 3.$$

Hence, we have

$$\mu = E(X) = 0, \mu_2 = E[X - E(X)]^2 = E(X^2) = 1$$

$$\mu_3 E[X - E(X)]^3 = E(X^3) = 0, \mu_4 = E[X - E(X)]^4 = E(X^4) = 3.$$

Thus, the skewness parameter $\beta_1 = 0$ and the kurtosis parameter,

$$\beta_2 = \frac{\mu_4}{\mu_2^2} = 3.$$

In general, we can see that the coefficients of $t$, $t^3$, $t^5$ are zero in the mgf (13.2). Hence all odd-ordered moments are zero for the standard normal distribution. ◄

We now state and prove the following theorem which allows us to find the mgf of a linear transformation of random variable.

**Theorem 13.2** *Let a random variable $X$ have the mgf $M_X(t)$; let $Y = aX + b$ where $a$ and $b$ are given constants. Let $M_Y(t)$ denote the mgf of $Y$. Then for any value of $t$ such that $M_X(at)$ exists,*

$$M_Y(t) = e^{bt} M_X(at).$$

*Proof*

$$M_Y(t) = E(e^{tY}) = E[e^{t(aX+b)}] = e^{bt} E(e^{atX}) = e^{bt} M_X(at). \qquad \Box$$

**Example 13.7 (Normal)**

Return to the above normal example. Let $Y = \sigma X + \mu$. The mgf of $Y$ is:

$$M_Y(t) = e^{\mu t + \sigma^2 \frac{t^2}{2}}.$$

◄

## 13.2.1  Uniqueness of the Moment Generating Functions

We now state the following theorem on the uniqueness of the moment generating functions without proof.

**Theorem 13.3** *If the mgf's of two random variables $X$ and $Y$ are identical for all values of $t$ in an interval around the point 0, then the probability distributions of $X$ and $Y$ must be identical.*

In brief the above theorem is stated as:

If two mgfs are identical in an interval containing 0 then the two distributions are identical.

This theorem allows us to identify a probability distribution just by examining its mgf which is merely an expectation. This is powerful because, in general an expected value, e.g. $\mu = E(X)$, does not uniquely identify a distribution. For example, $E(X) = 5$ cannot be used to claim that $X$ has the normal distribution. However, if a random variable $X$ has the mgf $e^{\mu t + \sigma^2 \frac{t^2}{2}}$ then the above uniqueness theorem of the moment generating functions guarantees that $X$ follows the normal distribution with mean $\mu$ and variance $\sigma^2$.

Thus, to establish a certain random variable, which can be a transformed one, has a certain particular distribution we may just find its mgf and see which standard distribution also possesses the same mgf. The matched distribution is then claimed to be the required distribution. An example follows.

### Example 13.8 ($\chi^2$ Distribution with 1 Degree of Freedom)

Let $X$ have the standard normal distribution. By obtaining the mgf, show that $Y = X^2$ has the $\chi^2$ distribution with 1 degree of freedom defined in Sect. 7.3.1. Here we have,

$$
\begin{aligned}
M_Y(t) &= E\left(e^{tY}\right) \\
&= E\left(e^{tX^2}\right) \\
&= \int_{-\infty}^{\infty} e^{tx^2} \frac{1}{\sqrt{2\pi}} e^{-\frac{1}{2}x^2} dx \\
&= \frac{1}{\sqrt{2\pi}} \int_{-\infty}^{\infty} e^{-\frac{1}{2}x^2(1-2t)} dx \quad [\text{now let } \sigma^2 = \frac{1}{1-2t}] \\
&= \sigma \int_{-\infty}^{\infty} \frac{1}{\sqrt{2\pi\sigma^2}} e^{-\frac{1}{2}\frac{x^2}{\sigma^2}} dx \\
&= \sigma \quad [\text{Since the last integral is 1 for } t < \tfrac{1}{2}] \\
&= \left(\frac{1}{1-2t}\right)^{\frac{1}{2}}.
\end{aligned}
$$

This $M_Y(t)$ matches the mgf of the $\chi^2$ distribution with 1 degree of freedom. Hence, by using the uniqueness theorem of the mgf we conclude that $Y$ has the $\chi^2$ distribution with 1 degree of freedom. We shall prove this result again by using the transformation technique in Example 14.7. This shows that:

Square of a standard normal random variable follows the $\chi^2$ distribution with 1 df.

◄

## 13.2.2  Using mgf to Prove Distribution of Sample Sum

Recall that in Sect. 8.7 we were not able to prove that the sum of two normally distributed random variables will also follow the normal distribution. However, we did prove the results for the sums of independent binomial and Poisson distributions using complicated techniques. The uniqueness theorem of the mgf allows us to prove those results in a straightforward manner.

**Theorem 13.4** *Suppose that $X$ and $Y$ are independent with $X$ having mgf $M_X(t)$ and $Y$ having the mgf $M_Y(t)$ then the mgf of $Z = X + Y$ is*

$$M_Z(t) = M_X(t) \times M_Y(t).$$

*Proof*

$$M_Z(t) = E\left(e^{tZ}\right) = E\left(e^{t(X+Y)}\right) = E\left(e^{tX}e^{tY}\right) = M_X(t) \times M_Y(t).$$

$\square$

The theorem lets us find the distribution of sum of two random variables.

---

**Example 13.9 (Normal)**

Let $X \sim N(\mu, \sigma^2)$ independently of $Y \sim N(\theta, \gamma^2)$. Then

$$M_X(t) = e^{\mu t + \sigma^2 \frac{t^2}{2}}, \text{ and } M_Y(t) = e^{\theta t + \gamma^2 \frac{t^2}{2}}.$$

Therefore $Z$ has the mgf

$$M_Z(t) = e^{(\mu+\theta)t + (\sigma^2+\gamma^2)\frac{t^2}{2}}.$$

We recognise that $M_Z(t)$ is the mgf of the normal distribution with mean $\mu + \theta$ and variance $\sigma^2 + \gamma^2$. Now we use the uniqueness theorem of the mgf, Theorem 13.3, to conclude that $Z = X + Y$ has the normal distribution with mean $\mu + \theta$ and variance $\sigma^2 + \gamma^2$. ◄

---

**Example 13.10**

Let $X \sim \text{gamma}(m, \beta)$ and $Y \sim \text{gamma}(n, \beta)$ independently. Show that $Z = X + Y$ is gamma$(m + n, \beta)$.

**Proof** We have

$$M_X(t) = \left( \frac{\beta}{\beta - t} \right)^m, \quad \text{and} \quad M_Y(t) = \left( \frac{\beta}{\beta - t} \right)^n, t < \beta.$$

Now

$$M_Z(t) = M_X(t) \, M_Y(t) = \left( \frac{\beta}{\beta - t} \right)^{m+n}, t < \beta.$$

Now $M_Z(t)$ is recognised to be the mgf of the gamma distribution with parameters $m+n$ and $\beta$. Hence the result follows by the uniqueness theorem. ☐

◄

---

**Example 13.11**

Let $X \sim \chi^2$ with $m$ degrees of freedom and $Y \sim \chi^2$ with $n$ degrees of freedom independently. Show that $Z = X + Y$ is $\chi^2$ with $m + n$ degrees of freedom.

**Proof** This is exactly as the proof in the gamma distribution case. We have

$$M_X(t) = \left( \frac{1}{1 - 2t} \right)^{\frac{m}{2}}, \quad \text{and} \quad M_Y(t) = \left( \frac{1}{1 - 2t} \right)^{\frac{n}{2}}, t < \tfrac{1}{2}.$$

Now

$$M_Z(t) = M_X(t) \, M_Y(t) = \left( \frac{1}{1 - 2t} \right)^{\frac{m+n}{2}}, t < \tfrac{1}{2}.$$

Now $M_Z(t)$ is recognised to be the mgf of the $\chi^2$ distribution with parameters $m + n$ degrees of freedom. Hence the result follows by the uniqueness theorem of the mgf. ☐

Thus we proved:

Sum of independent $\chi^2$s also follows the $\chi^2$ distribution.

◄

This technique can be extended to multiple independent random variables. For example suppose that $X_1, \ldots, X_n$ is a random sample from the exponential distribution with parameter $\beta$. Hence $M_{X_i}(t) = \frac{\beta}{\beta - t}$ for $t < \beta$. Now let $Y$ be the sample sum, i.e.

$$Y = X_1 + \cdots + X_n.$$

Then

$$M_Y(t) = E\left(e^{tY}\right) = \prod_{i=1}^{n} M_{X_i}(t) = \left(\frac{\beta}{\beta - t}\right)^n$$

for $t < \beta$. But $M_Y(t)$ is recognised as the mgf of the gamma distribution with parameters $n$ and $\beta$. Hence we have the result that the sum of i.i.d. exponential random variables follows a gamma distribution. This result gives a way to construct a gamma distribution with an integer value for the shape parameter from independently distributed exponential distributions. The example below clarifies this.

**Example 13.12 (Non-uniform Random Variate Generation)**

This example continues on from Example 14.6 above on transforming a set of uniformly distributed random variables to obtain another random variable. Suppose that $U_1, U_2, \ldots, U_n$ are independent random variables each with a Uniform distribution in the interval $(0, 1)$. Use them to construct random variables that have the following distributions. You do not need to use all the $U_i$'s in every case.

1. A binomial distribution with parameter $n$ and $p$ where $0 < p < 1$.
2. An exponential distribution with mean $\theta > 0$.
3. A gamma distribution with parameter $n$ and $\beta$.

These problems can be treated as exercises. But due to the importance of this topic, to provide a better understanding of the inter-relationships between the distributions we discuss the solutions below.

1. To obtain the binomial distribution we form $n$ Bernoulli random variables where the $i$th random variable $X_i$ takes the value one if $U_i < p$ and zero otherwise. The sum of these $X_i$'s follows the binomial distribution with parameters $n$ and $p$. In R we can generate a binomial random variable with parameters $n = 10$ and $p = 0.4$ (say) by issuing the command `rbinom(n=1, size=10, p=0.4)`. Here the argument $n$ is the number of samples we aim to obtain. This should not be confused with $n$, the total number of trials, for the binomial distribution which is the parameter `size` in the argument of `rbinom`.
2. To obtain an exponential random variable with mean $\beta$ we apply the probability integral transform mentioned in Example 14.6. The cdf of the exponential distribution is

$$F(x|\theta) = 1 - e^{-\frac{x}{\theta}}, \quad x > 0.$$

Now,

$$
\begin{aligned}
F(x|\theta) &= u \\
\implies \quad 1 - e^{-\frac{x}{\theta}} &= u \\
\implies \quad e^{-\frac{x}{\theta}} &= 1 - u \\
\implies \quad -\frac{x}{\theta} &= \log(1 - u) \\
\implies \quad x &= -\theta \log(1 - u).
\end{aligned}
$$

Because $F(X\theta)$ follows the distribution in $(0, 1)$m the above calculation shows that $X = -\theta \log(U)$ will follow the exponential distribution with mean $\theta$ where $U \sim U(0, 1)$.

In R we can generate an exponential random variable with mean $\theta = 5$ (say) by issuing the command `rexp`(n=1, rate=1/5). Note that the `rate` argument is the reciprocal of the mean.

3. Note that sum of $n$ i.i.d. exponential random variables with parameter $\beta$ follows the gamma distribution with parameters $n$ and $\beta$. Hence

$$
Y = -\frac{1}{\beta} \sum_{i=1}^{n} \log(U_i)
$$

follows the $G(n, \beta)$ distribution.

◀

## 13.3 Cumulant Generating Function

The cumulant generating function (cgf) of a random variable is the natural logarithm of its moment generating function. Thus the cgf, denoted by $K_X(t)$, is defined as:

$$
K_X(t) = \log\left(M_X(t)\right).
$$

The cgf generates the cumulants defined by:

$$
\kappa_n = \left.\frac{d^n}{dt^n} K_X(t)\right|_{t=0},
$$

where $n$ is a positive integer.

For example, if $X \sim N(\mu, \sigma^2)$, $K_X(t) = \mu t + \sigma^2 \frac{t^2}{2}$. Obviously, for this example we can see that

$$
\mu = \left.\frac{d}{dt} K_X(t)\right|_{t=0}, \quad \text{and} \quad \sigma^2 = \left.\frac{d^2}{dt^2} K_X(t)\right|_{t=0}.
$$

These results are true in general. The first cumulant is the mean since:

$$\frac{d}{dt}K_X(t) = \frac{d}{dt}\log\left(M_X(t)\right) = \frac{1}{M_X(t)}\frac{d}{dt}M_X(t),$$

and $M_X(0) = 1$ and $\frac{d}{dt}M_X(t)\big|_{t=0} = \mu$. Similarly it can be shown that the second cumulant is the variance and the third cumulant is the third central moment. But the relationship between the fourth central moment $\mu_4$ and the fourth cumulant $\kappa_4$ is:

$$\mu_4 = \kappa_4 + 3\kappa_2^2.$$

Thus $\kappa_4 \neq \mu_4$. Higher order moments and cumulants have interesting relationships which we do not explore here.

---

**Example 13.13 (Gamma Distribution)**

We can obtain the skewness and kurtosis of the gamma distribution by using its cgf. Suppose $X \sim G(\alpha, \beta)$. Then from Example 13.3 we have for $t < \beta$,

$$\begin{aligned}
K_X(t) &= \log\left(M_X(t)\right) \\
&= \log\left[\left(\frac{\beta}{\beta-t}\right)^\alpha\right] \\
&= -\alpha \log\left(1 - \frac{t}{\beta}\right) \\
&= \alpha\left(\frac{t}{\beta} + \frac{1}{2}\frac{t^2}{\beta^2} + \frac{1}{3}\frac{t^3}{\beta^3} + \frac{1}{4}\frac{t^4}{\beta^4} + \cdots\right)
\end{aligned}$$

using the series expansion

$$\log(1 - x) = -x - \frac{x^2}{2} - \frac{x^3}{3} - \cdots$$

when $|x| < 1$. From the above expansion for $K_X(t)$, we see that

$$\begin{aligned}
\kappa_1 &= \tfrac{d}{dt}K_X(t)\big|_{t=0} = \mu = \tfrac{\alpha}{\beta} \\
\kappa_2 &= \tfrac{d^2}{dt^2}K_X(t)\big|_{t=0} = \mu_2 = \tfrac{\alpha}{\beta^2} \\
\kappa_3 &= \tfrac{d^3}{dt^3}K_X(t)\big|_{t=0} = \mu_3 = \tfrac{2\alpha}{\beta^3} \\
\kappa_4 &= \tfrac{d^4}{dt^4}K_X(t)\big|_{t=0} = \mu_4 - 3\kappa_2^2 = \tfrac{6\alpha}{\beta^4}.
\end{aligned}$$

Thus

$$\mu_4 = \kappa_4 + 3\kappa_2^2 = \frac{6\alpha}{\beta^4} + 3\frac{\alpha^2}{\beta^4} = 3\frac{\alpha}{\beta^4}(2 + \alpha).$$

Hence the skewness measure is given by:

$$\beta_1 = \frac{\mu_3}{\mu_2^{\frac{3}{2}}} = \frac{2\alpha}{\beta^3}\left(\frac{\beta^2}{\alpha}\right)^{\frac{3}{2}} = \frac{2}{\sqrt{\alpha}},$$

which is always greater than zero. Thus the gamma distribution is positively skewed always. Now we obtain the measure of kurtosis as:

$$\beta_2 = \frac{\mu_4}{\mu_2^2} = 3\frac{\alpha}{\beta^4}(2+\alpha)\left(\frac{\beta^2}{\alpha}\right)^2 = 3 + \frac{6}{\alpha}.$$

Thus the gamma distribution has more peakedness than the normal distribution.
◀

---

## 13.4 Probability Generating Function

The *probability generating function* (pgf) is a useful function for analysing discrete probability distributions. Parallel to the mgf we can differentiate the mgf to obtain the factorial moments, defined by

$$E(X(X-1)\cdots(X-k+1))$$

for any positive integer value of $k$. This method is useful for finding mean and variances of discrete distributions. For example, recall the methods we used to find means and variances of the binomial and Poisson distributions in Chap. 5.

**Definition 13.3** The *probability generating function* (pgf), denoted by $H(t)$ is:

$$H(t) = E\left(t^X\right)$$

if the expectation exists.                                                    ◇

---

### Example 13.14 (Binomial)

Suppose that $X$ follows the binomial distribution with parameters $(n, p)$. Let us first find its pgf. Its pmf is $f(x) = \binom{n}{x}p^x q^{n-x}, 0 \le x \le n$, where $q = 1 - p$.

$$H(t) = E\left(t^X\right) = \sum_{x=0}^{n} t^x \binom{n}{x}p^x q^{n-x}$$
$$= \sum_{x=0}^{n} \binom{n}{x}(pt)^x q^{n-x}$$
$$= (pt + q)^n. \quad \text{[because } \sum_{x=0}^{n}\binom{n}{x}a^x b^{n-x} = (a+b)^n.\text{]}$$

◀

Sometimes the pgf is called a *factorial moment generating function* because of the following theorem.

**Theorem 13.5** *For any random variable X we have the results:*

*(a)* $\frac{d^k}{dt^k} H(t)|_{t=1} = E[X(X-1)\cdots(X-k+1)]$

*(b)* *If X is a discrete random variable,* $\frac{1}{k!}\frac{d^k}{dt^k} H(t)|_{t=0} = P(X = k).$

*(c)* $M(\log(t)) = H(t)$ *or* $M(t) = H(e^t).$

*Proof*

**Part (a):**  Here we provide the proof when $X$ is discrete random variable. The same proof works for the continuous case by using integration instead of summation. Assume that we can differentiate under the summation sign.

$$\frac{d}{dt} H(t) = \frac{d}{dt}\sum_{\text{All } x} t^x f(x)$$
$$= \sum_{\text{All } x} \left(\frac{d}{dt} t^x\right) f(x)$$
$$= \sum_{\text{All } x} (x t^{x-1}) f(x)$$
$$= E(X t^{X-1})$$

Differentiating twice:

$$\frac{d^2}{dt^2} H(t)|_{t=1} = E[X(X-1) t^{X-2}]|_{t=1} = E[X(X-1)].$$

Now we can see the result.

**Part (b):**  Here we assume that $X$ is a discrete random variable. Therefore:

$$H(t) = E\left(t^X\right)$$
$$= \sum_{\text{All } x} t^x f(x)$$
$$= \sum_{\text{All} x} t^x P(X = x)$$
$$= P(X = 0) + t P(X = 1) + t^2 P(X = 2) + \ldots + t^k P(X = k)$$
$$+ t^{k+1} P(X = k+1) + \ldots.$$

Hence,

$$\frac{d^k}{dt^k} H(t) = (k!) P(X = k) + \frac{(k+1)!}{1!} t P(X = k+1) + \ldots.$$

It is clear that $t$ will be there in each term except for the first on the right hand side of the above expression. Therefore the imposition $t = 0$ will get rid of those. Hence the result.

**Part (c):** Let $t = e^s$.

$$H(t) = E(t^X) = E(e^{sX}) = M(s) = M[\log(t)]$$

□

## Example 13.15 (Binomial)

Suppose that $X$ is binomial with parameters $(n, p)$. Its pgf is $H(t) = (pt + q)^n$. Hence

$$\frac{d}{dt}H(t) = n(pt + q)^{n-1}p \text{ and } \frac{d^2}{dt^2}H(t) = n(n-1)(pt + q)^{n-2}p^2.$$

Therefore

$$E(X) = \frac{d}{dt}H(t)|_{t=1} = np \text{ and } E[X(X-1)] = \frac{d^2}{dt^2}H(t)|_{t=1} = n(n-1)p^2.$$

Now

$$E(X^2) = E[X(X-1)] + E(X) = n(n-1)p^2 + np.$$

Hence we have:

$$\text{Var}(X) = E(X^2) - [E(X)]^2 = n(n-1)p^2 + np - (np)^2$$

$$= np - np^2 = np(1-p) = npq.$$

Its mgf is

$$M(t) = H(e^t) = (pe^t + q)^n.$$

◄

## Example 13.16 (Poisson)

Suppose that $X$ follows the Poisson distribution with parameter $\lambda$. Let us first find its pgf. Its pmf is

$$f(x) = \frac{e^{-\lambda}\lambda^x}{x!}, x = 0, 1, 2\ldots, \lambda > 0.$$

$$H(t) = E(t^X) = \sum_{x=0}^{\infty} t^x \frac{e^{-\lambda}\lambda^x}{x!}$$
$$= \sum_{x=0}^{\infty} \frac{e^{-\lambda}(\lambda t)^x}{x!}$$
$$= \sum_{x=0}^{\infty} \frac{e^{-\lambda t}e^{\lambda t - \lambda}(\lambda t)^x}{x!}$$
$$= e^{\lambda t - \lambda} \sum_{x=0}^{\infty} \frac{e^{-\lambda t}(\lambda t)^x}{x!}$$
$$= e^{\lambda t - \lambda},$$

since $\sum_{x=0}^{\infty} \frac{e^{-\lambda t}(\lambda t)^x}{x!} = 1$ because this is the sum of all the probabilities of a Poisson random variable with parameter $\lambda t$. To explain more, if $X \sim Poisson(\lambda t)$ then $P(X = x) = \frac{e^{-\lambda t}(\lambda t)^x}{x!}$ for $x = 0, 1, \ldots$, and $\sum_{x=0}^{\infty} P(X = x) = 1$.

Now we can find the mean and variance by differentiating $H(t)$ and then setting $t = 1$ as in the above binomial example. ◄

## 13.5  Exercises

### 13.1 (mgf and pgf)

1. Suppose that $X$ is a random variable with the mgf:

$$M(t) = \frac{1}{2}\left(1 + \frac{1}{2}e^t + \frac{1}{2}e^{-t}\right), \quad -\infty < t < \infty.$$

   (a) Find the mean and variance of $X$.
   (b) Find the standard deviation of $Y = 6 - 7X$.
   (c) Guess and write down the probability mass function of $X$.
      (**Hint**: It is a discrete random variable taking only three values.)

2. Does a distribution exist for which $M_X(t) = \frac{t}{1-t}$, $|t| < 1$? If yes, find it. If no, prove it.

3. Suppose that $X$ has the pmf $f(x) = q^{x-1}p, x = 1, 2, 3, \ldots, 0 < p < 1, p + q = 1$. Find the pgf and the mgf of $X$. Find the mean and the variance of $X$ using the pgf.

4. Suppose that $X$ follows the exponential distribution, i.e. $f(x) = e^{-x}, x > 0$. Obtain the mgf of this distribution and show that $E(X^k) = k!$ where $k$ is a positive integer. Verify this result using direct integration.

5. Suppose that a random variable $X$ has the mgf $M_X(t) = e^{at+bt^2}, -\infty < t < \infty$ where $a$ and $b > 0$ are known constants. Let $Y = c(X - a)$ where $c > 0$ is a given constant.
   (a) Recognise the random variable $X$ and write down its mean and variance.
   (b) Derive the mgf of $Y$. What is the distribution of $Y$ when $b = \frac{1}{2}$ and $c = 1$?

6. Find the mgf of $Y = aX + b$ where $a \neq 0$, and hence show that if $X \sim N(\mu, \sigma^2)$ then $Y = \frac{X-\mu}{\sigma} \sim N(0, 1)$.

7. Assume that $\int_{-\infty}^{\infty} \frac{1}{\sqrt{2\pi}b}e^{-\frac{1}{2b^2}(y-a)^2}dy = 1$ where $b > 0$. Evaluate the integral $\int_0^{\infty} e^{-3y^2}dy$.

8. Suppose that $X$ has the standard normal distribution, i.e.

$$f(x) = \frac{1}{\sqrt{2\pi}}e^{-\frac{1}{2}x^2}, \quad -\infty < x < \infty.$$

Derive the moment generating function of $Y = X^2$. What distribution does $Y$ follow?

9. Suppose that a random variable $X$ has the mgf

$$M(t) = e^{t^2+3t} \quad \text{for} \quad -\infty < t < \infty.$$

Find $E\{[X - E(X)]^r\}$, the $r$th central moment of $X$, for $r = 1, 2, \cdots$. Does $X$ have a normal distribution? Give your reasoning.

10. Suppose that a random variable $X$ has a gamma distribution with parameters $\alpha$ and $\beta$. Use the mgf of $X$ to show that if $Y$ is defined as $Y \equiv bX$, where $b$ is a constant, then $Y$ has a gamma distribution with parameters $\alpha$ and $\beta/b$.

# Transformation and Transformed Distributions

# 14

**Abstract**

Chapter 14 is devoted to deriving distributions of transformed random variables in one and multiple dimensions. These techniques are used to derive sampling distributions of quantities of statistical interests while sampling from the normal distribution. Three new distributions: chi-squared, t and F are derived and their properties are discussed.

This chapter introduces a few essential concepts in statistical distribution theory. The theory is needed to obtain distributions of transformed random variables in both one and more than one dimensions. For example, we may want to obtain the probability distribution of the transformed random variable $Z^2$ where we know that the random variable $Z$ follows the standard normal distribution. In multi-dimension, e.g. when we have a pair of jointly distributed random variables we may want to obtain the joint probability distribution of the transformed pair $X + Y$ and $X - Y$.

This chapter also derives the probability distributions of three commonly used generated distributions, $\chi^2$, $t$ and $F$ distributions, often used to perform statistical inference. We have used the $t$-distribution before in constructing confidence intervals in Sect. 11.4 and hypothesis testing in Chap. 12. These three distributions are generated by drawing random samples from the standard normal distribution. We also discuss the joint distribution of the sample mean and variance when sampling from the normal distribution.

## 14.1 Transformation of a Random Variable

Suppose that we know the pdf or pmf, $f(x)$, of a random variable $X$. Let $Y$ be a new random variable defined by $Y = r(X)$ where $r(X)$ is a function of $X$. What is the distribution of $Y$? We need this quite often in practice, for example suppose that we use the degree Celsius for temperature but we have to communicate to someone who is only familiar with temperature values in Fahrenheit. The motivating task here is convert the probability distribution of temperature in degree Celsius to a probability distribution of temperature in Fahrenheit. We first start with a simpler discrete example.

**Example 14.1**

Let $X$ denote the number of heads in three tosses of a fair coin. The probability distribution of $X$ is given by:

| $x$ | $P(X = x)$ |
|---|---|
| 0 | 1/8 |
| 1 | 3/8 |
| 2 | 3/8 |
| 3 | 1/8 |
| Total | 1 |

Let $Y$ be a 0-1 random variable such that $Y = 0$ if $X = 0$ and $Y = 1$ if $X > 1$. What is the probability distribution of $Y$?

We can answer the above problem easily. We just see that $Y$ can take two values, namely, 0 and 1. $P(Y = 0) = P(X = 0) = \frac{1}{8}$ and $P(Y = 1) = P(X = 1) + P(X = 2) + P(X = 3) = \frac{3}{8} + \frac{3}{8} + \frac{1}{8} = \frac{7}{8}$. Therefore we have the probability distribution:

| $y$ | $P(Y = y)$ |
|---|---|
| 0 | 1/8 |
| 1 | 7/8 |
| Total | 1 |

◀

## 14.1.1  Transformation of Discrete Random Variables

Assume that a discrete random variable $X$ has the probability function $f(x)$. Let $Y = r(X)$ for a suitable one to one function $r(\cdot)$. To find $P(Y = y)$ for any value of $y$ we sum the probabilities $f(x)$ over all values of $x$ such that $r(x) = y$. Thus, the pmf of $Y$ is given by:

$$g(y) = P(Y = y)$$
$$= P[r(X) = y]$$
$$= \sum_{x: \, r(x)=y} f(x),$$

for any possible value of $y$.

**Example 14.2 (Binomial)**

Let $X \sim B(n, p)$, i.e.,

$$f(x) = P(X = x) = \binom{n}{x} p^x (1 - p)^{n-x}, x = 0, 1, \ldots, n.$$

What is the pmf of $Y = n - X$?

$$\begin{aligned} g(y) &= \sum_{x: \, n-x=y} f(x) \\ &= \sum_{x: \, x=n-y} f(x) \\ &= f(n - y) \\ &= \binom{n}{n-y} p^{n-y} (1 - p)^{n-(n-y)} \\ &= \binom{n}{y} p^{n-y} (1 - p)^y. \end{aligned}$$

Thus, $Y \sim B(n, 1 - p)$. ◄

## 14.1.2 Transformation of Continuous Random Variables

The general, one dimensional, version of the transformation problem is that given a continuous random variable $X$ has the pdf $f(x)$ over a certain range, what is the pdf of the transformed random variable $Y$ where $Y = r(X)$ and $r(\cdot)$ is suitable function of interest? There are several competing methods to find the pdf of the transformed random variable, $Y$. As expected, not all methods will be applicable to all problems. Hence we learn all the available methods and use the most appropriate one for the problem at hand.

We first discuss a basic technique based on finding the cdf of a transformed random variable, which we denote by $Y$. The basic technique is pretty simple. We first find the cdf of the random variable $Y$ and then differentiate the cdf to obtain the pdf of $Y$.

**Example 14.3**

Suppose that $X$ has a uniform distribution on the interval $(-1, 1)$, so $f(x) = \frac{1}{2}$ for $-1 < x < 1$. What's the pdf of $Y = X^2$? First find the cdf of $Y$. Note that $0 \leq Y < 1$.

$$\begin{aligned} G(y) &= P(Y \leq y) = P(X^2 \leq y) \\ &= P(-\sqrt{y} \leq X \leq \sqrt{y}) \\ &= \int_{-\sqrt{y}}^{\sqrt{y}} f(x) dx \\ &= \int_{-\sqrt{y}}^{\sqrt{y}} \tfrac{1}{2} dx = \sqrt{y} \end{aligned}$$

The pdf $g(y)$ of $Y$ is obtained by differentiation:

$$g(y) = \frac{dG(y)}{dy} = \frac{1}{2\sqrt{y}}, 0 < y < 1.$$

◀

The above technique can be used when the integration required to obtain the cdf of $Y$ can be evaluated in closed form. However, exact integration may not always be possible. Hence we learn the following general method based on applying the substitution method of integration in ordinary calculus. Here is an example of this method.

---

**Example 14.4**

Suppose we have to evaluate the integral

$$\int_0^{\pi/2} (\sin^2(x) - 2\sin(x) + 1)\cos(x)\, dx.$$

To solve this problem, we apply the substitution $y = \sin(x)$ and evaluate $\frac{dy}{dx} = \cos(x)$, so that $dy = \cos(x)\, dx$. Now,

$$\int_0^{\pi/2} (\sin^2(x) - 2\sin(x) + 1)\cos(x)dx = \int_0^1 (y^2 - 2y + 1)dy,$$

which can be evaluated as standard. ◀

We essentially use the substitution method in integration to find pdf of a transformed random variable.

**Theorem 14.1** *Let $X$ be a continuous random variable with pdf $f(x)$ and $P(a < X < b) = 1$. Let $Y = r(X)$, and suppose that $r(x)$ is continuous and either strictly increasing or decreasing in $(a, b)$. Suppose also that $a < X < b$ if and only if $\alpha < Y < \beta$, and let $X = s(Y)$ be the inverse function for $\alpha < Y < \beta$. Then the pdf $g(y)$ of $Y$ is*

$$g(y) = f[s(y)]\left|\frac{ds(y)}{dy}\right| \text{ for } \alpha < y < \beta. \qquad (14.1)$$

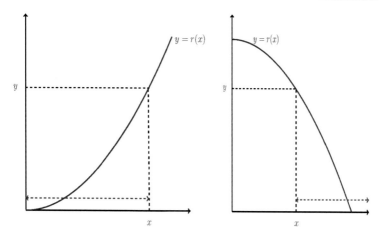

**Fig. 14.1** Transformation $y = r(x)$. Left panel plots an increasing function. Here $r(x) \leq y$ implies $x \leq r^{-1}(y)$. Right panel plots a decreasing function. Here $r(x) \leq y$ implies $x \geq r^{-1}(y)$

***Proof*** We use the above discussed cdf technique to find the pdf $g(y)$ of $Y$. To find the cdf $G(y)$ we use the substitution method of integration as illustrated above for a simple example. We consider two cases based on the nature of the transformation function $r(x)$.

First, suppose that $r(x)$ is strictly increasing as in the left panel of Fig. 14.1. Then for any $\alpha < y < \beta$,

$$G(y) = P(Y \leq y) = P[r(X) \leq y] = P[X \leq s(y)] = F[s(y)].$$

We obtain the pdf of $Y$ by differentiating $G(y)$. For $\alpha < y < \beta$,

$$g(y) = \frac{dG(y)}{dy} = \frac{dF[s(y)]}{dy} = f[s(y)]\frac{ds(y)}{dy}.$$

Now suppose that $r(x)$ is strictly decreasing as in the right panel of Fig. 14.1.

$$G(y) = P[r(X) \leq y] = P[X \geq s(y)] = 1 - F[s(y)]. \text{ Hence,}$$

$$\begin{aligned} g(y) &= \frac{dG(y)}{dy} \\ &= -f[s(y)]\frac{ds(y)}{dy} \\ &= f[s(y)]\left|\frac{ds(y)}{dy}\right|. \end{aligned}$$

In the last equality the negative sign has been absorbed in the absolute value of the derivative, $\frac{ds(y)}{dy}$. This is because, $\frac{ds(y)}{dy}$ is negative since the function $s(\cdot)$ is strictly decreasing. This negative value multiplied by the negative sign in the last but one line gives a positive value and the $\left|\frac{ds(y)}{dy}\right|$ guarantees that positive value.                □

*Jacobian* of the transformation: The derivative $\frac{ds(y)}{dy}$ in (14.1) is called the *Jacobian* of the transformation. Note that this is $\frac{dx}{dy}$ where $x$ is the old variable and $y$ is the new transformed variable. Hence the Jacobian is remembered as *"d old by d new"* to avoid confusion regarding the original and transformed variables. Note also that the Jacobian must be multiplied by its absolute value in the transformation pdf; otherwise a negative pdf $g(y)$ may be obtained incorrectly.

**Example 14.5**

Suppose that the random variable $X$ has the pdf $f(x) = 7e^{-7x}$ for $0 < x < \infty$. Find the pdf of $Y = 4X + 3$. Note that $\alpha = 3$ and $\beta = \infty$, so $3 < y < \infty$. Now we have $x = \frac{y-3}{4} = s(y)$. The Jacobian is $\frac{dx}{dy} = \frac{1}{4}$. Therefore $|\frac{dx}{dy}| = \frac{1}{4}$. We always remember to take absolute value of the Jacobian. Now,

$$g(y) = 7e^{-\frac{7}{4}(y-3)}\frac{1}{4} = \frac{7}{4}e^{-\frac{7}{4}(y-3)}, 3 < y < \infty.$$

◀

In general, we use the following the steps to obtain the pdf of a transformed random variable $Y$.

1. First find the range of $y$ values implied by the transformation.
2. See if the transformation is decreasing or increasing, draw a graph!
3. Invert the transformation.
4. Find the Jacobian remembering it to be *"d old by d new"*. Take its absolute value.
5. Re-write the pdf of the original random variable by substituting the inverse transformation.
6. Multiply the re-written pdf by the absolute value of the Jacobian.
7. The final expression written using the transformed random variable, $y$ with the range of $y$ is the pdf of $Y$.

**Example 14.6 (Probability Integral Transformation)**

Let $f(x) = e^{-x}, 0 < x < \infty$. What is the pdf of $Y = 1 - e^{-X}$?
Here $x = -\log(1 - y)$ and the transformation is one-to-one and $0 < y < 1$.
Hence, Jacobian is $\frac{dx}{dy} = \frac{1}{1-y}$. Hence, the pdf of $Y$ is

$$g(y) = \frac{1-y}{|1-y|} = 1, \quad \text{for } 0 < y < 1.$$

That is, $Y$ follows the uniform distribution $U(0, 1)$.

A more general result is true: *Let* $X$ *have a continuous cdf* $F(x)$. *Define* $Y =$ $F(X)$. *Then* $Y$ *is uniformly distributed on (0, 1)*. This is known as the *Probability integral transformation.* This result is very useful in drawing random samples, i.e. simulating, from the distribution of $X$ by first drawing a random sample, $Y$ from the uniform distribution $U(0, 1)$ and then setting $X = F^{-1}(Y)$ as a random sample from the distribution of $X$. Simulation, i.e. drawing random samples, from the uniform distribution is a much easier task which has been made even easier by the availability of routines to simulate uniform random variables in most computer programmes. See Example 13.12 below for simulating from few more standard distributions. ◄

**Many to One Transformation**
The above Jacobian based method only works for finding probability distributions only for one-to-one transformations as implicitly noted in Theorem 14.1 since the transformation should either be decreasing or increasing. Hence this method will not work to find pdf $Y = X^2$ where $X$ can take both positive and negative values, e.g. when $X$ is the standard normal random variable. A general strategy in this case is to divide the domain of $X$ values into separate intervals where within each domain the transformation is one-to-one. The next step is to find the pdf of the transformed random variable in each domain separately using the above method. The final pdf of the transformed random variable is the sum of the pdfs in the individual domains.

---

**Example 14.7 ($\chi^2$)**

Let $f(x) = \frac{1}{\sqrt{2\pi}}e^{-\frac{x^2}{2}}$ for $-\infty < x < \infty$. We will obtain the pdf of $Y = X^2$. Note that $\alpha = 0$ and $\beta = \infty$, so that $0 < y < \infty$. $x = \pm\sqrt{y} = s(y)$. Hence the transformation is not one-to-one but it is one-to-one in each of the domains $x < 0$ and $x > 0$.
When $x < 0$, the transformation is $x = -\sqrt{y}$ Jacobian is $\frac{dx}{dy} = -\frac{1}{2\sqrt{y}}$. Hence the pdf is

$$g_1(y) = \frac{1}{\sqrt{2\pi}}e^{-\frac{y}{2}}\left|-\frac{1}{2\sqrt{y}}\right| = \frac{1}{2}\frac{1}{\sqrt{2\pi}}y^{-\frac{1}{2}}e^{-\frac{y}{2}}, 0 < y < \infty. \qquad (14.2)$$

Similarly, when $x > 0$, the Jacobian is $\frac{dx}{dy} = \frac{1}{2\sqrt{y}}$. Hence the pdf is

$$g_2(y) = \frac{1}{\sqrt{2\pi}}e^{-\frac{y}{2}}\left|\frac{1}{2\sqrt{y}}\right| = \frac{1}{2}\frac{1}{\sqrt{2\pi}}y^{-\frac{1}{2}}e^{-\frac{y}{2}}, 0 < y < \infty. \qquad (14.3)$$

Hence, the pdf of $Y = X^2$ is the sum $g_1(y) + g_2(y)$ which is given by

$$g(y) = \frac{1}{\sqrt{2\pi}}y^{-\frac{1}{2}}e^{-\frac{y}{2}}, 0 < y < \infty.$$

This result is very important in statistics.

> Square of a standard normal random variable follows the $\chi^2$ (pronounced as chi-squared) distribution with one degree of freedom.

◄

---

## 14.2 Exercises

### 14.1 (Univariate Transformation)

1. Let $X$ be a random variable with the following cdf:

$$F(x) = \frac{x^2}{4} \quad \text{if } 0 < x < 2.$$

(a) Find the pdf of $X$.
(b) Find the pdf of $Y = 2 - X$.

2. Suppose that $X$ is uniformly distributed over the interval $(0,1)$, i.e. the pdf of $X$ is

$$f(x) = \begin{cases} 1, & \text{for } 0 < x < 1; \\ 0, & \text{otherwise.} \end{cases}$$

Find the pdf of $Y = -2\log(X)$.

3. Suppose that $X$ has a uniform distribution on the interval $(0,1)$. Show that the pdf of $Y = (8X)^{1/3}$ is given by

$$g(y) = \begin{cases} \frac{3}{8}y^2, & \text{for } 0 < y < 2; \\ 0, & \text{otherwise.} \end{cases}$$

4. Let $X$ have pdf $f(x) = 42x^5(1-x)$, $0 < x < 1$. Find the pdf of $Y = X^3$. Show that the pdf integrates to 1.

5. Let $X$ have pdf $f(x) = \frac{1}{2}(1+x)$, $-1 < x < 1$. Find the pdf of $Y = X^2$.

6. Suppose that $X$ follows the gamma random variable with parameters $\alpha$ and $\beta$. Find the pdf of $Y = \frac{1}{X}$.

7. Suppose that $X \sim N(\mu, \sigma^2)$. Find the pdf of $Y = e^X$. $Y$ is said to have the log-normal distribution.

## 14.3   Transformation Theorem for Joint Density

Assume that $X_1$ and $X_2$ have known joint pdf $f(x_1, x_2)$ and that two new random variables are defined by $Y_1 = r_1(X_1, X_2)$ and $Y_2 = r_2(X_1, X_2)$, where $r_1(\cdot, \cdot)$ and $r_2(\cdot, \cdot)$ are two given functions. The joint pdf of $Y_1$ and $Y_2$ is given in the following theorem.

**Theorem 14.2** *Suppose that*

**(a)** $(X_1, X_2)$ *take values only in S, a subset of $R^2$.*
**(b)** *The image of S under the given transformation,* $(Y_1, Y_2) = (r_1(X_1, X_2),$ $r_2(X_1, X_2))$, *is $T \subset R^2$, i.e. $(Y_1, Y_2)$ only takes values in T.*
**(c)** *the transformation from S to T is one-to-one, i.e. corresponding to each value of $(Y_1, Y_2)$ in T, there is only a unique value of $(X_1, X_2)$ in S such that*

$$\begin{cases} Y_1 = r_1(X_1, X_2) \\ Y_2 = r_2(X_1, X_2). \end{cases}$$

*We can, therefore, invert the equations above to get*

$$X_1 = s_1(Y_1, Y_2)$$
$$X_2 = s_2(Y_1, Y_2) \tag{14.4}$$

**(d)** *let*

$$J = \begin{vmatrix} \frac{\partial s_1(y_1, y_2)}{\partial y_1} & \frac{\partial s_1(y_1, y_2)}{\partial y_2} \\ \frac{\partial s_2(y_1, y_2)}{\partial y_1} & \frac{\partial s_2(y_1, y_2)}{\partial y_2} \end{vmatrix}$$

*which is the Jacobian of the transformation (14.4).*
*Then the joint pdf of $Y_1$ and $Y_2$ is given by*

$$g(y_1, y_2) = \begin{cases} f(s_1(y_1, y_2), s_2(y_1, y_2))|J|, & \text{for } (y_1, y_2) \in T, \\ 0, & \text{otherwise.} \end{cases}$$

The proof of this theorem follows directly from the result of changing variables in a bivariate integration. We also note that Theorem 14.2 also holds for any number of random variables $(X_1, X_2, \ldots, X_n)$ and the transformations: $(Y_1, Y_2, \ldots, Y_n)$, provided that the joint transformation is one-to-one. The Jacobian of the transformation must be evaluated as the determinant of the $n \times n$ matrix, which is defined similarly as above. This general theorem has been used to obtain the results in Sect. 14.6.

**Example 14.8 (Proof of $\Gamma\left(\frac{1}{2}\right) = \sqrt{\pi}$)**

This result has been used to prove many properties of the exponential and normal distributions in Chap. 7. There are many ways to prove this result. Here we apply the bivariate transformation technique stated above, but for ordinary integrals. ◄

***Proof*** First note that,

$$
\begin{aligned}
\Gamma\left(\tfrac{1}{2}\right) &= \int_0^\infty x^{\frac{1}{2}-1} e^{-x}\, dx \\
&= \int_0^\infty x^{\frac{1}{2}-1} e^{-x}\, dx \quad \text{[substitute } y^2 = x] \\
&= \int_0^\infty y^{-1} e^{-y^2}\, 2y\, dy \\
&= 2 \int_0^\infty e^{-y^2}\, dy.
\end{aligned}
$$

Now,

$$
\begin{aligned}
\Gamma\left(\tfrac{1}{2}\right) \times \Gamma\left(\tfrac{1}{2}\right) &= 4 \int_0^\infty e^{-x^2}\, dx \times \int_0^\infty e^{-y^2}\, dy \quad \text{[$x$ and $y$ are dummies]} \\
&= 4 \int_0^\infty \int_0^\infty e^{-(x^2+y^2)}\, dx\, dy.
\end{aligned}
$$

Now we will make the substitution

$$
\begin{aligned}
x &= r\cos(\theta) = s_1(r,\theta), \\
y &= r\sin(\theta) = s_2(r,\theta),
\end{aligned}
$$

where $r > 0$ and $0 \le \theta < \frac{\pi}{2}$ so that the constraints $0 \le x \le \infty$ and $0 \le y \le \infty$ are satisfied. Now the Jacobian of this transformation (remember $d$ old by $d$ new) is:

$$
J = \begin{vmatrix} \frac{\partial s_1(r,\theta)}{\partial r} & \frac{\partial s_1(r,\theta)}{\partial \theta} \\ \frac{\partial s_2(r,\theta)}{\partial r} & \frac{\partial s_2(r,\theta)}{\partial \theta} \end{vmatrix} = \begin{vmatrix} \cos(\theta) & -r\sin(\theta) \\ \sin(\theta) & r\cos(\theta) \end{vmatrix} = r\left[\cos^2(\theta) + \sin^2(\theta)\right] = r.
$$

Now,

$$
\begin{aligned}
\Gamma\left(\tfrac{1}{2}\right) \times \Gamma\left(\tfrac{1}{2}\right) &= 4 \int_0^\infty \int_0^\infty e^{-(x^2+y^2)}\, dx\, dy \\
&= 4 \int_0^\infty \int_0^{\frac{\pi}{2}} e^{-r^2}\, r\, dr\, d\theta \quad \text{[using the transformation method]} \\
&= 4 \left(\int_0^\infty r e^{-r^2}\, dr\right)\left(\int_0^{\frac{\pi}{2}} d\theta\right) \\
&= 4 \left(\int_0^\infty \tfrac{1}{2} e^{-u}\, du\right) \tfrac{\pi}{2} \\
&= \pi,
\end{aligned}
$$

which proves the result.                                                                    □

**Example 14.9 (Beta Function)**

Recall the *beta function,*

$$B(m, n) = \int_0^1 x^{m-1}(1-x)^{n-1}dx$$

and the gamma function

$$\Gamma(m) = \int_0^\infty x^{m-1}e^{-x}dx$$

◀

**Theorem 14.3** *The relationship between the beta function and the gamma function is*

$$B(m, n) = \frac{\Gamma(m)\Gamma(n)}{\Gamma(m+n)}.$$

**Proof** Note that

$$\Gamma(m)\Gamma(n) = \int_0^\infty x^{m-1}e^{-x}dx \int_0^\infty y^{n-1}e^{-y}dy$$
$$= \int_0^\infty \int_0^\infty x^{m-1}y^{n-1}e^{-(x+y)}dxdy$$

Now consider the transformation

$$u = \frac{x}{x+y} \quad \text{and } v = x + y$$

as in the second example of previous section. Note that $0 < u < 1$ and $0 < v < \infty$ and the Jacobian of the transformation is $v$. Hence

$$\Gamma(m)\Gamma(n) = \int_0^\infty \int_0^\infty x^{m-1}y^{n-1}e^{-(x+y)}dxdy$$
$$= \int_0^1 \int_0^\infty u^{m-1}(1-u)^{n-1}v^{m+n-1}e^{-v}dvdun$$
$$= \int_0^1 u^{m-1}(1-u)^{n-1}du \int_0^\infty v^{m+n-1}e^{-v}dv$$
$$= B(m, n) \ \Gamma(m+n)$$

□

## 14.4   Exercises

### 14.2 (Bivariate Transformation)

1. The joint pdf of $X$ and $Y$ is given by

$$f(x, y) = \frac{1}{\Gamma(m)\Gamma(n)} x^{m-1}(y-x)^{n-1}e^{-y}, \quad 0 < x < y < \infty$$

where $\Gamma(\cdot)$ is the gamma function.
(a) Using the transformation $w = x, z = y - x$, show that $Z = Y - X$ has a gamma distribution.
(b) Are $W$ and $Z$ independent? Give your reasoning.
(c) Find the mean and variance of $Y$.
2. Suppose that $U$ is uniform in the interval $(0, 2\pi)$ and $V$, independent of $U$, is exponential with parameter 1.
(a) Find the joint pdf of $X$ and $Y$ defined by

$$X = \sqrt{2V}\cos(U), \quad Y = \sqrt{2V}\sin(U).$$

(b) Show that $X$ and $Y$ are independent, each having a standard normal distribution.
(c) Derive the distribution of $\tan(U)$. Hence claim that the distribution of the ratio of two independent standard normal distributions is the Cauchy distribution.
3. Suppose that $X$ is standard normal and $Y$ is independent of $X$. The pdf of $Y$ is given by

$$f(y) = \sqrt{\frac{2}{\pi}} e^{-\frac{y^2}{2}}, \quad y > 0.$$

Define $U = X + \alpha Y$ and $V = Y$.
(a) Show that $E(Y) = \sqrt{\frac{2}{\pi}}$ and $\text{var}(Y) = 1 - \frac{2}{\pi}$.
(b) Obtain the mean and variance of $U$.
(c) Obtain the joint pdf of $U$ and $V$.
(d) Show that the marginal pdf of $U$ is given by

$$f(u) = 2\frac{1}{\sqrt{2\pi(1+\alpha^2)}} e^{-\frac{u^2}{2(1+\alpha^2)}} \Phi\left(\frac{u\alpha}{\sqrt{1+\alpha^2}}\right), \quad -\infty < u < \infty,$$

where $\Phi$ is the cdf of the standard normal distribution.
4. Suppose that $U_1$ and $U_2$ are an i.i.d. random sample from the uniform distribution in $(0, 1)$. Let

$$Z_1 = \sqrt{-2\log(U_1)}\cos(2\pi U_2), \quad Z_2 = \sqrt{-2\log(U_1)}\sin(2\pi U_2).$$

(a) Show that $Z_1$ and $Z_2$ are an i.i.d. random sample from the standard normal distribution. Thus this transformation, called the Box-Muller transformation allows us to simulate from the normal distribution based on a random sample from the uniform distribution.

(b) State a scheme for generating a random sample from the $\chi^2$ distribution with two degrees of freedom.

## 14.5 Generated Distributions: $\chi^2$, $t$ and $F$

### 14.5.1 $\chi^2$ Distribution

Recall from Example 7.9 in Sect. 7.3.1 that the $\chi^2$ distribution with $n$ degrees of freedom (df) is the same as the gamma distribution with parameter $\alpha = \frac{n}{2}$ and $\beta = \frac{1}{2}$, see Fig. 14.2 for some illustrations. Also recall that it has mean $n$ which is its df. From Example 7.9, we know that it has the pdf

$$f(x) = \frac{1}{2^{\frac{n}{2}} \Gamma\left(\frac{n}{2}\right)} x^{\frac{n}{2}-1} e^{-\frac{1}{2}x}, \quad x > 0.$$

Previously in this chapter we have obtained that the mgf of the $\chi^2$ is

$$M_X(t) = \left(\frac{1}{1-2t}\right)^{n/2}, \quad t < \frac{1}{2}.$$

Using the mgf it is easy to see that if we add independent $\chi^2$'s we get a $\chi^2$ whose df is the sum of the df's of the independent $\chi^2$'s. This is called the reproductive property of the $\chi^2$ distribution.

Another way to think of the $\chi^2$ is using the normal distributions. We proved earlier in this chapter that, if we square a standard normal then we get a $\chi^2$ with 1 df. Now using this and the reproductive property we see that if $X_1, X_2, \ldots, X_n$ are i.i.d. standard normal random variables then

$$Y = X_1^2 + X_2^2 + \cdots + X_n^2$$

is $\chi^2$ with n df.

### 14.5.2 $t$ Distribution

Suppose that $X \sim N(0, 1)$ and $Y \sim \chi^2$ with $n$ df and $X$ and $Y$ are independent. Then $U = \frac{X}{\sqrt{Y/n}}$ is said to have the $t$-distribution with $n$ df. It is easy to remember that the $t$ distribution is *standard normal divided by the square root of*

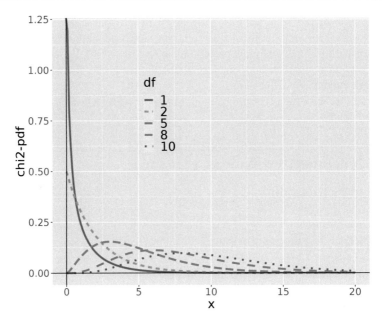

**Fig. 14.2** Illustration of the $\chi^2$ distribution for different degrees of freedom

*an independent $\chi^2$ divided by its df.* Let us find its pdf using the transformation approach.

Let $U = \frac{X}{\sqrt{Y/n}}$ and $V = Y$ and Now the pair $U$ and $V$ defines a transformation of $(X, Y)$. Note that $(X, Y)$ are the old variables and $U$ and $V$ are the new variables. Also $-\infty < U < \infty$ and $0 < V < \infty$.

The joint pdf of $X$ and $Y$ is (product because of independence):

$$
\begin{aligned}
f(x, y) &= \frac{1}{2^{n/2}\Gamma(n/2)} y^{\frac{n}{2}-1} e^{-\frac{1}{2}y} \frac{1}{\sqrt{2\pi}} e^{-\frac{1}{2}x^2} \\
&= \frac{1}{2^{(n+1)/2}\Gamma(n/2)\sqrt{\pi}} y^{\frac{n}{2}-1} e^{-\frac{1}{2}(y+x^2)}.
\end{aligned}
$$

By solving the pair of transformation equations $u = \frac{x}{\sqrt{y/n}}$ and $v = y$ we get $x = u\sqrt{\frac{v}{n}}$ and $y = v$. The Jacobian of the transformation is:

$$
J = \begin{vmatrix} \frac{\partial x}{\partial u} & \frac{\partial x}{\partial v} \\ \frac{\partial y}{\partial u} & \frac{\partial y}{\partial v} \end{vmatrix} = \begin{vmatrix} \sqrt{\frac{v}{n}} & \frac{u}{\sqrt{n}} \frac{1}{2} v^{\frac{1}{2}-1} \\ 0 & 1 \end{vmatrix} = \sqrt{\frac{v}{n}}.
$$

Now the joint pdf of $U$ and $V$ is:

$$f(u, v) = \frac{1}{2^{(n+1)/2}\Gamma(n/2)\sqrt{\pi}} v^{\frac{n}{2}-1} e^{-\frac{1}{2}(v+vu^2/n)} \sqrt{\frac{v}{n}}$$

$$= \frac{1}{2^{(n+1)/2}\Gamma(n/2)\sqrt{n\pi}} v^{\frac{n+1}{2}-1} e^{-v\frac{1}{2}(1+u^2/n)}.$$

Therefore the pdf of $U$ is

$$f(u) = \int_0^\infty f(u, v)dv$$

$$= \frac{1}{2^{(n+1)/2}\Gamma(n/2)\sqrt{n\pi}} \int_0^\infty v^{\frac{n+1}{2}-1} e^{-v\frac{1}{2}(1+u^2/n)} dv.$$

$$= \frac{1}{2^{(n+1)/2}\Gamma(n/2)\sqrt{n\pi}} \Gamma\left(\frac{n+1}{2}\right) \frac{1}{\left\{\frac{1}{2}(1+u^2/n)\right\}^{(n+1)/2}}$$

$$= \frac{\Gamma\left(\frac{n+1}{2}\right)}{\Gamma(n/2)\sqrt{n\pi}} \frac{1}{(1+u^2/n)^{(n+1)/2}}, \quad -\infty < u < \infty.$$

We have used the identity

$$\int_0^\infty x^{\alpha-1} e^{-x\beta} dx = \frac{\Gamma(\alpha)}{\beta^\alpha}, \quad \alpha > 0, \beta > 0$$

which follows from the definition of the gamma integral (7.1) in Sect. 7.1.

The $t$ distribution is symmetric about zero, has variance $\frac{n}{n-2}$ if $n > 2$, see the exercises. For $n = 1$ it is the Cauchy distribution and it approaches the normal distribution if $n \to \infty$.

## 14.5.3 $F$ Distribution

Suppose that $X \sim \chi^2$ with $n$ df and $Y \sim \chi^2$ with $m$ df and $X$ and $Y$ are independent. Then

$$U = \frac{X/n}{Y/m}$$

is said to follow the $F$ distribution with $n$ and $m$ df. Let us find its pdf using the transformation approach.

Let $U = \frac{X/n}{Y/m}$ and $V = Y$. Now the pair $U$ and $V$ defines a transformation of $(U, V)$. Note that $(X, Y)$ are the old variables and $U$ and $V$ are the new variables. Also $0 < U < \infty$ and $0 < V < \infty$.

The joint pdf of $X$ and $Y$ is (product because of independence):

$$f(x, y) = \frac{1}{2^{n/2}\Gamma(n/2)} x^{\frac{n}{2}-1} e^{-\frac{1}{2}x} \frac{1}{2^{m/2}\Gamma(m/2)} y^{\frac{m}{2}-1} e^{-\frac{1}{2}y}$$

$$= \frac{1}{2^{(n+m)/2}\Gamma(n/2)\Gamma(m/2)} x^{\frac{n}{2}-1} y^{\frac{m}{2}-1} e^{-\frac{1}{2}(x+y)}.$$

By solving we get $x = \frac{n}{m}uv$ and $y = v$. The Jacobian of the transformation is:

$$J = \begin{vmatrix} \frac{\partial x}{\partial u} & \frac{\partial x}{\partial v} \\ \frac{\partial y}{\partial u} & \frac{\partial y}{\partial v} \end{vmatrix} = \begin{vmatrix} \frac{n}{m}v & \frac{n}{m}u \\ 0 & 1 \end{vmatrix} = \frac{n}{m}v.$$

Now the joint pdf of $U$ and $V$ is:

$$f(u, v) = \frac{1}{2^{(n+m)/2}\Gamma(n/2)\Gamma(m/2)}\left(\frac{n}{m}vu\right)^{\frac{n}{2}-1} v^{\frac{m}{2}-1} e^{-\frac{1}{2}(\frac{n}{m}vu+v)} \frac{n}{m}v$$

$$= \frac{1}{2^{(n+m)/2}\Gamma(n/2)\Gamma(m/2)}\left(\frac{n}{m}\right)^{\frac{n}{2}} u^{\frac{n}{2}-1} v^{\frac{n+m}{2}-1} e^{-v\frac{1}{2}(1+\frac{n}{m}u)}.$$

To find the marginal pdf of $U$ we need to integrate out $v$ from the above joint density in the range $0 < v < \infty$. Hence, we need to obtain:

$$I = \int_0^\infty v^{\frac{n+m}{2}-1} e^{-v\frac{1}{2}(1+\frac{n}{m}u)} dv$$

$$= \frac{\Gamma(\frac{n+m}{2})}{\left\{\frac{1}{2}(1+\frac{n}{m}u)\right\}^{(n+m)/2}} \quad [\text{ using (7.9)}]$$

Now the pdf of $U$ is:

$$f(u) = \int_0^\infty f(u, v)dv$$

$$= \frac{1}{2^{(n+m)/2}\Gamma(n/2)\Gamma(m/2)}\left(\frac{n}{m}\right)^{\frac{n}{2}} u^{\frac{n}{2}-1} \frac{\Gamma(\frac{n+m}{2})}{\left\{\frac{1}{2}(1+\frac{n}{m}u)\right\}^{(n+m)/2}}$$

$$= \frac{\Gamma(\frac{n+m}{2})}{\Gamma(n/2)\Gamma(m/2)}\left(\frac{n}{m}\right)^{\frac{n}{2}} \frac{u^{\frac{n}{2}-1}}{(1+\frac{n}{m}u)^{\frac{n+m}{2}}}, \quad 0 < u < \infty.$$

Figures 14.3 and 14.4 provide a few illustrations of the F-distribution for various choice of degrees of freedom, $m$ and $n$.

Continue to assume that $X \sim \chi^2$ with $n$ df and $Y \sim \chi^2$ with $m$ df and $X$ and $Y$ are independent. Here are three results which we can prove easily.

1. It follows from definitions that, if $T$ has the $t$-distribution with $n$ df then $Y = T^2$ has the $F$ distribution with 1 and $n$ df.
2. By definition it follows that $Y = 1/U$ has the $F$ distribution with $m$ and $n$ df, where $U$ follows $F$ distribution with $n$ and $m$ df.
3. Note that $U = \frac{m}{n}\frac{X}{Y}$. Hence, $E(U) = \frac{m}{n}E(X)E\left(\frac{1}{Y}\right)$ because of independence of $X$ and $Y$. By using the Gamma function, it is easy to obtain $E\left(\frac{1}{Y}\right)$ (see the exercises) and then to show: $E(U) = \frac{m}{m-2}$ where $U$ follows the $F$ distribution with $n$ and $m$ df and $m > 2$. This technique of finding the mean of the $F$ distribution avoids having to calculate the $F$ integral, which is slightly more complicated than the Gamma integral.

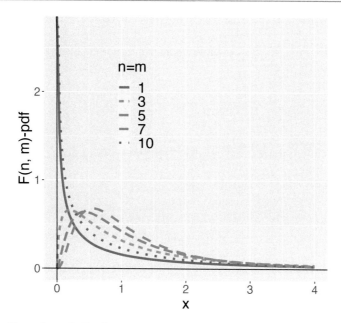

**Fig. 14.3** Illustrations of pdfs of the F($n.m$)distribution when $n = m$

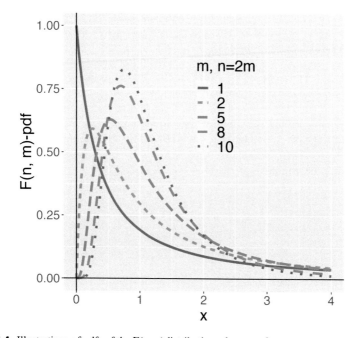

**Fig. 14.4** Illustrations of pdfs of the F($n.m$)distribution when $n = 2 \times m$

## 14.6    Sampling from the Normal Distribution

Suppose that $X_1, \ldots, X_n$ is a random sample from a normal distribution with mean $\mu$ and variance $\sigma^2$. Let

$$\bar{X} = \frac{1}{n} \sum_{i=1}^{n} X_i \quad \text{and} \quad S^2 = \frac{1}{n-1} \sum_{i=1}^{n} (X_i - \bar{X})^2.$$

Then we can prove the following results:

**R1** The sample mean $\bar{X}$ and the sample variance $S^2$ are independent random variables.

**R2** $\bar{X}$ has the normal distribution with mean $\mu$ and variance $\frac{\sigma^2}{n}$.

**R3** The random variable $\frac{1}{\sigma^2} \sum_{i=1}^{n} (X_i - \mu)^2$ has the $\chi^2$ distribution with $n$ degrees of freedom.

**R4** $Y = \frac{(n-1) S^2}{\sigma^2}$ has the $\chi^2$ distribution with $n - 1$ degrees of freedom.

**R5** $T = \frac{\sqrt{n}(\bar{X}-\mu)}{S}$ has the $t$ distribution with $n - 1$ degrees of freedom. This result has been used without proof in Sect. 11.4.

Before proving these results we first state the following lemma:

**Lemma 14.1** *If $Z_1, \ldots, Z_n$ are iid $N(0, 1)$ random variables, then*

*(i)* $\bar{Z} \sim N\left(0, \frac{1}{n}\right)$.

*(ii)* $\sum_{i=1}^{n} (Z_i - \bar{Z})^2 \sim \chi_{n-1}^2$.

*(iii)* $\bar{Z}$ and $\sum_{i=1}^{n} (Z_i - \bar{Z})^2$ are independent.

The proof of this lemma uses multivariate transformation technique, which may seem a bit difficult to comprehend at the first reading. Hence, the proof is given at the end of this section.

We now use the lemma to prove the results **R1** to **R5**. Let

$$Z_i = \frac{X_i - \mu}{\sigma}, \quad i = 1, \ldots, n.$$

Hence,

$$\bar{Z} = \frac{\bar{X} - \mu}{\sigma}, \quad \text{i.e.,} \quad \bar{X} = \sigma \bar{Z} + \mu$$

and

$$\sum_{i=1}^{n}(Z_i - \bar{Z})^2 = \sum_{i=1}^{n}\left(\frac{X_i-\mu}{\sigma} - \frac{\bar{X}-\mu}{\sigma}\right)^2$$
$$= \frac{1}{\sigma^2}\sum_{i=1}^{n}\left(X_i - \bar{X}\right)^2$$
$$= \frac{1}{\sigma^2}(n-1)S^2.$$

Thus we have proved $S^2 = \sigma^2 \frac{1}{n-1}\sum_{i=1}^{n}(Z_i - \bar{Z})^2$.
Now we provide arguments to prove the results **R1–R5**.

**R1** This result follows from part (iii) of the lemma. This is because we have $\bar{X} = \sigma\bar{Z} + \mu$ and $S^2 = \sigma^2 \frac{1}{n-1}\sum_{i=1}^{n}(Z_i - \bar{Z})^2$ as shown above.

**R2** This result follows from part (i) of the lemma since $\bar{X} = \sigma\bar{Z} + \mu$ is a linear transformation of the normally distributed random variable $\bar{Z}$.

**R3** This result follows since $\frac{1}{\sigma^2}\sum_{i=1}^{n}(X_i - \mu)^2 = \sum_{i=1}^{n} Z_i^2$, where $Z_i, i = 1, \ldots, n$ are i.i.d. $N(0, 1)$ random variables. See also the results in Sect. 14.5.1.

**R4** This result follows directly from part (ii) of the lemma.

**R5** To prove this, note that $\frac{\sqrt{n}(\bar{X}-\mu)}{\sigma} \sim N(0, 1)$ independently of $Y = \frac{(n-1)S^2}{\sigma^2}$, which follows the $\chi^2$ distribution with $n-1$ degrees of freedom. Then $T = \frac{\sqrt{n}(\bar{X}-\mu)}{S}$ follows the $t$-distribution using the definition of the $t$-distribution in Sect. 14.5.2.

Now we prove the lemma.

***Proof*** Define $n$ new random variables $Y_1, \ldots, Y_n$ by $\mathbf{Y} = (Y_1, \ldots, Y_n) = H\mathbf{Z}$, where $\mathbf{Z} = (Z_1, \ldots, Z_n)$ and the $n$ by $n$ matrix $H$ is given by

$$\begin{pmatrix} \frac{1}{\sqrt{n}} & \frac{1}{\sqrt{n}} & \frac{1}{\sqrt{n}} & \frac{1}{\sqrt{n}} & \cdots & \frac{1}{\sqrt{n}} & \frac{1}{\sqrt{n}} \\ \frac{-1}{\sqrt{2\times1}} & \frac{1}{\sqrt{2\times1}} & 0 & 0 & \cdots & 0 & 0 \\ \frac{-1}{\sqrt{3\times2}} & \frac{-1}{\sqrt{3\times2}} & \frac{2}{\sqrt{3\times2}} & 0 & \cdots & 0 & 0 \\ \cdots & \cdots & \cdots & \cdots & \cdots & \cdots & \cdots \\ \frac{-1}{\sqrt{n(n-1)}} & \frac{-1}{\sqrt{n(n-1)}} & \frac{-1}{\sqrt{n(n-1)}} & \frac{-1}{\sqrt{n(n-1)}} & \cdots & \frac{-1}{\sqrt{n(n-1)}} & \frac{n-1}{\sqrt{n(n-1)}} \end{pmatrix}.$$

This matrix is called the **Helmert** matrix; it is orthogonal and the $i$th row sum is equal to zero for $i > 1$. Check that $H'H = I$ for the $n = 2, 3, 4$ cases. Then orthogonality follows by induction.

Since $H$ is orthogonal, the determinant $|H| = \pm1$. Therefore the Jacobian of the transformation is 1. Since $Z_i$ are iid $N(0, 1)$ their joint pdf is:

$$f(z_1, \ldots, z_n) = \frac{1}{(2\pi)^{n/2}} e^{-\frac{1}{2}\sum_{i=1}^{n} z_i^2}.$$

Now

$$\sum_{i=1}^{n} Y_i^2 = \mathbf{Y}'\mathbf{Y} = \mathbf{Z}'H'H\mathbf{Z} = \mathbf{Z}'\mathbf{Z} = \sum_{i=1}^{n} Z_i^2$$

since $H'H = I$. Therefore, using the general transformation theorem, see Theorem 14.2 and discussion, the joint pdf of $Y_1, \ldots, Y_n$ is:

$$f(y_1, \ldots, y_n) = \frac{1}{(2\pi)^{n/2}} e^{-\frac{1}{2}\sum_{i=1}^{n} y_i^2}.$$

This shows that $Y_1, \ldots, Y_n$ are iid $N(0, 1)$ variables. Now, from the transformation $\mathbf{Y} = H\mathbf{Z}$ we have

$$Y_1 = \left(\frac{1}{\sqrt{n}}, \ldots, \frac{1}{\sqrt{n}}\right)\mathbf{Z}$$
$$= \frac{1}{\sqrt{n}}\sum_{i=1}^{n} Z_i$$
$$= \frac{1}{\sqrt{n}}n\bar{Z}$$
$$= \sqrt{n}\bar{Z}$$

Thus,

$$\bar{Z} = \frac{Y_1}{\sqrt{n}}, \tag{14.5}$$

and so

$$\sum_{i=1}^{n}(Z_i - \bar{Z})^2 = \sum_{i=1}^{n} Z_i^2 - n\bar{Z}^2$$
$$= \sum_{i=1}^{n} Y_i^2 - n\left(\frac{Y_1}{\sqrt{n}}\right)^2$$
$$= Y_2^2 + \cdots + Y_n^2. \tag{14.6}$$

Then parts (i) and (ii) of the lemma follow directly from (14.5) and (14.6), respectively. The independence of $\bar{Z}$ and $\sum_{i=1}^{n}(Z_i - \bar{Z})^2$ can be seen from (14.5) and (14.6), since $\bar{Z}$ depends only on $Y_1$ and $\sum_{i=1}^{n}(Z_i - \bar{Z})^2$ depends only on $Y_2, \cdots, Y_n$. The proof is thus completed.                                    □

---

**Example 14.10 (Confidence Interval for $\sigma^2$)**

Assume that we have an independent random sample, $X_1, \ldots, X_n$ from the normal distribution $N(\mu, \sigma^2)$. The result **R4** stated above can be used as a pivot statistic to derive a confidence interval for $\sigma^2$. **R4** states that $(n-1)S^2/\sigma^2$ has a $\chi_{n-1}^2$ distribution. Therefore, we can use R to find two numbers $h_1$ and $h_2$ such that

$$P\left(h_1 \leq \chi_{n-1}^2 = (n-1)\frac{S^2}{\sigma^2} \leq h_2\right) = 1 - \alpha$$

for a given value of $0 < \alpha < 1$. For example, if $\alpha = 0.05$ we may take $h_1$ and $h_2$ to be respectively the 2.5th and 97.5th percentiles of the $\chi_{n-1}^2$ distribution, which can be found using the R commands: `qchisq(p=0.025, df=n-1)` and `qchisq(p=0.0975, df=n-1)`.

The above probability inequality can be re-written as:

$$P\left((n-1)\frac{S^2}{h_2} < \sigma^2 < (n-1)\frac{S^2}{h_1}\right) = 1 - \alpha$$

Hence a $100(1 - \alpha)\%$ confidence interval for $\sigma^2$ will be

$$\left[(n-1)\frac{s^2}{h_2}, \ (n-1)\frac{s^2}{h_1}\right],$$

where $s^2$ is the observed value of $S^2$. ◄

---

## 14.7  Exercises

### 14.3 (Generated Distributions)

1. Suppose that a random variable $X$ has the pdf

$$f(x) = e^{-(x-\mu)}, x > \mu.$$

   Obtain the mean and variance of $X$.
2. The lifetime $X$ of an electronic component has an exponential distribution such that $P(X \leq 1000) = 0.75$. What is the expected lifetime of the component?
3. If $X \sim$ exponential($\beta$) then find the pdf of $Y = X^{1/\gamma}$. The random variable $Y$ is known as the Weibull random variable. See the list of distributions and from there write down its mean and variance.

4. Suppose that $X \sim \text{beta}(\alpha, \beta)$. Show that the pdf of $Y = \frac{1}{X} - 1$ is

$$
g(y|\alpha, \beta) = \begin{cases} \frac{1}{B(\alpha, \beta)} \frac{y^{\beta-1}}{(1+y)^{\alpha+\beta}}, & \text{for } y \geq 0, \\ 0, & \text{otherwise.} \end{cases}
$$

This distribution is called the *beta distribution of the second type*.

5. Let $X_1$ and $X_2$ be independent random variables where $X_1$ has a gamma$(\alpha_1, \beta)$ distribution and $X_2$ has a gamma$(\alpha_2, \beta)$ distribution.
   (a) Write down the joint probability density function of $(X_1, X_2)$.
   (b) Find the joint probability density function of $(Y_1, Y_2)$, where $Y_1 = X_1/X_2$ and $Y_2 = X_2$. Hence show that the marginal probability density function of $Y_1 = X_1/X_2$ is

   $$
   f_{Y_1}(y_1) = \frac{\Gamma(\alpha_1 + \alpha_2)}{\Gamma(\alpha_1)\Gamma(\alpha_2)} \frac{y_1^{\alpha_1-1}}{(1 + y_1)^{\alpha_1+\alpha_2}}, \qquad y_1 > 0.
   $$

   (c) Find the joint probability density function of $(Y_1, Y_2)$, where $Y_1 = X_1/(X_1 + X_2)$ and $Y_2 = X_1 + X_2$. Hence show that $X_1/(X_1 + X_2)$ and $X_1 + X_2$ are independent, and write down their marginal probability density functions.

6. If $X$ has a beta$(\alpha, \beta)$ distribution and $Y = \delta X$ for some positive constant $\delta$, show that the probability density function for $Y$ is given by

   $$
   f_Y(y) = \frac{\Gamma(\alpha + \beta)}{\Gamma(\alpha)\Gamma(\beta)\delta^{\alpha+\beta-1}} y^{\alpha-1}(\delta - y)^{\beta-1}, \qquad 0 < y < \delta
   $$

   Let $X_1$ and $X_2$ be independent random variables where $X_1$ has a beta$(\alpha_1, \beta_1)$ distribution and $X_2$ has a beta$(\alpha_2, \beta_2)$ distribution, and suppose that $\alpha_2+\beta_2 = \alpha_1$. Write down the joint probability density function of $(X_1, X_2)$. Find the joint probability density function of $(Y_1, Y_2)$, where $Y_1 = X_1 X_2$ and $Y_2 = X_1(1 - X_2)$. Hence find the marginal distributions of $Y_1$ and $Y_2$, and show that they are both beta distributions.
   Are $Y_1$ and $Y_2$ independent?

7. Let $X_1$ and $X_2$ be independent standard normally distributed random variables. Write down the joint probability density function of $(X_1, X_2)$. Find the joint probability density function of $(Y_1, Y_2)$, where $Y_1 = X_1$ and $Y_2 = X_2/X_1$. Are $Y_1$ and $Y_2$ independent? Hence show that $Y_2 = X_2/X_1$ has a Cauchy (or $t_1$) distribution, with marginal p.d.f.

$$
f_{Y_2}(y_2) = \frac{1}{\pi(1 + y_2^2)}, \qquad y_2 \in \mathcal{R}.
$$

(**Hint**: Use the fact that, if $g$ is an even function (satisfying $g(y) = g(-y)$ for all $y$), then $\int_{-\infty}^{\infty} |y| g(y) dy = 2\int_0^{\infty} y g(y) dy$.) Hence the ratio of two independent standard normal random variables is Cauchy.

## Abstract

Chapter 15 discusses bivariate and multivariate probability distributions. In particular, it discusses the marginal and conditional distributions associated with bivariate and multivariate normal distributions. It also discusses the joint moment generating function for the multivariate normal distribution. In the discrete case it introduces the multinomial distribution as a generalisation of the binomial distribution.

This chapter continues on from the discussion in Chap. 8 for continuous bivariate distributions. In that first part of the book we did not assume any knowledge of bivariate integration in Calculus. Hence we were not able to deeply explore various concepts and properties of the bivariate distributions. We now assume that the reader is able to perform bivariate integration and this chapter provides many such examples to enhance the reader's skills. This chapter also presents the bivariate normal distribution which is a generalisation of the univariate normal distribution. The chapter also presents a multivariate discrete distribution, called a multinomial distribution, is a generalisation of the binomial distribution where trials can have multiple outcomes.

## 15.1   An Example of a Bivariate Continuous Distribution

Suppose that the random variables, $X$ and $Y$, have the joint pdf given by

$$f(x, y) = \begin{cases} cx^2 y, & \text{if } x^2 \leq y \leq 1, \\ 0, & \text{otherwise.} \end{cases}$$

Our first task is to find $c$ so that $f(x, y)$ is a pdf. We then find the marginal pdfs of $X$ and $Y$. We also evaluate the probability $P(X \geq Y)$ and find the conditional pdf $f(y|X = x)$ of $Y$ given that the random variable $X$ takes the particular value $x$.

**Solution**

It is clear that $x^2 \leq 1$ if it is considered alone. Therefore, the marginal range of $x$ is $-1 \leq x \leq 1$. Now we see that $y \leq 1$. Also, it must be non-negative since it is obvious that $0 \leq x^2$ for any real $x$. So we have found the marginal ranges, $-1 \leq x \leq 1$ and $0 \leq y \leq 1$.

We now turn to find $c$. Note that the pdf is positive (by definition) in the region $A = \{(x, y) : x^2 \leq y \leq 1; -1 \leq x \leq 1; 0 \leq y \leq 1\}$. Therefore we can rewrite $A$ as: $A = \{(x, y) : -1 \leq x \leq 1; \ x^2 \leq y \leq 1\}$.

$$\int \int_A f(x, y)dxdy = \int \int_A cx^2 y \, dxdy$$
$$= c \int_{-1}^{1} x^2 dx \int_{x^2}^{1} y \, dy$$
$$= c \int_{-1}^{1} x^2 dx \, \frac{(1-x^4)}{2}$$
$$= \frac{c}{2} \int_{-1}^{1} x^2(1 - x^4)dx = c \, \frac{4}{21}.$$

Hence $c = \frac{21}{4}$. To obtain the marginal pdf of $X$ we integrate out $y$, just as we did above, treating $x$ as constant.

$$f_X(x) = \int_A f(x, y)dy$$
$$= \frac{21}{4} \int_{x^2}^{1} x^2 y \, dy$$
$$= \frac{21}{4} x^2 \int_{x^2}^{1} y \, dy$$
$$= \frac{21}{8} x^2(1 - x^4), -1 \leq x \leq 1.$$

Check that the above is a pdf, i.e. it is non-negative and integrates to 1.

Similarly, the marginal pdf of $Y$ is

$$f_Y(y) = \int_R f(x, y)dx$$
$$= \frac{21}{4} y \int_{-\sqrt{y}}^{\sqrt{y}} x^2 \, dx, \quad \text{[using } x^2 \leq y \text{ to find the range]}$$
$$= \frac{7}{2} y^{5/2}, 0 \leq y \leq 1.$$

Check that the above is a pdf. Find $P(X \geq Y)$. Here $A = \{(x, y) : x \geq y, x^2 \leq y \leq 1\}$. $A$ can be written as: $A = \{(x, y) : 0 \leq x \leq 1; \ x^2 \leq y \leq x\}$.

$$P(X \geq Y) = \int \int_A f(x, y)dxdy$$
$$= \frac{21}{4} \int_0^1 x^2 dx \int_{x^2}^{x} y \, dy$$
$$= \frac{21}{4} \int_0^1 x^2 dx \, \frac{x^2 - x^4}{2}$$
$$= \frac{21}{8} \int_0^1 x^2(x^2 - x^4)dx = \frac{3}{20}.$$

To find the conditional pdf note that

$$f(y|x) = \frac{f(x,y)}{f_X(x)}$$
$$= \frac{21}{4}x^2 y \frac{8}{21} \frac{1}{x^2(1-x^4)}$$
$$= \frac{2y}{1-x^4},$$

where $x^2 \le y \le 1$. Note that $f(y|x)$ is non-negative and it integrates to 1 within the range $x^2$ and 1 for any given $x^2$ in $(0, 1)$. Also, we emphasise that $f(y|x)$ is a pdf of $Y$ and $x$ has to be treated as a given constant. For example, when $x = \frac{1}{2}$,

$$f(y|X = \tfrac{1}{2}) = \frac{32}{15}y, \quad \frac{1}{4} \le y \le 1.$$

We can also find the conditional expectation $E(Y|X = x)$ in the usual way as illustrated in Sect. 8.5. Here

$$E(Y|X = x) = \int_{x^2}^{1} yf(y|x)dy$$
$$= \int_{x^2}^{1} y\frac{2y}{1-x^4}dy$$
$$= \frac{2}{3}\frac{1-x^6}{1-x^4}$$
$$= \frac{2}{3}\frac{1+x^2+x^4}{1+x^2}.$$

There are more examples of bivariate continuous distributions in the exercises.

## 15.2   Joint cdf

The joint cdf of a pair continuous random variables $X$ and $Y$ is defined to be

$$F(x, y) = P(X \le x, Y \le y) = \int_{-\infty}^{x} \int_{-\infty}^{y} f(s, t)dt ds.$$

The following is true (by the Fundamental Theorem of Calculus).

$$\frac{\partial^2 F(x, y)}{\partial x \partial y} = f(x, y).$$

**Example 15.1**

Let

$$F(x, y) = \frac{1}{16}xy(x + y), \ 0 \le x \le 2, \ 0 \le y \le 2.$$

By differentiating twice, $f(x, y) = \frac{1}{8}(x + y), \ 0 \le x \le 2, \ 0 \le y \le 2.$ ◄

## 15.3   Iterated Expectations

For two random variables $X$ and $Y$ the following is a very useful theorem which allows us to evaluate the marginal expectation and variance of one random variable from its expectation and variance conditional on the other. That is, we can obtain the marginal expectation $E(X)$ and marginal variance $\text{Var}(X)$ starting from the conditional expectation $E(X|Y)$ and conditional variance $\text{Var}(X|Y)$. Here is a theorem stating the results.

**Theorem 15.1** *If $X$ and $Y$ are any two random variables, then*

*(a) $E(X) = E[E(X|Y)]$.*
*(b) $\text{Var}(X) = E[\text{Var}(X|Y)] + \text{Var}[E(X|Y)]$.*

The above theorem is sometimes called the theorem of iterated expectations and variance. The marginal expectation $E(X)$ is sometimes called total expectation and the marginal variance $\text{Var}(X)$ is total variance of the random variable $X$. These results will be useful in learning about properties of compound distributions in Sect. 15.9.

Below we prove part (a) of the above theorem only. The proof of part (b) is slightly complicated and is omitted from here. The interested reader is referred to Theorem 4.4.2 in Casella and Berger [3].

***Proof*** Part (a)

$$\begin{aligned}
E(X) &= \int x \ f_X(x)dx \\
&= \int x \ \int f(x, y)dydx \quad [\text{definition of } f_X(x)] \\
&= \int x \ \int f(x|y)f_Y(y)dydx \quad [\text{definition of } f(x|y)] \\
&= \int \left[\int xf(x|y)dx\right] f_Y(y)dy \quad [\text{re-group terms}] \\
&= \int E(X|Y)f_Y(y)dy \\
&= E[E(X|Y)]
\end{aligned}$$

□

**Example 15.2**

Let $X|Y = y$ follows $N(a + by, \tau^2)$ and $Y \sim N(\mu, \sigma^2)$. Then

$$E(X) = E[E(X|Y = y)] = E(a + bY) = a + b\mu, \quad \text{and}$$

$$\begin{aligned}
\text{Var}(X) &= E[\text{Var}(X|Y)] + \text{Var}[E(X|Y)] \\
&= E[\tau^2] + \text{Var}[a + bY] \\
&= \tau^2 + b^2\text{Var}(Y) \\
&= \tau^2 + b^2\sigma^2.
\end{aligned}$$

◀

**Example 15.3 (Egg Laying)**

An insect lays a random number of eggs, $Y$, each surviving with probability $p$. Let $Y \sim \text{Poisson}(\lambda)$. On the average how many will survive? Let $X = $ number of survivors. Find $E(X)$. We have:

$$X|Y = y \sim \text{binomial}(y, p) \quad \text{and} \quad Y \sim \text{Poisson}(\lambda).$$

Here $E(X|Y) = Yp$ and $\text{Var}(X|Y) = Yp(1 - p)$. We know $E(X) = E[E(X|Y)] = E(Yp) = E(Y)p = \lambda p$.

$$\begin{aligned}
\text{Var}(X) &= E[\text{Var}(X|Y)] + \text{Var}[E(X|Y)] \\
&= E[Yp(1 - p)] + \text{Var}[Yp] \\
&= \lambda p(1 - p) + p^2\text{Var}[Y] \\
&= \lambda p(1 - p) + p^2\lambda = \lambda p.
\end{aligned}$$

◀

## 15.4 Exercises

### 15.1 (Bivariate Distributions)

1. A pdf is defined by $f(x, y) = 2x$, $0 \leq x \leq 1$, $0 \leq y \leq 1$. Show that $f(x, y)$ is a joint pdf. Find the marginal pdfs of $X$ and $Y$ and the conditional pdf of $Y|X = x$. Find $P(X^2 < Y < X)$.

2. A pdf is defined by $f(x, y) = x + y$, $0 \le x \le 1$, $0 \le y \le 1$. Find the marginal pdfs of $X$ and $Y$ and the conditional pdf of $Y|X = x$. Find $P(X > \sqrt{Y})$.

3. A pdf is defined by $f(x, y) = e^{-y}$, $0 < x < y < \infty$. Find the marginal pdfs of $X$ and $Y$ and the conditional pdf of $Y|X = x$.

4. A pdf is defined by

$$f(x, y) = \begin{cases} Cxy & \text{if } 0 \le x \le y \le 1, \\ 0 & \text{otherwise.} \end{cases}$$

(a) Show that $C = 8$.
(b) Find the marginal distributions. Are $X$ and $Y$ independent?

5. A pdf is defined by

$$f(x, y) = \begin{cases} C & \text{if } 0 \le x \le 1, \ 0 \le y \le 1, \ 0 \le x^2 + y^2 \le 1, \\ 0 & \text{otherwise.} \end{cases}$$

**Hint**: Assume that for a known constant $a > 0$,

$$\int \sqrt{a^2 - x^2}\, dx = \frac{1}{2}x\sqrt{a^2 - x^2} + \frac{1}{2}a^2 \sin^{-1}\frac{x}{a} + \text{Constant.}$$

(a) Show that $C = \frac{4}{\pi}$. Draw a graph of the region where $f(x, y) > 0$.
(b) Show that $P(X + Y \le 1) = \frac{2}{\pi}$. Justify this result geometrically.
(c) Find the marginal distributions. Are $X$ and $Y$ independent?
(d) Show that $E(X) = \frac{4}{3\pi}$ and $E(X^2) = \frac{1}{4}$.
   **Hint:** Use a simple substitution to evaluate the integral for $E(X)$. Again using a simple substitution you will be able to reduce the integral for $E(X^2)$ to a beta integral; then use the properties of the beta and gamma distributions.]
(e) Find the correlation $\rho_{XY}$ between $X$ and $Y$.
(f) Find the conditional pdf of $Y|X = x$. Verify that the pdf integrates to 1. Find $E(Y|X = x)$ and $\text{Var}(Y|X = x)$.
   **Hint:** The mean and variance can easily be found by recognising that $Y|X = x$ follows a standard distribution. To fix ideas set $x = \frac{\sqrt{3}}{2}$ and see what's the pdf of $Y|X = \frac{\sqrt{3}}{2}$. Do not forget to write the range.

## 15.5  Bivariate Normal Distribution

Bivariate normal is a very useful distribution. Below we derive this distribution as a particular transformation of two independently distributed standard normal random variables. As a result this exercise can also be seen as an example illustrating the bivariate transformation technique introduced in Sect. 14.3.

## 15.5.1 Derivation of the Joint Density

Suppose that $U$ and $V$ are independent standard normal random variables. Let $\mu_x, \mu_y, \sigma_x > 0, \sigma_y > 0$ and $-1 < \rho < 1$ be given constants. Define

$$X = \sigma_x U + \mu_x, \quad Y = \sigma_y [\rho U + \sqrt{1 - \rho^2} V] + \mu_y.$$

Find the joint pdf of $X$ and $Y$ using the transformation technique.

The joint pdf of $U$ and $V$ is:

$$f_{U,V}(u, v) = \frac{1}{2\pi} e^{-\frac{1}{2}(u^2 + v^2)}. \tag{15.1}$$

We want to find the joint pdf $f_{X,Y}(x, y)$ of $X$ and $Y$. Note that earlier we were using the notations $U$ and $V$ to denote the new random variables. But now the notations are reversed. Now $U$ and $V$ are old, $X$ and $Y$ are new.

Solving for $u$ and $v$ from the definition above, we have

$$u = \frac{x - \mu_x}{\sigma_x}, \quad v = \frac{1}{\sqrt{1 - \rho^2}} \left[ \frac{y - \mu_y}{\sigma_y} - \rho \frac{x - \mu_x}{\sigma_x} \right].$$

We note that $-\infty < x, y < \infty$ since $-\infty < u, v < \infty$

The Jacobian of the transformation $(u, v) \rightarrow (x, y)$ is

$$J = \begin{vmatrix} \frac{\partial u}{\partial x} & \frac{\partial u}{\partial y} \\ \frac{\partial v}{\partial x} & \frac{\partial v}{\partial y} \end{vmatrix} = \begin{vmatrix} \frac{1}{\sigma_x} & 0 \\ * & \frac{1}{\sigma_y \sqrt{1-\rho^2}} \end{vmatrix} = \frac{1}{\sigma_x \sigma_y \sqrt{1 - \rho^2}},$$

where $*$ does not need to be evaluated.

Now we have to replace $u$ and $v$ by $x$ and $y$ in $f_{U,V}(u, v)$. Note that only thing we should worry about is $u^2 + v^2$. But

$$u^2 + v^2 = \left( \frac{x - \mu_x}{\sigma_x} \right)^2 + \frac{1}{1 - \rho^2} \left[ \frac{y - \mu_y}{\sigma_y} - \rho \frac{x - \mu_x}{\sigma_x} \right]^2$$

$$= \left( \frac{x - \mu_x}{\sigma_x} \right)^2 + \frac{1}{1 - \rho^2} \left[ \left( \frac{y - \mu_y}{\sigma_y} \right)^2 + \rho^2 \left( \frac{x - \mu_x}{\sigma_x} \right)^2 \right.$$

$$\left. -2\rho \left( \frac{x - \mu_x}{\sigma_x} \right) \left( \frac{y - \mu_y}{\sigma_y} \right) \right]$$

$$= \frac{1}{1 - \rho^2} \left[ \left( \frac{x - \mu_x}{\sigma_x} \right)^2 + \left( \frac{y - \mu_y}{\sigma_y} \right)^2 - 2\rho \left( \frac{x - \mu_x}{\sigma_x} \right) \left( \frac{y - \mu_y}{\sigma_y} \right) \right]$$

$$\tag{15.2}$$

Therefore, the joint pdf of $X$ and $Y$ ($= f_{U,V}(u, v) \times |J|$) is:

$$f_{X,Y}(x, y) = \frac{1}{2\pi\sigma_x\sigma_y\sqrt{1-\rho^2}} \times$$
$$\exp\left[-\frac{1}{2(1-\rho^2)}\left\{\left(\frac{x-\mu_x}{\sigma_x}\right)^2 - 2\rho\left(\frac{x-\mu_x}{\sigma_x}\right)\left(\frac{y-\mu_y}{\sigma_y}\right) + \left(\frac{y-\mu_y}{\sigma_y}\right)^2\right\}\right]$$

where $x$ and $y$ both can take any values on the real line.                                  □

A pair of random variables $X$ and $Y$ having the above pdf is said to follow the *bivariate normal distribution* with parameters, $\mu_x$, $\mu_y$, $\sigma_x^2$, $\sigma_y^2$ and $\rho$. We sometimes use the **notation** $N_2(\mu_x, \mu_y, \sigma_x^2, \sigma_y^2, \rho)$ to denote the bivariate normal distribution.

The identity (15.2) helps us to understand the shape of the bivariate normal density. To fix ideas, note that $u^2 + v^2 = c$ is the equation of a circle on the $(u, v)$ plane. Hence, a plot of the bivariate density (15.1) will be a three-dimensional plot which sits on the $(u, v)$ plan with concentric circular rings stacked up corresponding to increasing values of $c$. The density (15.1) is at its maximum value when $c = u^2 + v^2 = 0$. The all possible solutions of the equation $u^2 + v^2 = c$ for a given value of $c$ are called the *level sets*. These level sets are clearly circular. The level sets define the *contours* of the bivariate density. The plot of the contours for different values of $c$ will be just concentric circles with the origin $(0, 0)$ as the centre, see Fig. 15.1.

Now we attempt to understand the shape of the general bivariate normal density. The equation $u^2 + v^2 =$ constant implies that

$$\left(\frac{x-\mu_x}{\sigma_x}\right)^2 - 2\rho\left(\frac{x-\mu_x}{\sigma_x}\right)\left(\frac{y-\mu_y}{\sigma_y}\right) + \left(\frac{y-\mu_y}{\sigma_y}\right)^2 = c$$

for another constant value $c$. For known values of the parameters, $\mu_x$, $\mu_y$, $\sigma_x^2$, $\sigma_y^2$, $\rho$, and $c$, the all possible solutions of the above identity will form an ellipse whose

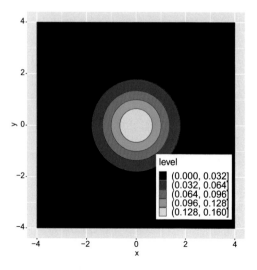

**Fig. 15.1** Illustration of the contours of the bivariate normal distribution with zero correlation

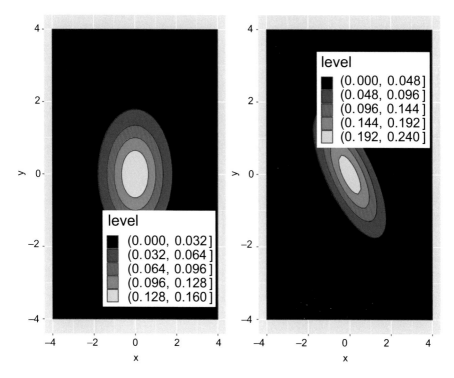

**Fig. 15.2** Illustration of the contours of the bivariate normal distribution. Left panel is for $\rho = 0.8$ and the right panel is for $\rho = -0.7$

major and minor axes can be obtained by changing the origin and rotating the axes. This is due to the transformation we have applied to define $x$ and $y$ from $u$ and $v$. Thus the contours for this distribution will be elliptical in shape. Indeed, see Fig. 15.2 where the plots are obtained for $\mu_x = \mu_y = 0$, $\sigma_x^2 = \sigma_y^2 = 1$ and two different values of $\rho$. The red lines show the contours for different particular values of $c$. Each of the areas between the two adjacent red lines is filled by one unique colour as there is very little change in the value of the density when $x$ and $y$ are in that particular region. Finally, contrasting the two plots in this figure, we see a positive and a negative relationship between $x$ and $y$, respectively for the positive and negative value of $\rho$ noted in the figure caption.

**Example 15.4**

The joint density $X$ and $Y$ when $\mu_x = \mu_y = 0$, and $\sigma_x = \sigma_y = 1$ is much easier:

$$f(x, y) = \frac{1}{2\pi\sqrt{1 - \rho^2}} e^{-\frac{1}{2(1-\rho^2)}\{x^2 - 2\rho xy + y^2\}}.$$

◀

## 15.5.2 Marginal Distributions

We obtain the marginal distributions from the basic relations:

$$X = \sigma_x U + \mu_x, \quad Y = \sigma_y [\rho U + \sqrt{1 - \rho^2} V] + \mu_y.$$

Since $U$ and $V$ are independent standard normal random variables. , $X$ and $Y$ must be normally distributed. Further, we have that $E(U) = E(V) = 0$ and $\text{Var}(U) = \text{Var}(V) = 1$ and $\text{Cov}(U, V) = 0$. Hence

$$E(X) = \sigma_x E(U) + \mu_x = \mu_x, \quad \text{Var}(X) = \sigma_x^2 \text{Var}(U) = \sigma_x^2.$$

Therefore, marginally $X$ follows $N(\mu_x, \sigma_x^2)$.

$$E(Y) = \sigma_y [\rho E(U) + \sqrt{1 - \rho^2} E(V)] + \mu_y = \mu_y, \quad \text{Var}(Y)$$
$$= \sigma_y^2 [\rho^2 \text{Var}(U) + (1 - \rho^2) \text{Var}(V)] = \sigma_y^2.$$

Therefore, marginally $Y$ follows $N(\mu_y, \sigma_y^2)$.

## 15.5.3 Covariance and Correlation

Note that $\text{Cov}(U, V) = 0$ since $U$ and $V$ are independent. Now the covariance of $X$ and $Y$ is:

$$\begin{aligned}
\text{Cov}(X, Y) &= \text{Cov}(\sigma_x U + \mu_x, \ \sigma_y [\rho U + \sqrt{1 - \rho^2} V] + \mu_y) \\
&= \text{Cov}(\sigma_x U, \ \sigma_y [\rho U + \sqrt{1 - \rho^2} V]) \\
&= \sigma_x \sigma_y \rho \text{Var}(U) + \sigma_x \sigma_y \sqrt{1 - \rho^2} \text{Cov}(U, V) \\
&= \sigma_x \sigma_y \rho + \sigma_x \sigma_y \sqrt{1 - \rho^2}(0) \\
&= \sigma_x \sigma_y \rho
\end{aligned}$$

Hence the correlation between $X$ and $Y$ is

$$\rho_{XY} = \frac{\text{Cov}(X, Y)}{\sigma_x \sigma_y} = \rho.$$

## 15.5.4 Independence

Recall that if $X$ and $Y$ are independent then $\rho_{XY} = 0$. Now only for this bivariate normal case the reverse is also true: Suppose that $\rho_{XY} = \rho = 0$, then the joint pdf of $X$ and $Y$ is:

$$f(x, y) = \frac{1}{2\pi\sigma_x\sigma_y} \exp\left[-\frac{1}{2}\left\{\left(\frac{x-\mu_x}{\sigma_x}\right)^2 + \left(\frac{y-\mu_y}{\sigma_y}\right)^2\right\}\right].$$
$$= \frac{1}{\sqrt{2\pi\sigma_x^2}} \exp\left[-\frac{1}{2\sigma_x^2}(x-\mu_x)^2\right] \times \frac{1}{\sqrt{2\pi\sigma_y^2}} \exp\left[-\frac{1}{2\sigma_y^2}(y-\mu_y)^2\right]$$
$$= f_X(x) \times f_Y(y)$$

Hence the above pdf factorises. Therefore $X$ and $Y$ are independent if $\rho = 0$. Hence we have the result: *Two random variables $X$ and $Y$ that have a bivariate normal distribution are independent if and only if they are uncorrelated.*

## 15.5.5 Conditional Distributions

The conditional distribution of $Y$ given that $X = x$ is found from the basic relations:

$$X = \sigma_x U + \mu_x, \quad Y = \sigma_y[\rho U + \sqrt{1-\rho^2} V] + \mu_y.$$

Given $X = x$, we have $U = \frac{x-\mu_x}{\sigma_x}$. Hence

$$Y = \sigma_y[\rho \frac{x-\mu_x}{\sigma_x} + \sqrt{1-\rho^2} V] + \mu_y,$$

given that $X = x$. The only random variable on the right hand side of the above is $V$, since we are assuming that $x$ is a constant. Therefore $Y|X = x$ is normally distributed (because $V \sim N(0, 1)$) with

$$E(Y|X = x) = \mu_y + \rho\frac{\sigma_y}{\sigma_x}(x-\mu_x), \quad \text{Var}(Y|X = x) = \sigma_y^2(1-\rho^2). \quad (15.3)$$

Alternatively, the pdf of the conditional distribution $Y$ given $X = x$, can be directly derived using the general formula

$$f(y|x) = \frac{f(x, y)}{f_X(x)},$$

see Sect. 8.5, where $f_X(x)$ is the marginal pdf of $X$.

The two results above (15.3) are very important in statistics. The conditional mean is often used in regression in the following form:

$$E(Y|X = x) = \mu_y + \frac{\text{Cov}(X, Y)}{\text{Var}(X)}(x - \mu_x).$$

This important formula is called the regression of $Y$ on $x$, see Chap. 17. The other conditional distribution $X|Y = y$ is obtained by switching the letters $x$ and $y$. $X|Y = y$ is normally distributed with

$$E(X|Y = y) = \mu_x + \rho \frac{\sigma_x}{\sigma_y}(y - \mu_y), \quad \text{Var}(X|Y = y) = \sigma_x^2(1 - \rho^2).$$

### 15.5.6  Linear Combinations

Distribution of $W = aX + bY + c$ for given constants $a$, $b$ and $c$ is normal because $Z$ can be expressed as a linear combination of $U$ and $V$ that we started with. The mean and variance of $W$ are:

$$E(W) = E(aX + bY + c) = a\mu_x + b\mu_y + c,$$

$$\begin{aligned}\text{Var}(W) &= \text{Var}(aX + bY + c) \\ &= a^2\text{Var}(X) + b^2\text{Var}(Y) + 2ab\text{Cov}(X, Y) \\ &= a^2\sigma_x^2 + b^2\sigma_y^2 + 2ab\rho\sigma_x\sigma_y.\end{aligned}$$

**Example 15.5 (Husbands and Wives)**

Let $X$ denote the height (in inches) of a wife and $Y$ denote the height of her husband. Suppose that the joint distribution of $X$ and $Y$ is bivariate normal with $\mu_x = 66.8$, $\sigma_x = 2$, $\mu_y = 70$, $\sigma_y = 2$ and $\rho = 0.68$. What is the probability that the wife will be taller than the husband for a randomly selected married couple? We want $P(X - Y > 0)$. Let $W = X - Y$.

$$E(W) = E(X - Y) = 66.8 - 70 = -3.2.$$

$$\text{Var}(W) = \text{Var}(X) + \text{Var}(Y) - 2\text{Cov}(X, Y) = 4 + 4 - 2(0.68)(2)(2) = 2.56.$$

Hence the standard deviation of $W$ is $\sqrt{2.56} = 1.6$. Therefore $Z = \frac{W - (-3.2)}{1.6}$ have a standard normal distribution and

$$P(W > 0) = P(Z > 2) = 1 - \Phi(2) = 0.0227$$

from the normal probability tables.

In this example let us answer another question. Suppose that the wife (of a randomly selected married couple) is 60 inches tall. What is the probability that her husband is taller than 66 inches?

We should use the conditional distribution of $Y|X = 60$ to find the probability. Note that

$$E(Y|X = 60) = \mu_y + \rho\frac{\sigma_y}{\sigma_x}(60 - \mu_x) = 65.376, \quad \text{Var}(Y|X = x)$$

$$= \sigma_y^2(1 - \rho^2) = 2.1504.$$

Let $W$ denote the random variable $Y|X = 60$. Now all that remains to find is $P(W > 66)$ where $W \sim N(65.376, 2.1504)$. Using the normal probability tables the required probability is $1 - \Phi\left(\frac{66-65.376}{\sqrt{2.1504}}\right) = 0.3352$. ◄

## 15.5.7 Exercises

**15.2 (Bivariate Normal Distribution)**

1. Let $X$ and $Y$ denote the scores of two class tests for a randomly selected student, called Miss T. Assume that $X$ and $Y$ is bivariate normal $N_2(\mu_x = 85, \mu_y = 90, \sigma_x = 10, \sigma_y = 16, \rho = 0.8)$.
   (a) What is the probability that sum of her score on the two tests will be greater than 200?
   (b) What is the probability that her score on the first test $(X)$ will be higher than her score on the second test?
   (c) If Miss T's score $X$ on the first test is 80, what is the probability that her score on the second test will be higher than 90?
2. Suppose that $X$ and $Y$ is $N_2(\mu_x, \mu_y, \sigma_x, \sigma_y, \rho)$ for which

$$E(X|Y = y) = 3.7 - 0.15y, \ E(Y|X = x) = 0.4 - 0.6x, \ \text{Var}(Y|X = x) = 3.64.$$

Find all the five parameters.
3. Consider the bivariate normal distribution $N_2(\mu_x, \mu_y, \sigma_x^2, \sigma_y^2, \rho)$ with pdf

$$f(x, y) = \frac{1}{2\pi\sigma_x\sigma_y\sqrt{1-\rho^2}} \times$$
$$\exp\left[-\frac{1}{2(1-\rho^2)}\left\{\left(\frac{x-\mu_x}{\sigma_x}\right)^2 - 2\rho\left(\frac{x-\mu_x}{\sigma_x}\right)\left(\frac{y-\mu_y}{\sigma_y}\right) + \left(\frac{y-\mu_y}{\sigma_y}\right)^2\right\}\right],$$

for $-\infty < x, y < \infty$. Assume the following.

$$E(Y|X = x) = \mu_y + \rho\frac{\sigma_y}{\sigma_x}(x - \mu_x) \text{ and } E(X|Y = y) = \mu_x + \rho\frac{\sigma_x}{\sigma_y}(y - \mu_y).$$

(a) Suppose that $\mu_x = \mu_y = 0$ and other three parameters are unknown. Is it possible to have $E(Y|X = x) = \frac{x}{2}$ and $E(X|Y = y) = 3y$? Give reasons.
(b) Suppose that $\mu_x = 0, \mu_y = 3, \sigma_x = 1, \sigma_y = 2$ and $\rho = 0.8$. Find $P(Y > 4|X = 1)$.

4. Suppose that $X$ and $Y$ is $N_2(\mu_x = 0, \mu_y = 0, \sigma_x = 1, \sigma_y = 1, \rho)$. Show that the sum $U = X + Y$ and the difference $V = X - Y$ are independent random variables. (**Hint**: Use the transformation technique.)

5. Consider the bivariate normal distribution $N_2(0, 0, 1, 1, \rho)$ with pdf

$$f(x, y) = \frac{1}{2\pi\sqrt{1 - \rho^2}} e^{\left[-\frac{1}{2(1-\rho^2)}\{x^2+y^2-2\rho xy\}\right]}, \quad -\infty < x, y < \infty$$

Consider the transformation $U = aX + bY$ and $V = aX - bY$.
(a) Obtain the covariance between $U$ and $V$. Hence conclude that $U$ and $V$ are independent if and only if $a^2 = b^2 (\neq 0)$.
(b) Obtain the joint pdf of $U$ and $V$.

6. Consider the bivariate normal distribution $N_2(\mu_x, \mu_y, \sigma_x^2, \sigma_y^2, \rho)$ with pdf

$$f(x, y) = \frac{1}{2\pi\sigma_x\sigma_y\sqrt{1-\rho^2}} \times$$
$$\exp\left[-\frac{1}{2(1-\rho^2)}\left\{\left(\frac{x-\mu_x}{\sigma_x}\right)^2 - 2\rho\left(\frac{x-\mu_x}{\sigma_x}\right)\left(\frac{y-\mu_y}{\sigma_y}\right) + \left(\frac{y-\mu_y}{\sigma_y}\right)^2\right\}\right],$$

for $-\infty < x, y < \infty$. Assume the following.

$$E(Y|X = x) = \mu_y + \rho\frac{\sigma_y}{\sigma_x}(x - \mu_x) \text{ and } E(X|Y = y) = \mu_x + \rho\frac{\sigma_x}{\sigma_y}(y - \mu_y).$$

(a) Suppose that $E(Y|X = x) = 3 + 1.6x$, $E(X|Y = y) = -1.2 + 0.4y$, and $\sigma_y = 2$. Find the remaining four parameters.
(b) For what value of $b$ is $\text{Var}(Y - bX)$ minimised? Show your derivation. Identify the value of $b$.
(c) Suppose that $\mu_x = \mu_y = 0$ and $\sigma_x = \sigma_y = 1$. Write down the joint pdf of $X$ and $Y$ for these specific values of the parameters. Let $U = X$ and $V = Y - \rho X$. Obtain the joint pdf of $U$ and $V$. Are $U$ and $V$ independent? Write down the marginal distribution of $V$.

## 15.6  Multivariate Probability Distributions

A multivariate probability distribution describes the joint variation of a collection of random variables.

Suppose that $X_1, X_2, \ldots, X_n$ are random variables. Then we write $\mathbf{X} = (X_1, X_2, \ldots, X_n)^T$ as the *vector* of random variables, and $S_{\mathbf{X}}$ as the sample space for $\mathbf{X}$.

**Discrete** multivariate probability distributions are defined by a probability (mass) function (p.f.)

$$f_{\mathbf{X}}(\mathbf{x}) \equiv P(\mathbf{X} = \mathbf{x})$$
$$= P(X_1 = x_1, X_2 = x_2, \ldots, X_n = x_n) \qquad \text{for } \mathbf{x} \in S_{\mathbf{X}}$$

where

$$\sum_{\mathbf{x} \in S_{\mathbf{X}}} f_{\mathbf{X}}(\mathbf{x}) = 1.$$

**Continuous** multivariate probability distributions are defined by a *joint* probability density function (p.d.f.) $f_{\mathbf{X}}(\mathbf{x})$ where

$$P(\mathbf{X} \in A) = \int_A f_{\mathbf{X}}(\mathbf{x})d\mathbf{x}, \qquad A \subseteq S_{\mathbf{X}}.$$

Hence

$$\int_{\mathcal{R}^n} f_{\mathbf{X}}(\mathbf{x})d\mathbf{x} = 1.$$

Henceforth, we shall restrict attention to continuous distributions (and assume that $S_{\mathbf{X}} = \mathcal{R}^n$). All of the properties we describe, can be applied to discrete distributions by replacing integration by summation.

The pdf of random variable $X_1$, a component of $\mathbf{X}$ is obtained from the joint pdf $f_{\mathbf{X}}$ by integrating out all the other variables

$$f_{X_1}(x_1) = \int_{-\infty}^{\infty} \int_{-\infty}^{\infty} \cdots \int_{-\infty}^{\infty} f_{\mathbf{X}}(x_1, x_2, \ldots, x_n)dx_2 \cdots dx_n.$$

The distribution of a component of a jointly distributed collection of variables is called the *marginal* distribution. This definition of a marginal distribution is also used for a selected set of components, or a function of selected set of components.

### 15.6.1 Expectation

The expectation $E(\mathbf{X})$ of $\mathbf{X} = (X_1, X_2, \ldots, X_n)^T$ is defined to be

$$E(\mathbf{X}) = [E(X_1), E(X_2), \ldots, E(X_n)]^T,$$

the vector containing the marginal expectations of each of the random variables. We often denote $E(\mathbf{X})$ by $\boldsymbol{\mu}$. It is clear that, for example,

$$E(X_1) = \int_{-\infty}^{\infty} \int_{-\infty}^{\infty} \cdots \int_{-\infty}^{\infty} x_1 f_{\mathbf{X}}(x_1, x_2, \ldots, x_n) dx_1 \cdots dx_n.$$

More generally, for any function $g(\mathbf{X})$ of $\mathbf{X}$, we define

$$E[g(\mathbf{X})] \equiv \int_{\mathcal{R}^n} g(\mathbf{x}) f_{\mathbf{X}}(\mathbf{x}) d\mathbf{x}.$$

It is immediately clear that

$$E(X_1 + X_2 + \ldots + X_n) = E(X_1) + E(X_2) + \ldots + E(X_n).$$

## 15.6.2 Variance and Covariance

The covariance of a pair of jointly distributed random variables, $X_1$ and $X_2$ is defined to be

$$\mathrm{Cov}(X_1, X_2) \equiv E([X_1 - E(X_1)][X_2 - E(X_2)])$$
$$= E(X_1 X_2) - E(X_1)E(X_2)$$

Note that $\mathrm{Cov}(X_1, X_1) = \mathrm{Var}(X_1)$, and

$$\mathrm{Var}(X_1 + X_2 + \ldots + X_n) = \sum_{i=1}^{n} \sum_{j=1}^{n} \mathrm{Cov}(X_i, X_j)$$
$$= \sum_{i=1}^{n} \mathrm{Var}(X_i) + \sum_{i \neq j} \mathrm{Cov}(X_i, X_j)$$
$$= \sum_{i=1}^{n} \mathrm{Var}(X_i) + \sum_{i < j} 2\mathrm{Cov}(X_i, X_j)$$

For a jointly distributed collection of random variables $\mathbf{X} = (X_1, X_2, \ldots, X_n)^T$, $\mathrm{Var}(\mathbf{X})$ is the *variance-covariance matrix*, denoted by $\boldsymbol{\Sigma}$, which is a $n \times n$ matrix whose entries are given by

$$\mathrm{Var}(\mathbf{X})_{ij} \equiv \mathrm{Cov}(X_i, X_j).$$

Thus we write,

$$\Sigma \equiv \text{Var}(\mathbf{X}) \equiv \begin{pmatrix} \text{Var}(X_1) & \text{Cov}(X_1, X_2) & \cdots & \text{Cov}(X_1, X_n) \\ \text{Cov}(X_2, X_1) & \text{Var}(X_2) & \cdots & \text{Cov}(X_2, X_n) \\ \vdots & \vdots & \ddots & \vdots \\ \text{Cov}(X_n, X_1) & \text{Cov}(X_n, X_2) & \cdots & \text{Var}(X_n) \end{pmatrix}.$$

The correlation of random variables $X_1$ and $X_2$ is defined as

$$\text{Corr}(X_1, X_2) = \frac{\text{Cov}(X_1, X_2)}{\sqrt{\text{Var}(X_1)\text{Var}(X_2)}}.$$

Note that $-1 \leq \text{Corr}(X_1, X_2) \leq 1$ as has been proved by the Cauchy-Schwarz inequality in Theorem 8.2.

### 15.6.3 Independence

A collection of continuous random variables $X_1, X_2, \ldots, X_n$ are said to be jointly *independent* if and only if their joint density is the product of their marginal densities *i.e.*

$$f_{\mathbf{X}}(\mathbf{x}) = f_{X_1}(x_1) f_{X_2}(x_2) \cdots f_{X_n}(x_n),$$

for all $\mathbf{x} \in S_{\mathbf{X}}$.

If $X_1, X_2, \ldots, X_n$ are independent then $E(X_1 X_2 \cdots X_n) = E(X_1)E(X_2) \cdots E(X_n)$, and hence for any pair of independent random variables

$$X_1 \text{ and } X_2 \text{ are independent} \Rightarrow \text{Cov}(X_1, X_2) = 0.$$

Independent random variables are uncorrelated, but uncorrelated random variables are not necessarily independent.

Other properties of independent random variables follow.

1. If the joint p.d.f. $f_{\mathbf{X}}(\mathbf{x})$ can be factorised into $n$ separate terms depending only on $x_1, x_2, \ldots, x_n$ respectively, then $X_1, X_2, \ldots, X_n$ are independent random variables.
2. (see also Sect. 13.2.) If $X_1, X_2, \ldots, X_n$ are independent random variables with m.g.f.s $m_{X_1}, m_{X_2}, \ldots, m_{X_n}$ respectively, and $Y = X_1 + X_2 + \ldots + X_n$, then the m.g.f. of $Y$ is given by

$$m_Y(t) = m_{X_1}(t) m_{X_2}(t) \cdots m_{X_n}(t).$$

326 15 Multivariate Distributions

3. If $X_1, X_2, \ldots, X_n$ are independent random variables and $Y = X_1 + X_2 + \ldots + X_n$, then

$$\text{Var}(Y) = \text{Var}(X_1) + \text{Var}(X_2) + \ldots + \text{Var}(X_n).$$

4. Pairwise independence does not imply joint independence.

A general property of mean vectors and variance matrices is that, if $\mathbf{B}$ is a $m \times n$ matrix of constants, $\mathbf{a}$ is a $m$-vector of constants, and $\mathbf{Z} = \mathbf{B}\mathbf{X} + \mathbf{a}$, so $\mathbf{Z}$ is a random $m$-vector, then

$$E(\mathbf{Z}) = \mathbf{B}E(\mathbf{X}) + \mathbf{a} = \mathbf{B}\boldsymbol{\mu} + \mathbf{a}$$

and

$$\text{Var}(\mathbf{Z}) = \mathbf{B}\text{Var}(\mathbf{X})\mathbf{B}^T = \mathbf{B}\boldsymbol{\Sigma}\mathbf{B}^T.$$

## 15.7 Joint Moment Generating Function

The mgf of a single random variable can be generalised to the joint mgf of more than one random variable, and it has the similar important properties as the mgf of one random variable. For example, the uniqueness theorem, see Sect. 13.2.1, still holds and it allows us to find the moments, e.g. mean and variance, by differentiation and also to find the joint pdfs of transformed random variables.

**Definition** For two random variables $X$ and $Y$ with joint pdf $f_{X,Y}(x, y)$, the joint mgf is given by

$$m_{X,Y}(t_1, t_2) = E[\exp\{t_1 X + t_2 Y\}] = E[\exp\{\mathbf{t}'\mathbf{X}\}]$$

where $\boldsymbol{\theta} = (t_1, t_2)'$ and $\mathbf{X} = (X, Y)'$. This definition of joint mgf for two random variables naturally extends to any number of random variables $n > 2$, say as follows:

$$m_{\mathbf{X}}(\mathbf{t}) = E[\exp\{t_1 X_1 + t_2 X_2 + \cdots + t_n X_n\}] = E[\exp\{\mathbf{t}'\mathbf{X}\}].$$

An example of a joint mgf is given in Sect. 15.7.1 below.

### 15.7.1 The Multivariate Normal Distribution

Suppose that $\mathbf{X} = (X_1, X_2, \ldots, X_n)^T$ is a collection of jointly distributed random variables. Then $\mathbf{X}$ is said to have a *multivariate normal distribution* if the p.d.f. of

**X** is given by

$$f_{\mathbf{X}}(\mathbf{x}) = (2\pi)^{-\frac{n}{2}} |\mathbf{\Sigma}|^{-\frac{1}{2}} \exp\left[-\tfrac{1}{2}(\mathbf{x} - \boldsymbol{\mu})^T \mathbf{\Sigma}^{-1}(\mathbf{x} - \boldsymbol{\mu})\right], \qquad \mathbf{x} \in \mathcal{R}^n.$$

where $\boldsymbol{\mu}$ is the mean vector, defined by

$$\boldsymbol{\mu} \equiv \begin{pmatrix} \mu_1 \\ \mu_2 \\ \vdots \\ \mu_n \end{pmatrix} = \begin{pmatrix} E(X_1) \\ E(X_2) \\ \vdots \\ E(X_n) \end{pmatrix} \equiv E(\mathbf{X})$$

and the $(i, j)$th element of $\mathbf{\Sigma}$ is $\mathrm{Cov}(X_i, X_j)$. We write $\mathbf{X} \sim N(\boldsymbol{\mu}, \mathbf{\Sigma})$.

The multivariate normal distribution has several appealing properties. If **X** has the multivariate normal distribution $N(\boldsymbol{\mu}, \mathbf{\Sigma})$ then:

1. The marginal distribution of any component of **X** is univariate normal. For example, $X_1$ is a $N(\mu_1, \Sigma_{11})$ random variable.
2. If **X** has a multivariate normal distribution then $\mathrm{Cov}(X_i, X_j) = 0 \Rightarrow X_i$ and $X_j$ are independent. However, this result is not true for other distributions, i.e. zero correlation (or covariance) between two random variables does not always mean independence between them.
3. If **X** has a multivariate normal distribution, then so does $\mathbf{Z} = \mathbf{BX} + \mathbf{a}$, so $\mathbf{Z} \sim N(\mathbf{B}\boldsymbol{\mu} + \mathbf{a}, \mathbf{B}\mathbf{\Sigma}\mathbf{B}^T)$. From this it follows [by considering $\mathbf{B} = (1\ 0 \ldots 0), \mathbf{a} = 0]$ that $X_1$ (and indeed every other component of **X**) has a univariate normal distribution.
4. If $\mathbf{\Sigma}$ is a diagonal matrix (all off-diagonal elements are 0, diagonal components are $\sigma_1^2, \ldots, \sigma_n^2$) then all components of **X** are uncorrelated.
5. Although zero correlation does not generally imply independence, it does for multivariate normal vectors, as then

$$f(\mathbf{x}) = (2\pi)^{-n/2} \prod_{i=1}^{n} \sigma_i \exp\left\{-\tfrac{1}{2} \sum_{i=1}^{n} \frac{(x_i - \mu_i)^2}{\sigma_i^2}\right\}$$

$$= \prod_{i=1}^{n} (2\pi \sigma_i^2)^{-1/2} \exp\left\{-\tfrac{1}{2} \frac{(x_i - \mu_i)^2}{\sigma_i^2}\right\}$$

so the joint density for **X** is the product of the $N(\mu_i, \sigma_i^2)$ marginal densities, which implies independence of $X_1, \ldots, X_n$.

The multivariate normal distribution and its properties above are used in practical modelling examples in Chap. 18.

## 15.7.2 Conditional Distribution

We now state the conditional distribution of a subset of the random variables $\mathbf{X}$ given the other random variables. Suppose that we partition the $n$-dimensional vector $X$ into one $n_1$ and another $n_2 = n - n_1$ dimensional random variable $\mathbf{X}_1$ and $\mathbf{X}_2$. Similarly partition $\boldsymbol{\mu}$ into two parts $\boldsymbol{\mu}_1$ and $\boldsymbol{\mu}_2$ so that we have:

$$\mathbf{X} = \begin{pmatrix} \mathbf{X}_1 \\ \mathbf{X}_2 \end{pmatrix}, \quad \boldsymbol{\mu} = \begin{pmatrix} \boldsymbol{\mu}_1 \\ \boldsymbol{\mu}_2 \end{pmatrix}.$$

Partition the $n \times n$ matrix $\Sigma$ into four matrices: $\Sigma_{11}$ having dimension $n_1 \times n_1$, $\Sigma_{12}$ having dimension $n_1 \times n_2$, $\Sigma_{21} = \Sigma_{12}'$ having dimension $n_2 \times n_1$, and $\Sigma_{22}$ having dimension $n_2 \times n_2$ so that we can write

$$\Sigma = \begin{pmatrix} \Sigma_{11} & \Sigma_{12} \\ \Sigma_{21} & \Sigma_{22} \end{pmatrix}.$$

The conditional distribution of $\mathbf{X}_1 | \mathbf{X}_2 = \mathbf{x}_2$ is the following normal distribution:

$$N\left( \boldsymbol{\mu}_1 + \Sigma_{12}\Sigma_{22}^{-1}(\mathbf{x}_2 - \boldsymbol{\mu}_2), \quad \Sigma_{11} - \Sigma_{12}\Sigma_{22}^{-1}\Sigma_{21} \right).$$

The marginal distribution of $\mathbf{X}_i$ is $N\left( \boldsymbol{\mu}_i, \Sigma_{ii} \right)$ for $i = 1, 2$.

## 15.7.3 Joint mgf of the Multivariate Normal Distribution

**Theorem 15.2** *If* $\mathbf{X} \sim N(\boldsymbol{\mu}, \Sigma)$ *then*

$$m_{\mathbf{X}}(\mathbf{t}) = e^{\boldsymbol{\mu}'\mathbf{t} + \frac{1}{2}\mathbf{t}'\Sigma\mathbf{t}}. \tag{15.4}$$

***Proof*** The proof this is a generalisation of the completing a square used in the proof in the univariate case in Example 13.4 for the standard normal distribution. We have

$$m_{\mathbf{X}}(\mathbf{t}) = \int_{-\infty}^{\infty} \cdots \int_{-\infty}^{\infty} e^{\mathbf{t}'\mathbf{x}} (2\pi)^{-\frac{n}{2}} |\Sigma|^{-\frac{1}{2}} e^{-\frac{1}{2}(\mathbf{x}-\boldsymbol{\mu})'\Sigma^{-1}(\mathbf{x}-\boldsymbol{\mu})} dx_1 \ldots dx_n$$
$$= (2\pi)^{-\frac{n}{2}} |\Sigma|^{-\frac{1}{2}} \int_{-\infty}^{\infty} \cdots \int_{-\infty}^{\infty} e^{-\frac{1}{2}\left\{(\mathbf{x}-\boldsymbol{\mu})'\Sigma^{-1}(\mathbf{x}-\boldsymbol{\mu}) - 2\mathbf{t}'\mathbf{x}\right\}} dx_1 \ldots dx_n.$$

Now

$$(x - \mu)'\Sigma^{-1}(x - \mu) - 2t'x = x'\Sigma^{-1}x + \mu'\Sigma^{-1}\mu - 2x'\Sigma^{-1}\mu - 2x't$$
$$= x'\Sigma^{-1}x + \mu'\Sigma^{-1}\mu - 2x'\Sigma^{-1}(\mu + \Sigma t)$$
$$= (x - \mu - \Sigma t)'\Sigma^{-1}(x - \mu - \Sigma t)' + \mu'\Sigma^{-1}\mu$$
$$\quad -(\mu + \Sigma t)\Sigma^{-1}(\mu + \Sigma t)$$
$$= (x - \mu - \Sigma t)'\Sigma^{-1}(x - \mu - \Sigma t)' - t'\Sigma t - 2\mu't$$

since

$$(\mu + \Sigma t)\Sigma^{-1}(\mu + \Sigma t) = \mu'\Sigma^1\mu + 2\mu'\Sigma^1\Sigma t + t'\Sigma\Sigma^{-1}\Sigma t$$
$$= \mu'\Sigma^1\mu + 2\mu't + t'\Sigma t.$$

Now

$$m_X(t)$$
$$= (2\pi)^{-\frac{n}{2}}|\Sigma|^{-\frac{1}{2}} \int_{-\infty}^{\infty} \cdots \int_{-\infty}^{\infty} e^{-\frac{1}{2}\{(x-\mu-\Sigma t)'\Sigma^{-1}(x-\mu-\Sigma t)' - t'\Sigma t - 2\mu't\}} dx_1 \ldots dx_n$$
$$= e^{\mu't+\frac{1}{2}t'\Sigma t} \int_{-\infty}^{\infty} \cdots \int_{-\infty}^{\infty} (2\pi)^{-\frac{n}{2}}|\Sigma|^{-\frac{1}{2}} e^{-\frac{1}{2}\{(x-\mu-\Sigma t)'\Sigma^{-1}(x-\mu-\Sigma t)'\}} dx_1 \ldots dx_n$$
$$= e^{\mu't+\frac{1}{2}t'\Sigma t}$$

since the last integral is 1 as it is the total probability integral for the multivariate normal distribution with mean $\mu + \Sigma t$ and covariance matrix $\Sigma$.

**Results Obtained Using the mgf**

1. Any (non-empty) subset of multivariate normal is also multivariate normal. In the joint mgf, we simply put $t_j = 0$ for all $j$ for which $X_j$ is not in the subset. The resulting mgf can be recognised as the mgf of a multivariate normal distribution. For example,

$$M_{X_1}(t_1) = M_X(t_1, 0, \ldots, 0) = e^{t_1\mu_1 + t_1^2\sigma_1^2/2}.$$

Hence, by the uniqueness theorem of the mgf, $X_1 \sim N\left(\mu_1, \sigma_1^2\right)$. Obviously, any index $i = 1, \ldots, n$, can be used in place of 1, to conclude that $X_i \sim N\left(\mu_i, \sigma_i^2\right)$ for any $i$. Also, if $n = 2$, we have:

$$M_{X_1,X_2}(t_1, t_2) = M_X(t_1, t_2, 0, \ldots, 0) = e^{t_1\mu_1 + + t_2\mu_2 + \frac{1}{2}\left(t_1^2\sigma_1^2 + 2\sigma_{12}t_1t_2 + \sigma_2^2t_2^2\right)},$$

which is the mgf of the bivariate normal distribution, we discussed in Sect. 15.5.

2. $X$ is a vector of independent random variables if and only if $\Sigma$ is a diagonal matrix.

If $X_1, \ldots, X_n$ are independent then $\sigma_{ij} = \text{Cov}(X_i, X_j) = 0$ for $i \neq j$, so that $\Sigma$ is diagonal.

If $\Sigma$ is diagonal then $\mathbf{t}'\Sigma\mathbf{t} = \sum_{i=1}^{n} \sigma_i^2 t_i^2$ and hence

$$M_{\mathbf{X}}(\mathbf{t}) = e^{\mu'\mathbf{t}+\frac{1}{2}\mathbf{t}'\Sigma\mathbf{t}} = e^{\sum_{i=1}^{n}\left(\mu_i t_i + \frac{1}{2}\sigma_i^2 t_i^2\right)} = \prod_{i=1}^{n} M_{X_i}(t_i),$$

which shows that $X_1, \ldots, X_n$ are independent by using the uniqueness theorem of the joint mgf.

3. If $\mathbf{Y} = \mathbf{a} + B\mathbf{X}$ where $\mathbf{a}$ a column vector of $n$ constants and $B$ is an $n \times n$ non-singular matrix of constants, then $Y \sim N\left(\mathbf{a} + B\mu, B\Sigma B'\right)$. This is easily proved using the joint mgf as follows:

$$m_{\mathbf{Y}}(\mathbf{t}) = E[\exp\{\mathbf{t}'\mathbf{Y}\}]$$

$$= E[\exp\{\mathbf{t}'(\mathbf{a} + B\mathbf{X})\}]$$

$$= e^{\mathbf{t}'\mathbf{a}} E[\exp\{(B'\mathbf{t})'\mathbf{X}\}]$$

$$= e^{\mathbf{t}'\mathbf{a}} m_{\mathbf{X}}(B'\mathbf{t})$$

$$= e^{\mathbf{t}'\mathbf{a}} \exp\left(\mu'(B'\mathbf{t}) + \frac{1}{2}(B'\mathbf{t})'\Sigma(B'\mathbf{t})\right)$$

$$= \exp\left((\mathbf{a} + B\mu)'\mathbf{t} + \frac{1}{2}\mathbf{t}'(B\Sigma B')\mathbf{t}\right),$$

which is recognised to be the mgf of $N\left(\mathbf{a} + B\mu, B\Sigma B'\right)$, and hence the result follows by using the uniqueness theorem of the mgf.

$\square$

## Example 15.6

Suppose $X_1$ and $X_2$ are independent $N(0, 1)$ random variables. Two new random variables $Y_1$ and $Y_2$ are defined as

$$Y_1 = \frac{1}{\sqrt{5}}X_1 + \frac{2}{\sqrt{5}}X_2, \quad Y_2 = \frac{2}{\sqrt{5}}X_1 - \frac{1}{\sqrt{5}}X_2.$$

Find the joint moment generating function of $Y_1$ and $Y_2$ and state their joint distribution.

Here we have:

$$m_{\mathbf{X}}(\mathbf{t}) = e^{\mathbf{t}'\mu\frac{1}{2}\mathbf{t}'\Sigma\mathbf{t}}$$

where

$$\mu = \begin{pmatrix} 0 \\ 0 \end{pmatrix}, \Sigma = \begin{pmatrix} 1 & 0 \\ 0 & 1 \end{pmatrix}.$$

Now

$$m_{Y_1, Y_2}(t_1, t_2) = E\left(e^{t_1 Y_1 + t_2 Y_2}\right)$$

$$= E\left(e^{t_1\left(\frac{1}{\sqrt{5}}X_1 + \frac{2}{\sqrt{5}}X_2\right) + t_2\left(\frac{2}{\sqrt{5}}X_1 - \frac{1}{\sqrt{5}}X_2\right)}\right)$$

$$= E\left(e^{\left(\frac{t_1}{\sqrt{5}} + \frac{2t_2}{\sqrt{5}}\right)X_1}\right) E\left(e^{\left(\frac{2t_1}{\sqrt{5}} - \frac{t_2}{\sqrt{5}}\right)X_2}\right)$$

$$= m_{X_1}\left(\frac{t_1}{\sqrt{5}} + \frac{2t_2}{\sqrt{5}}\right) m_{X_2}\left(\frac{2t_1}{\sqrt{5}} - \frac{t_2}{\sqrt{5}}\right)$$

$$= e^{\frac{1}{2}\left(\frac{t_1}{\sqrt{5}} + \frac{2t_2}{\sqrt{5}}\right)^2} e^{\frac{1}{2}\left(\frac{2t_1}{\sqrt{5}} - \frac{t_2}{\sqrt{5}}\right)^2}$$

$$= e^{\frac{1}{2}(t_1^2 + t_2^2)}$$

$$= e^{(t_1, t_2)\begin{pmatrix} 0 \\ 0 \end{pmatrix} + \frac{1}{2}(t_1, t_2)\begin{pmatrix} 1 & 0 \\ 0 & 1 \end{pmatrix}\begin{pmatrix} t_1 \\ t_2 \end{pmatrix}}$$

Hence,

$$\begin{pmatrix} Y_1 \\ Y_2 \end{pmatrix} \sim N\left(\begin{pmatrix} 0 \\ 0 \end{pmatrix}, \begin{pmatrix} 1 & 0 \\ 0 & 1 \end{pmatrix}\right).$$

◀

## 15.8  Multinomial Distribution

Suppose that a population contains items of $k$ different types ($k \geq 2$) and that the proportion of the items in the population that are of type $i$ is $p_i$ ($i = 1, 2, \ldots, k$). It is assumed that $p_i > 0$ for each $i$ and that $\sum_{i=1}^{k} p_i = 1$. Furthermore, suppose that $n$ items are selected at random from the population with replacement; and let $X_i$ denote the number of selected items that are of type $i$. Then it is said that the random vector $\mathbf{X} = (X_1, \ldots, X_k)$ has a *multinomial distribution* with parameters $n$ and $\mathbf{p} = (p_1, \ldots, p_k)$.

The pmf of $\mathbf{X}$ is:

$$P(X_1 = x_1, \ldots, X_k = x_k) = \frac{n!}{x_1! \cdots x_k!} p_1^{x_1} \cdots p_k^{x_k},$$

where $0 < p_i < 1$ and $0 < x_i < n$ for all $i = 1, \ldots, n$ and

$$\sum_{i=1}^{k} p_i = 1, \quad \text{and} \quad \sum_{i=1}^{k} x_i = n.$$

**Example 15.7**

Suppose that 23% of the people attending a certain baseball game live within 10 miles of the stadium; 59% live within 10 and 50 miles, and 18% live more than 50 miles from the stadium. What is the probability that, out of 20 randomly selected people attending the game, 7 live within 10 miles, 8 live between 10 and 50 miles and 5 live more than 50 miles from the stadium?

$$\text{Answer} = \frac{20!}{7!\,8!\,5!}(0.23)^7\,(0.59)^8\,(0.18)^5 = 0.0094$$

◄

### 15.8.1 Relation Between the Multinomial and Binomial Distributions

When the population being sampled contains only two different types of items, that is, when $k = 2$ the multinomial distribution reduces to the binomial distribution.

Let us return to the general case. Look at $X_i$ alone. Since $X_i$ can be regarded as the total number of items of type $i$ that are selected in $n$ Bernoulli trials, when the probability of selection on each trial is $p_i$, it follows that the *marginal distribution* of each variable $X_i$ must be a binomial distribution with parameters $n$ and $p_i$. Hence for every $i$,

$$E(X_i) = np_i \quad \text{and} \quad \text{Var}(X_i) = np_i(1 - p_i).$$

### 15.8.2 Means, Variances and Covariances

Consider two components $X_i$ and $X_j$ where $i \neq j$. The sum $X_i + X_j$ can be regarded as the total number of items of either type $i$ or type $j$ that are selected in $n$ Bernoulli trials, when the probability of selection on each trial is $p_i + p_j$, it follows that $X_i + X_j$ has a binomial distribution with parameters $n$ and $p_i + p_j$. Hence:

$$\text{Var}(X_i + X_j) = n(p_i + p_j)(1 - p_i - p_j).$$

However we proved earlier Sect. 8.6 that:

$$\text{Var}(X_i + X_j) = \text{Var}(X_i) + \text{Var}(X_j) + 2\text{Cov}(X_i, X_j).$$

To find $\text{Cov}(X_i, X_j)$, we set

$$\text{Var}(X_i) + \text{Var}(X_j) + 2\text{Cov}(X_i, X_j) = n(p_i + p_j)(1 - p_i - p_j),$$

We know $\text{Var}(X_i) = np_i(1 - p_i)$ and $\text{Var}(X_j) = np_j(1 - p_j)$. Therefore:

$$np_i(1 - p_i) + np_j(1 - p_j) + 2\text{Cov}(X_i, X_j) = n(p_i + p_j)(1 - p_i - p_j),$$

Hence we obtain that

$$\text{Cov}(X_i, X_j) = -np_i p_j.$$

### 15.8.3 Exercises

#### 15.3 (Multinomial Distributions)

1. Suppose that F is a continuous cdf on the real line; and let $\alpha_1$ and $\alpha_2$ be numbers such that $F(\alpha_1) = 0.3$ and $F(\alpha_2) = 0.8$. If 25 observations are selected at random from the distribution for which the cdf is $F$, what is the probability that 6 of the observed values will be less than $\alpha_1$, 10 of the observed values will be between $\alpha_1$ and $\alpha_2$, and 9 of the observed values will be greater than $\alpha_2$?
2. Suppose that 40% of the students in a large population are freshmen, 30% are sophomores, 20% are juniors and 10% are seniors. Suppose that 10 students are selected at random from the population; and let $X_1, X_2, X_3, X_4$ denote, respectively, the numbers of freshmen, sophomores, juniors, and seniors that are obtained.
   (a) Determine $\rho(X_i, X_j)$ for each pair of values $i$ and $j$ $(i < j)$.
   (b) For what values of $i$ and $j$ $(i < j)$ is $\rho(X_i, X_j)$ most negative?
   (c) For what values of $i$ and $j$ $(i < j)$ is $\rho(X_i, X_j)$ closest to 0?  □

## 15.9  Compound Distributions

A compound distribution is a model for a random sum $Y = X_1 + X_2 + \cdots + X_N$, where the number of terms $N$ is itself a random variable. To make the compound distribution more tractable, we assume that the variables $X_n$ are independent and identically distributed and that each $X_n$, for $n = 1, \ldots, N$, is independent of $N$. The random sum $Y$ can be interpreted the sum of all the measurements that are associated with certain events that occur during a fixed period of time. For example, we may be interested in the total amount of rainfall in a 24-hour period, during which a random number of showers are observed and each of the showers provides a measurement of an amount of rainfall. Another example of a compound distribution is the random variable of the aggregate claims generated by an insurance policy or a group of insurance policies during a fixed policy period. In this setting, $N$ is the number of claims generated by the portfolio of insurance policies and $X_1$ is the amount of the first claim and $X_2$ is the amount of the second claim and so on.

When $N$ follows the Poisson distribution, the random sum $Y$ is said to have a compound Poisson distribution. Similarly, when $N$ follows the Binomial distribution, the random sum $Y$ is said to have a compound Binomial distribution, and when $N$ follows the negative Binomial distribution, the random sum $Y$ is said to have a compound negative Binomial distribution.

A compound Poisson distribution is the probability distribution of the sum of a number of independent identically-distributed random variables, where the number of terms to be added is itself a Poisson-distributed variable. In the simplest cases, the result can be either a continuous or a discrete distribution.

**Definition**
Suppose that

$$N \sim \text{Poisson}(\lambda),$$

i.e., $N$ is a random variable whose distribution is the Poisson distribution with expected value $\lambda$, and that $X_1, X_2, X_3, \ldots$ are identically distributed random variables that are mutually independent and also independent of $N$. Then the probability distribution of the sum of $N$ i.i.d. random variables

$$Y = \sum_{n=1}^{N} X_n \tag{15.5}$$

is a compound Poisson distribution.

In the case $N = 0$, then this is a sum of 0 terms, so the value of $Y$ is 0. Hence the conditional distribution of $Y$ given that $N{=}0$ is a degenerate distribution.

The compound Poisson distribution is obtained by marginalising the joint distribution of $(Y,N)$ over $N$, and this joint distribution can be obtained by combining the conditional distribution $Y|N$ with the marginal distribution of $N$. To find the $E(Y|N)$ and $\text{Var}(Y|N)$ we exploit the sum of iid random variables in (15.5). Thus, $E(Y|N)$ is the expectation of $N$, where $N$ is fixed because of conditioning, iid random variables and hence $E(Y|N) = N\,E(X)$. Using the same arguments, we also have $\text{Var}(Y|N) = N\text{Var}(X)$.

The expected value and the variance of the compound distribution can be derived in a simple way from the law of total expectation and the law of total variance introduced in Sect. 15.3. Thus

$$\begin{aligned} E(Y) &= E[E(Y|N)] \\ &= E\,[N(E(X)] \\ &= E(N)E(X), \end{aligned}$$

and

$$
\begin{aligned}
\mathrm{Var}(Y) &= E\left[\mathrm{Var}(Y \mid N)\right] + \mathrm{Var}\left[E(Y \mid N)\right]\\
&= E\left[N\mathrm{Var}(X)\right] + \mathrm{Var}\left[NE(X)\right],\\
&= E(N)\mathrm{Var}(X) + (E(X))^2\,\mathrm{Var}(N).
\end{aligned}
$$

Then, since $E(N)=\mathrm{Var}(N)$ if $N$ is Poisson, this formula is reduced to

$$
\begin{aligned}
\mathrm{Var}(Y) &= E(N)\left[\mathrm{Var}(X) + E(X)^2\right]\\
&= E(N)E(X^2).
\end{aligned}
$$

The probability distribution of $Y$ can be determined in terms of the characteristic function in probability theory defined by:

$$
\begin{aligned}
\varphi_Y(t) &= E\left[e^{itY}\right]\\
&= E_N\left[E_{Y|N}\left(e^{itY}\right)\right]\\
&= E_N\left[E_{Y|N}\left(e^{it\sum_{n=1}^{N}X_n}\right)\right]\\
&= E_N\left[E_{Y|N}\left(e^{itX_1}\cdots e^{itX_N}\right)\right]\\
&= E_N\left[E_X\left(e^{itX}\right)^N\right]\\
&= E_N\left[(\varphi_X(t))^N\right]\\
&= E\left[(\varphi_X(t))^N\right],
\end{aligned}
$$

and hence, using the probability-generating function of the Poisson distribution, we have

$$
\varphi_Y(t) = e^{\lambda(\varphi_X(t)-1)}.
$$

An alternative approach is via the cumulant generating function:

$$
\begin{aligned}
K_Y(t) &= \log E[e^{tY}]\\
&= \log E[E[e^{tY} \mid N]]\\
&= \log E[e^{NK_X(t)}] = K_N(K_X(t)).
\end{aligned}
$$

It can be shown that, if the mean of the Poisson distribution $\lambda = 1$, the cumulants of $Y$ are the same as the moments of $X_1$.

# Convergence of Estimators

**16**

**Abstract**

Chapter 16 discusses asymptotic theories which are often required to guarantee good properties of statistical inference techniques. Three types of modes of convergence in statistics are discussed and illustrated with the help of simulation using R routines. Large sample properties of the maximum likelihood estimators are stated and so are the laws of large numbers.

This chapter studies the behaviour of the estimators as the sample size increases in mostly hypothetical situations. The main idea in statistics is to draw samples to make inference about unknown population quantities. The study of convergence of estimators is to check up on that in the sense that if we keep taking increasing number of independent samples from the population will the estimators get closer and closer to the target fixed and unknown population parameter? It is not always the case that we can afford to draw an ever increasing number of samples repeatedly. Instead, often the main aim of the study of convergence is to establish the behaviour of the estimators for large samples and then choose these same estimators even for smaller sample sizes with the hope that some of the large sample properties may continue to hold even for smaller sample sizes. This is especially relevant in cases where we are not able to compare and then select from among competing estimators using exact theoretical results for small sample sizes.

Historically, some form of convergence concepts were studied when statisticians were faced with difficulty in evaluating distributions and probabilities for large sample sizes. For example, imagine a situation where the CLT stated in Sect. 8.8 has been used to estimate a binomial probability when the number of trials $n$ is very large. In such a case the CLT provides an approximate practical solution in the absence of a modern computer software based method. Limiting distributions of estimators are also used to facilitate statistical inference, e.g., interval estimation and testing of hypothesis when the distribution of a pivot statistic (Sect. 11.1) cannot be determined exactly. In this chapter we discuss what is known as general convergence in distribution concepts underlying the CLT.

Another important convergence concept is when the sampling distribution of a sequence of random variables converges to a constant. Such a concept is called convergence in probability and is formally defined later in this chapter. We have encountered illustration of such type of convergence already. For example, in Sect. 9.4 we have established $E(\bar{X}) = \mu$ and $\text{Var}(\bar{X}) = \frac{\sigma^2}{n}$. It is plain to see that $\text{Var}(\bar{X}) \to 0$ as $n \to \infty$ when $\sigma^2$ is finite. Zero variance random variables are essentially constants and hence we can say that the sampling distribution of $\bar{X}$ converges to the constant $\mu$, the population mean under suitable conditions, e.g. finite variance and i.i.d. sampling.

This chapter only studies two broadly important modes of convergence: convergence in distribution and convergence in probability and the associated consequences on the properties of estimators such as the CLT, consistency and what is known as the weak law of large numbers. There are many other modes of convergence such as: almost sure convergence, strong law of large numbers etc. that require much higher levels of sophistication in mathematics. Hence those topics are omitted from our presentations in this chapter. We also exclude more rigorous statements on the two modes of convergence requiring much less stringent assumptions due to our assumption of a lower level of preparedness in mathematics.

A drawback of the study of convergence lies in the practical inability to provide an exact sample size that may guarantee the large sample properties claimed in the theorems. For example, a theorem such as the CLT may state asymptotic normality as $n \to \infty$ but then a valid practical question might be: "is $n = 200$ large enough?" A satisfying answer to such a question is generally problem specific. Theoretical investigations studying the speed of convergence may also be conducted but such a topic is beyond the scope of the current book. Instead, where possible, we shall illustrate convergence behaviours using simulations in R. A simulation study involves drawing random samples from a population and calculating the statistics of interest repeatedly for increasing sample sizes to discover patterns. Several simulation examples are provided in this chapter.

## 16.1  Convergence in Distribution

Suppose that we have a statistical model specified as a joint probability distribution for the random variables $X_1, \ldots, X_n$, and that a particular estimator $\tilde{\theta}_n(\mathbf{X})$ is proposed for estimating a particular parameter $\theta$ of interest. Here, we make explicit the dependence of the estimator $\tilde{\theta}_n$ on the sample size $n$. So $\tilde{\theta}_1, \tilde{\theta}_2, \ldots, \tilde{\theta}_n, \ldots$ is a sequence of random variables representing the estimator for different sample sizes. For example, $\tilde{\theta}_n = \tilde{\theta}_n(\mathbf{X})$, e.g. for a random sample $X_1, \ldots, X_n$,

$$\tilde{\theta}_n = \bar{X}_n = \frac{1}{n} \sum_{i=1}^{n} X_i.$$

where $X_1, \ldots, X_n$ form an iid sample. That is $\tilde{\theta}_n$ is a sequence of random variables.

Suppose that $Y_n$, e.g. $\bar{X}_n$, is a sequence of random variables depending on $n$, the sample size. Let $Y$ be another random variable whose probability distribution does not depend on $n$. We say that $Y_n$ *converges in distribution* to $Y$ if

$$\lim_{n \to \infty} P(Y_n \leq y) = P(Y \leq y) \tag{16.1}$$

for all $-\infty < y < \infty$. Notice that $P(Y_n \leq y)$ in the left hand side of (16.1) is the c.d.f. of the random variable $Y_n$ for each fixed value of $n$. Hence, convergence in distribution guarantees the convergence of the cdf to the c.d.f. , $P(Y \leq y)$, of another suitable random variable $Y$.

The most practical use of this concept of convergence lies in the ability to approximate $P(a \leq Y_n \leq b)$ by $P(a \leq Y \leq b)$ for any two real numbers $a \leq b$ when $n$ is large. Note that, in general, we may not be able to derive the probability function, p.m.f. if discrete or pdf if continuous, of $Y_n$ to evaluate $P(a \leq Y_n \leq b)$. But if $Y_n$ converges in distribution to $Y$, the required probability can be approximated by $P(a \leq Y \leq b)$. Better yet, the distribution of $Y$ may be continuous even if $Y_n$ is obtained for sequence of discrete valued random variables, $X_1, \ldots, X_n$.

The CLT is an example of this mode of convergence where $Y_n = \frac{\sqrt{n}(\bar{X}_n - \mu)}{\sigma}$ and $Y = N(0, 1)$ where $X_1, \ldots, X_n$ is a random sample from a population having mean $\mu$ and variance $\sigma^2$. For example, $X_1, \ldots, X_n$ may be a sequence of Bernoulli trials, inherently discrete, but the limiting distribution $Y_n$ is the standard normal distribution which is continuous.

Thus, starting from almost no assumptions on the population other than having finite variance, the CLT claims that the sample average of a random sample is approximately normally distributed. This remarkable central behaviour ensues for any population distribution having finite variance. The population distribution can be uniform or even extremely skewed to the left or right, e.g. $\chi^2$ with 1 degree of freedom or *multi-modal*.

Why should the CLT hold? In other words, why can we even expect to see the remarkable central behaviour? The intuition behind is that normality comes from the average (or sum) of a large number (as $n \to \infty$) of small, guaranteed by the finite variance assumption, independent shocks or disturbances. The finite variance assumption is thus essential to achieve limiting normality. However, this assumption can be relaxed in other alternative versions of the CLT which we do not discuss here. Below, we sketch a proof of the CLT.

***Proof*** The proof uses the uniqueness of the moment generating function noted in Theorem 13.3 in Chap. 13. The aim is to prove that the mgf of $Y_n = \frac{\sqrt{n}(\bar{X}_n - \mu)}{\sigma}$ is the mgf of the standard normal distribution $e^{t^2/2}$ as $n \to \infty$ for some $|t| < h$ for some $h > 0$.

Let $W_i = \frac{X_i - \mu}{\sigma}$ so that

$$E(W_i) = \left. \frac{dM_W(t)}{dt} \right|_{t=0} = 0, \quad \mathrm{Var}(W_i) = E(W_i^2) = \left. \frac{d^2 M_W(t)}{dt^2} \right|_{t=0} = 1,$$

$$(16.2)$$

where $M_W(t) = E(e^{tW})$ denotes the mgf of the random variable $W$ for $|t| < h$. Now,

$$Y_n = \frac{\sqrt{n}(\bar{X}_n - \mu)}{\sigma} = \frac{n(\bar{X}_n - \mu)}{\sqrt{n}\sigma} = \frac{1}{\sqrt{n}} \sum_{i=1}^{n} W_i.$$

Now,

$$M_{Y_n}(t) = E\left( e^{\frac{t}{\sqrt{n}} \sum_{i=1}^{n} W_i} \right)$$
$$= M_{W_1}\left( \frac{t}{\sqrt{n}} \right) \times \ldots \times M_{W_n}\left( \frac{t}{\sqrt{n}} \right)$$
$$= \left[ M_W\left( \frac{t}{\sqrt{n}} \right) \right]^n,$$

where the first equality follows from the definition and the last equality follows from the fact $W_1, \ldots, W_n$ are i.i.d. We now expand $M_W\left( \frac{t}{\sqrt{n}} \right)$ in a Taylor series around 0,

$$M_W\left( \frac{t}{\sqrt{n}} \right) = 1 + \left. \frac{dM_W}{dt}\left( \frac{t}{\sqrt{n}} \right) \right|_{t=0} \left( \frac{t}{\sqrt{n}} \right) + \left. \frac{d^2 M_W\left( \frac{t}{\sqrt{n}} \right)}{dt^2} \right|_{t=0} \frac{1}{2!}\left( \frac{t}{\sqrt{n}} \right)^2 + R\left( \frac{t}{\sqrt{n}} \right)$$
$$= 1 + \frac{t^2}{2n} + R\left( \frac{t}{\sqrt{n}} \right), \quad \text{[by applying (16.2)]},$$

where $R\left( \frac{t}{\sqrt{n}} \right)$ is the remainder term. Further non-trivial technical arguments using the Taylor series, will guarantee that the remainder term approaches 0 as $n \to \infty$, see for example, Section 5.3 of Casella and Berger [3]. These arguments are even more technical since the remainder term is inside the $n$th power in $\left[ M_W\left( \frac{t}{\sqrt{n}} \right) \right]^n$.

Now,

$$\lim_{n \to \infty} M_{Y_n}(t) = \lim_{n \to \infty} \left[ M_W \left( \frac{t}{\sqrt{n}} \right) \right]^n$$
$$= \lim_{n \to \infty} \left[ 1 + \frac{t^2}{2n} \right]^n$$
$$= e^{\frac{t^2}{2}},$$

for $|t| < a$ where $a$ is a positive number depending on $h$, $n$ and $\sigma$.  □

### Example 16.1

Sampling from the Uniform distribution. Suppose that $X_1, \ldots, X_n$ is a random sample from the uniform distribution taking values in the unit interval $(0, 1)$. Here $E(X_i) = 0.5 = \mu$ and $\text{Var}(X_i) = \sigma^2 = \frac{1}{12}$. Hence, by the CLT we can claim that

$$Y_n = \frac{\sqrt{n}(\bar{X}_n - 0.5)}{\sqrt{\frac{1}{12}}} = \sqrt{3n}(2\bar{X}_n - 1)$$

converges to the standard normal distribution as $n \to \infty$. Indeed, the convergence has been claimed to occur very quickly, e.g. for $n$ as low as 12. Thus it has been claimed that

$$Y_{12} = \sum_{i=1}^{12} X_i - 6,$$

follows the standard normal distribution, where $X_1, \ldots, X_{12}$ is a random sample from the uniform distribution $U(0, 1)$.

Hence, we expect that if we draw a sample of size 12 from the $U(0, 1)$ distribution and then form the statistic $Y_{12}$, then $Y_{12}$ will be approximately normally distributed with mean 0 and variance 1. Of-course, one single observation of $Y_{12}$ will not show a bell shaped histogram plot! Hence, we draw a fixed number, $R$ say, of replicated realised values of $Y_{12}$. Note that in order to get a single observed value $y_{12}$ of $Y_{12}$, we require $n = 12$ samples from $U(0, 1)$. In our experiment we take $R = 10, 000$ and evaluate $y_{12}$ for each of these $R$ samples. The resulting samples are used to draw the summary graphs in Fig. 16.1. The R package `ipsRdbs` provides the function `see_the_clt_for_uniform` to help draw the two panels in this figure. ◄

**Fig. 16.1** Illustration of the CLT for sampling from the uniform distribution. The plots can be replicated with example code seen by issuing the command ? `see_the_clt_for_uniform`

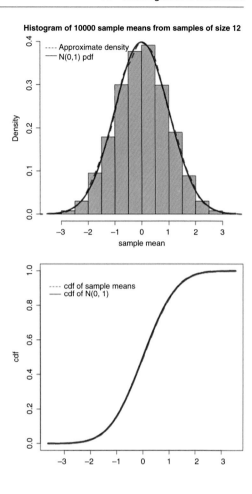

## 16.2    Convergence in Probability

The concept of convergence in probability is now discussed which allows us to discuss a mode of convergence of random variables to a constant. We say that $Y_n$ *converges in probability* to $a$, or $Y_n \xrightarrow{p} a$, if for any $\epsilon > 0$

$$P(|Y_n - a| > \epsilon) \to 0 \qquad \text{as } n \to \infty. \tag{16.3}$$

This definition states that the limit of the probability of divergence of $Y_n$ from $a$, $(|Y_n - a| > \epsilon)$, approaches zero as $n \to \infty$. The extent of divergence is measured by $\epsilon > 0$ for any $\epsilon > 0$. The specification of *any* $\epsilon$ assures that the extent of divergence can be taken to be negligible.

Convergence in probability in is used to prove the *Weak law of large numbers* (**WLLN**), stated as follows. Let $X_1, \ldots, X_n$ be i.i.d. random variables following the probability distribution of $X$ where $E(X) = \mu$ and finite variance $\sigma^2$. Let $\bar{X}_n = \frac{1}{n}\sum_{i=1}^{n} X_i$ denote the sample mean of the random sample. The WLLN states that

$$\bar{X}_n \xrightarrow{p} \mu, \qquad \text{as } n \to \infty.$$

Using the interpretation of convergence in probability, see (16.3), we see that the WLLN, if it holds, guarantees that the sample mean can be taken to be an arbitrarily close approximation of the population mean since the limit of the probability of divergence $|X_n - \mu| > \epsilon$ goes to zero for any $\epsilon > 0$. Note that we do not simply write that $\lim_{n\to\infty} \bar{X}_n = \mu$ since $\bar{X}_n$ is a random variable which has a probability (sampling) distribution, and as a result $\bar{X}_n$ does not represent a fixed sequence of numbers. The following proof of the WLLN uses the Chebychev's inequality which states that for any random variable, $Y$,

$$P\left(|Y - \mu| \geq k\sigma\right) \leq \frac{1}{k^2}$$

for any $k > 0$, where $\mu = E(Y)$ and $\sigma^2 = \text{Var}(Y)$. We do not prove this inequality here. The reader can find a proof from textbooks on mathematical statistics and probability, see e.g. Section 4.7.3 in Casella and Berger [3]. We now prove the WLLN as follows.

**_Proof_** Note that $E(\bar{X}_n) = \mu$ and $\text{Var}(\bar{X}_n) = \frac{\sigma^2}{n}$. Now we write down the Chebychev's inequality for $\bar{X}_n$,

$$P\left(|\bar{X}_n - \mu| \geq k\frac{\sigma}{\sqrt{n}}\right) \leq \frac{1}{k^2}$$

for any $k > 0$. Let $k = \epsilon\frac{\sqrt{n}}{\sigma}$ so that

$$P\left(|\bar{X}_n - \mu| \geq \epsilon\right) \leq \frac{\sigma^2}{n\epsilon^2}$$

for any $\epsilon > 0$. Now

$$\lim_{n\to\infty} \frac{\sigma^2}{n\epsilon^2} = 0$$

since $\sigma^2$ and $\epsilon > 0$ have been both assumed to be finite. $\qquad\square$

---

### Example 16.2

To illustrate the WLLN we generate $n$ i.i.d. uniform random variables on the interval $[0, 1]$: $X_i \sim U(0, 1)$ for $i = 1, \cdots, n$ and compute the mean $\bar{X}_n = \frac{1}{n}\sum_{i=1}^{n} X_i$.

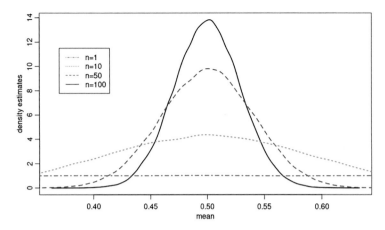

**Fig. 16.2** Illustration of the WLLN for sampling from the uniform distribution. This figure can be reproduced using the code provided in the helpfile `?see_the_wlln_for_uniform` in the R package `ipsRdbs`

This process is repeated $R$ times so that there are $R$ means $\bar{X}_n^{(1)}, \bar{X}_n^{(2)}, \cdots, \bar{X}_n^{(R)}$ for these we can plot a density estimate. Typical values for $R$ are 1000 or larger. We do this experiment for $n = 1, 10, 50, 100$. We expect to see that the densities become more and more concentrated around the population mean $E(X) = 0.5$, see Fig. 16.2. ◀

## 16.3  Consistent Estimator

The WLLN is used to define and prove the concept of consistency. We say that $\tilde{\theta}_n$ is a *consistent* estimator of $\theta$ if as $n \to \infty$,

$$\tilde{\theta}_n \xrightarrow{P} \theta, \quad \text{i.e.,} \quad P\left(\left|\tilde{\theta}_n - \theta\right| > \epsilon\right) \to 0$$

for any $\epsilon > 0$. Consistency is a highly desirable property of an estimator as it means that for any error level, the probability of the estimation error exceeding that level converges to zero as the sample size increases. Estimators that are not consistent are called inconsistent and are generally avoided. A biased estimator can be consistent, if its bias $\to 0$, as $n \to \infty$.

To prove consistency of $\tilde{\theta}_n$ for estimating $\theta$ we often make use of the following lemma.

**Lemma 16.1** *A sufficient condition for $Y_n$ to converge in probability to a is that*

$$E\left[(Y_n - a)^2\right] \to 0, \quad as\ n \to \infty. \tag{16.4}$$

The above mode of convergence of $Y_n$ to $a$ is called *convergence in mean square*. This lemma, proved below, establishes that convergence in mean square implies convergence in probability. Thus to prove convergence in probability it will suffice to prove convergence in mean square.

***Proof*** The proof of the lemma uses the following version of the Chebychev's inequality which states that for any $b$,

$$E\left[(Y_n - a)^2\right] \geq b^2 P\left(|Y_n - a| \geq b\right).$$

Hence,

$$P\left(|Y_n - a| \geq \epsilon\right) \leq \frac{1}{\epsilon^2} E\left[(Y_n - a)^2\right]$$

for any finite $\epsilon > 0$. Now, $E\left[(Y_n - a)^2\right] \to 0$ implies that $P\left(|Y_n - a| \geq \epsilon\right) \to 0$ since $\epsilon^2$ is positive but not equal to zero. Thus $Y_n$ converges to $a$ in probability. $\square$

To establish convergence in mean square, i.e. to have (16.4), we need two conditions,

$$E(Y_n) \to a \quad and \quad \text{Var}(Y_n) \to 0.$$

This is easily established by splitting the mean square error we have seen in Sect. 9.3.3. Here we have,

$$\begin{aligned} E\left[(Y_n - a)^2\right] &= E\left[(Y_n - E(Y_n) + E(Y_n) - a)^2\right] \\ &= E\left[(Y_n - E(Y_n))^2\right] + [E(Y_n) - a]^2 \\ &= \text{Var}(Y_n) + [E(Y_n) - a]^2. \end{aligned}$$

Note that $E(Y_n) \to a$ means that $Y_n$ is asymptotically unbiased for $a$ and $\text{Var}(Y_n) \to 0$ implies that the asymptotic variance of $Y_n$ is zero.

The above two conditions are used to prove consistency of estimators. The result is stated as: if, as $n \to \infty$ $E(\tilde{\theta}_n) \to \theta$, i.e. $\tilde{\theta}_n$ is asymptotically unbiased for $\theta$ and $\text{Var}(\tilde{\theta}_n) \to 0$, i.e. the asymptotic variance of $\tilde{\theta}_n$ is 0, then $\tilde{\theta}_n$ is a consistent estimator for $\theta$.

**Example 16.3**

Suppose that $\tilde{\theta}_n = \bar{X}_n$. We know from Sect. 9.4.1 that $E(\bar{X}_n) = \mu$ and $\text{Var}(\bar{X}_n) = \frac{\sigma^2}{n} \to 0$ as $n \to \infty$. Hence $\bar{X}_n$ is a consistent estimator of $\mu$. Hence, under any i.i.d. model, the sample mean $(\bar{X}_n)$ is a *consistent* estimator of the population mean, $\mu$ which is the expectation of the corresponding probability distribution). ◄

**Example 16.4**

Suppose that $X_1, \ldots, X_n$ is a random sample from a population with finite mean $\mu$ and variance $\sigma^2$. Define the sample variance

$$S_n^2 = \frac{1}{n-1} \sum_{i=1}^{n} (X_i - \bar{X}_n)^2.$$

Earlier in Sect. 9.4.1 we have proved that the sample variance $S_n^2$ is an unbiased estimator of population variance, i.e., $E\left[S_n^2\right] = \sigma^2$. Hence, the additional assumption that $\text{Var}(S_n^2) \to 0$ as $n \to \infty$ guarantees that $S_n^2$ converges in probability to $\sigma^2$. For example, when the underlying population is normal, the result that $\frac{(n-1)S_n^2}{\sigma^2} \sim \chi_{n-1}^2$, see Sect. 14.6, guarantees that

$$\text{Var}(S_n^2) = \frac{2\sigma^4}{n-1} \to 0$$

as $n \to \infty$. Thus it is possible to check that $\text{Var}(S_n^2)$ approaches zero on a case by case basis. However, this is not very elegant. In the next section, we provide an alternative approach. ◄

**Example 16.5**

Return to the uniform distribution Example 9.3 previously discussed in Sect. 9.3. In the example, $X_1, \ldots, X_n$ is a random sample from the uniform distribution on the interval $[0, \theta]$ where $\theta > 0$ is unknown. To estimate $\theta$ we proposed two estimators: (i) the sample mean, $2\bar{X}_n$ and (ii) the sample maximum

$$Y_n = \max\{X_1, \ldots, X_n\}.$$

For $\bar{X}_n$ we obtained the following results:

$$E(2\bar{X}_n) = \theta, \quad \text{Var}(2\bar{X}_n) = \frac{\theta^2}{3n}.$$

Clearly, $\text{Var}(2\bar{X}_n) \to 0$ as $n \to \infty$. Hence $2\bar{X}_n$ is a consistent estimator of $\theta$.

For the second estimator, we proved

$$E(Y_n) = \frac{n}{n+1}\theta, \quad \text{Var}(Y_n) = \frac{n\theta^2}{(n+2)(n+1)^2}.$$

Since $\lim_{n\to\infty}\frac{n}{n+1} = 1$, we have $\lim_{n\to\infty} E(Y_n) = \theta$. This implies that the $Y_n$ is an asymptotically unbiased estimator or $\theta$. We also have

$$\lim_{n\to\infty} \frac{n\theta^2}{(n+2)(n+1)^2} = 0,$$

since $\theta^2$ is finite. Hence, the second condition for consistency of $Y_n$ also holds. Hence $Y_n$ is a consistent estimator of $\theta$. ◄

## 16.4 Extending the CLT Using Slutsky's Theorem

Note that from a random sample $X_1, \ldots, X_n$, the CLT states that the probability distribution of $Y_n = \frac{\sqrt{n}(\bar{X}_n - \mu)}{\sigma}$ converges to the standard normal distribution as $n \to \infty$. In order to apply the CLT, we need to know the population variance $\sigma^2$ in addition to $\mu$. For example, to find an approximate confidence interval for $\mu$ using $Y_n$ as the pivot we need to know $\sigma^2$. However, this will not be known in general, especially without assuming the distribution of underlying random variable $X$. In such a situation, we may be tempted to estimate $\sigma$ by $S$, the sample standard deviation, and then use the new pivot $\tilde{Y}_n = \frac{\sqrt{n}(\bar{X}_n - \mu)}{S}$ for constructing a confidence interval. Indeed, this has been done in Sect. 11.3, but without any theoretical justification.

The **Slutsky's Theorem** stated below, but without proof, allows us to justify the above procedure. The theorem itself is much more general than the motivating application above. The theorem obtains the approximate distributions of the sum and product of two random variables where one converges in distribution to a random variable, $X$ say, and the other converges in probability to a constant, $a$ say. The main results are that the sum will converge to a new random variable $X + a$ and the product will converge to a new random variable $aX$. Here is the formal statement.

**Theorem 16.1 (Slutsky's Theorem)** *Suppose that $X_n \to X$ in distribution and another random variable $Y_n \to a$ where $a$ is a constant in probability. Then, it can be proven that:*

*1. $X_n + Y_n \to X + a$ in distribution.*
*2. $Y_n X_n \to aX$ in distribution.*  □

**Example 16.6**

Suppose that $X_1, \ldots, X_n$ is a random sample from a population with finite mean $\mu$ and variance $\sigma^2$. In Example 16.4, we have seen that $S_n^2 \to \sigma^2$ as $n \to \infty$ assuming $\text{Var}(S_n^2) \to 0$ as $n \to \infty$. Using further technical arguments (not provided here) it can be shown that $\frac{\sigma}{S_n} \to 1$ as $n \to \infty$. Hence the sequence of random variables,

$$\frac{\sqrt{n}(\bar{X}_n - \mu)}{S_n} = \frac{\sigma}{S_n} \frac{\sqrt{n}(\bar{X}_n - \mu)}{\sigma}$$

will approach the standard normal random variable by applying the Slutsky's Theorem. ◄

A result similar to Slutsky's theorem holds for convergence in probability. The result, see e.g. Lehman [10], is given as the following theorem.

**Theorem 16.2** *If $A_n$, $B_n$ and $Y_n$ converge in probability to $a$, $b$ and $c$ respectively, then $A_n + B_n Y_n$ converges to $a + bc$ in probability.*  □

Like the Slutsky's theorem, proof of this theorem is not included here. The interested reader can see Lehman ( This theorem provides a more elegant proof of the consistency of $S_n^2$ for $\sigma^2$, captured in the following example.

**Example 16.7**

Suppose that $X_1, \ldots, X_n$ is a random sample from a population with finite mean $\mu$ and variance $\sigma^2$. Also assume that $E(X_i - \mu)^4$ is finite. Define the sample variance

$$S_n^2 = \frac{1}{n-1} \sum_{i=1}^{n} (X_i - \bar{X}_n)^2.$$

We now prove that $S_n^2 \to \sigma^2$ in probability. ◄

***Proof*** Assume without loss if generality that $\mu = 0$. Otherwise, we can use the transformed variables $X_i - \mu$. With this assumption, $E(\bar{X}) = 0$ and $E(X_i^2) = \sigma^2$ for all $i = 1, \ldots, n$ and consequently,

$$E\left(\frac{1}{n} \sum_{i=1}^{n} X_i^2\right) = \sigma^2.$$

Thus, by writing $Y_i = X_i^2$, we can conclude that the sample mean $\bar{Y}_n$ converges in probability to the population mean $\sigma^2$ assuming $\text{Var}(Y_i)$ to be finite. Also, since $\bar{X}_n$

converges in probability to the population mean 0, we claim that $\bar{X}_n{}^2$ also converges in probability to 0. Now,

$$S_n^2 = \frac{n}{n-1}\left[\frac{1}{n}\sum_{i=1}^{n} X_i^2 - \bar{X}_n^2\right].$$

Since, $\frac{n}{n-1} \to 1$, we see that $S_n^2$ converges to $\sigma^2$ in probability by applying the above stated theorem.                                                                □

---

## 16.5   Score Function

Suppose that $x_1, \ldots, x_n$ are observations of $X_1, \ldots, X_n$, whose joint pdf $f_{\mathbf{X}}(\mathbf{x}; \boldsymbol{\theta})$ is completely specified except for the values of $p$ unknown parameters $\boldsymbol{\theta} = (\theta_1, \ldots, \theta_p)^T$. Let

$$u_i(\boldsymbol{\theta}) \equiv \frac{\partial}{\partial \theta_i}\log f_{\mathbf{X}}(\mathbf{x}; \boldsymbol{\theta}) \qquad i = 1, \ldots, p$$

and $\mathbf{u}(\boldsymbol{\theta}) \equiv [u_1(\boldsymbol{\theta}), \ldots, u_p(\boldsymbol{\theta})]^T$. Then we call $\mathbf{u}(\boldsymbol{\theta})$ the *vector of scores* or *score vector*. Where $p = 1$ and $\boldsymbol{\theta} = (\theta)$, the *score* is the scalar defined as

$$u(\theta) \equiv \frac{\partial}{\partial \theta} \log f_{\mathbf{X}}(\mathbf{x}; \theta).$$

The maximum likelihood estimate $\hat{\boldsymbol{\theta}}$ satisfies

$$u(\hat{\boldsymbol{\theta}}) = \mathbf{0} \qquad \Leftrightarrow \qquad u_i(\hat{\boldsymbol{\theta}}) = 0 \quad i = 1, \ldots, p.$$

Note that $\mathbf{u}(\boldsymbol{\theta})$ is a function of $\boldsymbol{\theta}$ for fixed (observed) $\mathbf{x}$. However, if we replace $x_1, \ldots, x_n$ in $\mathbf{u}(\boldsymbol{\theta})$, by the corresponding random variables $X_1, \ldots, X_n$ then we obtain a vector of random variables $\mathbf{U}(\boldsymbol{\theta}) \equiv [U_1(\boldsymbol{\theta}), \ldots, U_p(\boldsymbol{\theta})]^T$.

An important result in likelihood theory is that the expected score at the true (but unknown) value of $\boldsymbol{\theta}$ is zero, i.e.

$$E[\mathbf{U}(\boldsymbol{\theta})] = \mathbf{0} \qquad \Leftrightarrow \qquad E[U_i(\boldsymbol{\theta})] = 0 \quad i = 1, \ldots, p,$$

provided that

1. The expectation exists.
2. The sample space for $\mathbf{X}$ does not depend on $\boldsymbol{\theta}$.

***Proof*** Assume $\mathbf{x}$ to be continuous. If, instead, it is discrete, we replace $\int$ by $\sum$ in the following proof. Note that the integral sign ($\int$) below implies a multiple definite integral over the whole domain of $\mathbf{x}$. Also, the reader is alerted to the mathematical result in Appendix A.3 regarding differentiation of an integral. Lastly, note that $\int f_{\mathbf{X}}(\mathbf{x}; \boldsymbol{\theta}) d\mathbf{x} = 1$ since it is the total probability for the probability distribution of the random variable $\mathbf{X}$. For any $i = 1, \ldots, p$,

$$E[U_i(\boldsymbol{\theta})] = \int U_i(\boldsymbol{\theta}) f_{\mathbf{X}}(\mathbf{x}; \boldsymbol{\theta}) d\mathbf{x} \quad \left[\text{definition of expectation}\right]$$

$$= \int \frac{\partial}{\partial \theta_i} \log f_{\mathbf{X}}(\mathbf{x}; \boldsymbol{\theta}) f_{\mathbf{X}}(\mathbf{x}; \boldsymbol{\theta}) d\mathbf{x} \quad [\text{definition of } U_i(\boldsymbol{\theta})]$$

$$= \int \frac{\frac{\partial}{\partial \theta_i} f_{\mathbf{X}}(\mathbf{x}; \boldsymbol{\theta})}{f_{\mathbf{X}}(\mathbf{x}; \boldsymbol{\theta})} f_{\mathbf{X}}(\mathbf{x}; \boldsymbol{\theta}) d\mathbf{x} \quad [\text{chain rule of differentiation}]$$

$$= \int \frac{\partial}{\partial \theta_i} f_{\mathbf{X}}(\mathbf{x}; \boldsymbol{\theta}) d\mathbf{x} \quad \left[\text{cancels since } f_{\mathbf{X}}(\mathbf{x}; \boldsymbol{\theta}) > 0 \text{ for any } \mathbf{x}.\right]$$

$$= \frac{\partial}{\partial \theta_i} \int f_{\mathbf{X}}(\mathbf{x}; \boldsymbol{\theta}) d\mathbf{x} \quad \left[\text{See result in Appendix A.3}\right]$$

$$= \frac{\partial}{\partial \theta_i} 1 \quad \left[\text{total probability is 1}\right]$$

$$= 0.$$

$\square$

## Example 16.8

Suppose $x_1, \ldots, x_n$ are observations of $X_1, \ldots, X_n$, i.i.d. Bernoulli($p$) random variables. Here $\boldsymbol{\theta} = (p)$ and,

$$u(p) = n\overline{x}/p - n(1 - \overline{x})/(1 - p)$$

$$E[U(p)] = 0 \quad \Rightarrow \quad E[\overline{X}] = p.$$

◀

## Example 16.9

Suppose $x_1, \ldots, x_n$ are observations of $X_1, \ldots, X_n$, i.i.d. $N(\mu, \sigma^2)$ random variables. Here $\boldsymbol{\theta} = (\mu, \sigma^2)$ and,

$$u_1(\mu, \sigma^2) = n(\overline{x} - \mu)/\sigma^2$$

$$u_2(\mu, \sigma^2) = -\frac{n}{2\sigma^2} + \frac{1}{2(\sigma^2)^2} \sum_{i=1}^{n} (x_i - \mu)^2$$

$$E[\mathbf{U}(\mu, \sigma^2)] = \mathbf{0} \quad \Rightarrow \quad E[\overline{X}] = \mu \quad \text{and} \quad E\left[\frac{1}{n}\sum_{i=1}^{n}(X_i - \mu)^2\right] = \sigma^2.$$

◀

## 16.6 Information

Suppose that $x_1, \ldots, x_n$ are observations of $X_1, \ldots, X_n$, whose joint pdf $f_{\mathbf{X}}(\mathbf{x}; \boldsymbol{\theta})$ is completely specified except for the values of $p$ unknown parameters $\boldsymbol{\theta} = (\theta_1, \ldots, \theta_p)^T$. Previously, we defined the Hessian matrix $\mathbf{H}(\boldsymbol{\theta})$ to be the matrix with components

$$[\mathbf{H}(\boldsymbol{\theta})]_{ij} \equiv \frac{\partial^2}{\partial \theta_i \partial \theta_j} \log f_{\mathbf{X}}(\mathbf{x}; \boldsymbol{\theta}) \qquad i = 1, \ldots, p; \; j = 1, \ldots, p.$$

We call the matrix $-\mathbf{H}(\boldsymbol{\theta})$ the *observed information matrix*. Where $p = 1$ and $\boldsymbol{\theta} = (\theta)$, the *observed information* is a scalar defined as

$$-H(\theta) \equiv -\frac{\partial}{\partial \theta^2} \log f_{\mathbf{X}}(\mathbf{x}; \theta).$$

Here, we are interpreting $\boldsymbol{\theta}$ as the true (but unknown) value of the parameter. As with the score, if we replace $x_1, \ldots, x_n$ in $\mathbf{H}(\boldsymbol{\theta})$, by the corresponding random variables $X_1, \ldots, X_n$, we obtain a matrix of random variables. Then, we define the *expected information matrix* or *Fisher information matrix* to be $\mathcal{I}(\boldsymbol{\theta})$, where

$$[\mathcal{I}(\boldsymbol{\theta})]_{ij} = E(-[\mathbf{H}(\boldsymbol{\theta})]_{ij}) \qquad i = 1, \ldots, p; \; j = 1, \ldots, p.$$

An important result in likelihood theory is that the variance-covariance matrix of the score vector is equal to the expected information matrix i.e.

$$Var[\mathbf{U}(\boldsymbol{\theta})] = \mathcal{I}(\boldsymbol{\theta}) \quad \Leftrightarrow \quad Var[\mathbf{U}(\boldsymbol{\theta})]_{ij} = [\mathcal{I}(\boldsymbol{\theta})]_{ij} \quad i = 1, \ldots, p; \; j = 1, \ldots, p,$$

provided that

1. The variance exists.
2. The sample space for $\mathbf{X}$ does not depend on $\boldsymbol{\theta}$.

**Proof** (*Assume* $\mathbf{x}$ *Is Continuous. Replace* $\int$ *by* $\sum$ *in the Discrete Case*)

$$Var[\mathbf{U}(\boldsymbol{\theta})]_{ij} = E[U_i(\boldsymbol{\theta}) U_j(\boldsymbol{\theta})]$$

$$= \int \frac{\partial}{\partial \theta_i} \log f_{\mathbf{X}}(\mathbf{x}; \boldsymbol{\theta}) \frac{\partial}{\partial \theta_j} \log f_{\mathbf{X}}(\mathbf{x}; \boldsymbol{\theta}) f_{\mathbf{X}}(\mathbf{x}; \boldsymbol{\theta}) d\mathbf{x}$$

$$= \int \frac{\frac{\partial}{\partial \theta_i} f_{\mathbf{X}}(\mathbf{x}; \boldsymbol{\theta})}{f_{\mathbf{X}}(\mathbf{x}; \boldsymbol{\theta})} \frac{\frac{\partial}{\partial \theta_j} f_{\mathbf{X}}(\mathbf{x}; \boldsymbol{\theta})}{f_{\mathbf{X}}(\mathbf{x}; \boldsymbol{\theta})} f_{\mathbf{X}}(\mathbf{x}; \boldsymbol{\theta}) d\mathbf{x}$$

$$= \int \frac{1}{f_{\mathbf{X}}(\mathbf{x}; \boldsymbol{\theta})} \frac{\partial}{\partial \theta_i} f_{\mathbf{X}}(\mathbf{x}; \boldsymbol{\theta}) \frac{\partial}{\partial \theta_j} f_{\mathbf{X}}(\mathbf{x}; \boldsymbol{\theta}) d\mathbf{x}$$

$$i = 1, \ldots, p; \; j = 1, \ldots, p.$$

Now

$$[\mathcal{I}(\boldsymbol{\theta})]_{ij} = E\left[-\frac{\partial^2}{\partial\theta_i\partial\theta_j}\log f_{\mathbf{X}}(\mathbf{x};\boldsymbol{\theta})\right]$$

$$= \int -\frac{\partial^2}{\partial\theta_i\partial\theta_j}\log f_{\mathbf{X}}(\mathbf{x};\boldsymbol{\theta})f_{\mathbf{X}}(\mathbf{x};\boldsymbol{\theta})d\mathbf{x}$$

$$= \int -\frac{\partial}{\partial\theta_i}\left[\frac{\frac{\partial}{\partial\theta_j}f_{\mathbf{X}}(\mathbf{x};\boldsymbol{\theta})}{f_{\mathbf{X}}(\mathbf{x};\boldsymbol{\theta})}\right]f_{\mathbf{X}}(\mathbf{x};\boldsymbol{\theta})d\mathbf{x}$$

$$= \int \left[-\frac{\frac{\partial^2}{\partial\theta_i\partial\theta_j}f_{\mathbf{X}}(\mathbf{x};\boldsymbol{\theta})}{f_{\mathbf{X}}(\mathbf{x};\boldsymbol{\theta})} + \frac{\frac{\partial}{\partial\theta_i}f_{\mathbf{X}}(\mathbf{x};\boldsymbol{\theta})\frac{\partial}{\partial\theta_j}f_{\mathbf{X}}(\mathbf{x};\boldsymbol{\theta})}{f_{\mathbf{X}}(\mathbf{x};\boldsymbol{\theta})^2}\right]f_{\mathbf{X}}(\mathbf{x};\boldsymbol{\theta})d\mathbf{x}$$

$$= -\frac{\partial^2}{\partial\theta_i\partial\theta_j}\int f_{\mathbf{X}}(\mathbf{x};\boldsymbol{\theta})d\mathbf{x} + \int \frac{1}{f_{\mathbf{X}}(\mathbf{x};\boldsymbol{\theta})}\frac{\partial}{\partial\theta_i}f_{\mathbf{X}}(\mathbf{x};\boldsymbol{\theta})\frac{\partial}{\partial\theta_j}f_{\mathbf{X}}(\mathbf{x};\boldsymbol{\theta})d\mathbf{x}$$

$$= -\frac{\partial^2}{\partial\theta_i\partial\theta_j}1 + \int \frac{1}{f_{\mathbf{X}}(\mathbf{x};\boldsymbol{\theta})}\frac{\partial}{\partial\theta_i}f_{\mathbf{X}}(\mathbf{x};\boldsymbol{\theta})\frac{\partial}{\partial\theta_j}f_{\mathbf{X}}(\mathbf{x};\boldsymbol{\theta})d\mathbf{x}$$

$$= \int \frac{1}{f_{\mathbf{X}}(\mathbf{x};\boldsymbol{\theta})}\frac{\partial}{\partial\theta_i}f_{\mathbf{X}}(\mathbf{x};\boldsymbol{\theta})\frac{\partial}{\partial\theta_j}f_{\mathbf{X}}(\mathbf{x};\boldsymbol{\theta})d\mathbf{x}$$

$$= Var[\mathbf{U}(\boldsymbol{\theta})]_{ij} \qquad i = 1,\ldots,p; \;\; j = 1,\ldots,p.$$

□

---

### Example 16.10

Suppose $x_1,\ldots,x_n$ are observations of $X_1,\ldots,X_n$, i.i.d. Bernoulli($p$) random variables. Here $\boldsymbol{\theta} = (p)$ and

$$u(p) = \frac{n\overline{x}}{p} - \frac{n(1-\overline{x})}{(1-p)}$$

$$-H(p) = \frac{n\overline{x}}{p^2} + \frac{n(1-\overline{x})}{(1-p)^2}$$

$$\mathcal{I}(p) = \frac{n}{p} + \frac{n}{(1-p)} = \frac{n}{p(1-p)}.$$

◄

**Example 16.11**

Suppose $x_1, \ldots, x_n$ are observations of $X_1, \ldots, X_n$, i.i.d. $N(\mu, \sigma^2)$ random variables. Here $\boldsymbol{\theta} = (\mu, \sigma^2)$ and,

$$u_1(\mu, \sigma^2) = \frac{n(\overline{x} - \mu)}{\sigma^2}$$

$$u_2(\mu, \sigma^2) = -\frac{n}{2\sigma^2} + \frac{1}{2(\sigma^2)^2} \sum_{i=1}^{n} (x_i - \mu)^2$$

$$-\mathbf{H}(\mu, \sigma^2) = \begin{pmatrix} \frac{n}{\sigma^2} & \frac{n(\overline{x}-\mu)}{(\sigma^2)^2} \\ \frac{n(\overline{x}-\mu)}{(\sigma^2)^2} & \frac{1}{(\sigma^2)^3} \sum_{i=1}^{n} (x_i - \mu)^2 - \frac{n}{2(\sigma^2)^2} \end{pmatrix}$$

$$\mathcal{I}(\mu, \sigma^2) = -E\left(\mathbf{H}(\mu, \sigma^2)\right) = \begin{pmatrix} \frac{n}{\sigma^2} & 0 \\ 0 & \frac{n}{2(\sigma^2)^2} \end{pmatrix},$$

since

$$E(\overline{X}) = \mu, \quad \text{and } E\,(X_i - \mu)^2 = \sigma^2 \text{ for any } i = 1, \ldots, n.$$

◀

## 16.7 Asymptotic Distribution of the Maximum Likelihood Estimators

Maximum likelihood estimation, see Sect. 10.2, is an attractive method of estimation for a number of reasons. It is intuitively sensible (choosing $\boldsymbol{\theta}$ which makes the observed data most probable) and usually reasonably straightforward to carry out. Even when the simultaneous equations we obtain by differentiating the log likelihood function are impossible to solve directly, solution by numerical methods is usually feasible.

Perhaps the most compelling reason for considering maximum likelihood estimation is the asymptotic properties of maximum likelihood estimators.

Suppose that $x_1, \ldots, x_n$ are observations of independent random variables $X_1, \ldots, X_n$, whose joint pdf $f_{\mathbf{X}}(\mathbf{x}; \boldsymbol{\theta}) = \prod_{i=1}^{n} f_{Y_i}(y_i; \boldsymbol{\theta})$ is completely specified except for the values of an unknown parameter vector $\boldsymbol{\theta}$, and that $\hat{\boldsymbol{\theta}}$ is the maximum likelihood estimator of $\boldsymbol{\theta}$.

Then, as $n \to \infty$, the distribution of $\hat{\boldsymbol{\theta}}$ tends to a multivariate normal distribution with mean vector $\boldsymbol{\theta}$ and variance covariance matrix $\mathcal{I}(\boldsymbol{\theta})^{-1}$.

Where $p = 1$ and $\boldsymbol{\theta} = (\theta)$, the distribution of the m.l.e. $\hat{\theta}$ tends to $N[\theta, 1/\mathcal{I}(\theta)]$, where

$$I(\theta) = -E\left[\frac{\partial^2}{\partial\theta^2}\log f_{\mathbf{X}}(\mathbf{x};\theta)\right],$$

which is the Fisher Information number for $\theta$, see Sect. 16.6.

**Proof** Below we only sketch the proof for the one parameter case and identically distributed random variables $Y_1, \ldots, Y_n$.

Suppose that $x_1, \ldots, x_n$ are observations of independent identically distributed random variables $X_1, \ldots, X_n$, whose joint p.d.f. is therefore $f_{\mathbf{X}}(\mathbf{x};\theta) = \prod_{i=1}^{n} f_{\mathbf{X}}(x_i;\theta)$. We can write the score as

$$u(\theta) = \frac{\partial}{\partial\theta}\log f_{\mathbf{X}}(\mathbf{y};\theta) = \sum_{i=1}^{n}\frac{\partial}{\partial\theta}\log f_{\mathbf{X}}(x_i;\theta)$$

so $U(\theta)$ can be expressed as the sum of $n$ i.i.d. random variables. Therefore, asymptotically, as $n \to \infty$, by the central limit theorem , $U(\theta)$ is normally distributed. Furthermore, for the unknown true $\theta$ we know that $E[U(\theta)] = 0$ and $Var[U(\theta)] = I(\theta)$, so $U(\theta)$ is asymptotically $N[0, I(\theta)]$.

Now, a Taylor series expansion of $U(\hat{\theta})$ around the true $\theta$ gives

$$U(\hat{\theta}) = U(\theta) + (\hat{\theta} - \theta)U'(\theta) + \ldots$$

Now, $U(\hat{\theta}) = 0$, and if we approximate $U'(\theta) \equiv H(\theta)$ by $E[H(\theta)] \equiv -I(\theta)$, and also ignore higher order terms,[1] we obtain

$$\hat{\theta} = \theta + \frac{1}{I(\theta)}U(\theta)$$

As $U(\theta)$ is asymptotically $N[0, I(\theta)]$, $\hat{\theta}$ is asymptotically $N[\theta, I(\theta)^{-1}]$. □

For 'large enough $n$', we can treat the asymptotic distribution of the m.l.e. as an approximation. The fact that $E(\hat{\boldsymbol{\theta}}) \approx \boldsymbol{\theta}$ means that the maximum likelihood estimator is approximately *unbiased* (correct on average) for large samples. Furthermore, its variability, as measured by its variance $I(\boldsymbol{\theta})^{-1}$ is the smallest possible amongst unbiased estimators, so the maximum likelihood has good precision. Therefore the m.l.e. is a desirable estimator in large samples (and therefore presumably also reasonable in small samples).

---

[1] This requires that $\hat{\theta}$ is close to $\theta$ in large samples, which is true but we do not prove it here.

## 16.8   Exercises

### 16.1 (Convergence of Estimators)

1. Suppose that $X_1, \ldots, X_n$ is a random sample from a population with finite mean $\mu$ and variance $\sigma^2$. Define

$$Y_n = \frac{1}{n+1} \sum_{i=1}^{n} X_i.$$

   Show that $Y_n$ is asymptotically unbiased for $\mu$.

2. Suppose that $X_1, \ldots, X_n$ is a random sample from a population with finite mean $\mu$ and variance $\sigma^2$. Define

$$\tilde{S}_n^2 = \frac{1}{n} \sum_{i=1}^{n} (X_i - \bar{X}_n)^2.$$

   Show that $\tilde{S}_n^2$ is asymptotically unbiased for $\sigma^2$.

3. Suppose that $X_1, \ldots, x_n$ is a sequence of Bernoulli trials with success probability $\theta$. Show that $\bar{X}_n$ converges to $\theta$ in probability, i.e. $\bar{X}_n \xrightarrow{P} \theta$.

4. Suppose that we have a random sample of observations $X$ on the Poisson model which has marginal probability function

$$f(x; \lambda) = \frac{e^{-\lambda} \lambda^x}{x!}, \qquad x = 0, 1, \ldots, \qquad \lambda > 0.$$

(a) Show that

$$Z_n = \frac{\sqrt{n}(\bar{X}_n - \lambda)}{\sqrt{\lambda}} \to N(0, 1)$$

   in distribution as $n \to \infty$.

(b) Show that $\bar{X}_n$ converges to $\lambda$ in probability.

(c) Use the Slutsky's theorem to prove that

$$Y_n = \frac{\sqrt{n}(\bar{X}_n - \lambda)}{\sqrt{\bar{X}_n}} \to N(0, 1)$$

   in distribution as $n \to \infty$.

5. Suppose that $X_1, \ldots, X_n$ are a random sample from the negative binomial distribution with parameters $r$ and $p$ such that

$$E(X_i) = \frac{r(1-p)}{p} = \mu, \quad \mathrm{Var}(X_i) = \frac{r(1-p)}{p^2} = \sigma^2.$$

Suppose that $r = 10$, $p = 0.5$ and $n = 40$.
(a) State the CLT for the sample mean $\bar{X}_n$.
(b) Using the R command `pnbinom` find the exact value of $P(\bar{X}_n \leq 11)$.
(c) Using the CLT find an approximate value of this probability.

6. Suppose that $X_1, \ldots, X_n$ are a random sample from the exponential distribution with parameter $\theta > 0$ having the pdf:

$$f(x|\theta) = \begin{cases} \frac{1}{\theta} e^{-\frac{x}{\theta}} & \text{if } x > 0 \\ 0 & \text{if } x \leq 0 \end{cases}$$

(a) Show that the maximum likelihood estimator of $\theta$ is $\bar{X}_n$ the sample mean.
(b) Show that $\bar{X}_n$ is a consistent estimator of $\theta$.                          □

**16.2**

1. Suppose that $x_1, \ldots, x_n$ are observations of random variables $X_1, \ldots, X_n$ which are i.i.d. with p.d.f. (or p.m.f.) $f_X(x; \theta)$ for a scalar parameter $\theta$. For each of the distributions below, find the score function, and use the fact that the expected score is zero to derive an unbiased estimator of $\theta^{-1}$.
(a) $f_X(x; \theta) = \theta \exp(-\theta x)$, $x \in \mathcal{R}_+$ (exponential distribution).
(b) $f_X(x; \theta) = \theta x^{\theta - 1}$, $x \in (0, 1)$.
(c) $f_X(x; \theta) = \theta(1 - \theta)^{x-1}$, $x \in \{1, 2, 3, \ldots\}$ (geometric distribution).

2. Let $x_1, \ldots, x_n$ be independent observations of $X$, a geometrically distributed random variable with p.m.f.

$$f_X(x) = \theta(1 - \theta)^{x-1}, \quad x = 1, 2, \ldots,$$

where $0 < \theta < 1$.

Find the maximum likelihood estimator for $\theta$. Derive its asymptotic distribution.

Show that the score function, $U(\theta)$, is given by

$$U(\theta) = \frac{n}{\theta} - \frac{n(\bar{X} - 1)}{1 - \theta}$$

and, by considering $E[U(\theta)]$ and $\mathrm{Var}[U(\theta)]$, show that $E(\bar{X}) = 1/\theta$ and $\mathrm{Var}(\bar{X}) = (1 - \theta)/(n\theta^2)$.

3. Let $x_1, \ldots, x_n$ be independent observations of $X$, a continuous random variable with pdf

$$f_X(x; \theta) \;=\; \frac{1}{2\theta^3} x^2 \exp(-x/\theta), \quad x > 0.$$

(a) Derive the score function, $U(\theta)$, and use the fact that $E[U(\theta)] = 0$ to show that $E(\overline{X}) = 3\theta$.
(b) Find the maximum likelihood estimator $\hat{\theta}$ for $\theta$ and derive its asymptotic distribution. $\qquad\square$

# Part V
# Introduction to Statistical Modelling

This is Part V of the book where we introduce statistical modeling. It assumes some familiarity with basic linear algebra. Appendix A.4 provides a few revision exercises.

# Simple Linear Regression Model

# 17

**Abstract**

Chapter 17 kicks off Part V of the book on introduction to statistical modelling. It discusses the concepts related to simple regression modelling with many practical examples. The concepts of estimation, inference and predictions are discussed along with the required theoretical derivations. Simultaneously, illustrations are carried along with R code so that the reader can immediately transfer their skills into the practical domain.

## 17.1  Motivating Example: The Body Fat Data

As a motivating example we return to the body fat data set introduced in Example 1.5. The data set is `bodyfat` in the R package `ipsRdbs`. The mathematical problem here is to find a relationship between percentage of bodyfat and skinfold measurements so that it is possible to use the relationship to predict bodyfat percentage of a new athlete just by knowing their skinfold measurement.

The first step of any statistical modelling exercise is to examine the data graphically, or by calculating informative summaries. Here, a natural plot is a scatterplot of the main variable of interest, percentage of bodyfat, against the other variables, skinfold measurement here for all the observed values. The scatterplot shows both the structure and the variability in the data.

The left panel of Fig. 17.1 provides a scatter plot of the data set where the percentage of bodyfat for each of the 102 Australian athletes is plotted against their skinfold measurement. The scatterplot shows that the percent body fat increases linearly with the skinfold measurement and that the variation in percent body fat also increases with the skinfold measurement. There do not seem to be any unusual data points and there are no other obvious patterns to note.

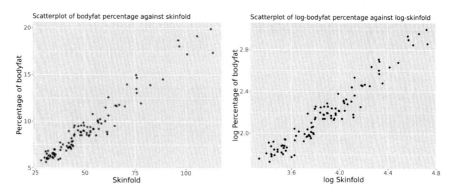

**Fig. 17.1** Scatter plots of the body fat data: (**a**) left panel on original scale, (**b**) right panel on log-transformed scale for both the axes

It is often the case that we can simplify the relationships exhibited by data by transforming one or more of the variables. In this example, we want to preserve the simple linear relationship noted in the above paragraph but stabilise the variability (i.e. make it constant) so that it is easier to describe. We therefore try transforming both variables. There is no theoretical argument to suggest a transformation here but empirical experience suggests that we try the log transformation.

The scatterplot on the log-scale for both the axes is shown in the right panel of Fig. 17.1. The plot confirms a reduction in variability in the log of the percentage variability for higher values of the log of the skinfold measurements. Thus the log transformation of both variables preserves the linear relationship and stabilises the variability.

To interpret a scatterplot, think of dividing the scatterplot into narrow vertical strips so that each vertical strip contains the observations on samples which have nearly the same skinfold measurement, denoted by $x$. Within each vertical strip, we see a distribution of percent bodyfat which is the conditional distribution of bodyfat percent given that the skinfold measurement is in the strip.

We need to describe how these distributions change as we move across the vertical strips. The most effective graphical way of seeing how the distributions change is to draw a boxplot for the data in each strip. Examining these boxplots in Fig. 17.2 broadly shows the features we noted above. In practice, we learn to interpret scatterplots without going through this formal exercise, but it does give useful insight into the conditional nature of the interpretation for building a regression model in Sect. 17.2.

## 17.2  Writing Down the Model

Let $y_i$ for $i = 1, \ldots, n$ denote the response or the main variable of interest that we would like to predict. This is the log percent body fat measurement for the $i$th athlete in the body fat example. Let $x_i$ denote the covariate or the explanatory variable that

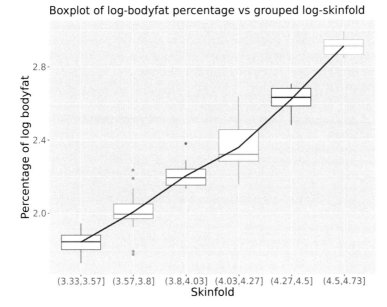

**Fig. 17.2** Boxplots for the log of the bodyfat percentage data for eight different intervals of skinfold data. A superimposed line joins the means of the log bodyfat percentage in the eight groups

we may use to explain the variation in the response variable $y_i$ for $i = 1, \ldots, n$. This is the log skinfold measurement in the above example.

Ideally we should continue to use the convention of writing uppercase letters, e.g. $Y_i$, $X_i$ etc. for the random variables and lower case letters, $y_i$, $x_i$ etc. for the observed values. However, in this chapter we treat $X$ as known (or given) and aim to study the conditional random variable $Y | X = x$. As a result, here and throughout our treatment of regression models, we will discuss the form of the conditional mean $E(Y | X = x)$ as a function of the covariate $x$. No attempt is made to model the variability in $X$, and hence the $X_i$ are not considered to be random variables—they are treated as constants, fixed at the observed values $x_i$. Henceforth, we shall use the notation $x_i$ to emphasise that these are known constants.

Therefore, given a pair of variables, it matters which we choose as the response and which as the explanatory variables. Often, this choice is clear, as for example when

- the aim of the modelling exercise is to predict one of the variables using the other (as in the above example).
- one of the variables is controlled (fixed by the experimenter) while the variability of the other is observed.

These notations are to be used generically in the sense that the notation $y$ will denote the dependent variable which we would like to predict and $x$ will denote the independent variable, which we can control for in the experiment. Sometimes these variables are also called using different terminologies. For example, in a dose-response study in medical statistics $y$ will be called the response and $x$ may be called a dose or a risk factor. Sometimes $x$ is also called an explanatory variable which is used to explain the variation in the random variable $Y$.

From the scatterplot, a plausible model for the **conditional** distribution of $Y$ given $X = x$ has

$$E(Y_i|x_i) = \beta_0 + \beta_1 x_i \quad \text{and} \quad \text{Var}(Y_i|x_i) = \sigma^2,$$

for $i = 1, \ldots, n$. There are three parameters in this model:

- **Conditional mean parameters**
  - **the intercept** $\beta_0$ is the value of mean percent log body fat when the log skinfold equals zero.
  - **the slope** $\beta_1$ is the effect on the mean log skinfold of a unit change in log skin fold.
- **Conditional variance** $\sigma^2$, sometimes called a nuisance parameter, as its value does not affect predictions (but does affect uncertainty concerning predictions).

To complete the model, we need to say more about the random variation.

- Provided the observations have been made from unrelated athletes, it seems reasonable to begin by assuming independence.
- It is difficult to say much about the shape of the conditional distribution of $Y_i$ given $x_i$: we usually start by making the optimistic assumption that it is normal. The discrepant points may make this somewhat implausible—it is important to check later.

Putting the assumptions together, we have the simple linear regression model

$$Y_i|x_i \overset{\text{ind}}{\sim} N(\beta_0 + \beta_1 x_i, \sigma^2),$$

for $i = 1, \ldots, n$. The structure of the underlying population can then be viewed graphically in Fig. 17.3. This figure is a theoretical analogue of Fig. 17.2; each vertical strip in Fig. 17.2 is represented by a density curve in Fig. 17.3. Moreover, the blue line through the middle of Fig. 17.2 is approximated by the theoretical (in blue) regression line in Fig. 17.3.

A useful alternative way to write the model is as

$$Y_i = \beta_0 + \beta_1 x_i + \epsilon_i$$

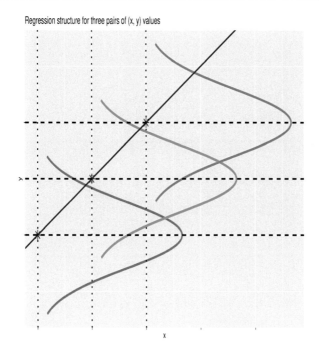

Regression structure for three pairs of (x, y) values

**Fig. 17.3** The structure of the underlying population

with $\epsilon_i \overset{\text{ind}}{\sim} N(0, \sigma^2)$ independent of $x_i$ for $i = 1, \ldots, n$.. Here $\epsilon_i$ is the random difference between the observation $Y_i$ and its conditional mean $\beta_0 + \beta_1 x_i$. Thus $\epsilon_i$ is a random variable called an error (not meaning a mistake, though) with a distribution called the error distribution. In this form, the model is a decomposition of the observation into a structural part (the regression line) plus random variation (the error), see Fig. 17.4.

If there was no variability, the points in the plot in Fig. 17.4 would fall exactly on a curve describing the exact relationship between $Y$ and $x$; the variability is reflected in the scatter of points about this curve.

An interesting question is what is the source of the variability? One explanation is that the variability is due to the measurement process and stochastic inaccuracies which arise in it. Alternatively, body fat may be affected by other variables which we have not held constant and their hidden variation induces variation in the data. An entirely pragmatic view is that the origin of the variability does not mater, variability is simply there and we must deal with it.

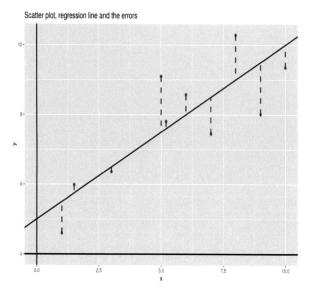

**Fig. 17.4** The structural and random components of the underlying population

## 17.3   Fitting the Model

Fitting a model to observed data refers to the act of finding estimates and their uncertainties for the unknown parameters. The choice of model fitting method depends on what we know about the distribution of the data, i.e., the error distribution assumed in the model. We do not actually know much at this stage, but we have made some optimistic assumptions, so we fit the model under these assumptions and then review their plausibility.

### 17.3.1   Least Squares Estimation

The method of **least squares estimation** works well when the error distribution is assumed to be normal. This entails estimating $(\beta_0, \beta_1)$ by $(\hat{\beta}_0, \hat{\beta}_1)$ which minimises the sum $S$ of squared vertical distances between the observed response values $y_i$ and the regression line $\beta_0 + \beta_1 x_i$, at the corresponding observed values of the explanatory variable, $x_i$, see Fig. 17.4. Clearly,

$$S = \sum_{i=1}^{n} (y_i - \beta_0 - \beta_1 x_i)^2 .$$

In the illustrative Fig. 17.4, $S$ is the sum of squares of the lengths of the dashed red lines. The least squares (LS) estimation method obtains estimates of $\beta_0$ and $\beta_1$ by minimising $S$ with respect to these two parameters. We use the derivative method as outlined in the exercise Sect. 1.4. Because there are two unknown parameters, we obtain the following partial derivatives:

$$\frac{\partial S}{\partial \beta_0} = -2\sum_{i=1}^{n}(y_i - \beta_0 - \beta_1 x_i)$$

and

$$\frac{\partial S}{\partial \beta_1} = -2\sum_{i=1}^{n}x_i(y_i - \beta_0 - \beta_1 x_i).$$

To minimise $S$ with respect to $\beta_0$ and $\beta_1$ we set

$$\frac{\partial S}{\partial \beta_0} = 0, \quad \frac{\partial S}{\partial \beta_1} = 0,$$

which are called the *normal equations*. So the least squares estimates $(\hat{\beta}_0, \hat{\beta}_1)$ satisfy the *normal equations*

$$\bar{y} = \hat{\beta}_0 + \hat{\beta}_1\bar{x} \quad \text{or} \quad \hat{\beta}_0 = \bar{y} - \hat{\beta}_1\bar{x}$$

and

$$\sum_{i=1}^{n}x_i y_i - n\bar{x}\hat{\beta}_0 - \hat{\beta}_1\sum_{i=1}^{n}x_i^2 = 0.$$

The first equation shows that the fitted line passes through the point $(\bar{x}, \bar{y})$. It follows that

$$0 = \sum_{i=1}^{n}x_i y_i - n\bar{y}\bar{x} - \hat{\beta}_1\left(\sum_{i=1}^{n}x_i^2 - n\bar{x}^2\right)$$
$$= \sum_{i=1}^{n}(x_i - \bar{x})y_i - \hat{\beta}_1\sum_{i=1}^{n}(x_i - \bar{x})^2$$
$$= \sum_{i=1}^{n}(x_i - \bar{x})y_i - \hat{\beta}_1(n-1)s_x^2,$$

say, where

$$s_x^2 = \frac{1}{(n-1)}\sum_{i=1}^{n}(x_i - \bar{x})^2.$$

We also define the sample variance of $y$ values,

$$s_y^2 = \frac{1}{(n-1)} \sum_{i=1}^{n} (y_i - \bar{y})^2,$$

and the sample standard deviations:

$$s_y = \sqrt{s_y^2}, \quad s_x = \sqrt{s_x^2}.$$

In order to proceed, let $r$ be the sample correlation coefficient defined by

$$r = \frac{s_{xy}}{s_y s_x}$$

where

$$s_{xy} = \text{Cov}(x, y) = \frac{1}{(n-1)} \sum_{i=1}^{n} (y_i - \bar{y})(x_i - \bar{x}),$$

is the sample covariance. Note that positive (negative) values of $r$ indicate that $y$ tends to increase (decrease) with increasing $x$. The statistic $r$ is called the Pearson or product-moment correlation coefficient.

With these additional notations, we have

$$\hat{\beta}_1 = \frac{\sum_{i=1}^{n} (x_i - \bar{x}) y_i}{(n-1) s_x^2}$$

$$= \frac{\sum_{i=1}^{n} (x_i - \bar{x}) y_i - \bar{y}(\sum_{i=1}^{n}(x_i - \bar{x}))}{(n-1)s_x^2} \quad [\text{since } \sum_{i=1}^{n}(x_i - \bar{x}) = 0]$$

$$= \frac{\sum_{i=1}^{n}(x_i - \bar{x})(y_i - \bar{y})}{(n-1)s_x^2}$$

$$= \frac{(n-1)s_{xy}}{(n-1)s_x^2}$$

$$= \frac{s_{xy}}{s_x^2} = \frac{r s_y s_x}{s_x^2} = r \frac{s_y}{s_x}.$$

Substituting the estimate $\hat{\beta}_1$ back into the first equation, we obtain

$$\hat{\beta}_0 = \bar{y} - \hat{\beta}_1 \bar{x}, \quad \text{where } \hat{\beta}_1 = r \frac{s_y}{s_x}.$$

Hence we can obtain $\hat{\beta}_0$ and $\hat{\beta}_1$ by calculating the summary statistics, $\bar{x}, \bar{y}, s_x, s_y$ and $r$ from the data.

### Example 17.1

For the body fat data with $x$ as the log of the skinfold values and $y$ as the log of the percentage body fat values we issue the following commands to obtain the various statistics we need to calculate $\hat{\beta}_0$ and $\hat{\beta}_1$.

```
bodyfat$logskin <- log(bodyfat$Skinfold)
bodyfat$logbfat <- log(bodyfat$Bodyfat)
n <- nrow(bodyfat)
x <- bodyfat$logskin
y <- bodyfat$logbfat
xbar <- mean(x)
ybar <- mean(y)
sx2 <- var(x)
sy2 <- var(y)
sxy <- cov(x, y)
r <- cor(x, y)
print(list(n=n, xbar=xbar, ybar=ybar, sx2=sx2, sy2=sy2, sxy=
    sxy, r=r))
```

These commands give us the following values:

$$n = 102, \bar{x} = 3.8831, \bar{y} = 2.1760, s_x^2 = 0.1084, s_y^2 = 0.0911, s_{xy}$$
$$= 0.0957, r = 0.9627.$$

Therefore,

$$\hat{\beta}_1 = r\frac{s_y}{s_x} = 0.9627\sqrt{\frac{0.0911}{0.1084}} = 0.8822,$$

$$\hat{\beta}_0 = 2.1760 - 0.8825(3.8831) = -1.250.$$

◀

Note that

$$\hat{\beta}_1 = \frac{\sum_{i=1}^{n}(x_i - \bar{x})y_i}{(n-1)s_x^2}, \quad \hat{\beta}_0 = \bar{y} - \hat{\beta}_1\bar{x}.$$

found above are particular realised values of the corresponding estimators

$$\hat{\beta}_1 = \frac{\sum_{i=1}^{n}(x_i - \bar{x})Y_i}{(n-1)s_x^2}, \quad \hat{\beta}_0 = \bar{Y} - \hat{\beta}_1\bar{x}.$$

Note the use of $Y_i$ and $\bar{Y}$ instead of $y_i$ and $\bar{y}$ to define the estimators. We do not use the upper case notation $X$ since we treat $x$-values as fixed numbers. Moreover, we do not use the upper case Greek letter $\hat{\beta}$ to denote the above two estimators. Now the above estimators can be re-written as:

$$\hat{\beta}_1 = \sum_{i=1}^{n} w_{1i} Y_i, \quad \hat{\beta}_0 = \sum_{i=1}^{n} w_{0i} Y_i, \tag{17.1}$$

where

$$w_{1i} = \frac{x_i - \bar{x}}{(n-1)s_x^2}, \quad w_{0i} = \frac{1}{n} - \bar{x}\, w_{1i} = \frac{1}{n} - \bar{x}\frac{x_i - \bar{x}}{(n-1)s_x^2}. \tag{17.2}$$

These definitions will be used to derive theoretical properties of the estimators in Sect. 17.5 below. It is easy to prove that

$$\sum_{i=1}^{n} w_{1i} = 0, \quad \sum_{i=1}^{n} w_{1i}^2 = 1, \sum_{i=1}^{n} w_{0i} = 1,$$

see the exercises.

### 17.3.2 Estimating Fitted Values

Once the estimates of the two unknown parameters have been found we obtain the *fitted values* as estimates:

$$\hat{y}_i = \hat{\beta}_0 + \hat{\beta}_1 x_i, \quad i = 1, 2, \ldots, n.$$

These fitted values all lie in the fitted straight line, see for example, Fig. 17.4. Furthermore, the fitted line is familiar as the theoretical conditional expectation of $Y$ given $X = x$ in Eq. (15.3) discussed in Sect. 15.5.
    For any $i = 1, \ldots, n$,

$$\begin{aligned}
\hat{y}_i &= \hat{\beta}_0 + \hat{\beta}_1 x_i \\
&= \bar{y} - \hat{\beta}_1 \bar{x} + \hat{\beta}_1 x_i \\
&= \bar{y} + \hat{\beta}_1 (x_i - \bar{x}).
\end{aligned}$$

Thus, we may write the fitted values as:

$$\hat{y}_i = \bar{y} + \hat{\beta}_1(x_i - \bar{x}) \equiv \bar{y} + r\frac{s_y}{s_x}(x_i - \bar{x}),$$

for any $i = 1, \ldots, n$. This proves that the fitted regression equation is a sample analogue of the conditional expectation (15.3),

$$E(Y|X = x) = \mu_y + \rho\frac{\sigma_y}{\sigma_x}(x - \mu_x),$$

where

$$\rho = \frac{E[(X - \mu_x)(Y - \mu_y)]}{\sigma_x \sigma_y},$$

as defined in (8.1) for jointly distributed random variables with respective means $\mu_x$ and $\mu_y$ and variances $\sigma_x^2$ and $\sigma_y^2$.

Thus, when $r = 0$ we have $\hat{\beta}_1 = 0$, i.e. there is no regression without correlation. However, this algebraic relationship ignores differences in context and interpretation.

- The regression model represents the error distribution as a normal distribution. This implies that the response has a normal distribution. The explanatory variable is treated as fixed and not modelled.
- Correlation on the other hand treats both variables as random and is most meaningful when the joint distribution of the response and explanatory variables is the bivariate normal distribution.

Further discussion regarding this is provided in Sect. 17.10.

---

**Example 17.2**

The fitted linear regression model for the body fat percent example is given by:

$$\hat{Y} = -1.25 + 0.88x$$

where $\hat{Y}$ is the fitted log-percent body fat and $x$ is the log-skinfold measurement. Here we have worked on the simpler log scale, it may be useful to make predictions on the raw scale. Note that we can transform the fitted model to show that on the raw scale, the fitted model is

$$\text{Fitted \% Body fat} = \exp(-1.25)(\text{Skinfold})^{0.88},$$

a multiplicative model which is often biologically interpretable. ◄

### 17.3.3  Defining Residuals

Residuals are the differences:

$$r_i = y_i - \hat{y}_i = y_i - \hat{\beta}_0 - \hat{\beta}_1 x_i$$

between the observed responses and the fitted values. Thus,

$$\text{Residual} = \text{Observed} - \text{Fitted}$$

The decomposition

$$y_i = \hat{\beta}_0 + \hat{\beta}_1 x_i + r_i,$$

is analogous to the model decomposition into a structural part (regression line) plus random variation (error) and shows that the residuals can be regarded as estimates of the errors.

Note that $r_i$ is an observation of the random variable

$$R_i = Y_i - \hat{Y}_i$$

where $\hat{Y}_i$ is also random variable obtained by using the **estimators** defined in (17.1).

## 17.4  Estimating the Variance

In addition to the parameters $\beta_0$ and $\beta_1$ which describe the conditional mean relationship, it is also useful to be able to estimate the error variance $\sigma^2$. We usually use the estimator

$$S^2 = \frac{1}{n-2} \sum_{i=1}^{n} (Y_i - \hat{\beta}_0 - \hat{\beta}_1 x_i)^2 = \frac{1}{n-2} \sum_{i=1}^{n} R_i^2,$$

where division by $n-2$ instead of $n$ or $n-1$ is used since two degrees of freedom have been 'lost' in estimating the conditional mean parameters. It can be shown that

$$\frac{(n-2)S^2}{\sigma^2} \sim \chi_{n-2}^2$$

independent of $\hat{\beta}_0$ and $\hat{\beta}_1$. Hence, $S^2$ is an unbiased estimator of $\sigma^2$. As usual, we use $s^2$ to denote the actual value of $S^2$ for our observed data.

The sampling (probability) distribution of $S^2$ can be used to construct confidence intervals for $\sigma^2$ in the same way as described for the simple normal models in Chap. 13.

---

**Example 17.3**

Continuing with the body fat data example, we issue the following commands to obtain $s^2$ as an estimate of $\sigma^2$.

```
hatbeta1 <- r * sqrt(sy2/sx2) # calculates estimate of the
    slope
hatbeta0 <- ybar - hatbeta1 * xbar # calculates estimate of
    the intercept
rs <- y - hatbeta0 - hatbeta1 * x # calculates residuals
s2 <- sum(rs^2)/(n-2) # calculates estimate of sigma2
s2
```

These command yield $s^2 = 0.00673$ as an estimate of $\sigma^2$. The residual standard error is $s = \sqrt{0.00673} = 0.0820.$ ◄

---

## 17.5   Quantifying Uncertainty

Consider the least squares estimator $\hat{\beta}_1$ of the slope parameter $\beta_1$. (This is usually of greater interest than the intercept, but the following concepts apply to any estimator of any model parameter). An estimate $\hat{\beta}_1$ of $\beta_1$ is never exactly equal to $\beta_1$ because different samples give different estimates. There is therefore uncertainty associated with an estimate. We can describe this uncertainty by the sampling distribution of the estimator which gives the distribution of the estimator across all possible samples of the same size from the population. The sampling distribution is the distribution of the estimator under the assumed model.

Recall that the regression model assumes $Y_i | x_i \sim N(\beta_0 + \beta_1 x_i, \sigma^2)$ independently for $i = 1, \ldots, n$. The least squares estimators of $\beta_0$ and $\beta_1$, see (17.1), are linear functions of the independently distributed random variables $Y_1 \ldots, Y_n$. This guarantees that the sampling distributions of $\hat{\beta}_0$ and $\hat{\beta}_1$ are both normal. We now derive the means and variances of these two normal distributions. We first prove

$$E(\hat{\beta}_1) = \beta_1, \quad \text{Var}(\hat{\beta}_1) = \frac{\sigma^2}{(n-1)s_x^2},$$

where $\hat{\beta}_1$ is the estimator defined in (17.1).

**Proof** Here we have

$$
E(\hat{\beta}_1) = \sum_{i=1}^{n} w_{1i} E(Y_i)
$$

$$
= \sum_{i=1}^{n} w_{1i} (\beta_0 + \beta_1 x_i) = \beta_1,
$$

as
$$
\sum_{i=1}^{n} w_{1i} = 0, \qquad \sum_{i=1}^{n} w_{1i} x_i = 1.
$$

so $\hat{\beta}_1$ is an *unbiased* estimator of $\beta_1$. We now obtain the variance:

$$
\mathrm{Var}(\hat{\beta}_1) = \sum_{i=1}^{n} w_{1i}^2 Var(Y_i)
$$

$$
= \sigma^2 \sum_{i=1}^{n} w_{1i}^2
$$

$$
= \frac{\sigma^2}{(n-1)s_x^2}
$$

as

$$
\sum_{i=1}^{n} w_{1i}^2 = \frac{1}{(n-1)s_x^2}.
$$

$\square$

Putting everything together,

$$
\hat{\beta}_1 \sim N\left(\beta_1, \frac{\sigma^2}{(n-1)s_x^2}\right).
$$

The variance of the slope estimator is made as small as possible for a fixed $n$ by making $s_x^2$ as large as possible. This can be achieved by putting the $X$s at either end of their feasible range. Note however that this optimal design leaves no possibility of checking the linearity of the model. The effect of discrepant explanatory variables is to increase $s_x^2$ and hence decrease $Var(\hat{\beta}_1)$. This will result in a misleadingly small variance if the corresponding $Y$ value is incorrectly measured.

From the variance, above, we obtain

$$
s.d.(\hat{\beta}_1) = \frac{\sigma}{\sqrt{(n-1)s_x^2}}
$$

as a measure of precision of the estimator $\hat{\beta}_1$. As this standard deviation depends on the unknown parameter $\sigma$, we use the *standard error*

$$s.e.(\hat{\beta}_1) = \frac{s}{\sqrt{(n-1)s_x^2}}$$

where $\sigma$ has been replaced by the estimate $s$ (see Sect. 17.4), to summarise the precision of $\hat{\beta}_1$.

The treatment of $\hat{\beta}_0$ is entirely analogous. We have

$$E(\hat{\beta}_0) = \sum_{i=1}^{n} w_{0i} E(Y_i)$$
$$= \sum_{i=1}^{n} w_{0i} (\beta_0 + \beta_1 x_i) = \beta_0$$

as  $\displaystyle\sum_{i=1}^{n} w_{0i} = 1, \qquad \sum_{i=1}^{n} w_{0i} x_i = \bar{x} - \bar{x} = 0$

and

$$Var(\hat{\beta}_0) = \sum_{i=1}^{n} w_{0i}^2 Var(Y_i)$$
$$= \sigma^2 \sum_{i=1}^{n} w_{0i}^2$$
$$= \sigma^2 \left( \frac{1}{n} + \frac{\bar{x}^2}{(n-1)s_x^2} \right)$$

as

$$\sum_{i=1}^{n} w_{0i}^2 = \sum_{i=1}^{n} \left( \frac{1}{n^2} - 2 \frac{1}{n} \frac{\bar{x}(x_i - \bar{x})}{(n-1)s_x^2} + \frac{\bar{x}^2(x_i - \bar{x})^2}{(n-1)^2 s_x^4} \right)$$
$$= \frac{1}{n} + \frac{\bar{x}^2}{(n-1)s_x^2}.$$

It follows that

$$\hat{\beta}_0 \sim N \left( \beta_0, \sigma^2 \left[ \frac{1}{n} + \frac{\bar{x}^2}{(n-1)s_x^2} \right] \right)$$

and

$$s.e.(\hat{\beta}_0) = s \left( \frac{1}{n} + \frac{\bar{x}^2}{(n-1)s_x^2} \right)^{1/2}.$$

Finally,

$$\sum_{i=1}^{n} w_{0i} w_{1i} = \sum_{i=1}^{n} \left( \frac{1}{n} - \frac{\bar{x}(x_i - \bar{x})}{(n-1)s_x^2} \right) \frac{x_i - \bar{x}}{(n-1)s_x^2} = -\frac{\bar{x}}{(n-1)s_x^2}$$

$$
\begin{aligned}
\text{Cov}(\hat{\beta}_0, \hat{\beta}_1) &= \text{Cov} \left( \sum_{i=1}^{n} w_{0i} Y_i, \sum_{i=1}^{n} w_{1i} Y_i \right) \\
&= \sum_{i=1}^{n} w_{0i} w_{1i} \text{Var}(Y_i) \\
&= \sigma^2 \sum_{i=1}^{n} w_{0i} w_{1i} \\
&= -\frac{\sigma^2 \bar{x}}{(n-1)s_x^2}.
\end{aligned}
$$

**Example 17.4**

Continuing with the body fat example, we have

$$
\begin{aligned}
s.e.(\hat{\beta}_0) &= s \left( \frac{1}{n} + \frac{\bar{x}^2}{(n-1)s_x^2} \right)^{1/2} \\
&= \sqrt{0.00673 \left( \frac{1}{102} + \frac{3.8831^2}{101(0.1084)} \right)} = 0.0966,
\end{aligned}
$$

and

$$s.e.(\hat{\beta}_1) = \frac{s}{\sqrt{(n-1)s_x^2}} = \sqrt{\frac{0.00673}{101(0.1084)}} = 0.0248.$$

◄

## 17.6   Obtaining Confidence Intervals

Recall from Chap. 11 that a confidence interval summarises the information jointly conveyed by a parameter estimate and its standard error, by providing an *interval* of values of the parameter which are plausible given the observed data. In Sect. 11.4, we derived confidence intervals for a parameter of a simple normal model by constructing a function involving the response variables $Y_1, \ldots, Y_n$ and the parameter of interest (but no other parameter) whose distribution was known.

The sampling distributions of $\hat{\beta}_0$ and $\hat{\beta}_1$ derived in Sect. 17.5, cannot be used directly in confidence interval construction, as they both depend on the unknown variance $\sigma^2$, as well as the parameter of interest. However, it can be proved that

$$\frac{(n-2)S^2}{\sigma^2} \sim \chi^2_{n-2}$$

independent of $\hat{\beta}_0$ and $\hat{\beta}_1$ using a multivariate transformation technique similar to the one used in Sect. 14.6.

As the sampling distribution of $S^2$ depends only on $\sigma^2$, so we can construct a confidence interval for $\sigma^2$ in a way which exactly parallels the construction of the confidence interval for the variance $\sigma^2$ as in Example 14.10. Indeed, following that example, a $100(1-\alpha)\%$ confidence interval for $\sigma^2$ will be

$$\left[ (n-2)\frac{s^2}{h_2} \ , \ (n-2)\frac{s^2}{h_1} \right],$$

where $s^2$ is the observed value of $S^2$. Here $h_1$ and $h_2$ are obtained from the $\chi^2$ distribution with $n-2$ degrees of freedom such that

$$P\left( h_1 \le \chi^2_{n-2} \le h_2 \right) = 1 - \alpha.$$

However, it is usually of greater interest to evaluate a confidence interval for one of the regression parameters $\beta_0$ or (particularly) $\beta_1$. Using these stated results it follows that,

$$t(\hat{\beta}_1) = \frac{\hat{\beta}_1 - \beta_1}{s.e.(\hat{\beta}_1)} \sim t_{n-2}$$

and

$$t(\hat{\beta}_0) = \frac{\hat{\beta}_0 - \beta_0}{s.e.(\hat{\beta}_0)} \sim t_{n-2}.$$

**Fig. 17.5** t-density function

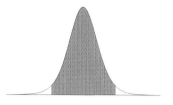

As, for any given $n$, the $t_{n-2}$ distribution is known, we can find $c$ such that

$$P(-c \leq t(\hat{\beta}_1) \leq c) = 1 - \alpha$$

for any given confidence level $1 - \alpha$, see Fig. 17.5.
Hence, as

$$-c \leq t(\hat{\beta}_1) \leq c \quad \Rightarrow \quad \hat{\beta}_1 - c \, s.e.(\hat{\beta}_1) \leq \beta_1 \leq \hat{\beta}_1 - c \, s.e.(\hat{\beta}_1)$$

we have constructed a confidence interval for $\beta_1$. Exactly the same argument can be applied for $\beta_0$. This is of the general form: *Estimate* $\pm$ *Critical Value* $\times$ *Standard Error*, mentioned in Sect. 11.2.

The confidence interval can be interpreted similarly as in Sect. 11.2. The random interval will contain the true value of $\beta_1$ with probability $1 - \alpha$. Any particular realised interval may or may not contain $\beta_1$; the probability is calculated over the set of possible intervals.

---

**Example 17.5**

Continuing with the body fat example, we have found earlier that:

$$\hat{\beta}_1 = 0.8822, \quad \text{and} \quad s.e.(\hat{\beta}_1) = 0.0248.$$

For a 95% confidence interval the critical value is `qt(0.975, df=100)` from the $t$-distribution with $n - 2 = 100$ degrees of freedom, which gives the result 1.9840. Hence the 95% confidence interval is evaluated to be $(0.833, 0.931)$ (upto three decimal places) using the R command

```
round(0.8822 + c(-1, 1) * qt(0.975, df=100) * 0.0248, 3).
```

◄

---

## 17.7   Performing Hypothesis Testing

Suppose that we conjecture that one of the parameters $\beta_0$ or $\beta_1$ takes a particular specified value. We can formally examine whether our conjectured value is consistent with the observed data by performing a *hypothesis test*. The null hypothesis (denoted $H_0$) corresponds to the conjecture of interest which here, without loss of generality, we shall assume involves $\beta_1$. Hence, we write

$$H_0 : \beta_1 = \beta_1^*.$$

where $\beta_1^*$ is the conjectured value.

The basic idea of testing is to take our estimator of $\beta_1$ and see how close it is to $\beta_1^*$. If it is close to $\beta_1^*$, then we conclude that the data supports $H_0$; if it is far from $\beta_1^*$, then we conclude there is evidence against $H_0$. As described in Sect. 12.2.4 we summarise the evidence against $H_0$, by computing a p-value which is the probability, under the model and assuming $H_0$ to be true, of observing a value of $\hat{\beta}_1$ as far from $\beta_1^*$ or further than that actually observed.

In order to calculate a p-value, we require a *test statistic* involving $\hat{\beta}_1$, the distribution of which is known under the null hypothesis. From Sect. 17.6, we know that

$$t(\hat{\beta}_1) = \frac{\hat{\beta}_1 - \beta_1}{s.e.(\hat{\beta}_1)} \sim t_{n-2}$$

Hence, the test statistic

$$t^*(\hat{\beta}_1) = \frac{\hat{\beta}_1 - \beta_1^*}{s.e.(\hat{\beta}_1)}$$

has a $t_{n-2}$ distribution *if* $\beta_1 = \beta_1^*$, that is, when $H_0$ is true. Note that $t^*(\hat{\beta}_1)$ is a statistic, as it can be calculated using data values alone. No knowledge of any unknown parameter is required.

We compute the p-value for the test by examining how probable it is that a $t_{n-2}$ distribution would generate a value of $t^*(\hat{\beta}_1)$ as, or more, extreme than the observed data value. We can write the calculation succinctly as

$$p = P(|T| \geq |t^*(\hat{\beta}_1)|) = 2P(T \leq -|t^*(\hat{\beta}_1)|).$$

where $T$ denotes a random variable with a $t_{n-2}$ distribution, and $t(\hat{\beta}_1)$ is the observed value of the test statistic.

This is a two-sided test. They are used when we are checking for departures from $H_0$ in either direction, see Fig. 17.6. Directional tests are possible but relatively uncommon in practical regression modelling.

**Fig. 17.6** Illustration of
p-value calculation for a two
sided test

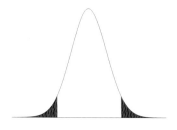

---

**Example 17.6**

Continuing with the body fat example, suppose that we would like to test

$$H_0 : \beta_1 = \beta_1^* = 0.$$

Now

$$t^*(\hat{\beta}_1) = \frac{\hat{\beta}_1}{s.e.(\hat{\beta}_1)} = \frac{0.8822}{0.0248} = 35.57,$$

which is much greater than the t-critical value of 1.9840 at 5% level of
significance. Hence the null hypothesis is rejected at 5% level of significance.
The p-value can be calculated as

$$P(|T_{100}| > 35.57) = 2 * (1 - pt(35.57, df = 100)) \approx 0.$$

To illustrate a one sided p-value calculation suppose that we test

$$H_0 : \beta_1 = \beta_1^* = 1$$

so that the regression line is a 45° line against the alternative

$$H_1 : \beta_1 < 1.$$

Here we calculate:

$$t^*(\hat{\beta}_1) = \frac{\hat{\beta}_1 - 1}{s.e.(\hat{\beta}_1)} = \frac{0.8822 - 1}{0.0248} = -4.75.$$

Hence the p-value is given by

$$P(T_{100} < -4.75) = 3.4 \times 10^{-6}.$$

This p-value is very small and hence we also reject this null hypothesis. ◄

## 17.8    Comparing Models

The most common test of interest for a simple linear regression model is the test of the hypothesis that $\beta_1 = 0$. The reason that this test is so important is that it provides a formal way of comparing the linear regression model

$$Y_i \overset{\text{ind}}{\sim} N(\beta_0 + \beta_1 x_i, \sigma^2), \quad i = 1, \ldots, n$$

where the distribution of the response variable $Y$ explicitly depends on the explanatory variable $x$ through a linear relationship in its conditional mean, with the simpler (more parsimonious) model

$$Y_i \overset{\text{ind}}{\sim} N(\beta_0, \sigma^2) \quad i = 1, \ldots, n$$

where there is no relationship between the variables $Y$ and $x$.

These models can be compared by testing the hypothesis $H_0 : \beta_1 = 0$ within the linear regression model. If this hypothesis is not rejected, then we can draw the conclusion that the simpler model is adequate for the data, and there is no reason to entertain the more complex model—there is no apparent relationship between $y$ and $x$ (or at least not one which can be explained by a simple linear regression).

Here, and more generally when using a hypothesis test to compare two models, the null hypothesis corresponds to the simpler model. Furthermore, the null hypothesis has special status in a hypothesis test—it is not rejected unless the data provide significant evidence against it. This is consistent with a desire for parsimony in modelling. We do not reject the simpler model in favour of the more complex model unless the data provide significant evidence that we should (the simper model a poor fit to the data in comparison with the more complex model).

The two models above can be compared from a different perspective which will be useful later (where we want to perform more general model comparisons) on.

- When we fit the model $Y_i \overset{\text{ind}}{\sim} N(\beta_0 + \beta_1 x_i, \sigma^2)$, $i = 1, \ldots, n$, the quality of the fit is reflected by the residual sum of squares (RSS) $(n - 2)s^2 = \sum_{i=1}^{n} r_i^2$.

- Under the simpler model (which we still denote $H_0$) $Y_i \overset{\text{ind}}{\sim} N(\beta_0, \sigma^2)$, $i = 1, \ldots, n$, the least squares estimator of $\beta_0$ is $\hat{\beta}_0 = \bar{Y}$ so the residual sum of squares (RSS) is $\sum_{i=1}^{n} (y_i - \bar{y})^2 = (n - 1)s_y^2$.

  – Although we use the same symbol $\beta_0$ in both models, we have $\beta_0 = E(Y_i | x_i = 0)$ or $\beta_0$ is the intercept in the simple regression model and $\beta_0 = E(Y_i)$ or $\beta_0$ is the mean in the reduced model. This raises a point which is very important in general: the interpretation of a parameter depends on the context, and this is not reflected in the notation.

- We can compare the two models by examining

$$\text{RSS under } H_0 - \text{RSS under regression model}$$
$$= (n-1)s_y^2 - (n-2)s^2$$
$$= \sum_{i=1}^{n}(y_i - \bar{y})^2 - \sum_{i=1}^{n} r_i^2.$$

## 17.9  Analysis of Variance (ANOVA) Decomposition

The above discussion on model comparison leads to a fundamental identity, which is known as *analysis of variance (ANOVA) decomposition*. We derive the identity as follows:

$$\sum_{i=1}^{n}(y_i - \bar{y})^2 = \sum_{i=1}^{n}(y_i - \hat{y}_i + \hat{y}_i - \hat{\beta}_0 - \hat{\beta}_1\bar{x})^2$$
$$= \sum_{i=1}^{n}(r_i + \hat{\beta}_1(x_i - \bar{x}))^2$$
$$= \sum_{i=1}^{n} r_i^2 + \hat{\beta}_1^2 \sum_{i=1}^{n}(x_i - \bar{x})^2 + 2\hat{\beta}_1 \sum_{i=1}^{n} r_i(x_i - \bar{x})$$
$$= \hat{\beta}_1^2(n-1)s_x^2 + \sum_{i=1}^{n} r_i^2$$

as the normal equations imply that $\bar{y} = \hat{\beta}_0 + \hat{\beta}_1\bar{x}$ and

$$\sum_{i=1}^{n} r_i(x_i - \bar{x}) = 0.$$

Hence,

$$(n-1)s_y^2 = \hat{\beta}_1^2(n-1)s_x^2 + (n-2)s^2$$

which is sometimes referred to as the *analysis of variance (ANOVA) decomposition*.
   We call the first term in the above identity the total sum of squares, the second the regression sum of squares and the third the residual sum of squares. Hence, we have

$$\text{Total Sum of Squares} = \text{Regression Sum of Squares} + \text{Residual Sum of Squares}.$$

Now, as suggested above, we can reformulate the comparison of the two models, by considering a test statistic based on the improvement in fit provided by the regression model, over the simpler model

$$\sum_{i=1}^{n}(y_i - \bar{y})^2 - \sum_{i=1}^{n}r_i^2 = \hat{\beta}_1^2(n-1)s_x^2 = \text{Regression SS}.$$

If the regression model fits the data significantly better than the simple (null) model $H_0$ then the regression sum of squares will be large. But how do we determine if the regression sum of squares is so large that we should reject $H_0$.

It is clear that, under $H_0 : \beta_1 = 0$, the regression sum of squares $\hat{\beta}_1^2(n-1)s_x^2$, divided by $\sigma^2$ has a $\chi_1^2$ distribution. Furthermore, it can be proved that, it is independent of $S^2$. Hence, we consider the test statistic

$$
\begin{aligned}
F &= \frac{\text{Regression Mean Square}}{\text{Residual Mean Square}} \\
&= \frac{\hat{\beta}_1^2(n-1)s_x^2}{S^2} = \frac{\hat{\beta}_1^2(n-1)s_x^2/\sigma^2}{S^2/\sigma^2} \\
&= \frac{\chi_1^2/1}{\chi_{n-2}^2/(n-2)}
\end{aligned}
$$

where $\chi_1^2$ has a $\chi_1^2$ distribution and $\chi_{n-2}^2$ has a $\chi_{n-2}^2$ distribution. The last ratio is said to have the *F distribution* with degrees of freedom 1 and $n-2$, see Sect. 14.5.3. Hence, we denote the $F$ distribution above by $F_{1,n-2}$. The F distribution is a 'known' distribution, properties of which can be obtained from statistical software such as R.

The comparison of the null model and regression models based on the F-ratio is identical to a test of the same hypothesis based on the t-statistic because

- $F = \frac{\hat{\beta}_1^2(n-1)s_x^2}{S^2} = \left( \frac{\hat{\beta}_1}{S/[(n-1)s_x^2]^{1/2}} \right)^2 = \left( \frac{\hat{\beta}_1}{s.e.(\hat{\beta}_1)} \right)^2$

- $F_{1,k} \sim \frac{\chi_1^2/1}{\chi_k^2/k} \sim \frac{N(0,1)^2}{\chi_k^2/k} \sim \left( \frac{N(0,1)}{[\chi_k^2/k]^{1/2}} \right)^2 \sim t_k^2$

The above discussion is summarised by what is known as the ANOVA table for comparing the simple linear regression model with the null model ($Y_i = \beta_0 + \epsilon_i$) and is given by:

| Source | df | Sum of squares | Mean squares | F value | P value |
|---|---|---|---|---|---|
| Regression | 1 | $\hat{\beta}_1^2(n-1)s_x^2$ | $\hat{\beta}_1^2(n-1)s_x^2$ | $\frac{\hat{\beta}_1^2(n-1)s_x^2}{s^2}$ | $Pr(F > F_{obs})$ |
| Residuals | $n-2$ | $(n-2)s^2$ | $s^2$ | | |
| Total | $n-1$ | $(n-1)s_y^2$ | | | |

---

**Example 17.7**

Continuing with the body fat example, so far we have obtained

$$n = 102, \hat{\beta}_1 = 0.8822, s^2 = 0.00673, s_x^2 = 0.1084, s_y^2 = 0.0911.$$

Hence we obtain the following ANOVA table:

| Source | df | Sum of squares | Mean squares | F value | P value |
|---|---|---|---|---|---|
| Regression | 1 | 8.524 | 8.524 | 1266.3 | $< 2.2 \times 10^{-16}$ |
| Residuals | 100 | 0.673 | 0.0067 | | |
| Total | 101 | 9.197 | | | |

◄

---

## 17.10  Assessing Correlation and Judging Goodness-of-Fit

The correlation coefficient is sometimes used in conjunction with a regression model to summarise the quality of the fit. A low value of $s^2$ implies small estimated error variance, and hence a good fit. It is more meaningful to look at $s^2$ relative to how well we could fit the data if we ignored the explanatory variable than to compare $s^2$ to the theoretical but unattainable 'ideal value' of zero. Without information about the explanatory variable, then we would have to model the response values as identically distributed about a common mean, and the error variance estimate would be $s_y^2$ the sample variance of $Y$ as in Sect. 9.4. We can therefore measure the fit by

$$\text{Adj } R^2 = \frac{s_y^2 - s^2}{s_y^2} = 1 - \frac{s^2}{s_y^2},$$

the proportionate reduction in error variance achieved by including the explanatory variable in the model.

The linear relationship is strong if Adj $R^2 \approx 1$ and weak if Adj $R^2 \approx 0$. Using the *analysis of variance decomposition*

$$(n - 1)s_y^2 = \hat{\beta}_1^2(n - 1)s_x^2 + (n - 2)s^2$$

and the relationship between $\hat{\beta}_1$ and $r$ (the sample correlation coefficient), we can write

$$s^2 = \frac{n - 1}{n - 2}(s_y^2 - \hat{\beta}_1^2 s_x^2) = \frac{n - 1}{n - 2}(1 - r^2)s_y^2$$

and hence

$$\text{Adj } R^2 = \frac{n-1}{n-2}r^2 - \frac{1}{n-2}.$$

It follows that when $n$ is large  $\text{Adj } R^2 \approx r^2$. In simple linear regression we may write $r^2 = R^2$, and hence

$$R^2 = \frac{n-2}{n-1}\text{Adj } R^2 + \frac{1}{n-1} = 1 - \frac{(n-2)s^2}{(n-1)s_y^2}$$

defines a form of squared correlation coefficient which generalises to more complex regression models.

---

**Example 17.8**

Continuing with the body fat example, we calculate:

$$R^2 = \frac{n-2}{n-1}\text{Adj } R^2 + \frac{1}{n-1} = 1 - \frac{(n-2)s^2}{(n-1)s_y^2} = 1 - \frac{0.673}{9.197} = 0.9268,$$

$$\text{Adj } R^2 = \frac{n-1}{n-2}\hat{\rho}^2 - \frac{1}{n-2} = 1 - \frac{0.00673}{0.0911} = 0.9261.$$

These high values confirm that the regression model is a very good fit for the data. ◄

The sample correlation coefficient $\hat{\rho}$ (and consequently  $\text{Adj } R^2$ and $R^2$) has limitations as a summary of the strength of relationships. In particular,

- $\hat{\rho}$ is sensitive to outliers
- $\hat{\rho}$ is sensitive to skewed data distributions
- $\hat{\rho}$ takes the same numerical value for very different scatterplots and
- a correlation of zero means **no linear relationship** but does not mean **no relationship at all**. See Fig. 17.7 for an illustration

Thus the summary provided by a correlation coefficient can only be meaningfully interpreted if we also have a scatterplot of the data. This problem can be overcome by taking the initial steps in the modelling process (plotting the data, choosing transformations etc.) but this is rarely done as the correlation coefficient is widely, and incorrectly, viewed as a statistic which alleviates the need to examine the data. In any case, once we have begun the modelling process, we may as well complete it properly.

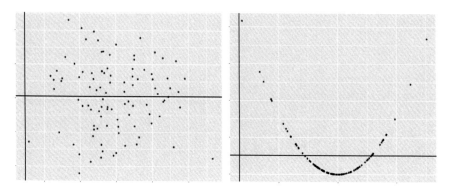

**Fig. 17.7**  Two examples of correlation $\approx 0$

A full linear regression model is far more useful in that it:

- provides a complete description,
- can be used to draw conclusions about how the response changes (or can be expected to change) as the explanatory variable changes,
- can be used to predict new values and,
- can be generalised in a variety of useful ways.

Calculation of a correlation coefficient complements this process, by summarising one particular aspect of goodness-of-fit, but the diagnostic plots provide a far more complete summary of the quality of the fit than $\hat{\rho}$ or Adj $R^2$ ever can.

## 17.11  Using the lm Command

The lm command in R can be used to fit linear regression models. Suppose that Assuming that there are two vectors x and y of equal length we can fit the above regression model simply by using the command lm(y~x). The model fitted object in R can be saved as any named object, say flm, and the summary of the fit can be obtained by issuing the commands:

```
flm <- lm(y~x)
summary(flm)
```

For the body fat data example, the last summary command produces the result:

```
Call:
lm(formula = y ~ x)

Residuals:
```

```
      Min        1Q     Median         3Q         Max
-0.223408 -0.055661  0.000052   0.056390   0.152984


Coefficients:
            Estimate Std. Error t value Pr(>|t|)
(Intercept) -1.24988    0.09661  -12.94   <2e-16 ***
x            0.88225    0.02479   35.59   <2e-16 ***
---
Signif. codes:  0 '***' 0.001 '**' 0.01 '*' 0.05 '.'
                0.1 ' ' 1


Residual standard error: 0.08205 on 100 degrees of
                         freedom
Multiple R-squared:  0.9268,Adjusted R-squared:  0.9261
F-statistic:  1266 on 1 and 100 DF,  p-value: < 2.2e-16
```

The 'Call' statement shows the formula used in fitting the model. This is followed by a summary of the residuals. All of the remaining numerical results have already been obtained by using hand calculations in previous sections. The last column in the 'Coefficients' table provides a star rating for each of the regression coefficients $\hat{\beta}_0$ and $\hat{\beta}_1$ and the rating scale is provided in the line just under the table. The residual standard error is $s$ which is square-root of the $s^2$ value we obtained earlier.

The hand calculated ANOVA table can be obtained by issuing the R command **anova(flm)**. This command produces the following output.

```
Analysis of Variance Table


Response: y
            Df Sum Sq Mean Sq F value      Pr(>F)
x            1 8.5240  8.5240  1266.3 < 2.2e-16 ***
Residuals 100 0.6731  0.0067
---
Signif. codes:  0 '***' 0.001 '**' 0.01 '*' 0.05 '.'
                0.1 ' ' 1
```

Again, this table matches with the ANOVA table obtained previously by using hand calculation.

## 17.12 Estimating the Conditional Mean

We sometimes need to estimate the conditional mean $\beta_0 + \beta_1 x_0$ at a new value $x_0$ of $x$ [which may, or may not, be one of the values of $X$ in our observed data]. A natural estimator of the conditional mean is $\hat{Y}_0 \equiv \hat{\beta}_0 + \hat{\beta}_1 x_0$. It follows from the sampling distributions of $\hat{\beta}_0$ and $\hat{\beta}_1$ that

$$E(\hat{Y}_0) = \beta_0 + \beta_1 x_0$$

so this estimator is unbiased, and

$$
\begin{aligned}
\mathrm{Var}(\hat{Y}_0) &= \mathrm{Var}(\hat{\beta}_0) + x_0^2 \mathrm{Var}(\hat{\beta}_1) + 2x_0 \, \mathrm{Cov}(\hat{\beta}_0, \hat{\beta}_1) \\
&= \sigma^2 \left( \frac{1}{n} + \frac{\bar{x}^2}{(n-1)s_x^2} + \frac{x_0^2}{(n-1)s_x^2} - \frac{2x_0\bar{x}}{(n-1)s_x^2} \right) \\
&= \sigma^2 \left( \frac{1}{n} + \frac{(x_0 - \bar{x})^2}{(n-1)s_x^2} \right) \\
&= \sigma^2 h_{00},
\end{aligned}
$$

with $h_{00} = \frac{1}{n} + \frac{(x_0-\bar{x})^2}{(n-1)s_x^2}$, so that

$$
\hat{\beta}_0 + \hat{\beta}_1 x_0 \sim N(\beta_0 + \beta_1 x_0, \sigma^2 h_{00})
$$

and

$$
\frac{\hat{\beta}_0 + \hat{\beta}_1 x_0 - (\beta_0 + \beta_1 x_0)}{s h_{00}^{1/2}} \sim t_{n-2}.
$$

Hence, we can obtain $100(1 - \alpha)\%$ confidence intervals of the form

$$
[\hat{\beta}_0 + \hat{\beta}_1 x_0 - c\, s\, h_{00}^{1/2}, \ \hat{\beta}_0 + \hat{\beta}_1 x_0 + c\, s\, h_{00}^{1/2}]
$$

where $P(t_{n-2} < -c) = \alpha/2$. Again, this is of the general form: *Estimate $\pm$ Critical Value $\times$ Standard Error.*

Notice that the width of these confidence intervals depends on $x_0$ and increases as $x_0$ gets further from the sample mean $\bar{x}$. That is, the best estimates are made in the centre of the observed explanatory variables.

The fitted values are estimates of the conditional mean function at the observed values of covariates so

$$
E(\hat{Y}_i) = \beta_0 + \beta_1 x_i, \qquad \mathrm{Var}(\hat{Y}_i) = \sigma^2 h_{ii}
$$

and

$$
\begin{aligned}
\mathrm{Cov}(\hat{Y}_i, \hat{Y}_j) &= \mathrm{Cov}(\hat{\beta}_0 + \hat{\beta}_1 x_i, \hat{\beta}_0 + \hat{\beta}_1 x_j) \\
&= \mathrm{Var}(\hat{\beta}_0) + (x_j + x_i)\mathrm{Cov}(\hat{\beta}_0, \hat{\beta}_1) + x_i x_j \mathrm{Var}(\hat{\beta}_1) \\
&= \sigma^2 \left( \frac{1}{n} + \frac{\bar{x}^2}{(n-1)s_x^2} - \frac{(x_j + x_i)\bar{x}}{(n-1)s_x^2} + \frac{x_i x_j}{(n-1)s_x^2} \right) \\
&= \sigma^2 \left( \frac{1}{n} + \frac{(x_i - \bar{x})(x_j - \bar{x})}{(n-1)s_x^2} \right) \\
&= \sigma^2 h_{ij}.
\end{aligned}
$$

This proves that the fitted variables $\hat{Y}_i$ and $\hat{Y}_j$ are correlated whereas $Y_i$ and $Y_j$ have been assumed to be independent.

---

**Example 17.9**

Continuing with the body fat data example, suppose that we want to predict the conditional mean at the skinfold value of 70. Here $x_0 = \log(70)$. In order to estimate at the conditional mean we can calculate $\hat{Y}_0 = \hat{\beta}_0 + \hat{\beta}_1 x_0$ by hand and then obtain $\exp(\hat{Y}_0)$. However, we can use the R command `predict` to obtain the standard error and the confidence intervals. In order to do this we need to create a new data frame in R containing the set of values of the predictor, $x$, where we want to predict the response and call the predict command with the fitted model object and the new data frame as arguments as below:

```
newx <- data.frame(x=log(70))
a <- predict(flm, newdata=newx, se.fit=T) # obtains the
    prediction and the standard error of predictions
a # These results are on log scale.
# Confidence interval for the mean of log bodyfat at
    skinfold=70
a <- predict(flm, newdata=newx, interval="confidence")
a # These results are on log scale. Need to exponentiate to
    get results on the original scale.
```

As noted in the above code lines the predictions are obtained on the log scale. We can exponentiate the prediction and the end points of the confidence interval to obtain approximate predictions on the original scale. Here, the last command produces the results:

|   | fit | lwr | upr |
|---|-----|-----|-----|
| 1 | 2.498339 | 2.474198 | 2.52248. |

Hence the conditional mean is estimated to be $e^{2.498339} = 12.16$ and a 95% confidence interval for the conditional mean is $(11.87, 12.46)$. ◄

---

## 17.13 Predicting Observations

Suppose that we want to predict the observation $Y_0 = \beta_0 + \beta_1 x_0 + \epsilon_0$ at a value $x_0$ of $X$, where $\epsilon_0$ is independent of the errors in the sample data observations. This is different from the inference problems we have considered so far because in those we estimated a fixed but unknown constant parameter whereas here we are 'estimating' a random variable. Nevertheless, the basic approach only needs minor modification.

Notice that $\hat{Y}_0 = \hat{\beta}_0 + \hat{\beta}_1 x_0$ satisfies

$$E(\hat{Y}_0 - Y_0) = 0$$

so $\hat{Y}_0$ is an unbiased predictor of $Y_0$. Making use of the fact that $Y_0$, and hence $\epsilon_0$, is independent of $Y_1, \ldots, Y_n$, the prediction variance is

$$
\begin{aligned}
E\left[(\hat{Y}_0 - Y_0)^2\right] &= E\left[(\hat{\beta}_0 + \hat{\beta}_1 x_0 - \beta_0 + \beta_1 x_0 - \epsilon_0)^2\right] \\
&= E\left[(\hat{\beta}_0 + \hat{\beta}_1 x_0 - \beta_0 + \beta_1 x_0)^2\right] + E\left[(\epsilon_0)^2\right] \\
&= \sigma^2(1 + h_{00})
\end{aligned}
$$

Using a similar approach to that in Sect. 17.12, we can obtain $100(1 - \alpha)\%$ *prediction* intervals of the form

$$[\hat{\beta}_0 + \hat{\beta}_1 x_0 - cs(1 + h_{00})^{1/2}, \; \hat{\beta}_0 + \hat{\beta}_1 x_0 + cs(1 + h_{00})^{1/2}]$$

where $P(t_{n-2} < -c) = \alpha/2$. Again, this is of the general form *Estimate* $\pm$ *Critical Value* $\times$ *Standard Error.*

The difference between estimating the conditional mean and prediction is important. If we are interested in predicting the percent body fat for an athlete with log-skinfold value $x_0 = \log(70)$, we are trying to predict the value for an individual athlete in the population. If on the other hand, we are going to examine a number of athletes with log-skinfold $x_0 = \log(70)$, we may be interested in their average percent body fat or conditional mean. Since it is easier to predict average properties than it is to predict individual properties, the intervals for $Y_0$ (a point on the scatterplot) are wider than those for $\beta_0 + \beta_1 x_0$ (the position of the regression line). Specifically, this is due to the fact that the prediction variance incorporates additional uncertainty in $\epsilon_0$. In practice, we are most often interested in the more difficult problem of making predictions for individuals.

Prediction can be thought of either as interpolation (prediction within the range of observed explanatory variables) or extrapolation (prediction beyond the range of observed explanatory variables). The above equations apply equally to both cases but it is important to note that extrapolation involves the implicit additional assumption that the model holds in a global sense beyond the range of the observed explanatory variable. This is a strong assumption which cannot be tested empirically and is often known to be untrue. It is therefore best to view the model as a local approximation and to avoid extrapolation unless there are special reasons to believe that the fitted model holds 'globally'.

---

**Example 17.10**

Continuing with the body fat data example, suppose that we want to predict percent body fat for a new athlete who has a skinfold value of 70. Here $x_0 =$

log(70). In order to do this we follow the above procedure, i.e. create a new data frame and then change the `predict` command by requesting prediction interval as follows:

```
newx <- data.frame(x=log(70))
# Prediction interval for skinfold=70
a <- predict(flm, newdata=newx, interval="prediction")
a # These results are on log scale. Need to exponentiate to
    get results on the original scale.
```

Thus a 95% prediction interval for the log body fat percentage of an athlete with a skinfold of 70 is (2.334, 2.663). We can convert this interval to the raw scale by exponentiating the endpoints. Exponentiation produces the prediction value 12.16, which is same as before but the 95% prediction interval is (10.32, 14.34), which is wider than the earlier interval (11.87, 12.46) for the conditional mean.
◀

## 17.14  Analysing Residuals

Note that the fitted values $\hat{y}_i$ and residuals $r_i$ are observed values of random variables

$$\hat{Y}_i = \sum_{j=1}^{n} w_{0j} Y_j + x_i \sum_{j=1}^{n} w_{1j} Y_j, \quad i = 1, \ldots, n$$

and

$$R_i = Y_i - \hat{Y}_i, \quad i = 1, \ldots, n$$

respectively.
We can also write down the properties of the residuals

$$R_i = Y_i - \hat{Y}_i = Y_i - \hat{\beta}_0 - \hat{\beta}_1 x_i.$$

This looks like, but is not the same as, the prediction problem because the observation $Y_i$ is in the sample and hence cannot be independent of $(\hat{\beta}_0, \hat{\beta}_1)$ which depend explicitly on $Y_1, \ldots, Y_n$. The mean and variance are

$$E(R_i) = 0$$

and

$$\text{Var}(R_i) = \text{Var}(Y_i) + \text{Var}(\hat{\beta}_0 + \hat{\beta}_1 x_i) - 2\text{Cov}(Y_i, \hat{\beta}_0 + \hat{\beta}_1 x_i).$$

Now

$$\text{Cov}(Y_i, \hat{\beta}_0 + \hat{\beta}_1 x_j) = Cov\left(Y_i, \sum_{k=1}^{n}(w_{0k} + w_{1k}x_j)Y_k\right)$$

$$= \left(w_{0i} + w_{1i}x_j\right)\text{Var}(Y_i)$$

$$[\text{since } Cov\,(Y_i, Y_k) = 0 \text{ for all } k \neq i]$$

$$= \sigma^2\left(\frac{1}{n} - \frac{\bar{x}(x_i - \bar{x})}{(n-1)s_x^2} + \frac{x_j(x_i - \bar{x})}{(n-1)s_x^2}\right)$$

$$= \sigma^2\left(\frac{1}{n} + \frac{(x_i - \bar{x})(x_j - \bar{x})}{(n-1)s_x^2}\right)$$

$$= \sigma^2 h_{ij}$$

so

$$\text{Var}(R_i) = \sigma^2(1 + h_{ii} - 2h_{ii}) = \sigma^2(1 - h_{ii})$$

and, for $i \neq j$,

$$\text{Cov}(R_i, R_j) = \text{Cov}(Y_i - \hat{\beta}_0 - \hat{\beta}_1 x_i, Y_j - \hat{\beta}_0 - \hat{\beta}_1 x_j)$$

$$= -\text{Cov}(Y_i, \hat{\beta}_0 + \hat{\beta}_1 x_j) - \text{Cov}(Y_j, \hat{\beta}_0 + \hat{\beta}_1 x_i)$$

$$+ \text{Cov}(\hat{\beta}_0 + \hat{\beta}_1 x_i, \hat{\beta}_0 + \hat{\beta}_1 x_j)$$

$$= -\sigma^2(h_{ij} + h_{ij} - h_{ij})$$

$$= -\sigma^2 h_{ij}.$$

Note that, in general $h_{ij} \neq 0$ which implies that possibly there is correlation between $R_i$ and $R_j$, even though we assumed that $\epsilon_i$ and $\epsilon_j$ are independent. This is because $R_i$ and $R_j$ are both dependent on $\hat{\beta}_0$ and $\hat{\beta}_1$. The correlation between $R_i$ and $R_j$ is defined by:

$$\text{Corr}(R_i, R_j) = \frac{\text{Cov}(R_i, R_j)}{\sqrt{\text{Var}(R_i)\,\text{Var}(R_j)}}$$

$$= \frac{-\sigma^2 h_{ij}}{\sqrt{\sigma^2(1-h_{ii})\,\sigma^2(1-h_{jj})}}$$

$$= -\frac{h_{ij}}{\sqrt{(1-h_{ii})(1-h_{jj})}}.$$

The residuals are useful for examining patterns and have nice simple distributional properties. It is sometimes useful to scale them so to the standardised residuals

$$R_i/s.$$

We expect 5% of $|R_i/s| > 2$ and very few of $|R_i/s| > 3$ in normally distributed data. The fact the variance of the residuals depends on $i$ does not usually matter (and this dependence diminishes as $n$ increases). When this is a concern, we may use the studentised residuals

$$R_i/[s(1 - h_{ii})^{1/2}]$$

instead.

## 17.15 Applying Diagnostics Techniques

Once we have fitted the model and computed the residuals from our fit, we can use the residuals to explore the quality of the fit and the properties of the error distribution. The most effective methods for carrying out this exploration are graphical. These are collectively known as diagnostic methods.

### 17.15.1 The Anscombe Residual Plot

The residual plot, credited to Anscombe, is a plot of the residuals $r_i$ against the fitted values $\hat{y}_i = \hat{\beta}_0 + \hat{\beta}_1 x_i$.

If the model is correct, the points in the residual plot should be randomly distributed within a horizontal strip centred about the horizontal axis. Significant deviations from the desirable shape indicate problems with the assumptions of the model.

1. **Curvature:** the band of points does not remain horizontal but curves as $X$ changes. This indicates that the conditional means are not linearly related to the explanatory variable (Fig. 17.8).
2. **Heteroscedasticity:** a fan shaped residual scatter indicates that the variability in the error term $\epsilon_i$ in the model is not constant (Fig. 17.9).

**Fig. 17.8** Curvature in a residual plot

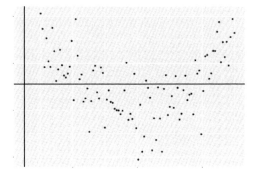

**Fig. 17.9** Heteroscedsaticity
in a residual plot

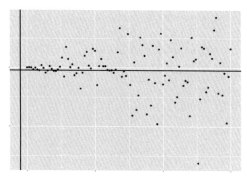

**Fig. 17.10** Discrepant
points in a residual plot

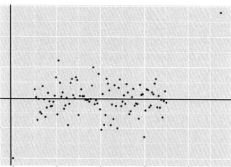

3. **Discrepant responses (outliers):** Y-values which do not follow the same linear
   model as the rest of the sample Y-values. The effect of such values can be to bias
   the sample estimates away from the parameters $\beta_0$ and $\beta_1$ of the linear model
   satisfied by the rest of the sample values and to inflate the error (Fig. 17.10).

   We can observe different forms of curvature, different forms of heteroscedasticity, more than one discrepant response and we can observe these departures
simultaneously.

   The information in the residual plot $(\hat{y}, y - \hat{y})$ is also contained in the scatterplot
$(x, y)$. Residual plots are graphically more effective than scatterplots because
the dominating linear relationship has been removed and it is easier to look for
deviations from the horizontal rather than from the fitted regression line.

   Residual plots are also useful in problems involving more than two variables
where scatterplots of pairs of variables are ineffective. In such cases, non-linearity
of the relationship between the response and an explanatory variable may not be
obvious from the Anscombe plot, and this plot should be supplemented by plotting
the residuals against each of the individual explanatory variables—more details
later.

## 17.15.2 Normal Probability Plots

Check the normality of the error distribution using a normal probability plot of the residuals. Order the residuals $r_1, \ldots, r_n$ in increasing size so $r_1^{\mathrm{ord}} \leq \ldots \leq r_n^{\mathrm{ord}}$. Suppose that $R_i \sim N(\mu, \sigma^2)$. It can be shown that:

$$E(R_i^{\mathrm{ord}}) \approx \mu + \sigma \Phi^{-1}\left(\frac{i - \frac{3}{8}}{n + \frac{1}{4}}\right), \quad i = 1, \ldots, n,$$

where $\Phi$ is the standard normal distribution function ($\Phi(x) = P(Z \leq x)$ where $Z \sim N(0, 1)$). Hence, a plot of $r_i^{\mathrm{ord}}$ against $\Phi^{-1}\left(\frac{i-\frac{3}{8}}{n+\frac{1}{4}}\right)$ should be approximately a straight line, if the normal model holds. Departures from linearity correspond to departures from the model.

Note that we always need to check which axis the observed residuals are on since this affects interpretation. R puts the observed residuals, $r_i^{\mathrm{ord}}$, on the $y$-axis and the theoretical quantiles, $\Phi^{-1}\left(\frac{i-\frac{3}{8}}{n+\frac{1}{4}}\right)$ on the $x$-axis. Thus, from a plot in R we can estimate $\sigma$ as the slope and estimate $\mu$ as the intercept, see Fig. 17.11.

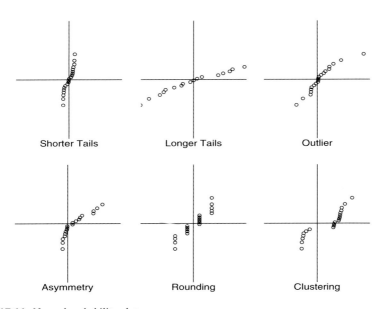

**Fig. 17.11** Normal probability plots

### 17.15.3 Dependence Plots

Dependence may be visible in the residual plot; it is more likely to be seen as patterns in plots of residuals against time and/or space variables or of residuals against adjacent residuals. For data collected sequentially in time, a relationship in a plot of the residual $R_i$ against the previous residual $R_{i-1}$ (called the lagged residual) often exhibits a linear relationship when the observations are dependent. Of course, in the absence of the required spatio-temporal information, we cannot construct such plots.

It is always important to think about the way in which the data were collected. Learning, repeated use of the same material, and arrangements in time and space can all induce dependence between the errors. Dependence-inducing structure in the data which is a result of the data collection process ordinarily needs to be reflected in the model. Failing to account for dependence, where it is present, typically leads to overconfident conclusions, which understate the amount of uncertainty. Models for dependent data are outside the scope of this course.

The beeswax data were collected through time (though we do not know the time structure or even the order) but there is no reason to suspect that this induced a dependence structure.

### 17.15.4 Discrepant Explanatory Variables

An X-value which is separated from the rest of the sample X-values is also important in modelling. The effect of such points is to attract the fitted line towards themselves and to increase the precision of the fitted line (see later). This means that discrepant explanatory variables can have a beneficial or a detrimental effect according to whether they are accurately observed or not.

**Leverage Points** In the simple regression framework, X-outliers are sometimes called leverage points. In more general models, leverage points have a more complicated definition than X-outliers but in the case of a simple regression model, a point $(x_i, y_i)$ is a leverage point if its leverage

$$h_{ii} = \frac{1}{n} + \frac{(x_i - \bar{x})^2}{(n-1)s_x^2}$$

is large, that is when $x_i$ is far from $\bar{x}$. This is basically, a definition of an X-outlier if we use $\bar{x}$ to represent the bulk of the X-values.

**Influential Points** Finally, X- and Y-outliers are sometimes referred to as influential points. Generally, influential points $(x_i, y_i)$ are observations that have an excessive 'influence' on the analysis in the sense that deleting them (or a set of them) results in large changes to the analysis. Clearly, large outliers are influential points and in the simple regression framework, there are no other kinds of influential points. However, when we work with more complicated models, it is possible to have influential points which are not also outliers. A measure of the influence of a point is **Cook's distance**

$$d_i = \frac{\sum_{j=1}^{n} \left( \hat{y}_j^{(i)} - \hat{y}_j \right)^2}{k\hat{\sigma}^2}$$

where $\hat{y}_j^{(i)}$ is the fitted value for observation $j$, calculated using the least squares estimates obtained from the modified data set with the $i$th observation deleted. In the denominator $k$ is the number of linear parameters estimated (2 for simple regression) and $\hat{\sigma}^2$ is an estimate of the error variance. A rule of thumb is that values of $d_i$ greater than 1 indicate influential points.

## 17.15.5 Suggested Remedies

**Transformation**  If the residual plot shows non-linearity and/or heteroscedasticity or the normal probability plot shows a long-tailed distribution, we may be able to find a transformation which linearises the relationship, stabilises the variability and makes the variability more normal. The most widely used transformations (in increasing order of strength) are the cube root, square root (for count data), logarithmic (for multiplicative relationships) or negative reciprocal (for time and rates). However, there is no guarantee that a transformation can be found to remedy any of these defects, never mind all simultaneously. Also, when a transformation has been used, the final model fitted is no longer in terms of the original variables. This latter objection is often unimportant but there are calibration problems in which it may matter.

**Reacting to Discrepant Observations**  First try to establish whether the points are mistakes, observations from a distinct, identifiable subgroup in the population or due to changes in the data collection process. Often, unless we have access to the people who collected the data, there is no additional information available. On the basis of whatever information we can collect, we can correct mistakes, build more complicated models (which either change the distributional assumptions or treat the subsets separately) or exclude the subsets from the analysis. The choice between these depends on whether we want to try to mix the possibly disparate structures into one model or we are interested in the pattern followed by the bulk (majority) of the data. We can exclude points in two ways.

- **Robust methods:** Use robust fitting methods to fit the model to the bulk of the data.

- **Exclusion:** Remove the points and then refit using least squares. (The disadvantage of this approach is that the identification of the points is not always straightforward and the inferences do not then reflect this facet of our analysis.)

Multiple analyses can be presented to show the range of possible scientific outcomes. It is important to identify discrepant points and then explain and justify the actions we take in relation to them.

## 17.16  Illustration of Diagnostics Plots for the Body Fat Data Example

The R command `plot` produces four important diagnostic plots as shown in Fig. 17.12. The first plot shows a slight curvature influenced by the three identified observations. The normal probability plot seems alright without much concern for any departure from normality. In practical real life data modelling it is natural to see slight departure of a few points from the middle straight line. The standardised residuals plots in the bottom row further diagnoses aspects of minor problems already identified in the two plots in the top row.

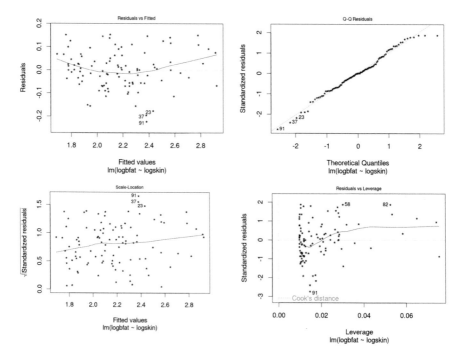

**Fig. 17.12**  Four diagnostic plots for the body fat data example

## 17.17   Remphasising the Paradigm of Regression Modelling

The basis for regression modelling is a strategy based on a clear sequence of steps. While these may be obvious from the example presented so far, it is helpful to make the strategy explicit. The steps are as follows:

1. Propose a plausible model
   (a) Look at the data
   (b) Use a scatterplot to try to choose scales (transformations) so a simple linear regression model is appropriate.
2. Ensure that the model is adequate
   (a) Fit the model (estimate the parameters)
   (b) Assess the quality of fit using the two basic diagnostic plots
   (c) If necessary in the light of step 3 (below), modify the model and repeat (a-b) until an appropriate model is obtained.
3. Present, interpret and use the model to make inferences and predictions, remembering always to appropriately summarise uncertainty.

The validity of the inferences depends on the representativeness of the sample and the validity of the model. It is vital when collecting data to ensure that it is representative of the population of interest and before making inferences to ensure that the model is appropriate. Strictly speaking, model assessment is a form of informal inference and has an impact on other inferences but it is not sensible to simply hope that a model is valid when empirical verification is available.

## 17.18   Summary of Linear Regression Methods

1. The simple linear regression model is given by:

$$Y_i = \beta_0 + \beta_1 x_i + \epsilon_i, \; \epsilon_i \overset{\text{ind}}{\sim} N(0, \sigma^2), \; i = 1, \ldots, n.$$

2. The least squares estimates are given by:

$$\hat{\beta}_0 = \bar{y} - \hat{\beta}_1 \bar{x}$$
$$\hat{\beta}_1 = \frac{\sum_{i=1}^{n}(x_i - \bar{x})(y_i - \bar{y})}{\sum_{i=1}^{n}(x_i - \bar{x})^2},$$
$$= \frac{\sum_{i=1}^{n}(x_i - \bar{x})y_i}{\sum_{i=1}^{n}(x_i - \bar{x})^2},$$
$$= \frac{\sum_{i=1}^{n} x_i y_i - n\bar{x}\bar{y}}{\sum_{i=1}^{n} x_i^2 - n\bar{x}^2}.$$

All these formulae are equivalent for $\hat{\beta}_1$.
3. Fitted values are given by $\hat{y}_i = \hat{\beta}_0 + \hat{\beta}_1 x_i$.

4. Residuals are given by: $r_i = y_i - \hat{y}_i$. It can be shown that:

$$\text{Residual SS} = \sum_{i=1}^{n} r_i^2 = \sum_{i=1}^{n} (y_i - \bar{y}^2) - \hat{\beta}_1^2 \sum_{i=1}^{n} (x_i - \bar{x})^2.$$

5. Residual df is the number of observations minus the number of parameters estimated $= n - 2$.
6. Can show that $\text{Var}(\hat{\beta}_1) = \frac{\sigma^2}{\sum_{i=1}^{n}(x_i - \bar{x})^2}$.
7. Can show that $\text{Var}(\hat{\beta}_0) = \sigma^2 \left( \frac{1}{n} + \frac{\bar{x}^2}{\sum_{i=1}^{n}(x_i - \bar{x})^2} \right)$.
8. Can show that $\text{Cov}(\hat{\beta}_0, \hat{\beta}_1) = -\sigma^2 \frac{\bar{x}}{\sum_{i=1}^{n}(x_i - \bar{x})^2}$.
9. We estimate $\sigma^2$ by $s^2 = \frac{\text{Residual SS}}{\text{Residual df}} = \frac{1}{n-2} \left( \sum_{i=1}^{n}(y_i - \bar{y}^2) - \hat{\beta}_1^2 \sum_{i=1}^{n} (x_i - \bar{x})^2 \right)$.
10. Can estimate the conditional mean of $Y_0$ for a new $x_0$ by $\hat{\beta}_0 + \hat{\beta}_1 x_0$. Its variance is $\sigma^2 \left( \frac{1}{n} + \frac{(x_0 - \bar{x})^2}{\sum_{i=1}^{n}(x_i - \bar{x})^2} \right) = \sigma^2 h_{00}$.
11. Can predict a future observation $Y_0$ for a new $x_0$ by $\hat{\beta}_0 + \hat{\beta}_1 x_0$. Its variance is $= \sigma^2 (1 + h_{00})$.
12. To construct a confidence interval, use the general formula: **Estimate $\pm$ Critical Value $\times$ Standard Error**. Use this to find intervals for any $\beta_0$ and $\beta_1$, the conditional mean and the future observation $Y_0$ for a new $x_0$.
13. The critical value is an appropriate quantile of the $t$-distribution with $n - 2$ df. For 95% intervals, the R command is qt(0.975, df=n-2).
14. **Standard Error** is the estimated value of the square root of the variance.
15. To test $H_0 : \beta_1 = 0$,
    (a) obtain $t_{obs} = \frac{\hat{\beta}_1}{\text{Standard Error}(\hat{\beta}_1)}$.
    (b) Calculate P-value $= 2P(t_{n-2} > |t_{obs}|)$. P-values are reported by R as the last column of the summary table for a fitted linear model.
    (c) **Reject $H_0$ if P-value is less than** $\alpha$, where $\alpha$ is the level of significance, usually 0.05.
16. The ANOVA table for comparing the simple linear regression model with the null model ($Y_i = \beta_0 + \epsilon_i$) is given by:

| Source | Df | Sum of squares | Mean squares | F value | P value |
|---|---|---|---|---|---|
| Regression | 1 | $\hat{\beta}_1^2(n-1)s_X^2$ | $\hat{\beta}_1^2(n-1)s_X^2$ | $\frac{\hat{\beta}_1^2(n-1)s_X^2}{s^2}$ | $Pr(F > F_{obs})$ |
| Residuals | $n-2$ | $(n-2)s^2$ | $s^2$ | | |
| Total | $n-1$ | $\sum_{i=1}^{n}(y_i - \bar{y}^2)$ | | | |

17. We use residual plots to check model assumptions.
    (a) We plot the residuals against the fitted values to check for homoscedasticity. Plot should look like a random scatter if this assumption holds.

(b) We order and plot the residuals against the normal order statistics to check for normality. This plot should be a straight line if the normality assumption holds.

18. We often report the $R^2$ :

$$R^2 = 1 - \frac{(n-2)S^2}{(n-1)S_y^2} = 1 - \frac{\text{Residual SS}}{\text{Total SS}} = \frac{\text{SS explained by the Model}}{\text{Total SS}}$$

19. **Adjusted** $R^2$ is the proportionate reduction in error variance achieved by the model.

$$\text{adj } R^2 = 1 - \frac{S^2}{S_y^2}.$$

## 17.19 Exercises

### 17.1 (Linear Regression Model)

1. Suppose that we have n observations on the simple regression model,

$$Y_i = \beta_0 + x_i\beta_1 + \epsilon_i, \quad i = 1, \ldots, n,$$

with $\epsilon_i \sim$ independent $N(0, \sigma^2)$.

(a) Let $\hat{\beta}_0$ and $\hat{\beta}_1$ represent the least squares estimators of the mean parameters in the model based on the observed data. Show how these estimators, the fitted values and residuals are affected if we replace each $Y_i$ by $aY_i + b$, where $a \neq 0$ and $b$ are constants.

(b) Repeat (a) but retaining $Y_i$ and replacing $x_i$ by $cx_i + d$, where $c \neq 0$ and $d$ are constants.

(c) Now suppose that instead of estimating $\beta_0$ and $\beta_1$ by minimising the vertical distance from the data to the fitted line, we estimate them by minimising the orthogonal distance

$$\sum_{i=1}^{n} \frac{(y_i - \beta_0 - \beta_1 x_i)^2}{1 + \beta_1^2}.$$

Show that $\tilde{\beta}_0 = \bar{Y} - \bar{X}\tilde{\beta}_1$ and

$$\tilde{\beta}_1 = \frac{S_{yy} - S_{xx} + \sqrt{(S_{yy} - S_{xx})^2 + 4S_{xy}^2}}{2S_{xy}},$$

where $S_{xx} = \sum_{i=1}^{n}(x_i - \bar{x})^2$, $S_{yy} = \sum_{i=1}^{n}(y_i - \bar{y})^2$ and $S_{xy} = \sum_{i=1}^{n}(x_i - \bar{x})(y_i - \bar{y})$, are solutions of the orthogonal distance estimating equations.

(d) Show how $\tilde{\beta}_1$ is affected if we replace each $Y_i$ by $aY_i + b$, where $a \neq 0$ and $b$ are constants.

(e) Show how $\tilde{\beta}_0$, $\tilde{\beta}_1$, the fitted values and residuals are affected if we replace both $Y_i$ and $x_i$ by $aY_i + b$ and $ax_i + d$, where $a \neq 0$, $b$ and $d$ are constants.

2. A model widely used in the analysis of sample surveys assumes that for $x_i > 0$ and $c$ known,

$$Y_i = x_i \beta + x_i^c \epsilon_i, \quad i = 1, \ldots, n,$$

with $\epsilon_i \sim$ independent $N(0, \sigma^2)$. Here the intercept is assumed to be zero and the conditional variance $Var(Y_i|x_i) = x_i^{2c}\sigma^2$ so that the responses are heteroscedastic.

(a) Let $\hat{\beta}_{LS}$ denote the least squares estimator of $\beta$. Show that

$$E(\hat{\beta}_{LS}) = \beta \quad \text{and} \quad Var(\hat{\beta}_{LS}) = \sigma^2 \frac{\sum_{i=1}^{n} x_i^{2(1+c)}}{(\sum_{i=1}^{n} x_i^2)^2}.$$

(b) Show that the variable $Y_i/x_i^c$ satisfies a homoscedastic linear regression model with slope $\beta$.

(c) Show that the least squares estimator of $\beta$ from the model in (b) satisfies

$$\hat{\beta} = \frac{\sum_{i=1}^{n} x_i^{1-2c} Y_i}{\sum_{i=1}^{n} x_i^{2(1-c)}}.$$

(d) Show that

$$E(\hat{\beta}) = \beta \quad \text{and} \quad Var(\hat{\beta}) = \frac{\sigma^2}{\sum_{i=1}^{n} x_i^{2(1-c)}}.$$

(e) Show that $Var(\hat{\beta}) \leq Var(\hat{\beta}_{LS})$.
    (**Hint**: use the Cauchy-Schwarz inequality.)

(f) The estimator

$$s_1^2 = \frac{1}{n-1} \sum_{i=1}^{n}(Y_i - x_i\hat{\beta}_{LS})^2$$

is estimating $n^{-1}\sum_{i=1}^{n} x_i^{2c}\sigma^2$. Show how to adjust $s_1^2$ to estimate $\sigma^2$. Construct an estimator of $\sigma^2$ based on the model from part (b).

3. The data given in the following table are the numbers of deaths from AIDS in Australia for 12 consecutive quarters starting from the second quarter of 1983.

| Quarter ($i$) | 1 | 2 | 3 | 4 | 5 | 6 | 7 | 8 | 9 | 10 | 11 | 12 |
|---|---|---|---|---|---|---|---|---|---|---|---|---|
| Number of deaths ($n_i$) | 1 | 2 | 3 | 1 | 4 | 9 | 18 | 23 | 31 | 20 | 25 | 37 |

(a) Draw a scatterplot of the data and comment on the nature of the relationship between the number of deaths and the quarter in this early phase of the epidemic.

(b) A statistician has suggested that a model of the form

$$E[N_i] = \gamma i^2$$

might be appropriate for these data, where $\gamma$ is a parameter to be estimated from the above data, and has proposed two methods for estimating $\gamma$ given below:

Show that the least squares estimate of $\gamma$, obtained by minimising $\sum_{i=1}^{12}(n_i - \gamma i^2)^2$ is given by

$$\hat{\gamma} = \frac{\sum_{i=1}^{12} n_i i^2}{\sum_{i=1}^{12} i^4}.$$

Show that an alternative (weighted) least squares estimate of , obtained by minimising $\sum_{i=1}^{12}(n_i - \gamma i^2)^2 / i^2$ is given by

$$\tilde{\gamma} = \frac{\sum_{i=1}^{12} n_i}{\sum_{i=1}^{12} i^2}.$$

(c) Calculate $\hat{\gamma}$ and $\tilde{\gamma}$ for the above data. [**Hint:** $\sum_{i=1}^{k} i^2 = \frac{1}{6}k(k+1)(2k+1)$ and $\sum_{i=1}^{k} i^4 = \frac{1}{30}k(k+1)(2k+1)(3k^2+3k-1)$]

(d) To assess whether the single parameter model which was used in part (b) is appropriate for the data, a two parameter model is now considered. The model is of the form

$$E[N_i] = \gamma i^\theta$$

To estimate the parameters $\gamma$ and $\theta$, a simple linear regression model

$$E[Y_i] = \alpha + \beta x_i$$

is used, where $x_i = \log(i)$ and $Y_i = \log(N_i)$ for $i = 1, \ldots, 12$. Relate the parameters $\gamma$ and $\theta$ to the regression parameters $\alpha$ and $\beta$.

The least squares estimates of $\alpha$ and $\beta$ are 0.6112 and 1.6008 with standard errors 0.4586 and 0.2525 respectively (you are not asked to verify these results). Using the value for the estimate of $\beta$, conduct a formal statistical test to assess whether the form of the model suggested in (b) is adequate.

4. Beanie Babies are toys, stuffed animals which have become valuable collectors' items. *Beanie World Magazine* has provided information on the age (in months) and value (in US$) of 50 Beanie Babies. The data is `beanie` in R package `ipsRdbs`. The data are also provided in the file `beanie.txt`. We want to model the relationship between the value (the response) and age (the explanatory variable).

(a) Obtain a scatter plot of the data and comment on the relationship between age and value. Suggest a suitable transformation for the response so that the methods of simple linear regression can be used to predict value. Provide the scatter plot on the transformed scale and comment on the relationship. On which scale would you fit a simple linear regression model?

(b) Fit a simple linear regression model on the chosen scale and explore whether it fits the data or not using two standard diagnostic plots. For this model, test the hypothesis of no relationship between age and value. Provide a 95% confidence interval for the slope parameter. Present an estimate of $\sigma^2$, the unknown variance parameter in your chosen regression model and state its degrees of freedom.

(c) Write down the fitted model on the original scale so that the model can be used to predict value in US$. Use the model to predict the mean value (in US$) of a Beanie Baby aged 35 months. Give a 95% confidence interval for this mean value (in US$). Using your model obtain a 97% prediction interval for the value of a Beanie Baby aged 45 months.

# Multiple Linear Regression Model

# 18

**Abstract**

This chapter generalises the simple regression techniques of the previous chapter to the case where there are multiple possible explanatory variables. This chapter describes the foundational basics for machine learning where the simple and multiple regression techniques are exploited heavily for practical problems. Again, the techniques are described both theoretically and using practical modelling examples in R so that the reader can easily form their own transferable skills.

This chapter generalises the simple linear regression methods of the previous chapter to include multiple explanatory variables (covariates). Indeed, the simple linear regression model will be seen as a particular special case of the multiple regression model to be introduced in this chapter. The methods for inference (estimation and hypothesis testing and prediction) and diagnostics plots outlined in the previous chapter will be generalised as required in this chapter. However, note that the topics remain largely the same as this chapter falls under the umbrella of statistical modelling.

## 18.1 Motivating Example: Optimising Rice Yields

Rice is often taken as a staple food source by millions of people worldwide. Rice is grown as the seeds of a semi-aquatic annual grass. It takes from three to six months to grow rice. Rice has three distinctive growth phases: the vegetative phase (germination, early seedling growth and reproduction), the reproductive phase (the time before heading) and the ripening phase (time after heading). The potential grain yield is determined largely in the reproductive phase while the ultimate yield which is based on the amount of starch produced is largely determined during

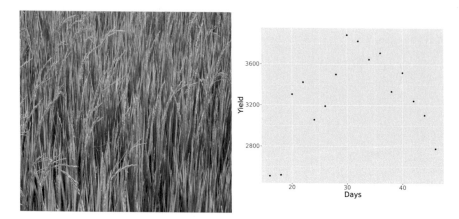

**Fig. 18.1** An image of Asian rice in Thailand (credit: Vyacheslav Argenberg / http://www.vascoplanet.com/ License: CC-BY-SA-4.0) and a scatter plot of the rice yield data

ripening. Experiments on rice development are usually conducted with a view towards optimising agricultural production.

Data from one such experiment, found in the data set `rice` in the R package `ipsRdbs`, consists of yields of rice and the number of days after flowering before the harvesting took place (days). This data set has been obtained via Prof Alan H. Welsh (author of Welsh [21]) from the research article, Bal and Ojha [1].

An initial scatterplot of the data (see Fig. 18.1) suggests that the relationship between yield and days is quadratic and the variability seems to be reasonably constant. It is useful (for numerical stability) to centre the days variable. The mean number of days in the experiment is 31 so we subtract 31 from each day. Then let $Y_i$ be the yield and $x_i$ be the number of days minus 31, $i = 1, \ldots, n$. Then a plausible model for the data is the quadratic regression model

$$Y_i = \beta_0 + \beta_1 x_i + \beta_2 x_i^2 + \epsilon_i, \qquad \epsilon_i \overset{\text{ind}}{\sim} N(0, \sigma^2).$$

The parameter $\beta_0$ is the intercept of the quadratic curve, $\beta_1$ is the linear coefficient and $\beta_2$ is the quadratic coefficient.

We fit this model using the least squares method of estimation, i.e., by minimising the sum of squares

$$S = \sum_{i=1}^{n} (y_i - \beta_0 - \beta_1 x_i - \beta_2 x_i^2)^2$$

to obtain estimates $\hat{\beta}_0$, $\hat{\beta}_1$ and $\hat{\beta}_2$ of $\beta_0$, $\beta_1$ and $\beta_2$ respectively. The fitted values are simply

$$\hat{y}_i = \hat{\beta}_0 + \hat{\beta}_1 x_i + \hat{\beta}_2 x_i^2$$

so the residuals from the model are

$$r_i = y_i - \hat{y}_i = y_i - \hat{\beta}_0 - \hat{\beta}_1 x_i - \hat{\beta}_2 x_i^2.$$

Once we have fitted the model and obtained the residuals, we can proceed to evaluate it using diagnostic plots. In particular, plots of residuals ($r_i$) against fitted (predicted) values ($\hat{y}_i$) and normal probability plots of the residuals should be checked. These plots are interpreted exactly as described previously in Sect. 17.15.

The residual plot (not included here) shows an even horizontal band of points with no particular pattern. The normal probability plot is linear showing that the error distribution is normal. The plots allow us to conclude that the quadratic model is adequate for the data.

The test in the R parameter estimates table corresponding to $X^2$ is testing the hypothesis that $\beta_2 = 0$. The p-value is zero (to 3 decimal places) so $\hat{\beta}_2$ is significantly different from zero, confirming that the linear model is not an adequate simplification of the quadratic model. Generally, it does not make sense to now test $\beta_1 = 0$. The F-test in the R analysis of variance table, obtained using the **anova** command in R, tests the hypothesis that $\beta_1 = \beta_2 = 0$ (the null model) and is now not a simple function of the two separate t-statistics.

We can also consider fitting the cubic model

$$Y_i = \beta_0 + \beta_1 x_i + \beta_2 x_i^2 + \beta_3 x_i^3 + \epsilon_i, \qquad \epsilon_i \overset{ind}{\sim} N(0, \sigma^2).$$

The diagnostics are similar to those obtained for the quadratic model so the cubic model also fits the data. However, the p-value for the test that $\beta_3 = 0$ is 0.6997 so the test is not significant and the cubic term adds little to the quadratic model. Note that the F-test (against the null model where $\beta_1 = \beta_2 = \beta_3 = 0$) is still highly significant.

Thus, our final fitted model is:

$$\text{Fitted Yield} = -1070.40 + 293.48\text{Days} - 4.54\text{Days}^2$$
$$\qquad\qquad\quad (617.25) \quad (42.18) \qquad\quad (0.67)$$

The estimate ($s$) of the error standard deviation ($\sigma$) is 203.88 on 13 degrees of freedom.

The optimal time to harvest is the value of X which maximises the fitted yield. This can be calculated analytically or it can be read off from the graph of yield against days with the fitted quadratic model. The approximate optimal value is 32.1465 days. A 95% prediction interval for the yield from a harvest of

$X = 32.1465$ days is

$$(3206.46, 4147.07).$$

## 18.2   Motivating Example: Cheese Testing

Cheese, see e.g. Fig. 18.2, is a popular food item which is made from milk by coagulating the milk solids into curd with acid, bacteria or rennet, removing the curds from the whey, salting and then leaving the cheese to mature. During maturation, ripening gases such as carbon dioxide and ammonia and other chemicals which affect the taste of the cheese are produced. In a study of cheddar cheese from the LaTrobe Valley in Victoria, Australia, samples of cheese were analysed for their chemical composition and subjected to taste tests. The response variable is the taste of the cheese and we would like to find the relationship between taste and the concentrations of acetic acid, hydrogen sulphide and lactic acid in the cheese. This data set is obtained from Prof Alan H. Welsh (author of Welsh [21]).

The difference between this problem and the preceding one on rice yield is that we have three explanatory variables rather than one. We can try modelling taste and each explanatory variable separately but it is possible that taste depends on the mix of chemicals rather than any single one of them and the one variable at a time approach cannot capture this kind of relationship. What we need is to try to relate taste to all three explanatory variables simultaneously.

If we examine the data, cheese in the R package ipsRdbs, we see that the range of values of hydrogen sulphide concentration is very large. This suggests that we may need to transform this variable. A plot may confirm the need to transform and enable us to evaluate different transformations. The first difficulty we encounter is that we cannot plot all four variables simultaneously to examine the four dimensional joint relationship. Essentially what we require are methods of reducing the dimensionality so that we can use graphical methods. A model, which, when fitted produces residuals, is one way of doing this; the residuals have dimension one whereas the data has dimension four (as there are four variables). Before we can fit a model, we need to formulate it. In the absence of theoretical reasons for adopting any particular model, we often begin with a linear model.

**Fig. 18.2**  Cheddar cheese cubes at the Public tasting event by Guillaume Paumier. Source: https://commons.wikimedia.org/ License: CC-BY-SA-3.0

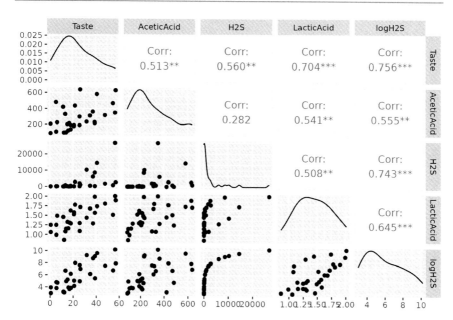

**Fig. 18.3**  Pairwise scatter plots for the cheese testing data

This is straightforward once we have determined plausible scales for the variables. We choose the scale (e.g. log scale) by thinking about the problem and by some exploratory analysis. The explanatory variables are all concentrations so perhaps they should all be on the same scale but there is no obvious scale to choose. The main tool for exploratory analysis is the scatterplot matrix which is an array of scatterplots of all pairs of variables, obtained using the **pairs** (or ggpairs using the ggplot library) command in R, see Fig. 18.3.

To examine relationships between taste and the four explanatory variables we inspect the four scatter plots in the first column of Fig. 18.3. There seems to be a fairly strong linear relationship between taste and lactic acid, there is a weaker but still linear relationship between taste and acetic acid and a highly nonlinear relationship between taste and hydrogen sulphide. This suggests possibly putting hydrogen sulphide on another scale. The choice of scale here is a matter of trial and error and in this circumstance we can start by trying the log scale.

The relationship between taste and log hydrogen sulphide is roughly linear so we will use this scale. Relationships between the explanatory variables are also of interest; these all seem to be linear though none are particularly strong. There are no obvious unusual points to cause concern. (The large hydrogen sulphide value on the raw scale does not look to be unusual on the log scale.)

We can now formulate an initial plausible model. Let $Y_i$ denote the taste, $x_{i1}$ the concentration of acetic acid, $x_{i2}$ the log of the concentration of hydrogen sulphide and $x_{i3}$ the concentration of lactic acid in the $i$th sample of cheese. Then a local linear approximation to the relationship between the response and the explanatory variables is

$$Y_i = \beta_0 + \beta_1 x_{i1} + \beta_2 x_{i2} + \beta_3 x_{i3} + \epsilon_i, \qquad \epsilon_i \overset{\text{ind}}{\sim} N(0, \sigma^2).$$

This is called a multiple regression model. If we think of the conditional mean response as a surface in four dimensional space, then this model approximates the conditional mean response by a hyperplane in that space. The parameter $\beta_0$ is the intercept of the plane (the value of the conditional mean response when the explanatory variables are all zero) and $\beta_1$, $\beta_2$ and $\beta_3$ are collectively referred to as the slopes. The parameter $\beta_1$ gives the effect of a unit change in the concentration of acetic acid $(x_1)$ on the conditional mean response when the concentration of log hydrogen sulphide and lactic acid $(x_2$ and $x_3)$ are held constant. Similar interpretations apply to the other slopes. The important point is that the interpretation of the slopes depends on the other explanatory variables in the model because these have to be held constant.

We fit this model by minimising the sum of squares

$$S = \sum_{i=1}^{n} (y_i - \beta_0 - \beta_1 x_{i1} - \beta_2 x_{i2} - \beta_3 x_{i3})^2$$

to obtain estimates $\hat{\beta}_0$, $\hat{\beta}_1$, $\hat{\beta}_2$ and $\hat{\beta}_3$ of $\beta_0$, $\beta_1$, $\beta_2$ and $\beta_3$ respectively. The fitted values are simply

$$\hat{y}_i = \hat{\beta}_0 + \hat{\beta}_1 x_{i1} + \hat{\beta}_2 x_{i2} + \hat{\beta}_3 x_{i3}$$

so the residuals from the model are

$$r_i = y_i - \hat{y}_i = y_i - \hat{\beta}_0 - \hat{\beta}_1 x_{i1} - \hat{\beta}_2 x_{i2} - \hat{\beta}_3 x_{i3}.$$

Once we have fitted the model and obtained the residuals, we can proceed to evaluate it using diagnostic plots. In particular, plots of residuals $(r_i)$ against fitted (predicted) values $(\hat{y}_i)$ and normal probability plots of the residuals should be checked when fitting a multiple linear regression model. These plots are interpreted exactly as described before. Plots of residuals against individual covariates are also important as these can help identify potential non-linear relationships which have not been modelled. Here the fitted values are a function of all the explanatory variables so the information in the diagnostic plots is not readily available from other representations of the data.

There is a slight appearance of heteroscedasticity in the residual plot though this is primarily due to one point which may be an outlier. The normal probability plot shows this point as a mild outlier so we will retain it in our analysis. There are no other obvious influential points of concern.

Since the model seems adequate, we can examine the parameter estimates (obtained using the `summary` command) and the ANOVA table (using the `anova` command):

| Parameter estimates | | | | |
|---|---|---|---|---|
| Term | Estimate | Std error | t ratio | Prob>|t| |
| Intercept | −27.142 | 9.278 | −2.93 | 0.007 |
| Acetic acid | 0.004 | 0.015 | 0.28 | 0.781 |
| $\log H_2 S$ | 3.84 | 1.220 | 3.15 | 0.004 |
| Lactic acid | 19.201 | 8.458 | 2.27 | 0.032 |

| Analysis of variance | | | | | |
|---|---|---|---|---|---|
| Source | DF | SS | MSE | F | P |
| Model | 3 | 5002.0 | 1667.3 | 16.29 | 0.000 |
| Error | 26 | 2660.9 | 102.3 | | |
| Total | 29 | 7662.9 | | | |

The root mean square error ($s$) is 10.12 on 26 degrees of freedom. (There are 30 observations and we have fitted 4 conditional mean parameters so the residual degrees of freedom is 26, =30-4.)

Can we simplify the model? The t-ratio for acetic acid is not significant (pvalue = 0.781) and the other two terms are significant so we can consider excluding acetic acid from the model. Formally, we have tested the hypothesis $H_0 : \beta_1 = 0$. That is, the coefficient $\beta_1$ of acetic acid is zero. The parameter estimates for the reduced model are given below

| Parameter estimates | | | | |
|---|---|---|---|---|
| Term | Estimate | Std error | t ratio | Prob>|t| |
| Intercept | −27.59 | 8.98 | −3.07 | 0.0048 |
| $\log H_2 S$ | 3.95 | 1.14 | 3.47 | 0.0017 |
| Lactic Acid | 19.89 | 7.96 | 2.50 | 0.0188 |

All the terms are significant so we stop trying to reduce the model. (If one of the slopes was not significant, then we could omit the corresponding variable, fit the reduced model and iterate the above process until the slopes remaining in the model are all significant.)

It is a good idea to check that our simplified model is still adequate by examining its diagnostics. This means that in spite of all the information provided, we cannot avoid fitting the reduced model. (Note also that the parameter estimates, standard errors, etc. change when we fit the reduced model; this is because their interpretation has changed with the omission of variables.)

The simplified model clearly still provides reasonable fit. The root mean square error is 9.94 on 27 degrees of freedom.
It is valuable to review the modelling process used in this example.

1. Propose an initial plausible model
   (a) Think about the data and scales
   (b) Use a scatterplot matrix to suggest linearising scales
2. Find an adequate model
   (a) Fit the plausible model
   (b) Check diagnostics
   (c) If the model is adequate, call it the full model and proceed to 3. Otherwise, modify the model and/or the data to obtain a new plausible model. Repeat steps 2(a)–2(c).
3. Find a simpler but still adequate model.
   (a) Examine the p-values for the effects. If they are all significant, skip to (c).
   (b) Fit the reduced model without the explanatory variable corresponding to the largest p-value and return to 3(a).
   (c) Check the diagnostics again to make sure the simplification has not reduced the adequacy of the fit. If it has, we may need to put some variables back into the model. Iterate until the simpler model is adequate.
4. Present and interpret the final model.

Notice that the strategy for simple regression fits in here; the only difference is that step 3 is usually not needed because the conditional mean structure in the simple linear regression model is already very simple. The scatterplot matrix is used in the same way as the scatterplot in a two variable problem but because we are only examining the relationship between two variables at a time, the scales we choose are less certain to work with than in the simple linear regression model. The basic difficulty is that looking at the relationship between two variables while ignoring the others is not necessarily informative about the relationship between these same two variables when the other variables are taken into account.

The model selection strategy for simplifying an adequate model is only one of a number of different possible approaches, see Sect. 18.8 below. Indeed, different approaches may yield different final models. In principle, there is nothing wrong with this, as we have to recognise that a model is at best an approximation to a data generating process, which may be useful in guiding understanding about the process and making predictions about future outcomes. It is common sense to acknowledge the possibility that more than one adequate such approximation may exist.

## 18.3   Matrix Formulation of the Model

Both of the previous two examples have illustrated extensions to the simple regression model we first considered in Sect. 17.2. Although they look different, the similarities are more important than the differences and suggest a unified approach.

Both models (and all the models we have considered so far are special cases of the general multiple regression model

$$Y_i = \beta_0 + \beta_1 x_{i1} + \beta_2 x_{i2} + \ldots + \beta_p x_{ip} + \epsilon_i, \qquad \epsilon_i \overset{\text{ind}}{\sim} N(0, \sigma^2). \qquad (18.1)$$

We have the special cases:

- **Simple Linear Regression** when $p = 1$
- **Quadratic Regression** when $p = 2$, $x_{i1} = x_i$ and $x_{i2} = x_i^2$
- **Multiple regression** when $p \geq 2$

With the extra cases and the flexibility achieved through transformation, this is a very flexible model. We can write the model in several different ways. One useful way is to represent the covariates and parameters as $(p + 1)$-vectors. We have

$$Y_i = \begin{pmatrix} 1 & x_{i1} & x_{i2} & \ldots & x_{ip} \end{pmatrix} \begin{pmatrix} \beta_0 \\ \beta_1 \\ \vdots \\ \beta_p \end{pmatrix} + \epsilon_i$$

$$= \mathbf{x}_i^T \boldsymbol{\beta} + \epsilon_i,$$

where $\mathbf{x}_i = \begin{pmatrix} 1 & x_{i1} & x_{i2} & \ldots & x_{ip} \end{pmatrix}^T$ is a $(p + 1)$-vector of covariates and $\boldsymbol{\beta} = \begin{pmatrix} \beta_0 & \beta_1 & \ldots & \beta_p \end{pmatrix}^T$ is a $(p + 1)$-vector of unknown (conditional mean) parameters. (All vectors are column vectors.) We can write

$$Y_i \overset{\text{ind}}{\sim} N(\mathbf{x}_i^T \boldsymbol{\beta}, \sigma^2),$$

for $i = 1, \ldots, n$. Then

$$\begin{pmatrix} Y_1 \\ Y_2 \\ \vdots \\ Y_n \end{pmatrix} = \begin{pmatrix} \beta_0 + \beta_1 x_{11} + \beta_2 x_{12} + \ldots + \beta_p x_{1p} \\ \beta_0 + \beta_1 x_{21} + \beta_2 x_{22} + \ldots + \beta_p x_{2p} \\ \vdots \\ \beta_0 + \beta_1 x_{n1} + \beta_2 x_{n2} + \ldots + \beta_p x_{np} \end{pmatrix} + \begin{pmatrix} \epsilon_1 \\ \epsilon_2 \\ \vdots \\ \epsilon_n \end{pmatrix}$$

$$= \begin{pmatrix} 1 & x_{11} & x_{12} & \cdots & x_{1p} \\ 1 & x_{21} & x_{22} & \cdots & x_{2p} \\ \vdots & \vdots & \vdots & \vdots & \vdots \\ 1 & x_{n1} & x_{n2} & \cdots & x_{np} \end{pmatrix} \begin{pmatrix} \beta_0 \\ \beta_1 \\ \vdots \\ \beta_p \end{pmatrix} + \begin{pmatrix} \epsilon_1 \\ \epsilon_2 \\ \vdots \\ \epsilon_n \end{pmatrix}$$

$$= \begin{pmatrix} \mathbf{x}_1^T \\ \mathbf{x}_2^T \\ \vdots \\ \mathbf{x}_n^T \end{pmatrix} \begin{pmatrix} \beta_0 \\ \beta_1 \\ \vdots \\ \beta_p \end{pmatrix} + \begin{pmatrix} \epsilon_1 \\ \epsilon_2 \\ \vdots \\ \epsilon_n \end{pmatrix}$$

Hence

$$\mathbf{Y} = \mathbf{X}\boldsymbol{\beta} + \boldsymbol{\epsilon}, \qquad \boldsymbol{\epsilon} \sim N(\mathbf{0}, \sigma^2 \mathbf{I}_n). \tag{18.2}$$

Here $\mathbf{y}$ is an $n$-vector of observations, $\mathbf{X}$ is an $n \times (p+1)$ matrix called the design matrix and $\boldsymbol{\epsilon}$ is an $n$-vector of errors which has a *multivariate* normal distribution with mean (vector) $\mathbf{0}$ and variance matrix $\sigma^2 \mathbf{I}_n$, see Sect. 15.7.1. It follows directly that

$$\mathbf{Y} \sim N(\mathbf{X}\boldsymbol{\beta}, \sigma^2 \mathbf{I}_n).$$

Hence, for multiple regression models, the assumption that the error terms $\epsilon_1, \ldots, \epsilon_n$ are independently normally distributed each with mean ) and variance $\sigma^2$ is equivalent to stating that the error vector $\boldsymbol{\epsilon} = (\epsilon_1 \ \epsilon_2 \ \ldots \ \epsilon_n)^T$ has a multivariate normal distribution with mean vector $\mathbf{0}$, and variance matrix $\sigma^2 \mathbf{I}_n$ where $\mathbf{I}_n$ is the $n \times n$ identity matrix.

## 18.4   Least Squares Estimation

Least squares estimation for the general linear model involves minimising the sum of squares

$$S = \sum_{i=1}^n (y_i - \mathbf{x}_i^T \boldsymbol{\beta})^2 = \sum_{i=1}^n \left( y_i - \sum_{j=0}^p x_{ij} \beta_j \right)^2$$

where we define $x_{i0} \equiv 1$. Then

$$\frac{\partial S}{\partial \beta_j} = -2 \sum_{i=1}^n x_{ij} (y_i - \mathbf{x}_i^T \boldsymbol{\beta})$$

and hence the vector of partial derivatives is given by

$$\frac{\partial S}{\partial \boldsymbol{\beta}} = -2\mathbf{X}^T (\mathbf{y} - \mathbf{X}\boldsymbol{\beta})$$

and the normal equations, $\frac{\partial S}{\partial \boldsymbol{\beta}} = \mathbf{0}$, are

$$\mathbf{X}^T \mathbf{X} \hat{\boldsymbol{\beta}} = \mathbf{X}^T \mathbf{y}.$$

Provided $\mathbf{X}$ is of full rank $p + 1$, the matrix $\mathbf{X}^T \mathbf{X}$ is nonsingular so

$$\hat{\boldsymbol{\beta}} = (\mathbf{X}^T \mathbf{X})^{-1} \mathbf{X}^T \mathbf{y}. \tag{18.3}$$

The fitted values are

$$\hat{\mathbf{y}} = \mathbf{X} \hat{\boldsymbol{\beta}} = \mathbf{X}(\mathbf{X}^T \mathbf{X})^{-1} \mathbf{X}^T \mathbf{y} = \mathbf{H} \mathbf{y}$$

where $\mathbf{H} = \mathbf{X}(\mathbf{X}^T \mathbf{X})^{-1} \mathbf{X}^T$ is called the hat matrix. Note that:

1. $\mathbf{H}$ is symmetric: $\mathbf{H}^T = \mathbf{H}$
2. $\mathbf{HX} = \mathbf{X}$ and
3. $\mathbf{H}$ is idempotent $\mathbf{H}^2 = \mathbf{H}$.

Also, for later use,

$$
\mathbf{H} = \begin{pmatrix} \mathbf{x}_1^T \\ \mathbf{x}_2^T \\ \vdots \\ \mathbf{x}_n^T \end{pmatrix} (\mathbf{X}^T \mathbf{X})^{-1} (\mathbf{x}_1 \ \mathbf{x}_2 \ \ldots \ \mathbf{x}_n)
$$

$$
= \begin{pmatrix} \mathbf{x}_1^T (\mathbf{X}^T \mathbf{X})^{-1} \\ \mathbf{x}_2^T (\mathbf{X}^T \mathbf{X})^{-1} \\ \vdots \\ \mathbf{x}_n^T (\mathbf{X}^T \mathbf{X})^{-1} \end{pmatrix} (\mathbf{x}_1 \ \mathbf{x}_2 \ \ldots \ \mathbf{x}_n)
$$

$$
= \begin{pmatrix} \mathbf{x}_1^T (\mathbf{X}^T \mathbf{X})^{-1} \mathbf{x}_1 & \mathbf{x}_1^T (\mathbf{X}^T \mathbf{X})^{-1} \mathbf{x}_2 & \cdots & \mathbf{x}_1^T (\mathbf{X}^T \mathbf{X})^{-1} \mathbf{x}_n \\ \mathbf{x}_2^T (\mathbf{X}^T \mathbf{X})^{-1} \mathbf{x}_1 & \mathbf{x}_2^T (\mathbf{X}^T \mathbf{X})^{-1} \mathbf{x}_2 & \cdots & \mathbf{x}_2^T (\mathbf{X}^T \mathbf{X})^{-1} \mathbf{x}_n \\ \vdots & \vdots & \vdots & \vdots \\ \mathbf{x}_n^T (\mathbf{X}^T \mathbf{X})^{-1} \mathbf{x}_1 & \mathbf{x}_n^T (\mathbf{X}^T \mathbf{X})^{-1} \mathbf{x}_2 & \cdots & \mathbf{x}_n^T (\mathbf{X}^T \mathbf{X})^{-1} \mathbf{x}_n \end{pmatrix}
$$

so

$$h_{ij} = \mathbf{x}_i^T (\mathbf{X}^T \mathbf{X})^{-1} \mathbf{x}_j.$$

The residuals are then obtained as:

$$\mathbf{r} = \mathbf{y} - \hat{\mathbf{y}} = \mathbf{y} - \mathbf{Hy} = (\mathbf{I}_n - \mathbf{H})\mathbf{y} \qquad (18.4)$$

and note that:

1. $\mathbf{I}_n - \mathbf{H}$ is symmetric: $(\mathbf{I}_n - \mathbf{H})^T = \mathbf{I}_n^T - \mathbf{H}^T = \mathbf{I}_n - \mathbf{H}$
2. $(\mathbf{I}_N - \mathbf{H})\mathbf{X} = \mathbf{X} - \mathbf{HX} = 0$ and
3. $\mathbf{I}_n - \mathbf{H}$ is idempotent $(\mathbf{I}_N - \mathbf{H})^2 = \mathbf{I}_n - \mathbf{H} - \mathbf{H} + \mathbf{H}^2 = \mathbf{I}_n - \mathbf{H}$.

Symmetric and idempotent matrices are also called projection matrices. Hence, least squares estimation can be given a geometric interpretation as follows. We have a vector $\mathbf{y} \in \mathcal{R}^n$ which we project (using $\mathbf{H}$) into the $(p + 1)$ dimensional subspace spanned by the columns of $\mathbf{X}$ to obtain the fitted values $\hat{\mathbf{y}}$. The component of $\mathbf{y}$ which is orthogonal (by 2. above) to $\hat{\mathbf{y}}$ is the residual vector $\mathbf{r}$.

The fitted value $\hat{\mathbf{y}}$ is interpreted to be a projection of the data vector $\mathbf{y}$ onto the column space of the design matrix $X$. The residual $\mathbf{r}$ is orthogonal to the fitted value $\hat{\mathbf{y}}$. We draw a representative diagram in Fig. 18.4 where there are only two covariates (including the intercept), but there are three observations so that $\mathbf{y}$ is three dimensional. In this diagram, the three axes are labelled as $u_1, u_2$ and $u_3$. The grided two dimensional plane $(u_1, u_2)$ represents the column space of the design matrix $X$ which has only two columns. To find the fitted value, a perpendicular is imagined to be dropped from the point $\mathbf{y}$ onto the column space of $X$. The fitted value $\hat{\mathbf{y}}$, is the point in the column space of $X$ where the perpendicular drops. (Recall that the fitted value is, $\hat{\mathbf{y}} = \mathbf{X}\hat{\boldsymbol{\beta}}$, which will lie in the column space of the design matrix $X$.) Notice that the dropping of the perpendicular is due to the method of the least squares estimation where the perpendicular represents the least distance between the data point $\mathbf{y}$ and the column space of $X$.

**Fig. 18.4** The geometric interpretation of least squares estimation

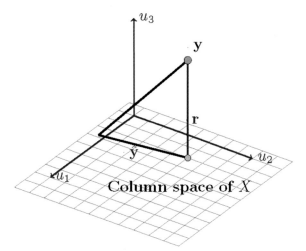

This geometric interpretation is not essential to understanding regression but can be helpful. One of its implications is that the analysis of variance decomposition, see Sect. 17.9, is simply Pythagoras's theorem.

## 18.5 Simple Linear Regression: A Special Case of Multiple Regression

In the simple regression model of Chap. 17,

$$
\mathbf{X} = \begin{pmatrix} 1 & x_1 \\ 1 & x_2 \\ \vdots & \vdots \\ 1 & x_n \end{pmatrix}
$$

It follows that

$$
\mathbf{X}^T\mathbf{X} = \begin{pmatrix} 1 & 1 & \cdots & 1 \\ x_1 & x_2 & \cdots & x_n \end{pmatrix} \begin{pmatrix} 1 & x_1 \\ 1 & x_2 \\ \vdots & \vdots \\ 1 & x_n \end{pmatrix}
$$

$$
= \begin{pmatrix} n & \sum_{i=1}^n x_i \\ \sum_{i=1}^n x_i & \sum_{i=1}^n x_i^2 \end{pmatrix}
$$

Then since

$$
|\mathbf{X}^T\mathbf{X}| = n\sum_{i=1}^n x_i^2 - \left(\sum_{i=1}^n x_i\right)^2 = n\sum_{i=1}^n (x_i - \bar{x})^2 = n(n-1)s_x^2,
$$

we obtain

$$
(\mathbf{X}^T\mathbf{X})^{-1} = \frac{1}{n(n-1)s_x^2} \begin{pmatrix} \sum_{i=1}^n x_i^2 & -\sum_{i=1}^n x_i \\ -\sum_{i=1}^n x_i & n \end{pmatrix}
$$

$$
= \frac{1}{(n-1)s_x^2} \begin{pmatrix} n^{-1}\sum_{i=1}^n x_i^2 & -\bar{x} \\ -\bar{x} & 1 \end{pmatrix}
$$

Since

$$
\mathbf{X}^T\mathbf{y} = \begin{pmatrix} 1 & 1 & \cdots & 1 \\ x_1 & x_2 & \cdots & x_n \end{pmatrix} \begin{pmatrix} y_1 \\ y_2 \\ \vdots \\ y_n \end{pmatrix} = \begin{pmatrix} \sum_{i=1}^n y_i \\ \sum_{i=1}^n x_i y_i \end{pmatrix}
$$

we have

$$
\begin{aligned}
\hat{\boldsymbol{\beta}} &= \frac{1}{(n-1)s_x^2} \begin{pmatrix} n^{-1}\sum_{i=1}^n x_i^2 & -\bar{x} \\ -\bar{x} & 1 \end{pmatrix} \begin{pmatrix} \sum_{i=1}^n y_i \\ \sum_{i=1}^n x_i y_i \end{pmatrix} \\
&= \frac{1}{(n-1)s_x^2} \begin{pmatrix} n^{-1}\sum_{i=1}^n x_i^2 \sum_{i=1}^n y_i - \bar{x}\sum_{i=1}^n x_i y_i \\ \sum_{i=1}^n x_i y_i - \bar{x}\sum_{i=1}^n y_i \end{pmatrix} \\
&= \frac{1}{(n-1)s_x^2} \begin{pmatrix} (\sum_{i=1}^n x_i^2 - n\bar{x}^2)\bar{y} + n\bar{x}^2\bar{y} - \bar{x}\sum_{i=1}^n x_i y_i \\ \sum_{i=1}^n (x_i - \bar{x})y_i \end{pmatrix} \\
&= \frac{1}{(n-1)s_x^2} \begin{pmatrix} (n-1)s_x^2\bar{y} - \bar{x}(\sum_{i=1}^n x_i y_i - \bar{x}\sum_{i=1}^n y_i) \\ (n-1)s_{xy} \end{pmatrix} \\
&= \frac{1}{(n-1)s_x^2} \begin{pmatrix} (n-1)s_x^2\bar{y} - \bar{x}(n-1)s_{xy} \\ (n-1)s_{xy} \end{pmatrix} \\
&= \begin{pmatrix} \bar{y} - \bar{x}s_{xy}/s_x^2 \\ s_{xy}/s_x^2 \end{pmatrix}
\end{aligned}
$$

which is equivalent to the expressions derived in Sect. 17.3. Similarly, the elements of the hat matrix are given by

$$
\begin{aligned}
h_{ij} &= (1 \; x_i)\frac{1}{(n-1)s_x^2} \begin{pmatrix} n^{-1}\sum_{k=1}^n x_k^2 & -\bar{x} \\ -\bar{x} & 1 \end{pmatrix} \begin{pmatrix} 1 \\ x_j \end{pmatrix} \\
&= \frac{1}{(n-1)s_x^2} \left( n^{-1}\sum_{k=1}^n x_k^2 - x_i\bar{x} - x_j\bar{x} + x_i x_j \right) \\
&= \frac{1}{(n-1)s_x^2} \left( n^{-1}\sum_{k=1}^n x_k^2 - \bar{x}^2 + \bar{x}^2 - x_i\bar{x} - x_j\bar{x} + x_i x_j \right) \\
&= \frac{1}{(n-1)s_x^2} \left\{ n^{-1}(n-1)s_x^2 + (x_i - \bar{x})(x_j - \bar{x}) \right\} \\
&= \frac{1}{n} + \frac{(x_i - \bar{x})(x_j - \bar{x})}{(n-1)s_x^2}.
\end{aligned}
$$

## 18.6    Inference for the Multiple Regression Model

The matrix notation really comes into its own in developing the properties of the least squares estimators. We have

$$E(\hat{\boldsymbol{\beta}}) = E[(\mathbf{X}^T\mathbf{X})^{-1}\mathbf{X}^T\mathbf{Y}] = (\mathbf{X}^T\mathbf{X})^{-1}\mathbf{X}^T E(\mathbf{Y})$$
$$= (\mathbf{X}^T\mathbf{X})^{-1}\mathbf{X}^T\mathbf{X}\boldsymbol{\beta} = \boldsymbol{\beta}$$

and

$$\mathrm{Var}(\hat{\boldsymbol{\beta}}) = \mathrm{Var}[(\mathbf{X}^T\mathbf{X})^{-1}\mathbf{X}^T\mathbf{Y}]$$
$$= (\mathbf{X}^T\mathbf{X})^{-1}\mathbf{X}^T \mathrm{Var}(\mathbf{Y})\mathbf{X}(\mathbf{X}^T\mathbf{X})^{-1}$$
$$= (\mathbf{X}^T\mathbf{X})^{-1}\mathbf{X}^T\sigma^2\mathbf{I}_n\mathbf{X}(\mathbf{X}^T\mathbf{X})^{-1}$$
$$= \sigma^2(\mathbf{X}^T\mathbf{X})^{-1}.$$

The sampling distribution of $\hat{\boldsymbol{\beta}}$ follows immediately as

$$\hat{\boldsymbol{\beta}} \sim N(\boldsymbol{\beta}, \sigma^2(\mathbf{X}^T\mathbf{X})^{-1})$$

and for an individual regression coefficient, $\beta_j$ for any $j$, we have

$$\hat{\beta}_j \sim N(\beta_j, \sigma^2[(\mathbf{X}^T\mathbf{X})^{-1}]_{jj}).$$

Hence

$$\frac{\hat{\beta}_j - \beta_j}{\sigma[(\mathbf{X}^T\mathbf{X})^{-1}]_{jj}^{1/2}} \sim N(0, 1).$$

However, this result cannot be used directly to construct confidence intervals for $\beta_j$, or to test hypotheses, because the error standard deviation $\sigma$ is unknown. As with simple linear regression, we need an estimator for $\sigma^2$ and to derive the sampling distribution of that estimator.

### 18.6.1  Estimating $\sigma^2$

As for simple linear regression, our estimator for $\sigma^2$ is based on the sum of squared residuals,

$$\sum_{i=1}^{n} r_i^2 = \mathbf{r}^T\mathbf{r}$$
$$= [(\mathbf{I}_n - \mathbf{H})\mathbf{y}]^T[(\mathbf{I}_n - \mathbf{H})\mathbf{y}], \quad \text{see } (18.4)$$
$$= \mathbf{y}^T(\mathbf{I}_n - \mathbf{H})^T(\mathbf{I}_n - \mathbf{H})\mathbf{y}$$

$$= \mathbf{y}^T (\mathbf{I}_n - \mathbf{H})(\mathbf{I}_n - \mathbf{H})\mathbf{y}$$
$$= \mathbf{y}^T (\mathbf{I}_n - \mathbf{H})\mathbf{y}.$$

The estimator we use for $\sigma^2$ is denoted $s^2$ and is given by

$$s^2 = \frac{1}{n-p-1} \sum_{i=1}^{n} r_i^2 = \frac{1}{n-p-1} \mathbf{y}^T (\mathbf{I}_n - \mathbf{H})\mathbf{y}.$$

We can now calculate standard errors for any $\hat{\beta}_j$, by replacing $\sigma$ in the standard deviation by its estimate $s$. Hence

$$s.e.(\hat{\beta}_j) = s[(\mathbf{X}^T \mathbf{X})^{-1}]_{jj}^{1/2}.$$

To derive the sampling distribution of $s^2$ (and other important properties for linear model inference), we require the following theorem (which is closely related to a famous theorem in Statistics known as the Cochran's theorem).

### 18.6.2  Cochran's Theorem

**Theorem 18.1** *Suppose that* $\mathbf{Y} \sim N(\boldsymbol{\mu}, \sigma^2 \mathbf{I}_n)$. *Let* $\mathbf{A}_1$ *and* $\mathbf{A}_2$ *be symmetric idempotent* $n \times n$ *(projection) matrices and define the random variables*

$$Q_1 \equiv \mathbf{Y}^T \mathbf{A}_1 \mathbf{Y} \quad and \quad Q_2 \equiv \mathbf{Y}^T \mathbf{A}_2 \mathbf{Y}.$$

*The following results hold:*

1. *$Q_1$ and $Q_2$ are independent if* $\mathbf{A}_1 \mathbf{A}_2 = \mathbf{0}$ ($= \mathbf{A}_2 \mathbf{A}_1$).
2. *If* $\mathbf{A}_1 \boldsymbol{\mu} = \mathbf{0}$ *then* $Q_1 / \sigma^2$ *has a* $\chi^2_{d_1}$ *distribution, where* $d_1 = trace\,(\mathbf{A}_1)$.
3. *Similarly, if* $\mathbf{A}_2 \boldsymbol{\mu} = \mathbf{0}$ *then* $Q_2 / \sigma^2$ *has a* $\chi^2_{d_2}$ *distribution, where* $d_2 = trace\,(\mathbf{A}_2)$.

The proof of the above theorem requires knowledge and familiarity of deep results in linear algebra and are omitted. The interested reader can find sketches of such proofs in textbooks such as Ravishanker et al. [16] and Searle [18].

### 18.6.3  The Sampling Distribution of $S^2$

The distribution of $S^2 = \frac{1}{n-p-1} \mathbf{Y}^T (\mathbf{I}_n - \mathbf{H})\mathbf{Y}$ follows from Cochran's Theorem in Sect. 18.6.2. We already know that $\mathbf{I}_n - \mathbf{H}$ is symmetric and idempotent. Now,

$$\text{trace}(\mathbf{I}_n - \mathbf{H}) = \text{trace}\,(\mathbf{I}_n) - \text{trace}(\mathbf{H})$$
$$= n - \text{rank}(\mathbf{H}),$$

since it can be proved that the trace of a projection matrix is equal to its rank. Here is a sketch of a proof. We use the facts that: (1) all eigenvalues of a projection matrix are either zero or one; hence (2) the sum of the eigenvalues of a projection matrix is equal to the rank; and (3) the trace of a matrix is sum of the eigenvalues.

The column space of $\mathbf{H} = \mathbf{X}(\mathbf{X}^T\mathbf{X})^{-1}\mathbf{X}^T$ and the column space of $\mathbf{X}$ are identical, so rank$(\mathbf{H})$ = rank$(\mathbf{X})$ = $p+1$ (as we assume that the $(p+1)$ columns of $\mathbf{X}$ are linearly independent), so trace $(\mathbf{I}_n - \mathbf{H}) = n - p - 1$. (The column space of a matrix is the vector space produced by all possible linear combination of its column vectors.)

Now $\mathbf{Y} \sim N(\boldsymbol{\mu}, \sigma^2\mathbf{I}_n)$, where $\boldsymbol{\mu} = \mathbf{X}\boldsymbol{\beta}$, so $(\mathbf{I}_n - \mathbf{H})\boldsymbol{\mu} = (\mathbf{I}_n - \mathbf{H})\mathbf{X}\boldsymbol{\beta} = \mathbf{0}$, as $(\mathbf{I}_n - \mathbf{H})\mathbf{X} = \mathbf{0}$. Hence, according to the Cochran's theorem

$$\frac{1}{\sigma^2}\mathbf{Y}^T(\mathbf{I}_n - \mathbf{H})\mathbf{Y} = \frac{n-p-1}{\sigma^2}S^2 \sim \chi^2_{n-p-1}.$$

An immediate consequence is that $E(S^2) = \sigma^2$ and hence that $S^2$ is an unbiased estimator of $\sigma^2$.

### 18.6.4  Inference for Regression Coefficients

As $(\mathbf{I}_n - \mathbf{H})\mathbf{H} = \mathbf{H} - \mathbf{H}^2 = \mathbf{0}$, the Cochran's theorem also tells us that $\mathbf{Y}^T(\mathbf{I}_n - \mathbf{H})\mathbf{Y}$ is independent of $\mathbf{H}\mathbf{Y}$. Hence $S^2$ is independent of $(\mathbf{X}^T\mathbf{X})^{-1}\mathbf{X}^T\mathbf{H}\mathbf{Y} = (\mathbf{X}^T\mathbf{X})^{-1}\mathbf{X}^T\mathbf{Y} = \hat{\boldsymbol{\beta}}$. Therefore, each $\hat{\beta}_j$ is independent of $S^2$, and

$$
\begin{aligned}
\frac{\hat{\beta}_j - \beta_j}{s.e.(\hat{\beta}_j)} &= \frac{\hat{\beta}_j - \beta_j}{s[(\mathbf{X}^T\mathbf{X})^{-1}]_{jj}^{1/2}} \\
&= \frac{\hat{\beta}_j - \beta_j}{\frac{s}{\sigma}\sigma[(\mathbf{X}^T\mathbf{X})^{-1}]_{jj}^{1/2}} \sim \frac{N(0,1)}{[\chi^2_{n-p-1}/(n-p-1)]^{1/2}} = t_{n-p-1},
\end{aligned}
$$

as the normal and chi-square distributions above are independent.

This is the result we require to construct confidence intervals for, or test hypotheses about, regression parameters.

### 18.6.5  Inference for Fitted Values and Residuals

We have

$$E(\hat{\mathbf{Y}}) = E(\mathbf{H}\mathbf{Y}) = \mathbf{H}E(\mathbf{Y}) = \mathbf{H}\mathbf{X}\boldsymbol{\beta} = \mathbf{X}\boldsymbol{\beta}$$

and

$$\mathrm{Var}(\hat{\mathbf{Y}}) = \mathrm{Var}(\mathbf{H}\mathbf{Y}) = \mathbf{H}\mathrm{Var}(\mathbf{Y})\mathbf{H} = \mathbf{H}\sigma^2\mathbf{I}_n\mathbf{H} = \sigma^2\mathbf{H}.$$

Furthermore

$$\hat{\mathbf{Y}} \sim N(\mathbf{X}\boldsymbol{\beta}, \sigma^2 \mathbf{H}).$$

For residuals, we have

$$E(\mathbf{R}) = E[(\mathbf{I}_n - \mathbf{H})\mathbf{Y}] = (\mathbf{I}_n - \mathbf{H})E(\mathbf{Y}) = (\mathbf{I}_n - \mathbf{H})\mathbf{X}\boldsymbol{\beta} = \mathbf{0}$$

and

$$\begin{aligned}
\mathrm{Var}(\mathbf{R}) &= \mathrm{Var}[(\mathbf{I}_n - \mathbf{H})\mathbf{Y}] \\
&= (\mathbf{I}_n - \mathbf{H})\mathrm{Var}(\mathbf{Y})(\mathbf{I}_n - \mathbf{H}) \\
&= (\mathbf{I}_n - \mathbf{H})\sigma^2 \mathbf{I}_n (\mathbf{I}_n - \mathbf{H}) \\
&= \sigma^2 (\mathbf{I}_n - \mathbf{H}).
\end{aligned}$$

Furthermore

$$\mathbf{R} \sim N(\mathbf{0}, \sigma^2 [\mathbf{I}_n - \mathbf{H}]).$$

This shows that, in general, the fitted residual values are not independent, since $\mathbf{H}$ is not a null matrix and as a result $\mathrm{Cor}(R_i, R_j) \neq 0$ for $i \neq j$.

## 18.6.6 Prediction

We estimate the conditional mean, $\mathbf{x}_0^T \boldsymbol{\beta}$, for the response variable at values of the explanatory variables given by $\mathbf{x}_0^T = (1 \; x_{01} \; x_{02} \; \ldots \; x_{0p})$, which may or may not match a set of values observed in the data, using

$$\hat{Y}_0 = \mathbf{x}_0^T \hat{\boldsymbol{\beta}}.$$

Then

$$E(\hat{Y}_0) = \mathbf{x}_0^T E(\hat{\boldsymbol{\beta}}) = \mathbf{x}_0^T \boldsymbol{\beta}$$

and

$$Var(\hat{Y}_0) = \mathbf{x}_0^T Var(\hat{\boldsymbol{\beta}})\mathbf{x}_0 = \mathbf{x}_0^T \sigma^2 (\mathbf{X}^T \mathbf{X})^{-1} \mathbf{x}_0 = \sigma^2 h_{00}.$$

We also have $s.e.(\hat{Y}_0) = s\sqrt{h_{00}}$, and

$$
\begin{aligned}
\frac{\hat{Y}_0 - \mathbf{x}_0^T \boldsymbol{\beta}}{s.e.(\hat{Y}_0)} &= \frac{\hat{Y}_0 - \mathbf{x}_0^T \boldsymbol{\beta}}{s\sqrt{h_{00}}} \\
&= \frac{\hat{Y}_0 - \mathbf{x}_0^T \boldsymbol{\beta}}{\frac{s}{\sigma}\sigma\sqrt{h_{00}}} \sim \frac{N(0, 1)}{[\chi_{n-p-1}^2/(n-p-1)]^{1/2}} = t_{n-p-1},
\end{aligned}
$$

as $S^2$ and $\hat{\boldsymbol{\beta}}$ are independent. Hence, we can derive confidence intervals for conditional mean estimates using the general form:

$$\textbf{Estimate} \pm \textbf{Critical Value} \times \textbf{Standard Error}.$$

as noted in Sect. 11.2.

For predicting the actual value $Y_0 = \mathbf{x}_0^T \boldsymbol{\beta} + \epsilon_0$, we also use $\hat{Y}_0$ as our predictor. Now

$$E(\hat{Y}_0 - Y_0) = E(\hat{Y}_0) - E(Y_0) = \mathbf{x}_0^T \boldsymbol{\beta} - \mathbf{x}_0^T \boldsymbol{\beta} = 0$$

and

$$
\begin{aligned}
Var(\hat{Y}_0 - Y_0) &= Var(\hat{Y}_0) + Var(Y_0) - 2Cov(\hat{Y}_0, Y_0) \\
&= \sigma^2 h_{00} + \sigma^2 = \sigma^2(1 + h_{00}).
\end{aligned}
$$

All of the above results hold generally for the multiple regression model, and hence of course in the special case of a simple linear regression model ($p = 1$). Hence, results we stated without proof for simple linear regression, (for example, see Sect. 17.6) have now been proved.

### 18.6.7  Model Comparison

One of the key problems in multiple regression model fitting is to compare competing models for data. This problem is tackled using the ANOVA based model comparison methods detailed in Sect. 17.9. The model comparison problem is to choose between the following two models $M$, called the full model and $M_0$, called the reduced model:

$$
\begin{aligned}
\text{M:} \quad & Y_i = \beta_0 + \beta_1 x_{i1} + \ldots + \beta_q x_{iq} + \beta_{q+1} x_{iq+1} \ldots + \beta_p x_{ip} + \epsilon_i \\
\text{M}_0: \quad & Y_i = \beta_0 + \beta_1 x_{i1} + \ldots + \beta_q x_{iq} + \epsilon_i
\end{aligned}
$$

424                                18 Multiple Linear Regression Model

where $q < p$ and $\epsilon_i \overset{\text{ind}}{\sim} N(0, \sigma^2)$ in both cases, $i = 1, \ldots, n$. In our statistical inference framework the equivalent problem is to test

$$H_0 : \beta_{q+1} = \beta_{q+2} = \ldots = \beta_p = 0$$

in Model M, against the alternative hypothesis that at least one equality does not hold in $H_0$. As a solution we employ the F-test as detailed in Sect. 17.9 for the simple linear regression model. The testing is performed with the help of an ANOVA table as constructed below.

We first fit Model M and find the fitted values

$$\hat{Y}_i = \hat{\beta}_0 + \hat{\beta}_1 x_{i1} + \ldots + \hat{\beta}_q x_{iq} + \ldots + \hat{\beta}_p x_{ip}$$

and then the Residual Sum of Squares (RSS)

$$\text{RSS under full model} = \text{RSS}_f = \sum_{i=1}^{n} \left( Y_i - \hat{Y}_i \right)^2$$

We then fit Model $M_0$ and find the fitted values

$$\tilde{Y}_i = \tilde{\beta}_0 + \tilde{\beta}_1 x_{i1} + \ldots + \tilde{\beta}_q x_{iq},$$

and the RSS,

$$\text{RSS under reduced model} = \text{RSS}_r = \sum_{i=1}^{n} \left( Y_i - \tilde{Y}_i \right)^2$$

Note that $\hat{\beta}_k \neq \tilde{\beta}_k, k = 0, \ldots, q$ since these estimates are obtained by minimising two different residual sums of squares for two different models. In fact, $\text{RSS}_r \geq \text{RSS}_f$ since on average the residuals under the more elaborate full model $M$ will be smaller in magnitude than the residuals under the reduced, i.e. restricted, model $M_0$. Now we perform the ANOVA decomposition as discussed previously in Sect. 17.9:

$$\begin{aligned}
\text{RSS}_r &= \sum_{i=1}^{n} \left( Y_i - \tilde{Y}_i \right)^2 \\
&= \sum_{i=1}^{n} \left( Y_i - \hat{Y}_i + \hat{Y}_i - \tilde{Y}_i \right)^2 \\
&= \sum_{i=1}^{n} \left( Y_i - \hat{Y}_i \right)^2 + \sum_{i=1}^{n} \left( \hat{Y}_i - \tilde{Y}_i \right)^2 + 2 \sum_{i=1}^{n} \left( Y_i - \hat{Y}_i \right) \left( \hat{Y}_i - \tilde{Y}_i \right) \\
&= \text{RSS}_f + \text{Difference in RSS}
\end{aligned}$$

where the Difference in RSS is simply calculated as the difference $\text{RSS}_r - \text{RSS}_f$, which is always non-negative as noted above. It can be shown that:

$$F = \frac{\text{Difference in RSS}/(p - q)}{\text{RSS}_f/(n - p - 1)} \sim F_{p-q,n-p-1}$$

when $H_0$ is true. Then the decision rule is to reject $H_0$ at $100\alpha\%$ level of significance if the observed value of $F$ is larger than the critical value, $F_{p-q,n-p-1}(1 - \alpha)$. The definition of this value is that the the area to the left of this under the $F$-density with $p - q$ and $n - p - 1$ degrees of freedom is $1 - \alpha$.

An alternative is to calculate the P-value:

$$P\left(F_{p-q,n-p-1} > F_{obs}\right),$$

which is area to the right of the observed value of $F$, $F_{obs}$ under the $F$-density with $p - q$ and $n - p - 1$ degrees of freedom. Now the decision rule is to reject $H_0$ if P-value is less than $\alpha$.

Now we construct the ANOVA table analogous to the one in Sect. 17.9:

| Source | Df | Sum of squares | Mean squares | F value | P value |
|---|---|---|---|---|---|
| Reduced model | $p - q$ | $\text{RSS}_r - \text{RSS}_f$ | $\frac{\text{RSS}_r - \text{RSS}_f}{p-q}$ | $\frac{\text{RSS}_r - \text{RSS}_f}{(p-q)\,S^2}$ | $Pr(F > F_{obs})$ |
| Residuals | $n - p - 1$ | $\text{RSS}_f$ | $S^2$ | | |
| Total | $n - q - 1$ | $\text{RSS}_r$ | | | |

As an example suppose $q = 0$ and $p = 1$. In this case, when $M_0$ is the true model, we have $Y_i \overset{\text{ind}}{\sim} N(\beta_0, \sigma^2)$, so that the response has no dependence on any explanatory variable. Then, we write

$$S^2 = \frac{\text{RSS under the full model}}{n - p - 1}, \quad S_Y^2 = \frac{\sum_{i=1}^n (Y_i - \bar{Y})^2}{n - 1},$$

and it can be shown that:

$$F = \frac{\frac{1}{p}[(n - 1)S_Y^2 - (n - p - 1)S^2]}{S^2}.$$

This test is performed routinely as default in most regression modelling software. If the explanatory variables are worthy of inclusion in the model, we want to reject the null hypothesis $H_0$.

As a second example suppose $q = p - 1$. In this case, model $M_0$ has exactly one less term than the full model M, and the F test derived above is testing the hypothesis $H_0$: $\beta_p = 0$. We can also test this hypothesis using the statistic

$$t = \frac{\hat{\beta}_p}{s.e.(\hat{\beta}_p)} \sim t_{n-p-1}$$

under $H_0$. In fact, it implies that this test is exactly equivalent to the ANOVA F-test, as $t^2 = F$, i.e., the distribution of square of a $t_{n-p-1}$ random variable is $F_{1,(n-p-1)}$, see Sect. 14.5.3.

## 18.7    Illustrative Example: Puffin Nesting

The above theory is now illustrated with a data from a study of nesting habits of common puffins see e.g. Fig. 18.5, on Great Island, Newfoundland reported in the article Nettleship [14]. The study explored the characteristics of desirable nesting sites by visiting 38 3 m × 3 m plots on the island and counting the number of nests and measuring the characteristics of each plot. The response variable is the number of burrows on each plot and the explanatory variables are the percentage grass cover, the mean soil depth (centimeter), the slope angle (degrees) and the distance from the cliff edge (meter). This data set has been obtained from Prof Alan H. Welsh (author of Welsh [21]).

The response variable shows a number of zero counts which may impact on modelling. As the response is a count, we expect to have to make a transformation. For large counts, taking logarithms often works well, for small counts taking square root is better. The largest count here is 25 which is moderate so we will try taking square root. Fitting the model with all the variables seems to fit reasonably well, see Fig. 18.6. The residual plot shows the zeros as a diagonal line across the left hand side of the plot and perhaps a suggestion of nonconstant variance for the non-zero counts. The normal probability plot shows a slightly short-tailed distribution. The data have been ordered by the size of the response so we cannot look for order effects. Also, there is no spatial information provided so we cannot explore spatial dependence.

We now have a model which seems to fit the data plausibly well. The question then is can we simplify the model by reducing the number of explanatory variables in the model?

**Fig. 18.5** Atlantic puffins, Isle of Lunga, Scotland by Steve Deger. Source: https://www.flickr.com/photos/ stevedeger/108316860/ License: CC BY 2.0

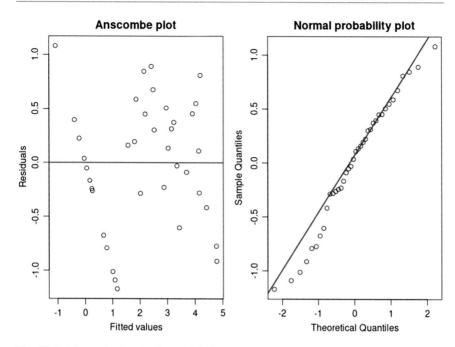

**Fig. 18.6** Diagnostic plots for the multiple linear regression model fitted to the puffin nesting data

```
pairs(puffin)
summary(puffin)
puffin$sqrtfreq <- sqrt(puffin$Nesting_Frequency)


puff.sqlm <- lm(sqrtfreq~ Grass_Cover + Mean_Soil_Depth +
    Slope_Angle
    + Distance_from_Edge, data=puffin)


#par(mfrow=c(2,1))
qqnorm(puff.sqlm$res,main="Normal probability plot")
qqline(puff.sqlm$res)
plot(puff.sqlm$fit, puff.sqlm$res,xlab="Fitted values",ylab="
    Residuals", main="Anscombe plot")
abline(h=0)

summary(puff.sqlm)
```

```
#####################################
# F test for two betas at the same time:
#####################################


puff.sqlm2 <- lm(sqrtfreq~ Mean_Soil_Depth + Distance_from_Edge,
    data=puffin)

anova(puff.sqlm)
anova(puff.sqlm2)

fval <- 1/2*(14.245-12.756)/0.387

# 1.924
qf(0.95, 2, 33)
# 3.28
1-pf(fval, 2, 33)
# 0.162

anova(puff.sqlm2, puff.sqlm)
```

## 18.8   Model Selection

Model selection refers to the process of choosing which explanatory variables to put into the model. We begin by considering the consequences of selecting a set of wrong covariates in the model.

### 18.8.1  Omitting a Variable Which Is Needed

This leads to bias in the coefficient estimates (and therefore in predictions) and a large estimated error variance $s^2$ reflecting the fact that the fit is poor.

As an example, suppose that we use the null model $Y_i \overset{\text{ind}}{\sim} N(\beta_0, \sigma^2)$ for prediction, when we should be using the simple linear model $Y_i \overset{\text{ind}}{\sim} N(\beta_0 + \beta_1 x_i, \sigma^2)$. Under the null model, our estimator $\hat{\beta}_0$ for $\beta_0$ is $\bar{Y}$ (the sample mean of the observed data). When predicting $Y_0$ a future value of the response for which the explanatory data take the value $x_0$, under the null model, we therefore use $\hat{Y}_0 = \bar{Y}$ . Then, under the correct model

$$
\begin{aligned}
E(\hat{Y}_0 - Y_0) &= E(\bar{Y} - Y_0) \\
&= (\beta_0 + \beta_1 \bar{x}) - (\beta_0 + \beta_1 x_0) \\
&= \beta_1(\bar{x} - x_0).
\end{aligned}
$$

Hence, our prediction is biased, with the bias getting worse, the further away our value of interest $x_0$ of the explanatory variable is from the mean of the explanatory variable in our data.

## 18.8.2 Including an Unnecessary Variable

This leads to an increase in the variance of the coefficient estimates (and therefore of predictions). An unnecessarily large model also requires more variables to measure (cost) and often gives poor predictions.

As an example, suppose that we use the simple linear model $Y_i \overset{\text{ind}}{\sim} N(\beta_0 + \beta_1 x_i, \sigma^2)$ for prediction, when in fact the null model $Y_i \overset{\text{ind}}{\sim} N(\beta_0, \sigma^2)$ is adequate (the explanatory variable does not effectively predict the response). Under both models $E(\hat{Y}_0 - Y_0) = 0$, as both models are 'correct' (the simple linear model has $\beta_1 = 0$), so we have unbiased prediction. However, under the null model

$$\text{Var}(\hat{Y}_0 - Y_0) = \text{Var}(\bar{Y}) + \text{Var}(Y_0)$$
$$= \sigma^2 \left(1 + \frac{1}{n}\right)$$

whereas, under the simple linear model, using the result of Sect. 17.13, we have

$$\text{Var}(\hat{Y}_0 - Y_0) = \sigma^2 (1 + h_{00})$$
$$= \sigma^2 \left(1 + \frac{1}{n} + \frac{(x_0 - \bar{x})^2}{(n-1)s_x^2}\right)$$

Therefore, the prediction variance (and hence expected magnitude of prediction error), is larger under the unnecessarily complex model, with the effect getting worse, the further away our value of interest $x_0$ of the explanatory variable is from the mean of the explanatory variable in our data.

A number of automatic procedures, such as backwards elimination and forward selection, see below, have been proposed for model selection. They all have strengths and weaknesses and should be used cautiously.

## 18.8.3 Backwards Elimination

1. Fit a more complicated model than we think we will need, and which fits the data well with satisfactory residual diagnostics.
2. Calculate the t-statistics for testing each coefficient equals zero separately
3. Drop the variable whose coefficient has the least non-significant t-statistic (highest *non-significant* p-value)
4. Return to step 2 until all variables are significant

This is quite a good method as we begin with a model that fits (and can always retrieve that model) so we do not have to fit a large number of models, the comparison of t-ratios is very straightforward and it works for large $p$.

We prefer to implement this and the other selection procedures explicitly rather than automatically because this gives more control, the ability to see where choices are based on borderline outcomes and because it avoids trying to set the software up correctly to do what we want.

For the puffin data, backwards elimination leads to a model with distance from the edge of the cliff (negative coefficient) and mean soil depth (positive coefficient) which makes biological sense. Forward selection, an opposite to the backward elimination process, is described below.

### 18.8.4  Forward Selection

This procedure proceeds as follows:

1. Fit the null model
2. Separately include each of the remaining variables in the current model and calculate the t-statistics for testing their slope equals zero
3. Add the variable whose coefficient has the highest significant t-statistic (smallest *significant* p-value)
4. Return to step 2 until no more variables can be included

Forward selection is less convenient that backwards elimination as there are a large number of models to fit, it is more difficult to keep track of the variables in each iteration, the t-ratios have to be extracted from separate output and the final model may not even fit (e.g. transformations may be needed). In addition, at each stage, the estimate of $\sigma^2$ may be upwardly biased. This can be handled by using $S^2$ based on all the variables but in this case we could equally well use backwards elimination.

For the puffin nesting data example, (data set `puffin` in the R package `ipsRdbs`, forward selection yields the same model as backwards elimination. However, in more complicated problems, this need not be the case.

We can modify forward selection to incorporate an elimination step, where if any variable becomes insignificant during the process of adding other variables, it is eliminated.

Backwards elimination and Forward selection are *stepwise methods.* An alternative, when the number of explanatory variables (including transformations, interactions etc.) is small, is *All Subsets*, where we explore fitting all possible models and then choosing the 'best'. Since there are $2^p$ possible models, this procedure is not feasible if $p$ is large.

## 18.8.5  Criteria Based Model Selection

The basic statistic for evaluating model fit is $S^2$ (or equivalently the RSS). Generally, $S^2$ decreases to a relatively stable minimum as the number of explanatory variables increases and then shows irregular behaviour as additional explanatory variables are added. Hence, any approach for comparing all possible models should be based on $S^2$. Measures based on $S^2$ include:

$R^2$, the multiple correlation coefficient.

$$R^2 = 1 - \frac{(n - p - 1)S^2}{(n - 1)S_Y^2}$$

is simple to calculate and almost always reported by model fitting software packages, but it is not a good measure because it can be made arbitrarily close to one by including a number of irrelevant explanatory variables.

**Adjusted** $R^2$, the adjusted multiple correlation coefficient.

$$\text{adj } R^2 = 1 - \frac{S^2}{S_Y^2}$$

Maximising adj $R^2$ is equivalent to minimising $S^2$. Note that $R^2$ and adj $R^2$ are equivalent to the expressions given in Sect. 17.10 when $p = 1$. Also, the interpretation of adj $R^2$, as the proportionate reduction in error variance (over the null model) achieved by the regression model clearly holds for multiple regression models.

**Mallows** $C_p$. For a model with $p + 1$ conditional mean parameters, Mallows proposed the statistic

$$C_{p+1} = \frac{(n - p - 1)S^2}{S_{\text{Large}}^2} - \{n - 2(p + 1)\},$$

where $S_{\text{Large}}^2$ is a 'reliable' estimate of $\sigma^2$ from a regression, with a large number of predictors (but not too large relative to $n$). In practice, this is often the model with all possible explanatory variables included. [The terminology $C_p$ comes from including the intercept in counting the number of covariates. As we let $p$ be the number of explanatory variables in addition to the intercept, $C_{p+1}$ is more natural for us.] When there is no lack of fit, $E(C_{p+1}) = (n - p - 1) - \{n - 2(p + 1)\} = p + 1$ so we choose a model with the smallest $p$ for which $C_{p+1} \approx p + 1$.

In addition to $C_p$ there are a number of model selection criteria, often called *information criteria*, that penalise a model both for lack of fit (high $S^2$) and complexity (large $p$).

**Information Criteria.** For a model with $p + 1$ conditional mean parameters, the information criteria are all of the form

$$IC(p) = (n - p - 1)S^2 + \alpha(n)(p + 1)S_{\text{Large}}^2,$$

where $\alpha(n)$ is a specified function of $n$. The two most common choices are

**Akaike's Information Criterion (AIC).** $\alpha(n) = 2$

**Bayesian Information Criterion (BIC).** $\alpha(n) = \log(n)$

We choose the model which minimises an information criterion. The information criteria penalise a model both for lack of fit (high $S^2$) and complexity (large $p$). AIC has a smaller penalty for extra complexity, so tends to choose large models where BIC tends to choose smaller ones.

## 18.9    Illustrative Example: Nitrous Oxide Emission

Data on the nitrous oxide content of exhaust emissions from a set of cars was collected by the Australian Traffic Accident Research Bureau to explore the relationship between several measures of nitrous oxide emissions. The data set is `emissions` in the R package `ipsRdbs`. The variables are the result of the Australian standard ADR37 test, a measurement taken while running the engine from a cold start for 505 seconds (CS505), a measurement taken while running the engine in transition from cold to hot for 867 seconds (T867), a measurement taken while running the hot engine for 505 seconds (H505) and a measurement from the superseded standard ADR27. The aim here is to find relationships between the ADR37 measurement and the other measurements. This data set has been obtained from Prof Alan H. Welsh (author of Welsh [21]).

The scatterplot matrix in Fig. 18.7 shows strong linear relationships between the variables which seem to be on an appropriate scale. There is a suggestion of heteroscedasticity in the relationship between ADR37 and T867 and a cluster of four unusual points in the plot of ADR37 and H505. These points turn out to be observations 14, 20, 49 and 50. Beyond the fact that they all have unusually high ADR37 readings for their H505 values, there is nothing in the data which particularly identifies these points.

If we fit a model to the data on the raw scale, the normal probability plot shows 3 distinct clusters in the data. These are less evident in the residual plot but can still be distinguished. There is no other variable or information in the data set which might explain these clusters or justify treating them separately.

The slight heteroscedasticity in the original scatterplot matrix suggests transforming ADR37. The linear relationships in that matrix suggest further that all the explanatory variables be put on the same scale. A reasonable start is to take the logarithms of all the variables.

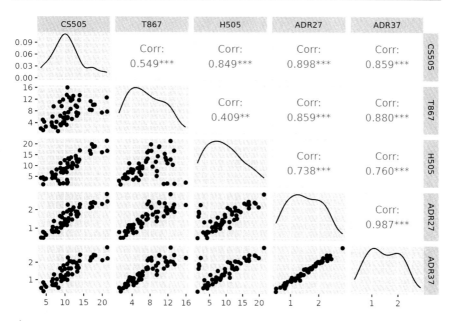

**Fig. 18.7**   Pairwise scatter plot of the nitrous oxide data on the original (raw) scale

The scatterplot matrix (not included here) on the log scale shows that taking logarithms does remove the heteroscedasticity while preserving the linear relationships. It does not take care of the cluster of 4 observations, however. The normal probability plot (not included) is reasonably linear and the residual plot looks broadly alright. Five points (33,5 on top and 39, 40, 44 below) stand out. Since they are not extreme on the normal probability plot, it is reasonable to simply include them in the analysis. Points 14 and 25 are identified as influential. (Incidentally, if we delete the original cluster, the fit is worse.)

The parameter estimates below show that log(CS505) has a negative coefficient. This is surprising because we expect positive relationships between the standard and the other emissions measurements. However, log(CS505) is the least significant non-significant variable so we omit it from the fit.

| Parameter estimates | | | | |
|---|---|---|---|---|
| Term | Estimate | Std error | t Ratio | Prob$_{>|t|}$ |
| Intercept | −0.1300 | 0.4307 | −0.30 | 0.76 |
| log(ADR27) | 0.9417 | 0.2414 | 3.90 | 0.00 |
| log(H505) | 0.1000 | 0.0186 | 5.38 | 0.00 |
| log(T867) | 0.1574 | 0.0899 | 1.75 | 0.09 |
| log(CS505) | −0.1679 | 0.1609 | −1.04 | 0.30 |

The parameter estimates then become

| Parameter estimates | | | | |
|---|---|---|---|---|
| Term | Estimate | Std error | t ratio | Prob>\|t\| |
| Intercept | −0.5748 | 0.0623 | −9.23 | 0.00 |
| log(ADR27) | 0.6959 | 0.0529 | 13.16 | 0.00 |
| log(H505) | 0.0903 | 0.0161 | 5.61 | 0.00 |
| log(T867) | 0.2447 | 0.0328 | 7.47 | 0.00 |

The diagnostic plots show that the quality of the fit has not changed by dropping log(CS505) from the model. However, the parameter estimates for log(T867) and log(ADR27) have changed and their standard errors have decreased dramatically. This means that log(T867) has moved from being not significant to highly significant. In a sense, given the other explanatory variables, log(CS505) is a not significant but highly influential variable.

## 18.10   Multicollinearity

In general, interpretation of the regression coefficients depends on the other variables in the model so it is not surprising that the estimates of the slopes change when we change the variables in the model. Substantial changes, such as those seen in the Nitrous Oxide emission example, indicate strong linear relationships between the explanatory variables. In particular, log(CS505) is strongly linearly related to log(T867) and log(ADR27). This is the phenomenon of *multicollinearity*; an explanatory variable is linearly related to one or more of the other explanatory variables. When log(CS505) is included in the model, it masks the contribution of log(T867) and to a lesser extent log(ADR27) and makes the slope estimates and their standard errors unstable. By this we mean that changing the terms in the model or when we take a number of similar datasets the estimates and their standard errors can change dramatically from dataset to dataset.

In the emissions example, we expect a relationship between the explanatory variables because they are all measuring the same quantity. When we recognise that log(CS505) is measuring the same quantity as log(T867) and log(ADR27), then we realise that it is redundant as an explanatory variable and feel confident about discarding it from the model.

In general, consider the two variable regression model

$$Y_i = \beta_0 + \beta_1 x_{i1} + \beta_2 x_{i2} + \epsilon_i, \qquad \epsilon_i \stackrel{ind}{\sim} N(0, \sigma^2).$$

Then we can show (see the exercises) that

$$\mathrm{Var}(\hat{\beta}_1) = \frac{\sigma^2}{(n-1)s_{X_1}^2} \frac{1}{1-r_{12}^2} \text{ and } \mathrm{Var}(\hat{\beta}_2) = \frac{\sigma^2}{(n-1)s_{X_2}^2} \frac{1}{1-r_{12}^2}$$

where $r_{12}$ is the sample correlation between $x_1$ and $x_2$. Clearly as $r_{12}^2$ nears 1, the variances increase. This corresponds to the relationship between $x_1$ and $x_2$ becoming more linear.

The general problem is analogous: near linear relationships between the explanatory variables (there exists a $\mathbf{b}$ such that $\mathbf{Xb}$ is close to $\mathbf{0}$) make the estimates unstable ($\mathbf{X}^T\mathbf{X}$ is close to singular) and their variances become large. There are two useful diagnostics for collinearity.

- **Variance inflation factors**. Let $R_k^2$ be the squared multiple correlation coefficient calculated from the regression of $x_k$ on the other explanatory variables. If $R_k^2$ is near 1, then we have approximate collinearity. In fact, $R_k^2$ plays the same role as $r_{12}^2$ in the above calculations so we have the equivalent formulation that we have approximate collinearity if the variance inflation factor

$$VIF_k = \frac{1}{1-R_k^2}$$

  is large.
- **Condition number**. A common measure of collinearity from numerical analysis is

$$\kappa = \left( \frac{\lambda_{\max}(\mathbf{X}^T\mathbf{X})}{\lambda_{\min}(\mathbf{X}^T\mathbf{X})} \right)^{1/2}$$

where $\lambda_{\max}$ and $\lambda_{\min}$ are the respectively the maximum and minimum eigenvalues. Large values of $\kappa$ imply collinearity; it has been suggested that a value of 30 be taken as large, but this has no real theoretical justification.

Multicollinearity is common in economic and social science data modelling problems where it is less clear that variables are linearly related. Often variables are introduced as proxies for quantities which are unmeasurable or difficult to measure and they may also turn out to be proxies for some of the other explanatory variables in the model. In the case when we measure the same quantity in different ways, we may comfortably discard the redundant variables. Sometimes it happens that the observations in the sample are multicollinear even though the variables are not globally collinear. In this case, prediction into regions where the multicollinearity does not hold is obviously problematic and the model can only be used in a limited way.

Multicollinearity does not in general affect our ability to obtain good fit in a particular dataset but it does complicate the interpretation of the parameter estimates.

## 18.11    Summary of Multiple Regression Modelling

1. The multiple linear regression model is given by:

$$Y_i = \beta_0 + \beta_1 x_{i1} + \ldots + \beta_p x_{ip} + \epsilon_i, \quad \epsilon_i \overset{\text{ind}}{\sim} N(0, \sigma^2), \quad i = 1, \ldots, n.$$

2. The model can be written as: $\mathbf{Y} = X\boldsymbol{\beta} + \boldsymbol{\epsilon}$. [Recall that notations in **boldface** denote vectors.]
3. The least squares estimates are given by: $\hat{\boldsymbol{\beta}} = (X^T X)^{-1} X^T \mathbf{y}$.
4. Fitted values are given by $\hat{\mathbf{y}} = X\hat{\boldsymbol{\beta}} = X(X^T X)^{-1} X^T \mathbf{y} = H\mathbf{y}$.
5. Residuals are given by: $\mathbf{r} = \mathbf{y} - \hat{\mathbf{y}} = (I_n - H)\mathbf{y}$. Residual SS $= \mathbf{r}^T \mathbf{r} = \mathbf{y}^T (I_n - H)\mathbf{y}$.
6. Residual df is the number of observations minus the number of parameters estimated $= n - (p + 1)$.
7. Can show that $\hat{\boldsymbol{\beta}}$ follows the multivariate normal distribution with mean vector $\boldsymbol{\beta}$ and variance matrix $\sigma^2 (X^T X)^{-1}$.
8. Can show that $Var(\hat{\beta}_j) = \sigma^2$ times the $j$th diagonal element of $(X^T X)^{-1}$.
9. We estimate $\sigma^2$ by $s^2 = \dfrac{\text{Residual SS}}{\text{Residual df}} = \dfrac{\mathbf{y}^T (I_n - H)\mathbf{y}}{n - p - 1}$.
10. Can estimate the conditional mean of $Y_0$ for a new $\mathbf{x}_0$ by $\mathbf{x}_0^T \hat{\boldsymbol{\beta}}$. Its variance is $\sigma^2 \mathbf{x}_0^T (X^T X)^{-1} \mathbf{x}_0 = \sigma^2 h_{00}$.
11. Can predict a future observation $Y_0$ for a new $\mathbf{x}_0$ by $\mathbf{x}_0^T \hat{\boldsymbol{\beta}}$. Its variance is $\sigma^2 (1 + \mathbf{x}_0^T (X^T X)^{-1} \mathbf{x}_0) = \sigma^2 (1 + h_{00})$.
12. To construct a confidence interval, use the general formula: **Estimate** $\pm$ **Critical Value** $\times$ **Standard Error**. Use this to find intervals for any $\beta_j$, the conditional mean and the future observation $Y_0$ for a new $\mathbf{x}_0$.
13. **Standard Error** is the estimated value of the square root of the variance.
14. For hypothesis testing we use the rule: **Reject $H_0$ if P-value is less than** $\alpha$ from Chap. 12.
15. To compare two models, where one model is a reduced version of the other full model, we use the $F$-test and draw up an ANOVA table. We find two residual sums of squares, calculate the difference and the F-ratio is the difference by its df divided by the mean residual SS for the more complex model.
16. Although often reported, multiple correlation is not a good selection criterion.

$$R^2 = 1 - \frac{(n - p - 1)S^2}{(n - 1)S_Y^2} = 1 - \frac{\text{Residual SS}}{\text{Total SS}} = \frac{\text{SS explained by the Model}}{\text{Total SS}}$$

17. To compare arbitrary linear models we can use the AIC criterion which penalises a model both for lack of fit (high $S^2$) and complexity (large $p$).

$$AIC = n \log \left\{ \frac{(n-p-1)S^2}{n} \right\} + 2 \times (p+1)$$

$$= n \log \left\{ \frac{\text{Residual SS}}{n} \right\} + 2 \times (\text{number of parameters}).$$

18. To search for the best model we can use stepwise procedures such as backwards elimination and forward selection.
19. Discussed two diagnostics (variance inflation factor and condition number) for detecting multicollinearity.
20. We use residual plots to check model assumptions.
    (a) We plot the residuals against the fitted values $(X\hat{\boldsymbol{\beta}})_i$ to check for homoscedasticity. Plot should look like a random scatter if this assumption holds.
    (b) We order and plot the residuals against the normal order statistics to check for normality. This plot should be a straight line if the normality assumption holds.

## 18.12  Exercises

### 18.1 (Multiple Linear Regression Model)

1. Consider the linear model

$$y_i = \beta_0 + \beta_1 x_i + \beta_2 (3x_i^2 - 2) + \epsilon_i, \quad i = 1, 2, 3.$$

where $x_1 = -1$, $x_2 = 0$, $x_3 = 1$ and the $\epsilon_i$ are independent and normally distributed with zero mean and common variance $\sigma^2$.
   (i) Write down the design matrix and hence evaluate the least squares estimators for $\beta_0$, $\beta_1$ and $\beta_2$.
   (ii) Show that the least squares estimates of $\beta_0$ and $\beta_1$ are unchanged if the simpler model

$$y_i = \beta_0 + \beta_1 x_i + \epsilon_i, \quad i = 1, 2, 3.$$

is fitted (i.e. $\beta_2$ is set equal to 0 in the more complex model).
Now consider the linear model

$$y_i = \beta_0 + \beta_1 (x_{i1} - \bar{x}_1) + \beta_2 (x_{i2} - \bar{x}_2) + \epsilon_i, \quad i = 1, \ldots, n,$$

where $(x_{i1}, x_{i2}), i = 1, \ldots, n$ are the observed values of two explanatory variables, for which $\overline{x_1}$ and $\overline{x_1}$ are the sample means . Show that

$$Var(\hat{\beta}_1) = \frac{\sigma^2}{(n-1)s_{x_1}^2(1 - r_{12}^2)}, \qquad Var(\hat{\beta}_2) = \frac{\sigma^2}{(n-1)s_{x_2}^2(1 - r_{12}^2)}$$

where $s_{x_1}$ and $s_{x_2}$ are the sample standard deviations of the two explanatory variables and

$$r_{12} = \frac{\sum_{i=1}^{n}(x_{i1} - \overline{x}_1)(x_{i2} - \overline{x}_2)}{(n-1)s_{x_1}s_{x_2}}$$

is their sample correlation. What are the implications of this result for two highly correlated explanatory variables?

2. For the general linear model $\mathbf{Y} = X\boldsymbol{\beta} + \boldsymbol{\epsilon}$, prove the following properties of the residual vector $\mathbf{r} = (r_1, \ldots, r_n)^T = \mathbf{y} - X\hat{\boldsymbol{\beta}}$.
   (i) $\mathbf{r} = (I - H)\mathbf{y}$
   (ii) $\mathbf{r}^T\mathbf{r} = \mathbf{r}^T\mathbf{y}$, where $\mathbf{r}^T\mathbf{r} \equiv \sum_{i=1}^{n} r_i^2$ and $\mathbf{r}^T\mathbf{y} \equiv \sum_{i=1}^{n} r_i y_i$.
   (iii) For models where the design matrix includes a column of 1s, $\overline{r} \equiv \frac{1}{n}\sum_{i=1}^{n} r_i = 0$. [Assume that this is the case from here on].
   (iv) $\sum_{i=1}^{n}(r_i - \overline{r})(y_i - \overline{y}) = \mathbf{r}^T\mathbf{r}$.
   (v) The sample correlation coefficient between residuals and observed values is $(1 - R^2)^{1/2}$, where $R^2$ is the multiple correlation coefficient, defined by

$$R^2 - \frac{(n-1)s_y^2 - \mathbf{r}^T\mathbf{r}}{(n-1)s_y^2}$$

where $s_y$ is the sample standard deviation of the $y$s.
   (vi) The population correlation coefficient between residuals $\mathbf{R} = (R_1, \ldots, R_n)^T$ and fitted values $\hat{\mathbf{Y}}$ is zero.
   (vii) $Var(\mathbf{R}) = \sigma^2(I - H)$
   (viii) Why should (v), (vi), and (vii) be remembered when plotting residuals?

3. Suppose that we have $n$ observations $(Y_i, x_{i1}, x_{i2})$ which satisfy the regression model

$$Y_i = \beta_0 + x_{i1}\beta_1 + x_{i2}\beta_2 + \epsilon_i, \quad i = 1, \ldots, n,$$

with $\epsilon_i \sim$ independent $N(0, \sigma^2)$. Assume that $\sum_{i=1}^{n} x_{i1} = \sum_{i=1}^{n} x_{i2} = 0$ and $\sum_{i=1}^{n} x_{i1}^2 = \sum_{i=1}^{n} x_{i2}^2 = 1$. Let $X_1$ be the $n$-vector with $i$th component $x_{i1}$ so $X_1 = (x_{11}, \ldots, x_{n1})^T$, and similarly for $X_2$, $y$ and $\epsilon$. We will explore what happens if we ignore $x_{i2}$ and simply use least squares to fit the regression of $Y_i$ on $x_{i1}$. The intercept and slope estimator in this case is

$$\hat{\boldsymbol{\alpha}} = (D_1^T D_1)^{-1} D_1^T \mathbf{Y},$$

where $\boldsymbol{\alpha} = (\alpha_0, \alpha_1)^T$ and $D_1 = (1, X_1)$ is the $n \times 2$ design matrix for the model without $X_2$.

(i) Express the model for $y$ in terms of the covariate matrix $D_1$ and the vector $X_2$.

(ii) Show that

$$E(\hat{\boldsymbol{\alpha}}) = \begin{pmatrix} \beta_0 \\ \beta_1 \end{pmatrix} + (D_1^T D_1)^{-1} D_1^T X_2 \beta_2 \quad \text{and} \quad Var(\hat{\boldsymbol{\alpha}}) = \sigma^2 (D_1^T D_1)^{-1}.$$

(iii) Interpret the results in (ii), paying careful attention to the cases $X_1^T X_2 = 0$ (orthogonality) and $X_1^T X_2 = 1$ (collinearity).

(iv) Show that the residuals $\mathbf{r} = \mathbf{Y} - D_1 \hat{\boldsymbol{\alpha}}$ satisfy

$$\mathbf{R} = (I_n - H_1)\boldsymbol{\epsilon} + (I_n - H_1)X_2 \beta_2,$$

where $H_1 = D_1 (D_1^T D_1)^{-1} D_1^T$ is the hat matrix for the regression of $Y_i$ on $x_{i1}$ ignoring $x_{i2}$. Hence or otherwise show that

$$E(\mathbf{R}) = (I_n - H_1)X_2 \beta_2 \quad \text{and} \quad Var(\mathbf{R}) = \sigma^2 (I_n - H_1).$$

(v) Show that the predictor of an independent new observation $Y_z$ at $(z_1, z_2)$ given by $\hat{Y}_z = (1, z_1)\hat{\boldsymbol{\alpha}}$ satisfies

$$E(\hat{Y}_z - Y_z) = \{(1, z_1)(D_1^T D_1)^{-1} D_1^T X_2 - z_2\}\beta_2$$

and

$$Var(\hat{Y}_z - Y_z) = \sigma^2 (1, z_1)(D_1^T D_1)^{-1} \begin{pmatrix} 1 \\ z_1 \end{pmatrix} + \sigma^2.$$

4. The following linear model is proposed for random variables $Y_1, Y_2, Y_3$:

$$Y_i = a \left( \frac{3 + 5x_i}{3 + x_i} \right) + b \left( \frac{3 - 5x_i}{3 - x_i} \right) + \epsilon_i$$

where $x_1 = -1$, $x_2 = 0$, $x_3 = 1$ and the $\epsilon_i$ are independent and normally distributed with zero mean and common variance $\sigma^2$.

(i) Write down the design matrix and hence evaluate the least squares estimators for $a$ and $b$.

(ii) Show that the $F$-statistic for the test of the null hypothesis $H_0 : a = 0$ against the alternative $H_1 : a \neq 0$ is given by

$$F = \frac{3(Y_2 + Y_3)^2}{2(Y_1 - Y_2 + Y_3)^2}.$$

What is the distribution of this statistic under $H_0$?

(iii) Describe briefly how residual plots can be used to check the distributional assumptions made when a linear model is fitted.

5. The data `crime` are crime-related and demographic statistics for 47 US states in 1960. The data were collected from the FBI's Uniform Crime Report and other government agencies to determine how the variable crime rate depends on the other variables measured in the study.

(a) Use R to fit linear models, explaining the relationship between the response and explanatory variables. First, include all the 13 covariates and then remove variables using the `drop1` command, until you can drop no further. Explain why further terms cannot be dropped, use R output as evidence.

(b) Write down the model equation and interpret parameter estimates of this final model. Do you see any difficulty in interpretation?

(c) Present residual plots for the model written down in part b. to check your model assumptions. What all concerns you have about the model assumptions? Would you say your model was a good fit?

# Analysis of Variance

# 19

**Abstract**

Finally, this chapter introduces the concepts of analysis of variance which is a seen as a general model comparison technique where there are categorical explanatory variables. Theoretical generalisation of the techniques from the two preceding chapters are included and so are illustrations using R. In particular, the one way analysis of variance technique is illustrated by using an ecological example on modelling body weights of brushtail possums—a nocturnal animal only native to Australia.

## 19.1 The Problem

Consider a generalisation of the two-sample t-test which compares means of two normal distributions, discussed in Sect. 12.5. Here the generalisation is regarding testing equality of the means for more than two populations. Let $g(\geq 2)$ be the number of populations and assume that a simple random sample of size $n_i$ is taken from the $i$th normally distributed population where $i = 1, \ldots, g$. Thus we have,

$$Y_{ij} \sim N(\mu_i, \sigma^2), \quad j = 1, \ldots, n_i, \ i = 1, \cdots, g, \qquad (19.1)$$

independently. In the above model, we assume that the $g$ normal distributions are independent and have the same unknown variance $\sigma^2$ but may have different unknown mean values $\mu_i$, $i = 1, \ldots, g$.

---

The original version of this chapter has been revised. A correction to this chapter can be found at
https://doi.org/10.1007/978-3-031-37865-2_21

© The Author(s), under exclusive license to Springer Nature Switzerland AG 2024,
corrected publication 2024
S. K. Sahu, *Introduction to Probability, Statistics & R*,
https://doi.org/10.1007/978-3-031-37865-2_19

The main problem is to assess whether the $g$ unknown mean values are equal (so that the $g$ populations are identical) by using the $n$ $(= \sum_{i=1}^{g} n_i)$ observed $Y$-values. This is equivalent to testing the hypotheses:

$$H_0 : \quad \mu_1 = \cdots = \mu_g \quad \text{against} \quad H_a : \quad \text{not } H_0.$$

It turns out that this testing can be done by using the methodologies already studied in Chap. 18 and the ANOVA table first introduced in Sect. 17.9. In order to exploit these theories, it is necessary to express model (19.1) equivalently using a slightly different but more general notation so that we can express model (19.1) as a special case of the multiple regression model (18.1). We start by reparameterising

$$\mu_i = \beta_0 + \beta_i$$

for $i = 1, \ldots, g$ so that $\beta_0$ represents the overall mean of the $g$ populations and $\beta_i$ is the incremental mean specific for population $i$ over and above the grand mean $\beta_0$. Clearly, the hypothesis $H_0$ then reduces to

$$\beta_1 = \beta_2 = \cdots = \beta_g = 0,$$

and model (19.1) is re-expressed as:

$$Y_{ij} \sim N(\beta_0 + \beta_i, \sigma^2), \quad j = 1, \ldots, n_i, \ i = 1, \cdots, g. \tag{19.2}$$

This model is called the one way anova model as it uses just one factor to group the observations. Let $\boldsymbol{\beta} = (\beta_0, \ldots, \beta_g)$ denote the $g + 1$ parameters and $\mathbf{Y}$ denote the collection of $n(= \sum_{i=1}^{g} n_i)$ observations $(Y_{11}, \ldots, Y_{1n_1}, Y_{21}, \ldots, Y_{2n_2}, \ldots, Y_{gn_g})$. In $\mathbf{Y}$, samples from the first population are collected first and the process continues with the other populations. Note that the population labels (1 to $g$) are chosen without loss of generality. Now our problem is to define the $n \times (g+1)$ design matrix $X$ so that we can write model (19.2) as $\mathbf{Y} = X\boldsymbol{\beta} + \boldsymbol{\epsilon}$ analogous to model (18.2). Before we do this in Sect. 19.3 below, we introduce a motivating practical data set which will also be used as a running example in this chapter to illustrate various concepts.

## 19.2 Motivating Example: Possum Data

**Example 19.1**

Variation in the external morphology of animals through their geographic distribution is of considerable scientific interest. Scientists, Lindenmayer et al. [13], collected data on brushtail possums (See Fig. 19.1), tree living furry animals who are mostly nocturnal (marsupial) at a number of different locations in

**Fig. 19.1**  Common brushtail
possum by GregTheBusker.
License: CC BY 2.0. Source:
https://flickr.com/photos/
17004938@N00/5653697137

Australia. For our analysis the locations are classified into 7 regions: (1) Western
Australia (W.A), (2) South Australia (S.A), (3) Northern Territory (N.T), (4)
Queensland (QuL), (5) New South Wales (NSW), (6) Victoria (Vic) and (7)
Tasmania (Tas).

The data set (obtained from Prof Alan Welsh, author of Welsh, 1996) is `possum`
in the R package `ipsRdbs` and the response variable is $y_{ij}$ = body weight of
the $j$th possum ($j = 1, \ldots, n_i$) caught in the $i$th location, ($i = 1, \ldots, 7$), and
the explanatory variables are ($z_{ij}$) body length, and location $x_{ij}$. One of the main
aims here is to explore whether the relationship between body length, $y_{ij}$ and
weight $z_{ij}$ changes with the different locations $x_{ij}$.

The explanatory variable length $z$ is interval scaled but the location variable $x$
is nominal although it has associated numerical labels 1 to 7. Such nominal
explanatory variables are often referred to as *factors* in regression modelling,
while interval scaled explanatory variables like length are often referred to as
covariates. Now our aims are to answer questions such as:

1. Ignoring length $z$, does weight $y$ vary between different levels of location $x$?
2. Does the relationship between length and weight differ between the locations?

These questions can be answered by using summary statistics and exploratory
graphical displays. For example, we can obtain side by side boxplots, see
Fig. 19.2, by issuing the R commands:

```
summary(possum)
boxplot(Body_Weight ~ region, data=possum)
meanwts <- tapply(possum$Body_Weight, possum$region, FUN=
    mean)
round(meanwts, 3)
varwts <- tapply(possum$IEq29}Body_Weight, possum$region,
    FUN=var)
round(varwts, 3)
```

The boxplots show the full range of variability in the body weights in each of
the seven regions. Table 19.1 provides the summary statistics. The mean weights
vary in a narrow range between 2.7 (Tasmania) to 3.5 (Queensland).

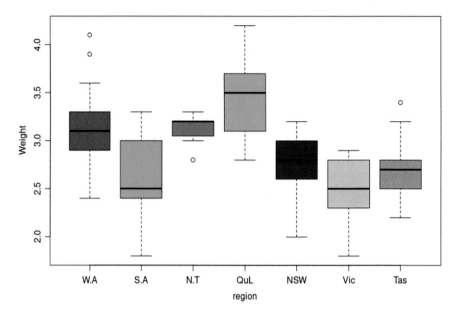

**Fig. 19.2** Side by side boxplots of body weights of the caught possums in 7 regions in Australia

**Table 19.1** Summary statistics of the body weight data of the caught possums. The overall mean is 2.88

| Region | W.A | S.A | N.T | QuL | NSW | Vic | Tas |
|---|---|---|---|---|---|---|---|
| Location ($i$) | 1 | 2 | 3 | 4 | 5 | 6 | 7 |
| $n_i$ | 33 | 12 | 7 | 6 | 13 | 13 | 17 |
| mean ($\bar{y}_i$) | 3.133 | 2.642 | 3.114 | 3.467 | 2.754 | 2.446 | 2.682 |
| var ($s_i^2$) | 0.149 | 0.183 | 0.028 | 0.239 | 0.133 | 0.128 | 0.090 |

Figure 19.3 provides a scatter plot to study the relationships between body weight and length in the presence of variation due to location. However, statistically significant differences cannot be judged from these summaries alone and we cannot decide whether to have seven different regression lines for the seven regions instead of just one straight line as shown in Fig. 19.3. Hence, we proceed to modelling. ◄

# 19.3    Formulating a Multiple Regression Model

Often, as in the possum example, at least one explanatory variable is a *factor*, a variable whose values are not numerical, but labels classifying the units of observation into different categories, e.g. sex, region, type etc. For the possum example, numerical values 1, 2, . . . , 7 for locations are used to denote the regions, but these values are just being used as labels and have no numerical significance. It is meaningless to perform any kind of arithmetic on these values, e.g. there is no interpretation for the region number 4.2.

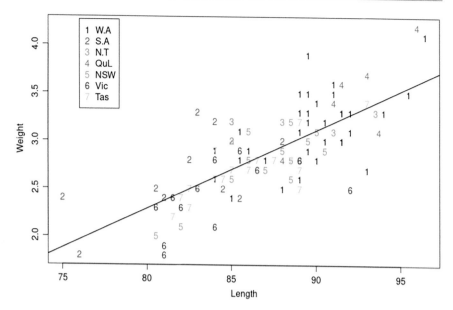

**Fig. 19.3** Scatter plot of body weights against body length of the caught possums in 7 regions in Australia. The superimposed black line shows the fitted simple linear regression model for the response variable weight and explanatory variable length

We incorporate factors such as region in the possum example in a general multiple regression model (18.1) using a set of *dummy variables*, which can take only two values, 0 and 1. In order to do this we first write the response $Y_{ij}$ using just one subscript, say $\ell$, because (18.1) has been written using just one subscript $i$. Here we use $\ell$ instrad of $i$ since $i$ has already been used to denote the groups in (19.2). Thus we write an individual response $Y_\ell$ instead of $Y_{ij}$. Corresponding to each $Y_\ell$ we have the factor valiable $X_\ell$, which identifies the location in the possum example.

With $g$ possible levels for the factor $X$, we define the $g$ dummy variables $\mathbf{X}_1, \ldots, \mathbf{X}_g$ where each of these vecor variables has $n$ entries corresponding to $n$ observations $Y_1, \ldots, Y_n$.

$$X_{\ell 1} = I(X_\ell = 1) = \begin{cases} 1 & \text{if } X_\ell = 1 \\ 0 & \text{if } X_\ell \neq 1 \end{cases}$$

$$X_{\ell 2} = I(X_\ell = 2) = \begin{cases} 1 & \text{if } X_\ell = 2 \\ 0 & \text{if } X_\ell \neq 2 \end{cases}$$

$$\vdots$$

$$X_{\ell g} = I(X_\ell = g) = \begin{cases} 1 & \text{if } X_\ell = g \\ 0 & \text{if } X_\ell \neq g \end{cases}$$

where $I(\cdot)$ is the indicator function, $I(A) = 1$ if $A$ is true and $I(A) = 0$, otherwise.

For data unit $\ell$ with $Y_\ell$ as the response, exactly one of the dummy variables takes the value 1 (corresponding to the value of factor $X$ for unit $\ell$), with the rest taking the value 0. Hence,

$$Y_\ell = \begin{cases} \beta_0 + \beta_1 + \epsilon_\ell & \text{if } \ell \text{ is in group } 1 \\ \beta_0 + \beta_2 + \epsilon_\ell & \text{if } \ell \text{ is in group } 2 \\ \vdots \\ \beta_0 + \beta_g + \epsilon_\ell & \text{if } \ell \text{ is in group } g \end{cases} \qquad \epsilon_\ell \overset{\text{ind}}{\sim} N(0, \sigma^2),$$

is equivalent to the model (19.2). Thus, $\beta_0 + \beta_i$ is the conditional mean response for a unit classified in group $i$ by factor $X$. Using the dummy variables we can re-write the above model equivalently as

$$Y_\ell = \beta_0 + \sum_{i=1}^{g} \beta_i X_{\ell i} + \epsilon_\ell, \qquad \epsilon_\ell \sim N(0, \sigma^2), \qquad (19.3)$$

independently for $\ell = 1, \ldots, n$. In effect, a single factor variable $X$ with $g$ possible labels (or values) is represented by the $g$ dummy (0-1) variables, $X_1, \ldots, X_g$. In order to estimate $\boldsymbol{\beta} = (\beta_0, \beta_1, \ldots, \beta_g)$, we construct the design matrix $\mathbf{X}$ as follows.

$$\mathbf{Y} = \mathbf{X}\boldsymbol{\beta} + \boldsymbol{\epsilon}, \qquad \boldsymbol{\epsilon} \sim N(\mathbf{0}, \sigma^2 \mathbf{I}_n)$$

where $\boldsymbol{\beta} = (\beta_0, \beta_1, \ldots, \beta_g)^T$ and

$$\mathbf{X} = \begin{pmatrix} \mathbf{1}_{n_1} & \mathbf{1}_{n_1} & \mathbf{0}_{n_1} & \cdots & \mathbf{0}_{n_1} \\ \mathbf{1}_{n_2} & \mathbf{0}_{n_2} & \mathbf{1}_{n_2} & & \mathbf{0}_{n_2} \\ \vdots & \vdots & & \ddots & \\ \mathbf{1}_{n_g} & \mathbf{0}_{n_g} & \mathbf{0}_{n_g} & & \mathbf{1}_{n_g} \end{pmatrix}$$

and $\mathbf{1}_n$ and $\mathbf{0}_n$ represent $n$-vectors of ones and zeros respectively. Therefore $\mathbf{X}$ is a $n \times (g + 1)$ matrix which is *not of full rank*, as column 1 is equal to the sum of columns $2, \ldots, g + 1$. Hence $\mathbf{X}^T \mathbf{X}$ is not invertible, and there is not a unique solution to the normal equations:

$$\mathbf{X}^T \mathbf{X} \hat{\boldsymbol{\beta}} = \mathbf{X}^T \mathbf{Y}.$$

We can see that

$$\mathbf{X}^T\mathbf{X} = \begin{pmatrix} n & n_1 & n_2 & \dots & n_g \\ n_1 & n_1 & 0 & \dots & 0 \\ n_2 & 0 & n_2 & \dots & 0 \\ \vdots & \vdots & & \ddots & \\ n_g & 0 & 0 & & n_g \end{pmatrix} \quad \text{and} \quad \mathbf{X}^T\mathbf{Y} = \begin{pmatrix} n\bar{Y} \\ n_1\bar{Y}_1 \\ \vdots \\ n_g\bar{Y}_g \end{pmatrix}$$

where $\bar{Y}_i$ is the sample mean of the response variable for those units classified in group $i$ by $X$. i.e. $\bar{Y}_i = \frac{1}{n_i}\sum_{j=1}^{n_i} Y_{ij}$. Hence the normal equations are the system of linear equations

$$n\hat{\beta}_0 + \sum_{i=1}^{g} n_i\hat{\beta}_i = n\bar{Y}$$
$$\hat{\beta}_0 + \hat{\beta}_1 = \bar{Y}_1$$
$$\hat{\beta}_0 + \hat{\beta}_2 = \bar{Y}_2$$
$$\vdots$$
$$\hat{\beta}_0 + \hat{\beta}_g = \bar{Y}_g$$

The first of these equations is clearly redundant since it is given by the sum of the remaining $g$ equations which are a linearly independent system of $g$ equations in $g + 1$ unknowns. Hence there is no unique solution. In this case, we say that the parameters are not identifiable in this model or equivalently the model (19.3) is overparameterised. For example, adding a fixed constant to $\beta_0$ and subtracting the same constant from $\beta_1, \dots, \beta_g$ does not change the model for the response variable $Y$. We fix this problem by imposing an *identifiability constraint* on the parameters. Here, we choose to fix $\beta_1 = 0$, so under the model $\beta_0$ is the mean response in group 1, and $\beta_i$ ($i \geq 2$) is the difference in mean response between group $i$ and group 1.

With this constraint, the normal equations have the unique solution:

$$\hat{\beta}_0 = \bar{Y}_1$$
$$\hat{\beta}_2 = \bar{Y}_2 - \bar{Y}_1$$
$$\vdots$$
$$\hat{\beta}_g = \bar{Y}_g - \bar{Y}_1.$$

Clearly, $\beta_0$ is estimated to be the mean of the first group and each $\beta_i$ for $i \geq 2$ is estimated to be the difference in mean between the $i$th group and the first group. This observation is very useful in interpreting the estimates of the incremental means $\beta_i$, $i = 2, \dots, g$ in practical problems.

The adopted identifiability constraint, $\beta_1 = 0$, does not affect calculation and interpretation of the fitted values, residuals and model based predictions, since the model (19.1) or the equivalent model (19.2) implies

$$\hat{Y}_{ij} = \hat{\beta}_0 + \hat{\beta}_i = \bar{Y}_i$$

directly from the normal equations. Hence, the residuals are given by

$$R_{ij} = Y_{ij} - \hat{Y}_{ij} = Y_{ij} - \bar{Y}_i$$

and the residual sum of squares (RSS) under model (19.2), is given by

$$\text{RSS}_f = \sum_{i=1}^{g} \sum_{j=1}^{n_i} R_{ij}^2 = \sum_{i=1}^{g} \sum_{j=1}^{n_i} (Y_{ij} - \bar{Y}_i)^2, \tag{19.4}$$

which is required for testing significance of the factor $X$ as described below.

## 19.4 Testing Significance of a Factor—One-Way Anova

To test significance of a factor $X$ (with no other explanatory variables present), we compare the full model:

$$M : Y_{ij} = \beta_0 + \beta_i + \epsilon_{ij}$$

with the reduced model

$$M_0 : Y_{ij} = \beta_0 + \epsilon_{ij},$$

where $\epsilon_{ij} \overset{\text{ind}}{\sim} N(0, \sigma^2)$ under both the models. That is, we test $H_0 : \beta_1 = \beta_2 = \ldots = \beta_g = 0$ in the full model M. Failure to reject the hypothesis indicates that the null model, where all groups have the same mean ($\beta_0$), is an adequate simplification of the full model M, which assumes a separate mean response for each group.

Under the reduced model $M_0$, the single mean parameter $\beta_0$ is estimated by the mean of all the data, i.e.,

$$\hat{\beta}_0 = \bar{Y} = \frac{1}{n} \sum_{i=1}^{g} \sum_{j=1}^{n_i} Y_{ij}.$$

As a result the residuals are $Y_{ij} - \bar{Y}$ for $i = 1, \ldots, g$, $j = 1, \ldots n_i$. The residual sum of squares under this reduced model is:

$$\text{RSS}_r = \sum_{i=1}^{g} \sum_{j=1}^{n_i} (Y_{ij} - \bar{Y})^2.$$

The full model $M$ and the reduced model $M_0$ are now compared using the general model comparison methodology outlined in Sect. 18.6.7. The crucial quantity in the model comparison is the difference in residual sum of squares $\text{RSS}_r - \text{RSS}_f$. To derive an expression for this we write:

$$
\begin{aligned}
\text{RSS}_r &= \sum_{i=1}^{g} \sum_{j=1}^{n_i} (Y_{ij} - \bar{Y})^2 \\
&= \sum_{i=1}^{g} \sum_{j=1}^{n_i} [(Y_{ij} - \bar{Y}_i]) + (\bar{Y}_i - \bar{Y})]^2 \\
&= \sum_{i=1}^{g} \sum_{j=1}^{n_i} (Y_{ij} - \bar{Y}_i)^2 + \sum_{i=1}^{g} \sum_{j=1}^{n_i} (\bar{Y}_i - \bar{Y})^2 \\
&\quad + \sum_{i=1}^{g} \sum_{j=1}^{n_i} (Y_{ij} - \bar{Y}_i)(\bar{Y}_i - \bar{Y}) \\
&= \sum_{i=1}^{g} \sum_{j=1}^{n_i} (Y_{ij} - \bar{Y}_i)^2 + \sum_{i=1}^{g} n_i (\bar{Y}_i - \bar{Y})^2 \\
&\quad + \sum_{i=1}^{g} (\bar{Y}_i - \bar{Y}) \sum_{j=1}^{n_i} (Y_{ij} - \bar{Y}_i) \\
&= \sum_{i=1}^{g} \sum_{j=1}^{n_i} (Y_{ij} - \bar{Y}_i)^2 + \sum_{i=1}^{g} n_i (\bar{Y}_i - \bar{Y})^2 \\
&= \text{RSS}_f + \sum_{i=1}^{g} n_i (\bar{Y}_i - \bar{Y})^2
\end{aligned}
$$

since $\sum_{j=1}^{n_i} (Y_{ij} - \bar{Y}_i) = 0$ for all $i = 1, \ldots, g$, recall the identity (1.1). The necessary analysis of variance decomposition is given by:

$$\text{RSS under } M_0 = \text{RSS under } M + \text{difference in RSS}$$

i.e., $\qquad \text{RSS}_r = \text{RSS}_f + \text{RSS}_r - \text{RSS}_f$

i.e., $\sum_{i=1}^{g} \sum_{j=1}^{n_i} (Y_{ij} - \bar{Y})^2 = \sum_{i=1}^{g} \sum_{j=1}^{n_i} (Y_{ij} - \bar{Y}_i)^2 + \sum_{i=1}^{g} n_i (\bar{Y}_i - \bar{Y})^2$.

An alternative expression for the difference in RSS is given by:

$$\sum_{i=1}^{g} n_i (\bar{Y}_i - \bar{Y})^2 = \sum_{i=1}^{g} n_i \bar{Y}_i^2 - n \bar{Y}^2.$$

Under model $M_0$, using the Cochran's Theorem 18.1, we know that

$$
\left.
\begin{aligned}
\frac{1}{\sigma^2} \sum_{i=1}^{g} n_i (\bar{Y}_i - \bar{Y})^2 &\sim \chi_{g-1}^2 \\
\frac{1}{\sigma^2} \sum_{i=1}^{g} \sum_{j=1}^{n_i} (Y_{ij} - \bar{Y}_i)^2 &\sim \chi_{n-g}^2
\end{aligned}
\right\} , \text{ independently}
$$

and hence

$$
F = \frac{\frac{1}{g-1} \sum_{i=1}^{g} n_i (\bar{Y}_i - \bar{Y})^2}{\frac{1}{n-g} \sum_{i=1}^{g} \sum_{j=1}^{n_i} (Y_{ij} - \bar{Y}_i)^2} \sim F_{(g-1),(n-g)},
$$

by the definition of the $F$-distribution, see Sect. 14.5.3. Note that the degrees of freedom here are $g - 1$ and $n - g$, rather than $g$ and $n - g - 1$ which is what we would expect if model $M$ was represented by a design matrix with $g + 1$ linearly independent columns. Effectively, we have removed the column corresponding to the constrained parameter $\beta_1$ from consideration (Table 19.2).

We refer to this analysis, testing the significance of a single factor, as a one-way analysis-of-variance (or one-way **ANOVA**, for short) by following the discussion

**Table 19.2**  Theoretical ANOVA table for the possum data

| Source | df | Sum of squares (SS) | Mean SS | F-value | P-value |
|--------|-----|---------------------|---------|---------|---------|
| Group | $g-1$ | $\sum_{i=1}^{g} n_i(\bar{y}_i - \bar{y})^2$ | $\frac{1}{g-1}$ Group SS | $\frac{\text{Mean Group SS}}{\text{Mean Error SS}}$ | $Pr(F_{g-1,n-g} > F_{\text{obs}})$ |
| Error | $n-g$ | $\sum_{i=1}^{g} \sum_{j=1}^{n_i}(y_{ij} - \bar{y}_i)^2$ | $\frac{1}{n-g}$ Error SS | – | – |
| Total | $n-1$ | $\sum_{i=1}^{g} \sum_{j=1}^{n_i}(y_{ij} - \bar{y})^2$ | | | |

**Table 19.3**  ANOVA table using sums of squares

| Source | df | Sum of squares (SS) | Mean SS | F-value | P-value |
|--------|-----|---------------------|---------|---------|---------|
| Group | $g-1$ | $RSS_f - RSS_r$ | $\frac{1}{g-1}$ Group SS | $\frac{\text{Mean Group SS}}{\text{Mean Error SS}}$ | $Pr(F_{g-1,n-g} > F_{\text{obs}})$ |
| Error | $n-g$ | $RSS_r$ | $\frac{1}{n-g}$ Error SS | – | – |
| Total | $n-1$ | $RSS_f$ | | | |

in Sect. 17.9. We write down the theoretical table as follows. where $F_{\text{obs}}$ denotes the observed value of the $F$-statistic. It is also sometimes helpful to note that the denominator of the F-statistic can be calculated by noting:

$$RSS_f = \sum_{i=1}^{g} \sum_{j=1}^{n_i}(y_{ij} - \bar{y}_i)^2 = \sum_{i=1}^{g}(n_i - 1)s_i^2$$

where

$$s_i^2 = \frac{1}{n_i - 1} \sum_{j=1}^{n_i}(y_{ij} - \bar{y}_i)^2,$$

is the sample variance of the observations from the $i$th group. This observation allows us to calculate the Error SS from the sample variances $s_i^2$ of the groups, $i = 1, \ldots, g$. Also we can easily see that:

$$\sum_{i=1}^{g} \sum_{j=1}^{n_i}(y_{ij} - \bar{y})^2 = \sum_{i=1}^{g} \sum_{j=1}^{n_i} y_{ij}^2 - n\bar{y}^2.$$

Thus, the total SS can be calculated by squaring and summing each observation $y_{ij}$ and then subtracting $n$ times the square of the grand mean $\bar{y}$. The group SS is then simply calculated as the difference between Total and Error SS. These calculations are illustrated using the gas mileage data Example 19.2 introduced below. For this example most of the calculations can be done using a calculator, or even the command prompt in R.

Using the above notations for the sum of squares, we re-write the ANOVA table as in Table 19.3:

**Table 19.4** Gas mileage data and summaries

| Model ($i$) | $A(1)$ | $B(2)$ | $C(3)$ | $D(4)$ |
|---|---|---|---|---|
| Mileage ($y_{ij}$) | 22, 26 | 28, 24, 29, 27 | 29, 32, 28 | 23, 24 |
| Count ($n_i$) | 2 | 4 | 3 | 2 |
| Mean ($\bar{y}_i$) | 24 | 27 | 29.67 | 23.5 |
| Variance ($s_i^2$) | 8 | 7 | 4.33 | 0.5 |

### Example 19.2

Table 19.4 above gives the gas mileages recorded during a series of road tests on four different models of Japanese luxury cars. This is data set gasmileage in the R package ipsRdbs. Assuming that the one-way ANOVA model holds for the observations, we want to test the null hypothesis that the four models give, on average, the same mileage at the level of significance $\alpha = 0.05$.
Here $g = 4$, $n = 10$ and the overall mean $\bar{y} = \frac{1}{10}\sum_{i=1}^{4} n_i \bar{y}_i = 26.5$, $\text{RSS}_f = \sum_{i=1}^{4}(n_i - 1)s_i^2 = 31.17$. We also have

$$\sum_{i=1}^{4}\sum_{j=1}^{n_i} y_{ij}^2 = 7115.$$

Hence,

$$\text{RSS}_r = \sum_{i=1}^{4}\sum_{j=1}^{n_i} y_{ij}^2 - n\bar{y}^2 = 7115 - 10(26.5)^2 = 92.5.$$

Now the F test statistic is given by

$$\frac{(\text{RSS}_r - \text{RSS}_f)/(4-1)}{\text{RSS}_f/(10-4)} = 3.935.$$

The upper 5% critical value for the $F$-distribution with degrees of freedom $g - 1 = 4 - 1 = 3$ and $n - g = 10 - 4 = 6$ is obtained from the R command qf(0.95, df1=3, df2=6) and turns out to be 4.757. Hence the null hypothesis $H_0$ of equal mean mileages is not rejected, that is, the observed data do not suggest that the four models have significantly different mean mileages at $\alpha = 5\%$ level of significance. The p-value of this test is calculated by using the command 1-pf (3.935, df1=3, df2=6), which evaluates to be 0.072. With these calculations we are able to complete the following ANOVA table (Table 19.5):
◀

**R Code for the Gas Mileage Data Example 19.2** These codelines can also be found in the R package ipsRdbs by issuing the command ?gasmileage.

**Table 19.5** ANOVA table for the gas mileage data

| Source | df | SS | Mean SS | F-value | P-value |
|--------|----|-----|---------|---------|---------|
| Group | 3 | 61.33 | $\frac{61.33}{3} = 20.44$ | $\frac{20.44}{5.194} = 3.935$ | $P(F_{3,6} > 3.936) = 0.072$ |
| Error | 6 | 31.17 | $\frac{31.17}{6} = 5.194$ | – | – |
| Total | 9 | 92.5 | | | |

```
###################################################
####This example is on one-way ANOVA for the ####
#### gas mileage data. #############################
###################################################


#####################
####Read in data####
#####################
modelA <- c(22, 26)
modelB <- c(28, 24, 29)
modelC <- c(29, 32, 28)
modelD <- c(23, 24)

#Create the Y vector
y <- c(modelA, modelB, modelC, modelD)
xx <- c(1,1,2,2,2,3,3,3,4,4)
# Plot the observations
plot(xx, y, col="red", pch="*", xlab="Model", ylab="Mileage")


# Method1: Hand calculation
ni <- c(2, 3, 3, 2)
means <- tapply(y, xx, mean)
vars <- tapply(y, xx, var)
round(rbind(means, vars), 2)


sum(y^2) # gives 7115


totalSS <- sum(y^2) - 10 * (mean(y))^2 # gives 92.5
RSSf <- sum(vars*(ni-1)) # gives 31.17
groupSS <- totalSS - RSSf # gives 61.3331.17/6


meangroupSS <- groupSS/3 # gives 20.44
meanErrorSS <- RSSf/6 # gives 5.194


Fvalue <- meangroupSS/meanErrorSS # gives 3.936
pvalue <- 1-pf(Fvalue, df1=3, df2=6)



#### Method 2: Illustrate using dummy variables
##################################
#Create the design matrix X for the full regression model
g <- 4
```

```
n1 <- length(modelA)
n2 <- length(modelB)
n3 <- length(modelC)
n4 <- length(modelD)
n <- n1+n2+n3+n4
X <- matrix(0, ncol=g, nrow=n) #Set X as a zero matrix initially
X[1:n1,1] <- 1 #Determine the first column of X
X[(n1+1):(n1+n2),2] <- 1 #the 2nd column
X[(n1+n2+1):(n1+n2+n3),3] <- 1 #the 3rd
X[(n1+n2+n3+1):(n1+n2+n3+n4),4] <- 1 #the 4th


#################################
####Fitting the full model####
#################################
#Estimation
XtXinv <- solve(t(X)%*%X)
betahat <- XtXinv %*%t(X)%*%y #Estimation of the coefficients
Yhat <- X%*%betahat #Fitted Y values
Resids <- y - Yhat #Residuals
SSE <- sum(Resids^2) #Error sum of squares
S2hat <- SSE/(n-g) #Estimation of sigma-square; mean square for
    error
Sigmahat <- sqrt(S2hat)

###############################################################
####Fitting the reduced model -- the 4 means are equal #####
###############################################################
Xr <- matrix(1, ncol=1, nrow=n)
kr <- dim(Xr)[2]
#Estimation
Varr <- solve(t(Xr)%*%Xr)
hbetar <- solve(t(Xr)%*%Xr)%*%t(Xr)%*% y #Estimation of the
    coefficients
hYr = Xr%*%hbetar #Fitted Y values
Resir <- y - hYr #Residuals
SSEr <- sum(Resir^2) #Total sum of squares

###############################################################
####F-test for comparing the reduced model with the full model ####
###############################################################
FStat <- ((SSEr-SSE)/(g-kr))/(SSE/(n-g)) #The test statistic of the
    F-test
alpha <- 0.05
Critical_value_F <- qf(1-alpha, g-kr,n-g) #The critical constant of
    F-test
pvalue_F <- 1-pf(FStat,g-kr, n-g) #p-value of F-test


SSerror = sum( (modelA-mean(modelA))^2 ) + sum( (modelB-mean(modelB)
    )^2 ) + sum( (modelC-mean(modelC))^2 ) + sum( (modelD-mean(
    modelD))^2 )
SStotal <- sum( (y-mean(y))^2 )
```

```
SSgroup <- SStotal-SSerror

####
#### Method 3: Use the built-in function lm directly
####################################

aa <- "modelA"
bb <- "modelB"
cc <- "modelC"
dd <- "modelD"
Expl <- c(aa,aa,bb,bb,bb,cc,cc,cc,dd,dd)
is.factor(Expl)
Expl <- factor(Expl)
model1 <- lm(y~Expl)
summary(model1)
anova(model1)

###Alternatively ###
xxf <- factor(xx)
is.factor(xxf)
model2 <- lm(y~xxf)
summary(model2)
anova(model2)
```

## 19.5   Modelling Interaction Effects

Suppose that we now have three variables: a continuous response variable $Y$, a factor $X$ with $g$ levels and a further continuous explanatory variable $Z$. For the possum data example, $Y$ is the body weight, $X$ is region and $Z$ is body length. The multiple regression model (19.3) is extended to include $Z$:

$$Y_\ell = \beta_0 + \beta_1 x_{\ell 1} + \beta_2 x_{\ell 2} + \ldots + \beta_g x_{\ell g} + \beta_{g+1} z_\ell + \epsilon_\ell, \qquad \epsilon_\ell \overset{\text{ind}}{\sim} N(0, \sigma^2), \qquad (19.5)$$

where $x_{\ell i}$ are dummy variables defined in Sect. 19.3. A simpler representation of this model is obtained by using the two subscripts notation $Y_{ij}$ to denote individual responses, index $i$ representing the group and $j$ indexing observations within group, where $j = 1, \ldots, n_i$, $i = 1, \ldots, g$. Then, the model (19.5) can be written as

$$Y_{ij} = \beta_0 + \beta_i + \gamma z_{ij} + \epsilon_{ij}, \qquad \epsilon_{ij} \overset{\text{ind}}{\sim} N(0, \sigma^2),$$

for all $i$ and $j$, where we write $\gamma$ in place of $\beta_{g+1}$. This model implies that $Y$ is linearly related to $Z$ with slope $\gamma$, but the intercepts of the linear regression lines depends on $X$, the group of the observation, see Fig. 19.4 for an illustration with $g = 3$ groups.

**Fig. 19.4** The model
$E(Y_{ij}) = \beta_0 + \beta_i + \gamma z_{ij}$ for
$i = 1, 2, 3$

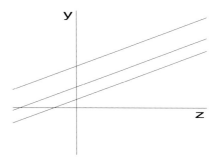

**Fig. 19.5** The general model
$E(Y_{ij}) = \beta_0 + \beta_i + (\gamma_0 + \gamma_i)z_{ij}$
for three groups

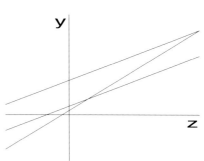

An important thing to note about Fig. 19.4 is that the three regression lines are parallel, which implies that the difference in $E(Y)$ for two different $Z$ values is the same for every group (value of $X$). Alternatively, the difference in $E(Y)$ between two particular groups (values of $X$) is the same for every value of $Z$. As a result, we can colnclude that the distribution of $Y$ depends on both $X$ and $Z$ but there is *no interaction* between $X$ and $Z$.

A linear model *with interaction* allows non-parallel regression lines for the groups, see Fig. 19.5 for an illustration. We can write the model *with interaction* as

$$Y_{ij} = \beta_0 + \beta_i + (\gamma_0 + \gamma_i)z_{ij} + \epsilon_{ij}, \qquad \epsilon_{ij} \overset{\text{ind}}{\sim} N(0, \sigma^2), \qquad (19.6)$$

for $j = 1, \ldots, n_i$,  $i = 1, \ldots, g$, where $\beta$ and $\gamma$ are unknown parameters. There are eight possible models as illustrated below. Figure 19.6 provides a schematic diagram of the eight models.

(a) $\alpha_i = 0, \beta = 0, \gamma_i = 0, \forall i$. Model: $E(Y_{ij}) = \mu$. Dependence on neither x nor z.

(b) $\alpha_i = 0, \beta \neq 0, \gamma_i = 0, \forall i$. Model: $E(Y_{ij}) = \mu + \beta z_{ij}$. Dependence on z only.

(c) $\alpha_i \neq 0, \beta = 0, \gamma_i = 0, \forall i$. Model: $E(Y_{ij}) = \mu + \alpha_i$. Dependence on x only.

(d) $\alpha_i \neq 0, \beta \neq 0, \gamma_i = 0, \forall i$. Model: $E(Y_{ij}) = \mu + \alpha_i + \beta z_{ij}$. Dependence on both x and z but no interaction.

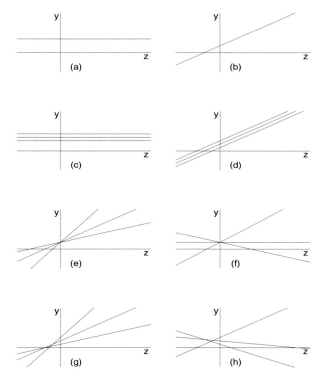

**Fig. 19.6** Eight possible graphs of the model: $E(Y_{ij}) = \mu + \alpha_i + (\beta + \gamma_i)z_{ij}$, $j = 1, \ldots, n_i$, $i = 1, 2, 3$

(e) $\alpha_i = 0$, $\beta = 0$, $\gamma_i \neq 0$, $\forall i$. Model: $E(Y_{ij}) = \mu + \gamma_i z_{ij}$. Dependence on z only, qualitatively same as (f) below.

(f) $\alpha_i = 0$, $\beta \neq 0$, $\gamma_i \neq 0$, $\forall i$. Model: $E(Y_{ij}) = \mu + (\beta + \gamma_i)z_{ij}$. Dependence on z only, qualitatively same as (e) above.

(g) $\alpha_i \neq 0$, $\beta = 0$, $\gamma_i \neq 0$, $\forall i$. Model: $E(Y_{ij}) = \mu + \alpha_i + \gamma_i z_{ij}$. Dependence on x and z with interaction, same as (h) below.

(h) $\alpha_i \neq 0$, $\beta \neq 0$, $\gamma_i \neq 0$, $\forall i$. Model: $E(Y_{ij}) = \mu + \alpha_i + (\beta + \gamma_i)z_{ij}$. Dependence on x and z with interaction, same as (g) above.

The interaction model (19.6) can be expressed in multiple regression from as

$$Y_\ell = \beta_0 + \beta_1 x_{\ell 1} + \beta_2 x_{\ell 2} + \ldots + \beta_g x_{\ell g} +$$
$$\beta_{g+1} z_\ell + \beta_{g+2} x_{\ell 1} z_\ell + \beta_{g+3} x_{\ell 2} z_\ell + \ldots + \beta_{2g+1} x_{\ell g} z_\ell + \epsilon_\ell,$$
$$\epsilon_\ell \overset{\text{ind}}{\sim} N(0, \sigma^2), \ \ell = 1, \ldots, n.$$

Note that the columns of the design matrix for interactions are the *products* of the corresponding dummy variable and the continuous explanatory variable $z$. We can formally compare models with and without interaction. We simply test the hypothesis $\beta_{g+2} = \beta_{g+3} = \ldots = \beta_{2g+1}$ (or equivalently $\gamma_1 = \gamma_2 = \ldots = \gamma_g = 0$).

## 19.6  Possum Data Example Revisited

Return to the possum data example discussed in Sect. 19.2 along with the body weight data $y_{ij}$, region $x_{ij}$ and body length $z_{ij}$ for the $j$th possum caught in the $i$ location, $j = 1, \ldots, n_i$ and $i = 1, \ldots, 7$. For this data we consider five primary models of interest:

$M_0$:  $Y_{ij} = \beta_0 + \epsilon_{ij}$, which states that there are no relationships between $Y$ and the covariate $Z$ and the factor region $X$.

$M_1$:  $Y_{ij} = \beta_0 + \beta_i + \epsilon_{ij}$, which states that there are no relationships between body weight $Y$ and the covariate length but the factor $X$ has an effect in changing the intercept of the regression line $Y$ on $Z$.

$M_2$:  $Y_{ij} = \beta_0 + \gamma z_{ij} + \epsilon_{ij}$, which states that there is a single regression line which is not affected by location.

$M_3$:  $Y_{ij} = \beta_0 + \beta_i + \gamma z_{ij} + \epsilon_{ij}$, which states that the regression lines are parallel with different intercepts due to differences in location.

$M_4$:  $Y_{ij} = \beta_0 + \beta_i + (\gamma_0 + \gamma_i) z_{ij} + \epsilon_{ij}$, which states that there are interaction effects between region $X$ and covariate $Z$.

The ANOVA table for comparing $M_1$ with the reduced model $M_0$ is easily completed by using the summaey statistics provided in Table 19.1. For this data set $n = 101$ (total number of Possums caught), $g = 7$ (number of different possible locations or regions). Now:

$$F = \frac{\frac{1}{6} \sum_{i=1}^{7} n_i (\bar{y}_i - \bar{y})^2}{\frac{1}{94} \sum_{i=1}^{7} (n_i - 1) s_i^2} = \frac{8.567/6}{12.714/94} = 10.56$$

Here the ANOVA table is given by (Table 19.6):

The critical value for a test at the 5% level of significance is 2.2, so there is strong evidence (p-value very close to zero) that the null hypothesis (simpler model) should

**Table 19.6**  ANOVA table for the possum data

| Source | df | Sum of squares (SS) | Mean SS | F-value | P-value |
|---|---|---|---|---|---|
| Location | 6 | 8.567 | 1.428 | $\frac{1.428}{0.135} = 10.56$ | $Pr(F_{6,94} > 10.56) = 6.2 \times 10^{-9}$. |
| Error | 94 | 12.714 | 0.135 | – | – |
| Total | 100 | 21.281 | | | |

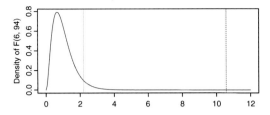

**Fig. 19.7**   F density function and the critical and observed values for the possum example

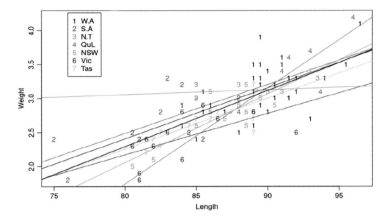

**Fig. 19.8**   Scatter plot of the possum data. Superimposed is the fitted full interaction model as coloured lines. The solid black line is the fitted reduced model

be rejected in favour of the alternative that the mean weights of Possums do depend on their region of origin (Fig. 19.7).

The body lengths of the individual possums are required to compare models $M_2$ and $M_3$ with $M_1$ and $M_0$. Also the required computations are cumbersome using a calculator. Hence, we use R to fit and compare these models. Note that in R an interaction effect between $x$ and $z$ is specified by   $x:z$ in the model formula. The reader is invited to try out the accompanying R code to self study all the model fitting and comparison results. Below we just provide a summary of the results for brevity.

The fitted full interaction model $M_4$ is illustrated in Fig. 19.8, where there are seven different regression lines with different slopes and intercepts for seven different regions. A summary of the fitted model however does not show the interaction term to be significant. Hence, we drop the interaction term and fit model $M_3$ instead. A summary of the fitted model $M_3$ shows that both $x$ and $z$ are significant and, hence $M_3$ is our preferred model for this data. A simpler model $M_2$ does not provide an adequate fit for the data.

**The following code lines can be found in the package `ipsRdbs` by issuing the command** `?possum`.

```r
n <- length(possum$Body_Weight)
# Wrong model since Location is not a numeric covariate
wrong.lm <- lm(Body_Weight ~ Location, data=possum)
summary(wrong.lm)


nis <- table(possum$Location)
datasums <- data.frame(nis=nis, mean=meanwts, var=varwts)
modelss <- sum(datasums[,2] * (meanwts - mean(meanwts))^2)
residss <- sum( (datasums[,2] - 1) * varwts)


fvalue <- (modelss/6) / (residss/94)
fcritical <- qf(0.95, df1= 6, df2=94)
x <- seq(from=0, to=12, length=200)
y <- df(x, df1=6, df2=94)

#postscript("fdensity.eps", horiz=F)
plot(x, y, type="l", xlab="", ylab="Density of F(6, 94)", col=4)

abline(v=fcritical, lty=3, col=3)
abline(v=fvalue, lty=2, col=2)

pvalue <- 1-pf(fvalue, df1=6, df2=94)

### Doing the above in R

# Convert the Location column to a factor
possum$Location <- as.factor(possum$Location)
summary(possum) Now Location is not a factor

# Put the identifiability constraint:
options(contrasts=c("contr.treatment", "contr.poly"))


colnames(possum) <- c("y", "z", "x", "r")

# Fit model M1
possum.lm1 <- lm(y ~ x, data=possum)
summary(possum.lm1)
anova(possum.lm1)

possum.lm2 <- lm(y ~ z, data=possum)
summary(possum.lm2)
anova(possum.lm2)

# Include both location and length but no interaction
possum.lm3 <- lm(y ~ x+z, data=possum)
summary(possum.lm3)
```

```
anova(possum.lm3)

# Include interaction effect
possum.lm4 <- lm(y~x+z+x:z, data=possum)
summary(possum.lm4)
anova(possum.lm4)

anova(possum.lm2, possum.lm3)


#Check the diagnostics for M3
plot(possum.lm3$fit, possum.lm3$res,xlab="Fitted values",ylab="
    Residuals", main="Anscombe plot")
abline(h=0)
qqnorm(possum.lm3$res,main="Normal probability plot")
qqline(possum.lm3$res)
```

## 19.7  Exercises

### 19.1 (Problems of Chap. 19)

1. The following table contains 10 claim amounts for repair costs arising from a particular type of storm damage to private houses, for each of four different postcode regions (denoted A, B, C and D):

|                  |     A    |     B    |     C    |     D    |
|------------------|----------|----------|----------|----------|
|                  |   961    |  1507    |  1303    |  1022    |
|                  |  1263    |  1349    |   959    |   997    |
|                  |  1304    |  1521    |  1297    |  1335    |
|                  |  1532    |  1134    |  1051    |  1216    |
|                  |  1294    |  1293    |  1163    |  1277    |
|                  |  1605    |   993    |   993    |  1135    |
|                  |  1308    |  1126    |   978    |  1273    |
|                  |  1393    |  1140    |   891    |  1244    |
|                  |  1255    |  1305    |  1177    |  1105    |
|                  |  1131    |  1224    |  1153    |  1524    |
| Sums             | 13046    | 12592    | 10965    | 12128    |
| Sums of squares  | 17322090 | 16116822 | 12208021 | 14929994 |

(a) Do the data provide significant evidence (at the 5% level of significance) of a difference between the regions in the magnitude of claims?
(b) Based on your answer to (a) provide predictions and associated 95% confidence intervals for (i) the mean claim level in region A, (ii) a single future claim in region C.

2. The data set `cement` in `ipsRdbs` represents the results of an experiment to investigate sources of variability in testing the strength (in pounds per square inch) of Portland cement. The cement was 'gauged' (mixed with water and worked for a fixed time) by three different gaugers, and then cast into cubes. Each gauger gauged 12 cubes which were then divided into three sets of 4, for testing by three 'breakers'.

(a) By fitting linear models to the data, investigate how variability in the test results is influenced by differences between gaugers and between breakers. What is your conclusion?

(b) Predict the result of a future test (i.e. future observation) carried out by gauger 2 and breaker 3, and give a 95% confidence interval for your prediction.

|  | Breaker 1 | | Breaker 2 | | Breaker 3 | |
|---|---|---|---|---|---|---|
| Gauger 1 | 5280 | 5520 | 4340 | 4400 | 4160 | 5180 |
| | 4760 | 5800 | 5020 | 6200 | 5320 | 4600 |
| Gauger 2 | 4420 | 5280 | 5340 | 4880 | 4180 | 4800 |
| | 5580 | 4900 | 4960 | 6200 | 4600 | 4480 |
| Gauger 3 | 5360 | 6160 | 5720 | 4760 | 4460 | 4930 |
| | 5680 | 5500 | 5620 | 5560 | 4680 | 5600 |

3. Suppose that we have recorded the time to death $Y_i, i = 1, \ldots, 6$ of six rats exposed to three different types of poison, I, II and III. Without loss of generality, assume that the first two rats received poison type I, rats 3 and 4 received poison type II and the last two rats received the third poison type. Consider the linear model

$$Y_i = \mu + \alpha(p_i) + \epsilon_i \qquad i = 1, \ldots, 6,$$

where

$$\alpha(p_i) = \begin{cases} \alpha_1 & \text{if the } i\text{th rat received poison type I,} \\ \alpha_2 & \text{if the } i\text{th rat received poison type II,} \\ \alpha_3 & \text{if the } i\text{th rat received poison type III.} \end{cases}$$

(a) Write down the above linear model as

$$\mathbf{Y} = X\boldsymbol{\beta} + \boldsymbol{\epsilon}$$

where $X$ is a matrix order $6 \times 4$, $\boldsymbol{\beta} = (\mu, \alpha_1, \alpha_2, \alpha_3)^T$. Write down the $X$ matrix and obtain the $X^T X$ matrix. Provide arguments for the fact that $X^T X$ is singular.

(b) Assume $\alpha_1 = 0$. Remove the second column of the $X$ matrix, call the resulting $6 \times 3$ matrix $X_1$ and write $\boldsymbol{\beta}_1 = (\mu, \alpha_2, \alpha_3)^T$. Obtain $\hat{\boldsymbol{\beta}}_1 =$

$(X_1^T X_1)^{-1} X_1^T \mathbf{y}$ explicitly as linear combinations of $y_1, y_2, \ldots, y_6$. Obtain a simplified expression for the residual sum of squares.

(c) Assume $\alpha_2 = 0$. Remove the third column of the $X$ matrix, call the resulting $6 \times 3$ matrix $X_2$ and write $\boldsymbol{\beta}_2 = (\mu, \alpha_1, \alpha_3)^T$. Obtain $\hat{\boldsymbol{\beta}}_2 = (X_2^T X_2)^{-1} X_2^T \mathbf{y}$ explicitly as linear combinations of $y_1, y_2, \ldots, y_6$. Obtain a simplified expression for the residual sum of squares and show that this is same as the one you obtained in part (b). Explain why this can be expected to happen.                                                                                    □

# Solutions to Selected Exercises

# 20

This chapter provides solutions to selected exercises. The abbreviation NSP next to an exercise implies that No Solutions have been Provided for that exercise.

## Problems of Chap. 1

### 1.1 Learning to add with the summation sign $\sum$

1. Prove that $\sum_{i=1}^{n} k\, x_i = k \sum_{i=1}^{n} x_i$.

$$\sum_{i=1}^{n} k\, x_i = k\, x_1 + k\, x_2 + \cdots + k\, x_n$$
$$= k(x_1 + x_2 + \cdots + x_n)$$
$$= k \sum_{i=1}^{n} x_i.$$

2. NSP
3. NSP
4. Prove that $\sum_{i=1}^{n} (x_i - \bar{x})^2 = \sum_{i=1}^{n} x_i^2 - n\bar{x}^2$.

$$\sum_{i=1}^{n} (x_i - \bar{x})^2 = \sum_{i=1}^{n} \left( x_i^2 - 2x_i\bar{x} + \bar{x}^2 \right)$$
$$= \sum_{i=1}^{n} x_i^2 - 2\bar{x} \left( \sum_{i=1}^{n} x_i \right) + n\bar{x}^2$$
$$= \sum_{i=1}^{n} x_i^2 - 2\bar{x}\, (n\bar{x}) + n\bar{x}^2$$
$$= \sum_{i=1}^{n} x_i^2 - n\bar{x}^2.$$

The original version of this chapter has been revised. A correction to this chapter can be found at https://doi.org/10.1007/978-3-031-37865-2_21

5. NSP

**1.2**

1. NSP
2. NSP

**1.3**

1.

$$\sum_{i=1}^{n}(x_i - \overline{x})^2 = \sum_{i=1}^{n} x_i^2 - n\overline{x}^2$$

$$= \sum_{i=1}^{n} x_i^2 - n\left[\frac{1}{n}\sum_{i=1}^{n} x_i\right]^2$$

$$= \sum_{i=1}^{n} x_i^2 - \frac{1}{n}\left[\sum_{i=1}^{n} x_i\right]^2$$

Thus $\sum_{i=1}^{n}(x_i - \overline{x})^2 \geq 0$ implies

$$\sum_{i=1}^{n} x_i^2 \geq \frac{1}{n}\left[\sum_{i=1}^{n} x_i\right]^2.$$

**1.4**

1. Here

$$\overline{x}_{n+1} = \frac{x_1 + \cdots + x_n + x_{n+1}}{n+1}$$

$$= \frac{n\overline{x}_n + x_{n+1}}{n+1}.$$

This proves the first part. Now,

$$nS_{n+1}^2 = \sum_{i=1}^{n+1}(x_i - \overline{x}_{n+1})^2$$

$$= \sum_{i=1}^{n+1}\left(x_i - \frac{n\overline{x}_n + x_{n+1}}{n+1}\right)^2$$

$$= \frac{1}{(n+1)^2} \sum_{i=1}^{n+1} ((n+1)x_i - n\bar{x}_n - x_{n+1})^2$$

$$= \frac{1}{(n+1)^2} \sum_{i=1}^{n+1} ((n+1)x_i - (n+1)\bar{x}_n + \bar{x}_n - x_{n+1})^2$$

$$= \frac{1}{(n+1)^2} \sum_{i=1}^{n+1} \Big((n+1)^2(x_i - \bar{x}_n)^2 + (\bar{x}_n - x_{n+1})^2$$

$$+2(n+1)(x_i - \bar{x}_n)(\bar{x}_n - x_{n+1}))$$

$$= \sum_{i=1}^{n+1} (x_i - \bar{x}_n)^2 + \frac{n+1}{(n+1)^2}(\bar{x}_n - x_{n+1})^2$$

$$+ \frac{2(n+1)}{(n+1)^2}(\bar{x}_n - x_{n+1}) \sum_{i=1}^{n+1}(x_i - \bar{x}_n)$$

$$= \sum_{i=1}^{n+1} (x_i - \bar{x}_n)^2 + \frac{1}{n+1}(\bar{x}_n - x_{n+1})^2 + \frac{2}{n+1}(\bar{x}_n - x_{n+1})(x_{n+1} - \bar{x}_n)$$

$$= \sum_{i=1}^{n+1} (x_i - \bar{x}_n)^2 + \frac{1}{n+1}(\bar{x}_n - x_{n+1})^2 - \frac{2}{n+1}(\bar{x}_n - x_{n+1})^2$$

$$= \sum_{i=1}^{n} (x_i - \bar{x}_n)^2 + (x_{n+1} - \bar{x}_n)^2 - \frac{1}{n+1}(\bar{x}_n - x_{n+1})^2$$

$$= (n-1)S_n^2 + \frac{n}{n+1}(x_{n+1} - \bar{x}_n)^2,$$

as required.

2. Note that: $\sum_{i=1}^{m} x_i = m\bar{x}$ and $\sum_{j=1}^{n} y_j = n\bar{y}$. Now:

$$\bar{z} = \frac{1}{m+n}(z_1 + z_2 + \cdots + z_{m+n})$$

$$= \frac{1}{m+n}(x_1 + x_2 + \cdots x_m + y_1 + \cdots + y_n)$$

$$= \frac{1}{m+n}(m\bar{x} + n\bar{y})$$

3.

$$(m+n-1)s_z^2 = \sum_{k=1}^{m+n} (z_k - \bar{z})^2$$

$$= \sum_{i=1}^{m} (x_i - \bar{z})^2 + \sum_{j=1}^{n} (y_j - \bar{z})^2$$

Now substitute $\bar{z} = \frac{1}{m+n}(m\bar{x} + n\bar{y})$ in the above and simplify.

**1.5**

1. NSP
2. Note that

$$\sum_{i=1}^{n}(a_i - b_i)^2 = \sum_{i=1}^{n} a_i^2 + \sum_{i=1}^{n} b_i^2 - 2 \sum_{i=1}^{n} a_i b_i$$

With the given hint we have,

$$\sum_{i=1}^{n} a_i^2 = \sum_{i=1}^{n} \left( \frac{x_i}{\sqrt{\sum_{i=1}^{n} x_i^2}} \right)^2$$

$$= \sum_{i=1}^{n} \frac{x_i^2}{\sum_{i=1}^{n} x_i^2}$$

$$= \frac{1}{\sum_{i=1}^{n} x_i^2} \sum_{i=1}^{n} x_i^2 \quad \text{since} \sum_{i=1}^{n} x_i^2 \text{ is free of } i$$

$$= 1.$$

Similarly, $\sum_{i=1}^{n} b_i^2 = 1$. Now,

$$\sum_{i=1}^{n} a_i b_i = \sum_{i=1}^{n} \frac{x_i}{\sqrt{\sum_{i=1}^{n} x_i^2}} \frac{y_i}{\sqrt{\sum_{i=1}^{n} y_i^2}}$$

$$= \frac{\sum_{i=1}^{n} x_i y_i}{\sqrt{\sum_{i=1}^{n} x_i^2} \sqrt{\sum_{i=1}^{n} y_i^2}}.$$

Now:

$$\sum_{i=1}^{n}(a_i - b_i)^2 = 2 - 2 \frac{\sum_{i=1}^{n} x_i y_i}{\sqrt{\sum_{i=1}^{n} x_i^2} \sqrt{\sum_{i=1}^{n} y_i^2}} \geq 0$$

implies

$$\frac{\sum_{i=1}^{n} x_i y_i}{\sqrt{\sum_{i=1}^{n} x_i^2} \sqrt{\sum_{i=1}^{n} y_i^2}} \leq 1$$

$$\implies \left(\sum_{i=1}^{n} x_i y_i\right)^2 \leq \left(\sum_{i=1}^{n} x_i^2\right)\left(\sum_{i=1}^{n} y_i^2\right).$$

This proves the stated version of the famous Cauchy-Schwarz inequality in Mathematics.

# Problems of Chap. 2

## 2.1

1.(a) The following are the R code lines for the data set `cfail`.

```
summary(cfail)
mean(cfail) ## 3.75
median(cfail) ## 3
# First find the frequency table.
table(cfail)
# Find the mode by inspection.
# Mode is the number which has the maximum frequency.
var(cfail) ## 11.43204
table(cfail)
hist(cfail)
boxplot(cfail)
```

The histogram shows a long right tail for the data with an isolated observation at 17, The box plot indicates a fairly symmetric central box, i.e. the upper and lower quartiles are equidistant from the median. But it also shows a long right tail with three observations which are beyond the top whisker.

(b)
```
plot(cfail, type="l", xlab="Week", ylab="Number of weekly
    computer failure
s in two years")
title("Time series plot of weekly number of computer
    failure over two year
s")
```

The plot shows few excursions to large number of failures in particular weeks. There may well be a time pattern in these excursions but we need further calendar time information to make any such conclusion.

(c) The following are the R codes for the two years in cfail.

```
year <- rep(c(1,2), each = length(cfail)/2)
tapply(X=cfail, INDEX=year, FUN=mean)
## Year 1 = 3.826923; Year 2 = 3.673077
tapply(X=cfail, INDEX=year, FUN=var)
## Year 1 = 14.930241; Year 2 = 8.145928
```

2. The following are the R code lines to explore the billionaires data set.
   (a) `summary(bill)`

   Summary on the column wealth reveals that the wealth varies from 1 to 37 billion dollars with the mean at 2.726 and median at 1.8. Summary on age reveals that the billionaires were between the ages of 7 and 102 with a mean of about 64. The table command on region shows that the number of billionaires were 37, 76, 22, 28 and 62 respectively in Asia, Europe, Middle-East, Other regions and USA respectively.

(b)

```
attach(bill)
tapply(X = wealth, INDEX = region, FUN = mean)
## A = 2.651; E = 2.258; M = 4.264; O = 2.279; U = 3.000
tapply(X = wealth, INDEX = region, FUN = var)
## A = 4.808; E = 2.635; M = 58.632; O = 1.601; U = 13.395
## Answers are rounded to 3 d.p. using the round function,
    e.g.
round(tapply(X = wealth, INDEX = region, FUN = mean), 3
```

(c) Include output of the command: **boxplot**(wealth region, **data**
=bill).

(d) Include output of the command: **plot**(bill$age, bill$wealth,
ylab="wealth in billions of US dollars").

**2.2** Here are the commands:

```
errors <- read.csv("2019ageguess.csv", head=T)
head(errors)
tail(errors)
errors
# Q1: How many rows and columns are there in the data set?
dim(errors)

# Q2: Number of students in class
sum(errors$size)/10 ## Is the total number of students
sum(errors$females)/10 ## Is the total number of femals
sum(errors$size)/10 - sum(errors$females)/10 # number of male
    students

# Q3: Note down the number of photographed person for each
    unique value of age.
table(errors$tru_age)/55
## divided by 55 since each of the 55 groups guessed the
    photos

# Q4: Cross tabulation
table(errors$sex, errors$race)/55
## divided by 55 since each of the 55 groups guessed the
    photos

# Q5: What are the minimum and maximum true ages of the
    photographed mathematicians?
# alternatively can find the minimums and maximums this way
min(errors$tru_age)
max(errors$tru_age)

# Q6. Obtain a barplot of the true age distribution.
barplot(table(errors$tru_age)/55)

# Q7: Obtain a histogram of the estimated age column and
    compare this with the true age distribution seen in the
    barplot drawn above.
hist(errors$est_age)
```

```r
# Q8: Plot estimated age against true ages

plot(errors$tru_age, errors$est_age)
plot(errors$tru_age, errors$est_age, pch="*", las=1)
abline(a=0, b=1, col=2)
abline(v=41, col=5)
# plot(x, y)
# Q9. What are the means and standard deviations for the
#     columns: size, females, est age, tru age, error and abs
#     error?
summary(errors)

# or alternatively
mean(errors$size)
mean(errors$females)
mean(errors$est_age)
mean(errors$tru_age)
mean(errors$error)
mean(errors$abs_error)
# For sd s
sd(errors$size)
sd(errors$females)
sd(errors$est_age)
sd(errors$tru_age)
sd(errors$error)
sd(errors$abs_error)

# Q10: What is the mean number of males in each group? What
#     is the mean number of females in each group?
# mean number of males in each group - first, create a new
#     variable containing the number of males in a group
errors$males <- errors$size - errors$female
head(errors) # to see the first few rows of the data set
#     containing the new variable
mean(errors$males)

# mean number of females in each group
mean(errors$females)

# Q11. How many of the photographs were of each race?
table(errors$race)/55

# Q12. Note down the frequency table of the sign of the
#     errors.
errors$sign <- sign(errors$error)
table(errors$sign)

# Q13. Obtain a histogram for the errors and another for the
#     absolute errors. Which one is bellshaped and why?
hist(errors$error)
hist(errors$abs_error)
```

```
# Q14. Obtain a histogram for the square-root of absolute
    errors.
hist(sqrt(errors$abs_error))


# Q15. Draw a boxplot
boxplot(errors$abs_error)
# IQR = Q3 - Q1
summary(errors$abs_error)

# Q16. Is it easier to guess the ages of female
    mathematicians?
errors$sex <- factor(errors$sex) # make sex a factor variable
levels(errors$sex)
tapply(X=errors$abs_error, INDEX=errors$sex, FUN =mean)


#Also draw a side by side boxplot of the absolute errors for
    the two groups of mathematicians: males and females.
boxplot(data=errors, abs_error~sex, col=c(2,4) )
tapply(X=errors$abs_error, INDEX=errors$sex, FUN =median)


# Q17. Is it easier to guess the ages of black mathematicians
    ?
tapply(X=errors$abs_error, INDEX=errors$race, FUN =mean)

# Q18 How would you order the mean absolute error by race?
boxplot(data=errors, abs_error~race, col=c(7, 8, 5, 0) )

# Q19. Is it easier to guess the ages of younger
    mathematicians?
tapply(X=errors$abs_error, INDEX=errors$tru_age, FUN =mean)

# Q20. Which person's age is the most difficult to guess?
boxplot(data=errors, abs_error~photo, col=heat.colors(8))

errors$morefemale <- sign(errors$females - errors$males)
boxplot(data=errors, abs_error~morefemale, col=heat.colors(8)
    )
table(errors$morefemale)
```

# Problems of Chap. 3

## 3.1

1. b.
2. NSP
3. Answer: $\frac{2}{3}$

4. Answer: 0.3.
5. Answer: 0.2.
6. Answer: 0.3.

**3.2**

1. NSP
2. • PASS: Answer: $\frac{4!}{2!} = 12$.
   • STATISTICS: Answer: $\frac{10!}{3!3!2!}$.
   • EXAM: Answer: $4!$.
3. $12 \times 9 = 108$.
4. (a) $\binom{25}{6} = 177100$
   (b) Number of ways of selecting 3 from the 10 Japanese cars $= \binom{10}{3}$.
   Number of ways of selecting 3 from the 15 European cars $= \binom{15}{3}$.
   By product rule, number of ways of selecting 3 Japanese and 3 European
   $= \binom{10}{3}\binom{15}{3} = 120 \times 455 = 54600$.
   (c) Probability of event is $\frac{54600}{177100} = 0.308$ (to 3 d.p)
5. NSP
6. A hand of 7 cards is dealt from a pack of 52. Find the probability that the pack
   contains:
   (a) 4 spades ♠ and 3 hearts ♡.
   Answer $= \frac{^{13}C_4 \times ^{13}C_3}{^{52}C_7}$.
   (b) 3 spades ♠ and 1 heart ♡.
   Answer $= \frac{^{13}C_3 \times ^{13}C_1 \times ^{26}C_3}{^{52}C_7}$.
   (c) 2 Aces.
   Answer $= \frac{^4C_2 \times ^{48}C_5}{^{52}C_7}$.
   (d) exactly one Ace and one King of the same suit.
   Answer $= \frac{^4C_1 \times ^{44}C_5}{^{52}C_7}$.
7. NSP
8.

$$\text{Answer:} ^9C_6 \times ^7C_4 \times ^2C_1.$$

9. Simplify: (i) $\frac{(n+1)!}{(n-1)!} = \frac{(n+1)n(n-1)!!}{(n-1)!} = (n+1)n$.
   (ii) $\frac{(n-2)!}{n!} = \frac{1}{n(n-1)}$.
10. Factorise: $n! + (n+1)! = n! + (n+1)n! = n!(n+2)$.
11. (a) $\binom{n}{n-k} = \frac{n!}{(n-k)!(n-[n-k])!} = \frac{n!}{(n-k)!k!} = \binom{n}{k}$ Explanation: The
    number of selections (without replacement) of $k$ objects from $n$ is exactly the
    same as the number of selections of $(n-k)$ objects from $n$.

(b)

$$RHS = \frac{n!}{k!(n-k)!} + \frac{n!}{(k-1)!(n-[k-1])!}$$

$$= \frac{n!}{k!(n-[k-1])!}[n-(k-1)+k]$$

$$= \frac{n![n+1]}{k!(n-[k-1])!}$$

$$= \frac{(n+1)!}{k!(n+1-k)!}$$

$$= LHS$$

Explanation: The number of selections of $k$ items from $(n+1)$ consists of:

- The number of selections that include the $(n+1)^{th}$ item. There are $\binom{n}{k-1}$ of these.
- The number of selections that exclude the $(n+1)^{th}$ item. There are $\binom{n}{k}$ of these.

12. NSP

13.

$$P(\text{more boys than girls}) = P(3 \text{ boys}) + P(4 \text{ boys})$$

$$= \frac{^5C_3 \times {}^4C_1 + {}^5C_4 \times {}^4C_0}{^9C_4}$$

$$= \frac{40+5}{126} = \frac{45}{126}.$$

## Problems of Chap. 4

### 4.1

1. Here we have $P(A) = 0.8$ and $P(B|A) = 0.5 = \frac{P(B \cap A)}{P(A)}$. Hence, $P(A \cap B) = P(A) \times P(B \cap A) = 0.4$.

2. $P(A|B) == \frac{P(A \cap B)}{P(B)} = \frac{0.1}{0.6} = \frac{1}{6}$.

3. NSP

4.

5.

$$P(B|A) = 4P(A) \Rightarrow P(A \text{ and } B) = P(B|A)P(A) = 4[P(A)]^2$$

$$P(A|B) = 9P(B) \Rightarrow P(A \text{ and } B) = P(A|B)P(B) = 9[P(B)]^2$$

$$\therefore 4[P(A)]^2 = 9[P(B)]^2$$
$$2P(A) = 3P(B)$$
$$P(B) = \frac{2}{3}P(A)$$

$$P(A \text{ or } B) = \frac{7}{48} = P(A) + P(B) - P(A \text{ and } B)$$
$$= P(A) + \frac{2}{3}P(A) - 4[P(A)]^2$$
$$4[P(A)]^2 - \frac{5}{3}P(A) + \frac{7}{48} = 0$$
$$\left(4P(A) - \frac{7}{6}\right)\left(P(A) - \frac{1}{8}\right) = 0$$
$$\Rightarrow P(A) = \frac{7}{24} \text{ or } P(A) = \frac{1}{8}$$

To decide which $P(A)$ value is correct, notice that

$$P(A \text{ and } B) = 4[P(A)]^2 \le P(A)$$
$$\frac{[P(A)]^2}{P(A)} \le \frac{1}{4}$$
$$P(A) \le \frac{1}{4}$$

Hence, $P(A)$ must be $\frac{1}{8}$.

[There are other ways of identifying the correct P(A) solution E.g. If $P(A) = \frac{7}{24}$, then $P(B) = \frac{2}{3} \times P(A) = \frac{7}{36}$ and $P(A|B) = 9 \times P(B) = 9 \times \frac{7}{36} \ge 1$ - This is impossible! So this contradiction shows us that $P(A) = \frac{1}{8}$.]

**4.2**

1. Define events

$GG$ : the chosen drawer has two gold coins,

$GS$ : the chosen drawer has one gold and one silver coin,

$SS$ : the chosen drawer has two silver coins,

$S$ : the coin chosen is silver.

We require $P(GS|S)$. By the Bayes Theorem

$$P(GS|S) = \frac{P(S|GS)P(GS)}{P(S|GG)P(GG) + P(S|GS)P(GS) + P(S|SS)P(SS)}$$

$$= \frac{\frac{1}{2} \times \frac{1}{3}}{0 \times \frac{1}{3} + \frac{1}{2} \times \frac{1}{3} + 1 \times \frac{1}{3}} = \frac{1}{3}.$$

2. NSP
3. We assume that the sexes of the children are independent.
   (a)

$$P(\text{both boys}|\text{older is a boy}) = \frac{P(\text{both boys and older is a boy})}{P(\text{older is a boy})}$$

$$= \frac{P(\text{both boys})}{P(\text{older is a boy})} = \frac{1/4}{1/2} = \frac{1}{2}.$$

   (b)

$$P(\text{both boys}|\text{at least one boy}) = \frac{P(\text{both boys and at least one boy})}{P(\text{at least one boy})}$$

$$= \frac{P(\text{both boys})}{1 - P(\text{both girls})} = \frac{1/4}{1 - 1/4} = \frac{1}{3}.$$

4. NSP
5. NSP
6. Here $k = 2$ in the Bayes theorem and let $B_1$ be the event that a randomly chosen person has the disease and $B_2$ is the complement of $B_1$. Let $A$ be the event that a randomly chosen person has the symptom. The problem is to determine $P(B_1|A)$.
   We have $Pr(B_1) = 0.01$ since 1% of the population has the disease, and $P(A|B_1) = 0.98$.
7. Also $P(B_2) = 0.99$ and $Pr(A|B_2) = 0.001$, this is the probability of having the symptom without having the disease. Now $P(\text{disease} \mid \text{symptom}) =$

$$P(\text{D} \mid \text{S}) = Pr(B_1|A) = \frac{Pr(A|B_1)\,Pr(B_1)}{Pr(A|B_1)\,Pr(B_1) + Pr(A|B_2)\,Pr(B_2)}$$

$$= \frac{0.98 \times 0.01}{0.98 \times 0.01 + 0.001 \times 0.99}$$

$$= 0.9082.$$

Thus, $Pr(\text{disease}) = Pr(B_1) = 0.01$ gets revised to: $P(\text{disease} \mid \text{symptom}) = Pr(B_1|A) = 0.9082$.
8. NSP

**4.3**

1.

$$P(\text{problem not solved}) = P(\text{Jane fails and Alice fails})$$
$$= P(\text{Jane fails})\,P(\text{Alice fails}) \text{ (by independence)}$$
$$= (1 - 0.4)(1 - 0.3)$$
$$= 0.42.$$

Hence $P(\text{problem solved}) = 0.58$.

2. Let $S$ be a sample space and $A, B \subset S$ two events. The events $A$ and $B$ are independent iff $P(A \cap B) = P(A)P(B)$.

(a) If $P(A) = 1$ then $P(A \cap B) = P(B)$.
From the axioms of probability we know if $P(A) = 1$ then $P(A') = 0$. Now

$$P(B) = P(B \cap S) = P(B \cap (A \cup A')) = P((B \cap A) \cup (B \cap A')).$$

The events $B \cap A$ and $B \cap A'$ are mutually exclusive and hence by the axiom of probability (A3)

$$P((B \cap A) \cup (B \cap A')) = P(B \cap A) + P(B \cap A').$$

Finally, since $P(A') = 0$ and $A' \cap B \subseteq A'$, it follows $P(A' \cap B) = 0$ and therefore

$$P(B) = P(A \cap B).$$

(b) Any event $A$ with $P(A) = 0$ or $P(A) = 1$ is independent of every event $B$.

If $P(A) = 1$ then from above $P(A \cap B) = P(B)$ and hence $P(A \cap B) = P(A)P(B)$.
If $P(A) = 0$ then since $A \cap B \subseteq A$, we have $P(A \cap B) = 0$ and therefore $P(A \cap B) = P(A)P(B)$.

(c) If the events $A$ and $B$ are independent, then every pair of events $A'$ and $B$; $A$ and $B'$; $A'$ and $B'$ are independent.
Assume $A$ and $B$ are independent events, i.e. $P(A \cap B) = P(A)P(B)$. We want to show that $A'$ and $B$ are independent.
Consider $P(B) = P(B \cap (A \cup A')) = P((B \cap A) \cup (B \cap A'))$. The events $B \cap A$ and $B \cap A'$ are mutually exclusive and hence by the axiom of probability (A3)

$$P((B \cap A) \cup (B \cap A')) = P(B \cap A) + P(B \cap A'),$$

leading to $P(B) = P(B \cap A) + P(B \cap A')$. Using the independence of the events $A$ and $B$ we have $P(B) = P(B)P(A) + P(B \cap A')$. Now rearranging the equation we get

$$P(B \cap A') = P(B) - P(B)P(A) = (1 - P(A))P(B) = P(A')P(B)$$

as required. The other claims follow applying the proved result, using symmetry of the definition of independence and $(A')' = A$.

3. Water system operates if $A$ and $B$ operate or $C$ operates or all three operate.

P(A operates) = 0.90, P(B operates) = 0.90, P(C operates) = 0.95,
P(AB path operates) = P(A and B operate) = 0.9 × 0.9 = 0.81
P(AB path fails to operate) = 1–0.81 = 0.19
P(C fails to operate) = 1–0.95 = 0.05
P(system fails to operate)= P(AB path fails and C fails)
= P(AB path fails) P(C fails)
= 0.19 × 0.05
= 0.0095
P(system operates) = 1–0.0095 = 0.9905
Reliability of water system is 0.9905 or 99.05%.

4. NSP

## Problems of Chap. 5

### 5.1

1. There are six words in the sentence and there is exactly one word having $X = i$ number of words for $i = 1, \ldots, 6$. Hence the probability distribution is:

| $x$ | 1 | 2 | 3 | 4 | 5 | 6 | Total |
|-----|---|---|---|---|---|---|-------|
| $f(x)$ | $\frac{1}{6}$ | $\frac{1}{6}$ | $\frac{1}{6}$ | $\frac{1}{6}$ | $\frac{1}{6}$ | $\frac{1}{6}$ | 1 |

2. NSP
3. The sample space is

$$(1, 1)\ (1, 2)\ (1, 3)\ (1, 4)\ (1, 5)\ (1, 6)$$
$$(2, 1)\ (2, 2)\ (2, 3)\ (2, 4)\ (2, 5)\ (2, 6)$$
$$(3, 1)\ (3, 2)\ (3, 3)\ (3, 4)\ (3, 5)\ (3, 6)$$
$$(4, 1)\ (4, 2)\ (4, 3)\ (4, 4)\ (4, 5)\ (4, 6)$$
$$(5, 1)\ (5, 2)\ (5, 3)\ (5, 4)\ (5, 5)\ (5, 6)$$
$$(6, 1)\ (6, 2)\ (6, 3)\ (6, 4)\ (6, 5)\ (6, 6)$$

(a) Working along the cross-diagonals we find by enumeration that $X$ has the following probability function

| $x$ | 2 | 3 | 4 | 5 | 6 | 7 | 8 | 9 | 10 | 11 | 12 |
|---|---|---|---|---|---|---|---|---|---|---|---|
| $p_x$ | $\frac{1}{36}$ | $\frac{2}{36}$ | $\frac{3}{36}$ | $\frac{4}{36}$ | $\frac{5}{36}$ | $\frac{6}{36}$ | $\frac{5}{36}$ | $\frac{4}{36}$ | $\frac{3}{36}$ | $\frac{2}{36}$ | $\frac{1}{36}$ |

More concisely,

$$p_x = \begin{cases} \frac{6-|x-7|}{36} & \text{if } x = 2, \ldots, 12 \\ 0 & \text{otherwise.} \end{cases}$$

(b)

$$F(x) = \begin{cases} 0 & \text{if } x < 2 \\ \frac{1}{36} & \text{if } 2 \leq x < 3 \\ \frac{3}{36} & \text{if } 3 \leq x < 4, \text{ etc.} \end{cases}$$

Using the formula $\sum_{i=1}^{n} i = \frac{1}{2}n(n+1)$ we find that the cumulative distribution function $F(x)$ can be written concisely in the form

$$F(x) = \begin{cases} 0 & \text{if } x < 2 \\ \frac{(6+[x-7])(7+[x-7])}{72} & \text{if } 2 \leq x < 7 \\ \frac{21}{36} + \frac{[x-7](11-[x-7])}{72} & \text{if } 7 \leq x < 12 \\ 1 & \text{if } x \geq 12, \end{cases}$$

where $[x]$ denotes the integral part of $x$.

For example, $F(3.5) = \frac{(6-4)(7-4)}{72} = \frac{3}{36}$, $F(10.5) = \frac{21}{36} + \frac{3(11-3)}{72} = \frac{33}{36}$.

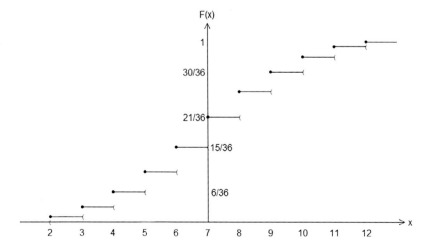

4. NSP

5. To find $k$ we have:

$$
\begin{aligned}
F(2) &= 1 \\
k(2^4 + 2^2 - 2) &= 1 \\
\implies k &= \tfrac{1}{18}
\end{aligned}
$$

6.

$$
\int_0^1 kx^2 dx = \frac{k}{3} = 1 \implies k = 3.
$$

7. (a) $k \int_0^1 x^2(1-x)\,dx = 1$, which implies that $k = 12$.

(b) $P(0 < X < \tfrac{1}{2}) = 12 \int_0^{1/2} x^2(1-x)\,dx = \tfrac{5}{16}$.

8. (a) We must show that $\int_{-\infty}^{\infty} f(x)\,dx = 1$ and $f(x) \geq 0$ for all $x$.

Now $\int_{-\infty}^{\infty} \frac{dx}{\pi(1+x^2)} = \frac{1}{\pi}[\tan^{-1}(x)]_{-\infty}^{\infty} = \frac{1}{\pi}(\frac{\pi}{2} - (\frac{-\pi}{2})) = 1$.

Also $\frac{1}{\pi(1+x^2)} \geq 0$ for all $x$.

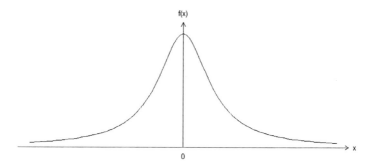

f(x)

0                                              x

(b) $F(x) = \frac{1}{\pi} \int_{-\infty}^{x} \frac{dy}{1+y^2} = \frac{1}{\pi}[\tan^{-1}(y)]_{-\infty}^{x} = \frac{1}{\pi}(\tan^{-1}(x) + \frac{\pi}{2})$.

(c) $P(-1 \leq X \leq 1) = F(1) - F(-1) = \frac{1}{\pi}(\tan^{-1}(1) + \frac{\pi}{2}) - \frac{1}{\pi}(\tan^{-1}(-1) + \frac{\pi}{2}) = \frac{1}{2}$.

9. (a) Since $f(x) = \frac{dF(x)}{dx}$, we have

$$
f(x) = \begin{cases} e^{x-\theta} & \text{for } x \leq \theta \\ 0 & \text{for } x > \theta. \end{cases}
$$

(b)  When $X \leq \theta$ we have $Y \geq 0$. The cdf of $Y = \theta - X$ is given by:

$$
\begin{aligned}
G(y) &= P(Y \leq y) \\
&= P(\theta - X \leq y) \\
&= P(X \geq \theta - y) \\
&= 1 - P(X < \theta - y) \\
&= 1 - F(\theta - y) \\
&= 1 - e^{\theta - y - \theta} \\
&= 1 - e^{-y},
\end{aligned}
$$

for $y \geq 0$ and $G(y) = 0$ if $y < 0$. Hence, the pdf of $Y$ is:

$$
g(y) = \begin{cases} e^{-y} & \text{for } y \geq 0 \\ 0 & \text{for } y < 0. \end{cases}
$$

**5.2**

1. There are six words in the sentence and there is exactly one word having $X = i$ number of words for $i = 1, \ldots, 6$. Hence the probability distribution is:

| $x$ | 1 | 2 | 3 | 4 | 5 | 6 | Total |
|---|---|---|---|---|---|---|---|
| $f(x)$ | $\frac{1}{6}$ | $\frac{1}{6}$ | $\frac{1}{6}$ | $\frac{1}{6}$ | $\frac{1}{6}$ | $\frac{1}{6}$ | 1 |

Here

$$
\begin{aligned}
E(X) &= \tfrac{1}{6}(1 + 2 + 3 + 4 + 5 + 6) = \tfrac{7}{2} \\
E(X^2) &= \tfrac{1}{6}(1^2 + 2^2 + 3^2 + 4^2 + 5^+ 6^2) \\
&= \tfrac{1}{6} \tfrac{6 \times 7 \times 13}{6} = \tfrac{91}{6} \\
\text{Var}(X) &= E(X^2) - (E(X))^2 \\
&= \tfrac{91}{6} - \tfrac{49}{4} = \tfrac{35}{12}.
\end{aligned}
$$

2. NSP
3. We find that $X$ has probability function

| $x$ | 1 | 2 | 3 | 4 | 5 | 6 |
|---|---|---|---|---|---|---|
| $p_x$ | $\frac{1}{36}$ | $\frac{3}{36}$ | $\frac{5}{36}$ | $\frac{7}{36}$ | $\frac{9}{36}$ | $\frac{11}{36}$ |

$$
\begin{aligned}
E(X) &= \frac{1}{36} \times 1 + \frac{3}{36} \times 2 + \frac{5}{36} \times 3 + \frac{7}{36} \times 4 + \frac{9}{36} \times 5 + \frac{11}{36} \times 6 \\
&= \frac{161}{36}.
\end{aligned}
$$

4.

$$E(X) = \sum_x x p_x = \frac{1}{2n+1} \sum_{x=-n}^{n} x = 0.$$

$$E(X^2) = \sum_x x^2 p_x = \frac{1}{2n+1} \sum_{x=-n}^{n} x^2$$

$$= \frac{2}{(2n+1)} \frac{n(n+1)(2n+1)}{6} = \frac{n(n+1)}{3}.$$

Therefore $\mathrm{Var}(X) = E(X^2) - [E(X)]^2 = \dfrac{n(n+1)}{3}.$

If $\mathrm{Var}(X) = 10,$

then $\dfrac{n(n+1)}{3} = 10$

$$n^2 + n - 30 = 0.$$

Therefore $n = 5$ (rejecting $-6$).

5. Solutions to the trianglular distribution problem.
   (a) The graph will be a triangle.
   (b) $P(X \leq -1) = 0.$
   (c)

$$P\left(|X| > \tfrac{1}{2}\right) = P\left(X < -\tfrac{1}{2}\right) + P\left(X > \tfrac{1}{2}\right)$$
$$= \int_{-1}^{-0.5} (1+x)dx + \int_{0.5}^{1} (1-x)dx$$
$$= \tfrac{1}{4}$$

   (d)

$$F(x) = \begin{cases} 0 & x < -1 \\ \int_{-1}^{x}(1+u)du = \frac{x^2}{2} + x + \frac{1}{2}, & -1 \leq x < 0 \\ \int_{-1}^{0}(1+u)du + \int_{0}^{x}(1-u)du = -\frac{x^2}{2} + x + \frac{1}{2}, & 0 \leq x < 1 \\ 1 & x \leq 1 \end{cases}$$

   (e)

$$E(X) = \int_{-1}^{1} x f(x)dx = \int_{-1}^{0} x(1+x)dx + \int_{0}^{1} x(1-x)dx = 0.$$

(f)

$$E(X^2) = \int_{-1}^{1} x^2 f(x)dx = \int_{-1}^{0} x^2(1+x)dx + \int_{0}^{1} x^2(1-x)dx = \frac{1}{6}.$$

$$\text{Var}(X) = E(X^2) - (E(X))^2 = \frac{1}{6}.$$

6. NSP
7.

$$E[(X-a)^2] = E\left[(X-\mu+\mu-a)^2\right]$$

$$= E\left[(X-\mu)^2 + (\mu-a)^2 + 2(X-\mu)(\mu-a)\right]$$

Expand the square

$$= E[(X-\mu)^2] + E[(\mu-a)^2] + E[2(X-\mu)(\mu-a)]$$

distribute expectations

$$= E[(X-\mu)^2] + (\mu-a)^2 + 2(\mu-a)E[X-\mu]$$

$$= E[(X-\mu)^2] + (\mu-a)^2 \quad \text{because } E(X) = \mu$$

$$= \text{Var}(X) + (\mu-a)^2.$$

We note that the first term $\text{Var}(X)$ is free of $a$ and the second term is non-negative and is excatly zero when $a = \mu = E(X)$. Thus we take $a = \mu = E(X)$ to minimise $E[(X-a)^2$ and in that case minimum value of $E[(X-a)^2$ is the variance of $X$.

8. NSP

## Problems of Chap. 6

### 6.1

1. (d) Here $\text{Var}(X) = p(1-p)$ which is maximised at $p = \frac{1}{2}$. $\frac{d(p(1-p))}{dp} = 1-2p$ and $\frac{d^2(p(1-p))}{dp^2} = -2$.

2. Let $X$ = number of defectives in a randonly selected box. Here $X$ follows binomial with parameters $n = 10$ and $p = 0.02$.

$$P(X=0) + P(X=1) = {}^{10}C_0 p^0(1-p)^{10} + {}^{10}C_1 p^1(1-p)^9 = 0.984$$

3. NSP

4. NSP
5. We know $\mathrm{Var}(Y) = \mu$, which implies $E(Y^2) = \mathrm{Var}(Y) + (E(Y))^2 = \mu + \mu^2$. Hence,

$$E(Y(Y-1)) = E(Y^2) - E(Y) = \mu + \mu^2 - \mu = \mu^2.$$

6. NSP
7. Let $X$ be the number of calls arriving in an hour and let $P(X \geq 15) = p$. Then $Y$, the number of times out of 10 that $X \geq 15$, is $B(n, p)$ with $n = 10$ and $p = 1 - 0.98274 = 0.01726$.

Therefore $P(Y \geq 2) = 1 - P(Y \leq 1)$

$$= 1 - ((0.98274)^{10} + 10(0.01726)(0.98274)^9)$$

$$= 0.01223.$$

8. NSP
9. We know

$$
\begin{aligned}
r\frac{(1-p)}{p} &= \mu \\
\implies \frac{(1-p)}{p} &= \frac{\mu}{r} \\
\implies \frac{(1-p)}{p} + 1 &= \frac{\mu}{r} + 1 \\
\implies \frac{1}{p} &= \frac{\mu+r}{r} \\
\implies p &= \frac{r}{\mu+r}.
\end{aligned}
$$

Thus, $1 - p = \frac{\mu}{\mu+r}$. Hence,

$$
\begin{aligned}
P(X = x) &= \binom{r+x-1}{x} p^r (1-p)^x \\
&= \binom{r+x-1}{x} \left(\frac{r}{\mu+r}\right)^r \left(\frac{\mu}{\mu+r}\right)^x, \quad x = 0, 1, \ldots,
\end{aligned}
$$

as required.
10. Answer: (c)
11. Let $X$ be the number of errors made. Then $X$ is binomially distributed with $n = 10^{15}$ and $p = 10^{-14}$. Thus

$$P(X \leq 9) = \sum_{x=0}^{9} \binom{10^{15}}{x} (10^{-14})^x (1 - 10^{-14})^{10^{15}-x}.$$

Using the Poisson approximation with $\lambda = np = 10$,

$$P(X \leq 9) = \sum_{x=0}^{9} e^{-10} \cdot \frac{10^x}{x!} = 0.45793$$

using the R command **ppois**$(9, \text{lambda}=10)$.

12. If the number of voters in the district is much larger than 100, then $X$ may be taken to be $B(100, 0.3)$ even though sampling is without replacement, since the proportion of Conservative voters will hardly change.

(a)

$$E(X) = 100 \times 0.3 = 30.$$

$$Var(X) = 100 \times 0.3 \times 0.7 = 21.$$

(b) We must therefore calculate

(i) $P(30-\sqrt{21} \leq X \leq 30+\sqrt{21})$ or $P(26 \leq X \leq 34)$, since $X$ can only take integral values. $P(26 \leq X \leq 34) = P(X \leq 34) - P(X \leq 25) = 0.67401$ (**pbinom**(34,size=100,prob=0.3)-pbinom(25,size=100, prob=0.3) in R.

(ii) $P(30 - 2\sqrt{21} \leq X \leq 30 + 2\sqrt{21})$ or $P(21 \leq X \leq 39)$. $P(21 \leq X \leq 39) = P(X \leq 39) - P(X \leq 20) = 0.96255$ **pbinom**(39,size=100,prob=0.3)-pbinom(20,size=100, prob=0.3) in R.

13. We find the variance of the negative binomial distribution.

We note that $Var(Y) = Var(X)$ where $X = Y - r$. Also, $E(X) = \frac{r(1-p)}{p}$. We first find $E\left[(X(X-1)\right]$.

***Proof***

$$E\left[X(X-1)\right] = \sum_{x=0}^{\infty} x(x-1)\binom{r+x-1}{x}p^r(1-p)^x$$

$$= p^r \sum_{x=2}^{\infty} x(x-1)\frac{(r+x-1)!}{x!(r-1)!}1-p)^x$$

$$= p^r(1-p)^2 \sum_{x=2}^{\infty} \frac{(r+x-1)!}{(x-2)!(r-1)!}(1-p)^{x-2}$$

$$= p^r r(r+1)(1-p)^2 \sum_{y=0}^{\infty} \frac{(r+y+2-1)!}{y!(r+1)!}(1-p)^y$$

$$= p^r r(r+1)(1-p)^2 \sum_{y=0}^{\infty} \binom{r+1+y}{y}(1-p)^y$$

$$= p^r r(r+1)(1-p)^2 \sum_{y=0}^{\infty} \binom{r+2+y-1}{y}(1-p)^y$$

$$= p^r r(r+1)(1-p)^2 \{1 - (1-p)\}^{-(r+2)}$$

$$= r(r+1)\frac{(1-p)^2}{p^2}.$$

Now

$$\mathrm{Var}(X) = E\,[X(X-1)] + E(X) - [E(X)]^2$$

$$= r(r+1)\frac{(1-p)^2}{p^2} + \frac{r(1-p)}{p} - \left[\frac{r(1-p)}{p}\right]^2$$

$$= \frac{r(1-p)}{p^2}\,\{(r+1)(1-p) + p - r(1-p)\}$$

$$= \frac{r(1-p)}{p^2}$$

14. NSP

15. (a) Suppose $x > 0$ is a positive integer. It will take $x$ trials to obtain the first success if and only if the first $x-1$ Bernoulli trails result in failure and the $x$th trial results in success. By definition the Bernoulli trials are independent and it is given that the success probability is $p$. Hence

$$P(X=x) = P(F)P(F)\cdots P(F)P(S) \quad \text{where there are } x-1 \text{ F's.}$$
$$= (1-p)^{x-1}p \text{ for } x = 1, 2, \ldots,$$

(b)

$$\sum_{x=1}^{\infty} P(X=x) = \sum_{x=1}^{\infty}(1-p)^{x-1}p$$

$$= p\sum_{y=0}^{\infty}(1-p)^y \quad [\text{substitute } y = x-1]$$

$$= p\,\frac{1}{1-(1-p)} \quad [\text{Geometric series }]$$

$$= 1.$$

(c) NSP

(d) We know:

$$P\{A|B\} = \frac{P\{A \cap B\}}{P\{B\}}.$$

Hence,

$$P(X > s+k|X > k) = \frac{P(X > s+k, X > k)}{P(X > k)}$$

$$= \frac{P(X > s + k)}{P(X > k)}$$

$$= \frac{(1 - p)^{s+k}}{(1 - p)^k}$$

$$= (1 - p)^s$$

$$= P(X > s).$$

(e)

$$E(X) = \sum_{x=1}^{\infty} x P(X = x)$$

$$= \sum_{x=1}^{\infty} x p (1 - p)^{x-1}$$

$$= p \left[ 1 + 2(1 - p) + 3(1 - p)^2 + 4(1 - p)^3 + \ldots \right]$$

For $n > 0$ and $|y| < 1$, the negative binomial series is given by:

$$(1 - y)^{-n} = 1 + ny + \frac{1}{2}n(n + 1)y^2 + \frac{1}{6}n(n + 1)(n + 2)y^3 + \cdots$$

With $n = 2$ and $y = 1 - p$ the coefficient of the general term $y^k$ is given by:

$$\frac{n(n + 1)(n + 2)(n + k - 1)}{k!} = \frac{2 \times 3 \times 4 \times \cdots \times (2 + k - 1)}{k!} = k + 1.$$

Thus $E(X) = p(1 - 1 + p)^{-2} = 1/p$.

16. NSP

---

# Problems of Chap. 7

## 7.1

1. NSP
2. The cdf of the exponential distribution is

$$F(x) = P(X \leq x) = \int_0^x \lambda e^{-\lambda u} du$$

$$= \left[ -e^{-\lambda u} \right]_0^x$$

$$= 1 - \exp(-\lambda x).$$

(a) The bulb is still functioning after a time $x$ if $X > x$.

$$P(X > x) = 1 - F(x) = \exp(-\lambda x)$$

Hence the answer to the first part is: $e^{-\lambda \frac{2}{\lambda}} = e^{-2}$.
(b) Let $p$ be the probability that the bulb is still functioning after time $1/\lambda$.

$$p = e^{-\lambda \frac{1}{\lambda}} = e^{-1}.$$

The required probability is the binomial probability of $Y = 1$ when $Y \sim B(n = 100, p)$.

$$^{100}C_1 \left(e^{-1}\right) \left(1 - e^{-1}\right)^{99} = 7 \times 10^{-19}.$$

3. NSP
4.(a) Since total probability is 1, we have,

$$1 = \int_0^\infty kt^2 e^{-\theta t^3} dt$$

$$= k \int_0^\infty t^2 e^{-\theta t^3} dt$$

$$= k \int_0^\infty \frac{1}{3\theta} e^{-u} du \text{ [substituting } \theta t^3 = u]$$

$$= k \frac{1}{3\theta} \text{ since } \int_0^\infty e^{-u} du = 1.$$

This proves, $k = 3\theta$.
(b) We have,

$$F(t) = \int_0^t 3\theta v^2 e^{-\theta v^3} dv$$

$$= \int_0^{\theta t^3} e^{-u} du \text{ [substituting } \theta v^3 = u]$$

$$= 1 - e^{\theta t^3}.$$

(c) We have, $S(10) = e^{-\theta 10^3} = 0.9$. Hence, $-1000\theta = \log(0.9)$ and then we have $\theta = -\log(0.9)/1000 = 0.10536 \times 10^{-3}$. Now the reliability at $t = 20$ is given by

$$S(20) = e^{-\theta 20^3} = e^{-0.10536 \times 10^{-3} \times 20^3} = e^{-0.10536 \times 8} = 0.4305.$$

(d) Here $R$ follows the binomial distribution with parameters $n = 100$ and $p = 1 - S(20)$ since we are interested in the proportion if failed bulbs. Here $p = 1 - 0.4305 = 0.5695$. Hence the mean of $R$ is $np = 100 \times 0.5695 = 56.95$ and variance $np(1 - p) = 100 \times 0.5695 \times 0.4305 = 24.52$.

5. NSP
6. NSP

## 7.2

1. NSP
2. Find $a$ such that $P(X > a) = 0.10$ when $X \sim N(185, 17^2)$. From tables, if $Z \sim N(0, 1)$, then $P(Z > 1.28) = 0.10$. But $Z = \frac{X-\mu}{\sigma}$ implies $X = \mu + \sigma Z$. Thus $a = 185 + 17 * 1.28 = 206.8$
3. We have $X \sim N(\mu = 3, \sigma^2 = 4)$.

$$P(X > -1) = P\left(\frac{X-\mu}{\sigma} > \frac{-1-3}{2}\right)$$
$$= P(Z > -2), \quad Z \sim N(0, 1)$$
$$= P(Z < 2).$$

Draw a cartoon of the standard distribution and note that $P(Z < -2) = P(Z > 2)$ since the distribution of $Z$ is symmetric about 0.
4. Here $X \sim N(0, 1)$. Hence $\text{Var}(X) = 1$. We have to find the variance of $Y$.

$$E(Y) = \int_0^1 y\,dy$$

$$= \frac{1}{2}.$$

$$E(Y^2) = \int_0^1 y^2\,dy$$

$$= \frac{1}{3}.$$

$$\text{Var}(Y) = E(Y^2) - (E(Y))^2$$

$$= \frac{1}{3} - \left(\frac{1}{2}\right)^2$$

$$= \frac{1}{12}.$$

Hence (i) is true since $\text{Var}(X) = 12 \times \text{Var}(Y)$.

Here $X \sim N(0, 1)$. Hence

$$P\left(X < \frac{1}{2}\right) = \texttt{pnorm(0.5)} = 0.6915$$

$$P\left(Y < \frac{1}{2}\right) = 0.5.$$

Hence $1.383 \times P\left(Y < \frac{1}{2}\right) = 0.6915$.

Hence (ii) is true as well. So the answer is (A).

5. NSP

6. The standardised variable $Z = \frac{X-4}{4} \sim N(0, 1)$.

(a) $P(|X - 4| \le a) = P(|Z| \le \frac{a}{4}) = P(-\frac{a}{4} \le Z \le \frac{a}{4}) = 2\Phi(\frac{a}{4}) - 1$.
We require the $a$ for which $2\Phi(\frac{a}{4}) - 1 = 0.95$, i.e. $\Phi(\frac{a}{4}) = 0.975$.
Now $\Phi(1.96) = 0.975$ ($\texttt{qnorm(0.975)}$ in R). Therefore $\frac{a}{4} = 1.96$, i.e. $a = 7.84$.

(b) The normal probability density function is symmetric about its mean.
Therefore, median = mean = 4.

(c) The 90th percentile (or 9th decile) $x_{90}$ satisfies

$$0.9 = P(X \le x_{90}) = P\left(Z \le \frac{x_{90} - 4}{4}\right) = \Phi\left(\frac{x_{90} - 4}{4}\right).$$

$\frac{x_{90} - 4}{4} = 1.2816$ ($\texttt{qnorm(0.9)}$ in R), which implies that $x_{90} = 9.126$.

(d) The upper quartile $u$ satisfies

$$0.75 = P(X \le u) = P\left(Z \le \frac{u - 4}{4}\right) = \Phi\left(\frac{u - 4}{4}\right).$$

$$\frac{u - 4}{4} = 0.6745 \ (\texttt{qnorm(0.75)} \ \text{in R}), \ \text{which implies that} \ u = 6.698.$$

Using symmetry, the interquartile range = 2(6.698–4) = 5.396.

7.(a) To find $C$ we note that

$$\int_a^\infty f(x)dx = 1$$
$$\implies \int_a^\infty \frac{C}{\sqrt{2\pi}} e^{-\frac{x^2}{2}} dx = 1$$
$$\implies C(1 - \Phi(a)) = 1$$
$$\implies C = \frac{1}{1-\Phi(a)}.$$

(b) Here

$$E(X) = \int_a^\infty \frac{1}{1 - \Phi(a)} \frac{x}{\sqrt{2\pi}} e^{-\frac{x^2}{2}} dx$$
$$= \frac{1}{\sqrt{2\pi}(1 - \Phi(a))} \int_a^\infty x e^{-\frac{x^2}{2}} dx$$
$$= \frac{1}{1 - \Phi(a)} \frac{1}{\sqrt{2\pi}} \int_{\frac{a^2}{2}}^\infty e^{-u} du \quad [\text{substitute} u = \frac{x^2}{2}]$$
$$= \frac{1}{1 - \Phi(a)} \frac{1}{\sqrt{2\pi}} e^{-\frac{a^2}{2}}$$
$$= \frac{1}{1 - \Phi(a)} \phi(a).$$

Now $\phi(0) = \frac{1}{\sqrt{2\pi}}$ and $\Phi(0) = \frac{1}{2}$, hence

$$E(X) = \frac{\phi(a)}{1 - \Phi(a)} = \frac{2}{\sqrt{2\pi}} = \sqrt{\frac{2}{\pi}}.$$

(c) The cdf is given by:

$$F(x) = \frac{1}{1 - \Phi(a)} \int_a^x \frac{1}{\sqrt{2\pi}} e^{-\frac{u^2}{2}} du$$
$$= \frac{1}{1 - \Phi(a)} (\Phi(x) - \Phi(a))$$
$$= \frac{\Phi(x) - \Phi(a)}{1 - \Phi(a)}, \quad x > a.$$

The cdf takes the value 0 if $x \leq a$.

(d) We are given

$$
\begin{aligned}
F(x) &= 0.95 \\
\frac{\Phi(x) - \Phi(0)}{1 - \Phi(0)} &= 0.95 \\
2\Phi(x) - 1 &= 0.95 \\
\Phi(x) &= 0.975 \\
x &= \Phi^{-1}(0.975) = 1.96.
\end{aligned}
$$

## 7.3

1. NSP
2. NSP
3. It matches with the $\chi^2$ distribution for $p = 1$. Since $\Gamma\left(\frac{1}{2}\right) = \sqrt{\pi}$

   It matches with the gamma distribution with $\alpha = \frac{1}{2}$ and $\beta = 2$.
4. The pdf can be written as:

$$
f(x) = C x^{\frac{1}{2} - 1} (1 - x)^{3 - 1}, \quad 0 < x < 1
$$

which is recognised as the beta distribution with parameters $\alpha = \frac{1}{2}$ and $\beta = 3$. Hence

$$
\begin{aligned}
C &= B\left(\tfrac{1}{2}, 3\right) \\
&= \frac{\Gamma(3)\Gamma\left(\frac{1}{2}\right)}{\Gamma\left(\frac{7}{2}\right)} \\
&= \frac{2! \sqrt{\pi}}{\frac{5}{2}\frac{3}{2}\frac{1}{2}\Gamma\left(\frac{1}{2}\right)} \\
&= \frac{16}{15}.
\end{aligned}
$$

# Problems of Chap. 8

## 8.1

1. NSP
2. Here

$$E(X) = 1 \times 0.3 + 2 \times 0.5 + 3 \times 0.2 = 1.9.$$

$$E(Y) = -1 \times 0.5 + 0 \times 0.3 + 1 \times 0.2 = -0.3.$$

$$E(XY) = (1)(-1)0.2 + (2)(-1)0.2 + (3)(-1)0.1$$
$$+ (2)(1)0.1 + (3)(1)0.1 = -0.4.$$

$$\text{Cov}(X.Y) = E(XY) - E(X)E(Y) = -0.4 - 1.9(-0.3) = 0.17.$$

3. First construct the following table calculating $x + y$ for each cell.

|   | $y$ |
|---|---|
|   | $-1$ $0$ $1$ |
| $1$ | $0$ $1$ $2$ |
| $x$ $2$ | $1$ $2$ $3$ |
| $3$ | $2$ $3$ $4$ |

Now the probability distribution of $Z = X + Y$ is calculated as:

| $z$ | 0 | 1 | 2 | 3 | 4 | total |
|---|---|---|---|---|---|---|
| $P(Z = z)$ | 0.2 | 0.3 | 0.3 | 0.1 | 0.1 | 1 |

Hence $E(Z) = 0(0.2) + 1(0.3) + 2(0.3) + 3(0.1) + 4(0.1) = 1.6$. Similarly, $E(Z^2) = 4$ and then $\text{Var}(Z) = 4 - 1.6^2 = 1.44$.

**Alternative Solution:**

$$\text{Var}(X) = E(X^2) - (E(X))^2 = 1 \times 0.3 + 4 \times 0.5 + 9 \times 0.2 - 1.9^2 = 0.49.$$

$$\text{Var}(Y) = E(Y^2) - (E(Y))^2 = 1 \times 0.5 + 0 \times 0.3 + 1 \times 0.2 - 0.3^2 = 0.61.$$

In general we have,

$$\text{Var}(aX + bY) = a^2 \text{Var}(X) + b^2 \text{Var}(Y) + 2ab\text{Cov}(X, Y).$$

$$\text{Var}(X + Y) = \text{Var}(X) + \text{Var}(Y) + 2\text{Cov}(X, Y) = 0.49 + 0.61 + 2 \times 0.17 = 1.44.$$

4. NSP
5. The marginal distributions of $X$ and $Y$ are found by summing respectively the rows and columns of the joint probability table. This gives

| $x$ | 1 | 2 | 3 |
|---|---|---|---|
| $p_x$ | $\frac{1}{4}$ | $\frac{3}{8}$ | $\frac{3}{8}$ |

| $y$ | 1 | 2 | 3 |
|---|---|---|---|
| $p_y$ | $\frac{1}{4}$ | $\frac{1}{2}$ | $\frac{1}{4}$ |

It follows that $P(X = 1)P(Y = 2) = \frac{1}{4} \times \frac{1}{2} = \frac{1}{8} = P(X = 1, Y = 2)$, as required.

It cannot be concluded that $X$ and $Y$ are independent since other values $x$ and $y$ can be found such that $P(X = x, Y = y) \neq P(X = x)P(Y = y)$.

For example, $P(X = 2, Y = 2) = \frac{1}{8}$, while $P(X = 2)P(Y = 2) = \frac{3}{16}$.

6. NSP

7.(a) Here is the complete table.

|  | | $y$ | | |
|---|---|---|---|---|
|  | | 0 | 1 | 2 | Total |
| $x$ | 0 | $\frac{10}{66}$ | $\frac{20}{66}$ | $\frac{6}{66}$ | $\frac{36}{66}$ |
|  | 1 | $\frac{15}{66}$ | $\frac{12}{66}$ | 0 | $\frac{27}{66}$ |
|  | 2 | $\frac{3}{66}$ | 0 | 0 | $\frac{3}{66}$ |
| Total | | $\frac{28}{66}$ | $\frac{32}{66}$ | $\frac{6}{66}$ | 1 |

$X$ and $Y$ not independent since $P(X = 1, Y = 2) \neq P(X = 1) \times P(Y = 2)$.

(b)

$$P(X + Y = 2) = \frac{6}{66} + \frac{12}{66} + \frac{3}{66} = \frac{21}{66} = \frac{7}{22}$$

$$P(Y = 2|X = 0) = \frac{P(X = 0, Y = 2)}{P(X = 0)} = \frac{\frac{6}{66}}{\frac{36}{66}} = \frac{1}{6}.$$

8. NSP

## 8.2

1. Here $n\lambda = 50 \times 0.1 = 5$. Let $Y = \sum_{i=1}^{n} X_i$. Thus $Y \sim$ Poisson(5). $P(Y = 0) = e^{-5} = 0.0067$.

2. NSP

3. Here $X =$ the number of sixes obtained follows the binomial distribution with $n = 600$ and $p = 1/6$ if the dice is fair. We have $E(X) = 100$ and $\text{Var}(X) = 100(5/6) = 250/3$. Using the CLT, this distribution is approximated by $N(100, 250/3)$. Let $Y$ denote this normal distribution.

(a)

$$
\begin{aligned}
P(90 \leq X < 100) &= P(89.5 \leq Y \leq 100.5) \\
&= \Phi\left(\frac{100.5 - 100}{\sqrt{250/3}}\right) - \Phi\left(\frac{89.5 - 100}{\sqrt{250/3}}\right) \\
&= \Phi(0.0548) - \Phi(-1.1502) \\
&= 0.3968
\end{aligned}
$$

(b) We wish to find a positive integer $N$ such that

$$P(100 - N \leq X \leq 100 + N) = 0.95.$$

Using the continuity correction, we need:

$$P(99.5 - N \leq Y \leq 100.5 + N) = 0.95.$$

That is,

$$\text{i.e.}\, \Phi\left(\frac{N + 0.5}{\sqrt{250/3}}\right) - \Phi\left(-\frac{N + 0.5}{\sqrt{250/3}}\right) = 0.95$$

$$\text{or}\, \Phi\left(\frac{N + 0.5}{\sqrt{250/3}}\right) = 0.975$$

$$\text{thus}\, \frac{N + 0.5}{\sqrt{250/3}} = 1.96.$$

This equation does not have integer solution but the integer most nearly satisfies is $N = 17$. You can check that $P(83 \leq X \leq 117) = 0.945$.
(c) He might conclude that that there is some evidence of that the dice is unfair since 120 is an unexpectedly large number of sixes if the die is fair.
4. NSP
5. (B) since $n = 40$ is large and $p = 0.01$ is small but $np = 0.4$ is finite.
6. NSP

---

## Problems of Chap. 9

### 9.1

1.(a) An unbiased estimate of $\mu_b$ is given by the mean weight of the boys,

$$\hat{\mu}_b = \frac{1}{10}(77 + 67 + \ldots + 81) = 67.3.$$

An unbiased estimate of $\sigma_b^2$ is the sample variance of the weights of the boys,

$$\hat{\sigma}_b^2 = ((77^2 + 67^2 + \ldots + 81^2) - 10\hat{\mu}_b^2)/9 = 52.6\dot{7}.$$

(b) Similarly, unbiased estimates of $\mu_g$ and $\sigma_g^2$ are

$$\hat{\mu}_g = \frac{1}{10}(42 + 57 + \ldots + 59) = 52.4,$$

$$\hat{\sigma}_g^2 = ((42^2 + 57^2 + \ldots + 59^2) - 10\hat{\mu}_g^2)/9 = 56.7\dot{1}.$$

(c) An unbiased estimate of $\mu_b - \mu_g$ is

$$\hat{\mu}_b - \hat{\mu}_g = 67.3 - 52.4 = 14.9.$$

$E(\hat{\sigma}_b^2) = E(\hat{\sigma}_g^2) = \sigma^2$ and so

$$E\left(\frac{\hat{\sigma}_b^2 + \hat{\sigma}_g^2}{2}\right) = \sigma^2.$$

Therefore an unbiased estimate of $\sigma^2$ which uses both sets of weights is

$$\frac{1}{2}(\hat{\sigma}_b^2 + \hat{\sigma}_g^2) = \frac{1}{2}(52.6\dot{7} + 56.7\dot{1})$$

$$= 54.69\dot{4}.$$

2. NSP

3.

$$E(Y) = \sum_{i=1}^{n} E(a_i X_i)$$

$$= \sum_{i=1}^{n} a_i E(X_i) = \sum_{i=1}^{n} a_i \mu$$

$$= \mu \sum_{i=1}^{n} a_i.$$

Therefore $Y$ is unbiased if $\sum_{i=1}^{n} a_i = 1$.

$$\text{Var}(Y) = \sum_{i=1}^{n} a_i^2 \text{Var}(X_i) = \sigma^2 \sum_{i=1}^{n} a_i^2.$$

Applying the criterion that $\sum_{i=1}^{n} a_i$ must be 1 for an unbiased estimator, we can see that $Y_1$, $Y_3$ and $Y_4$ are unbiased estimators. Using the formula just derived,

$$\text{Var}(Y_1) = \frac{\sigma^2}{4}, \quad \text{Var}(Y_3) = \frac{13\sigma^2}{25}, \quad \text{Var}(Y_4) = \frac{\sigma^2}{2}.$$

Therefore $Y_1$ is the most efficient of the unbiased estimators.

## Problems of Chap. 10

### 10.1

1. By definition, method of moments estimators $\tilde{\alpha}$ and $\tilde{\beta}$ satisfy

$$\overline{X} = \frac{\tilde{\alpha}}{\tilde{\alpha}+\tilde{\beta}} \quad (1)$$

$$\overline{X^2} = \frac{\tilde{\alpha}(\tilde{\alpha}+1)}{(\tilde{\alpha}+\tilde{\beta})(\tilde{\alpha}+\tilde{\beta}+1)}. \quad (2)$$

From (2), we have

$$\overline{X^2} = \frac{\tilde{\alpha}}{\tilde{\alpha}+\tilde{\beta}} \frac{\frac{\alpha}{\tilde{\alpha}+\tilde{\beta}} + \frac{1}{\tilde{\alpha}+\tilde{\beta}}}{1 + \frac{1}{\tilde{\alpha}+\tilde{\beta}}}. \quad (3)$$

Now, substituting (1) in (3), we have

$$\overline{X^2} = \overline{X}\frac{\overline{X}+\frac{1}{\tilde{\alpha}+\tilde{\beta}}}{1+\frac{1}{\tilde{\alpha}+\tilde{\beta}}}$$

$$\Rightarrow \overline{X^2} + \overline{X^2}\frac{1}{\tilde{\alpha}+\tilde{\beta}} = \overline{X}^2 + \overline{X}\frac{1}{\tilde{\alpha}+\tilde{\beta}}$$

$$\Rightarrow \frac{1}{\tilde{\alpha}+\tilde{\beta}} = \frac{\overline{X^2}-\overline{X}^2}{\overline{X}-\overline{X^2}}. \quad (4)$$

Now, substituing (4) in (1), we have

$$\tilde{\alpha} = \overline{X}\frac{\overline{X} - \overline{X^2}}{\overline{X^2} - \overline{X}^2}$$

and [using the fact that (1) $\Rightarrow \tilde{\beta} = \tilde{\alpha}(1 - \overline{X})/\overline{X}$],

$$\tilde{\beta} = (1 - \overline{X})\frac{\overline{X} - \overline{X^2}}{\overline{X^2} - \overline{X}^2}.$$

2. NSP
3. Note that the sample space depends on $\alpha$ so this has to be incorporated into the likelihood through an indicator function, so

$$f_{\mathbf{X}}(\mathbf{x}; \alpha, \theta) = \prod_{i=1}^{n}(1 - \theta)\theta^{x_i-\alpha}I[x_i \geq \alpha]$$
$$= (1 - \theta)^n\theta^{\sum_{i=1}^{n}(x_i-\alpha)}\prod_{i=1}^{n}I[x_i \geq \alpha]$$
$$= (1 - \theta)^n\theta^{\sum_{i=1}^{n}x_i-n\alpha}I[\min\{x_i\} \geq \alpha]$$

Now, as a function of $\alpha$,

$$f_{\mathbf{X}}(\mathbf{x}; \alpha, \theta) \propto \theta^{-n\alpha}I[\min\{x_i\} \geq \alpha] = \left(\theta^{-n}\right)^{\alpha}I[\min\{x_i\} \geq \alpha]$$

and hence, as $\theta \in (0, 1)$, $\theta^{-n} > 1$ and hence $\left(\theta^{-n}\right)^\alpha$ is an increasing function of $\alpha$. Hence, regardless of the value of $\theta$ $\left(\theta^{-n}\right)^\alpha I[\min\{x_i\} \geq \alpha]$ is maximised as a function of $\alpha$ at

$$\hat{\alpha} = \min\{x_i\}$$

(as for larger values of $\alpha$ the indicator function ensures that the likelihood is zero.) Now, for $\hat{\theta}$, we have

$$\log f_{\mathbf{X}}(\mathbf{x}; \alpha, \theta) = n \log(1 - \theta) + \log \theta \left(\sum_{i=1}^n x_i - n\alpha\right)$$
$$\Rightarrow \frac{\partial}{\partial \theta} \log f_{\mathbf{X}}(\mathbf{x}; \alpha, \theta) = -\frac{n}{1-\theta} + \frac{1}{\theta} \left(\sum_{i=1}^n x_i - n\alpha\right)$$

Hence, the m.l.e. $\hat{\theta}$ satisfies

$$-\frac{n}{1-\hat{\theta}} + \frac{1}{\hat{\theta}} \left(\sum_{i=1}^n x_i - n\hat{\alpha}\right) = 0$$
$$\Rightarrow n\hat{\theta} = (1 - \hat{\theta}) \left(\sum_{i=1}^n x_i - n \min\{x_i\}\right)$$
$$\Rightarrow \hat{\theta} = (1 - \hat{\theta}) \left(\overline{x} - \min\{x_i\}\right)$$
$$\Rightarrow \hat{\theta} = \frac{\overline{x} - \min\{x_i\}}{1 + \overline{x} - \min\{x_i\}}.$$

4. NSP
5.

$$f_{\mathbf{X}}(\mathbf{x}; \theta) = \prod_{i=1}^2 \left(\frac{\theta}{2}\right)^{|x_i|} (1 - \theta)^{1-|x_i|}$$
$$= \left(\frac{\theta}{2}\right)^{|x_1|+|x_2|} (1 - \theta)^{2-|x_1|-|x_2|}$$

$$\Rightarrow \log f_{\mathbf{X}}(\mathbf{x}; \theta) = (|x_1| + |x_2|)(\log \theta - \log 2) + (2 - |x_1| - |x_2|) \log(1 - \theta)$$
$$\Rightarrow \frac{\partial}{\partial \theta} \log f_{\mathbf{X}}(\mathbf{x}; \theta) = \frac{|x_1|+|x_2|}{\theta} - \frac{2-|x_1|-|x_2|}{1-\theta}$$

Hence, the m.l.e. $\hat{\theta}$ satisfies

$$\frac{|x_1|+|x_2|}{\hat{\theta}} - \frac{2-|x_1|-|x_2|}{1-\hat{\theta}} = 0$$
$$\Rightarrow (|x_1| + |x_2|)(1 - \hat{\theta}) = \hat{\theta}(2 - |x_1| - |x_2|)$$
$$\Rightarrow \hat{\theta} = \frac{1}{2}(|x_1| + |x_2|).$$

Now

$$f(x; \theta) = \begin{cases} \theta/2 & x = -1, 1 \\ 1 - \theta & x = 0 \end{cases}$$

and therefore

$$f(x_1, x_2; \theta) = \begin{cases} \theta^2/4 & (x_1, x_2) = (-1, -1), (-1, 1), (1, -1), (1, 1) \\ \theta(1-\theta)/2 & (x_1, x_2) = (0, -1), (0, 1), (-1, 0), (1, 0) \\ (1-\theta)^2 & (x_1, x_2) = (0, 0) \end{cases}$$

and

$$f(\hat{\theta}; \theta) = \begin{cases} \theta^2 & \hat{\theta} = 1 \\ 2\theta(1-\theta) & \hat{\theta} = 1/2 \\ (1-\theta)^2 & \hat{\theta} = 0. \end{cases}$$

Hence

$$E(\hat{\theta}) = \theta^2 \cdot 1 + 2\theta(1-\theta) \cdot \tfrac{1}{2} = \theta$$

so $\hat{\theta}$ is unbiased for $\theta$. As

$$E(\hat{\theta}^2) = \theta^2 \cdot 1 + 2\theta(1-\theta) \cdot 1/4 = \frac{\theta^2}{2} + \frac{\theta}{2}$$

we have

$$\text{Var}(\hat{\theta}) = \frac{\theta^2}{2} + \frac{\theta}{2} - \theta^2 = \tfrac{1}{2}\theta(1-\theta).$$

Now, for $T = I(X_1 = 1) + I(X_2 = 1)$, we have

$$f(T; \theta) = \begin{cases} \frac{\theta^2}{4} & T = 2 \\ \frac{\theta^2}{2} + \theta(1-\theta) & T = 1 \\ (1-\theta)^2 + \theta(1-\theta) + \frac{\theta^2}{4} & T = 0 \end{cases}$$

and therefore

$$E(T) = \frac{\theta^2}{2} + \frac{\theta^2}{2} + \theta(1-\theta) = \theta$$

so $T$ is unbiased for $\theta$. As

$$E(T^2) = \theta^2 + \frac{\theta^2}{2} + \theta(1-\theta) = \frac{\theta^2}{2} + \theta$$

we have

$$\text{Var}(T) = \frac{\theta^2}{2} + \theta - \theta^2 = \tfrac{1}{2}\theta(2 - \theta).$$

and hence $\text{Var}(T) = \text{Var}(\hat{\theta}) + \frac{\theta}{2}$ and so the variance of $T$ exceeds that of $\hat{\theta}$ unless $\theta = 0$. As $T = 2$ with probability $\frac{\theta^2}{4}$, then (unless $\theta = 0$) there is a positive probability that $T$ exceeds 1, and hence lies outside the parameter space $0 \le \theta \le 1$.

6. NSP
7.

$$\text{Likelihood: } f(\mathbf{x}|\theta) = \theta^n (x_1 \cdots x_n)^{\theta-1}$$

$$\text{Prior: } \pi(\theta) = \frac{\beta^\alpha}{\Gamma(\alpha)} \theta^{\alpha-1} e^{-\beta\theta}$$

$$\pi(\theta|\mathbf{x}) \propto \theta^{n+\alpha-1} e^{-\beta\theta + \theta \sum \log x_i}$$
$$= \theta^{n+\alpha-1} e^{-\theta(\beta - \sum \log x_i)}$$

Hence $\theta|\mathbf{x} \sim \text{Gamma}(n + \alpha, \beta - \sum \log x_i)$.

Therefore the Bayes estimator under squared error loss is

$$E(\theta|\mathbf{x}) = \frac{n+\alpha}{\beta - \sum \log x_i}.$$

8. NSP

---

## Problems of Chap. 11

### 11.1

1. The $100(1 - \alpha)\%$ symmetric confidence interval is

$$\left[\bar{x} - z_\gamma \times \frac{4}{10}, \bar{x} + z_\gamma \times \frac{4}{10}\right] \quad (\gamma = 1 - \frac{\alpha}{2})$$

and its width is $0.8z_\gamma$. The width of the quoted confidence interval is 1.316. Therefore, assuming that the quoted interval is symmetric,

$$0.8z_\gamma = 1.316 \Rightarrow z_\gamma = 1.645 \Rightarrow \gamma = 0.95 \ (\texttt{pnorm(1.645)} \text{ in R}).$$

This implies that $\alpha = 0.1$ and hence $100(1 - \alpha) = 90$, i.e. the confidence level is 90%.

2. NSP
3. Assuming that although the 972 homeowners are all insured within the same company they constitute a random sample from the population of all homeowners in the city, the 95% interval is given approximately by

$$\left[\hat{p} - 1.96\sqrt{\frac{\hat{p}(1-\hat{p})}{n}}, \ \hat{p} + 1.96\sqrt{\frac{\hat{p}(1-\hat{p})}{n}}\right],$$

where $n = 972$ and $\hat{p} = 357/972$. The interval is therefore $[0.337, 0.398]$.
4. NSP
5. In the usual notation,

$$\sum_{i=1}^{n} x_i = 13.676, \quad \sum_{i=1}^{n} x_i^2 = 18.703628.$$

These lead to

$$\bar{x} = 1.3676, \quad s = 0.00606.$$

Also, for the 95% CI, critical value $= 2.262$ (qt(0.975, df = 9) in R). Thus a 95% confidence interval for the mean is

$$\left[1.3676 - 2.262 \times \frac{0.00606}{\sqrt{10}}, \ 1.3676 + 2.262 \times \frac{0.00606}{\sqrt{10}}\right],$$

i.e. $[1.3633, 1.3719]$.
A 95% confidence interval for the mean error is obtained by subtracting the true wavelength of 1.372 from each endpoint. This gives $[-0.0087, -0.0001]$. As this contains negative values only, we conclude that the device tends to underestimate the true value.
6. NSP
7. Let $d_1, d_2, \ldots, d_{10}$ denote the differences in levels before and after treatment. Their values are

$$0.32, 0.16, -0.18, 0.11, 0.22, 0.00, 0.47, 0.31, 0.49, -0.12.$$

Then $\sum_{i=1}^{10} d_i = 1.78$ and $\sum_{i=1}^{10} d_i^2 = 0.7924$ so that $\bar{d} = 0.178$, $s_d = 0.2299$. A 95% confidence interval for the mean difference $\mu_1 - \mu_2$ is

$$\left[\bar{d} \pm \text{critical value} \ \frac{s_d}{\sqrt{10}}\right],$$

i.e. $\left[0.178 - 2.262 \times \frac{0.2299}{\sqrt{10}}, 0.178 + 2.262 \times \frac{0.2299}{\sqrt{10}}\right]$ or $[0.014, 0.342]$, as critical value $= 2.262$ (qt(0.975, df = 9) in R).

Note that the two samples are not independent. Thus the standard method of finding a confidence interval for $\mu_1 - \mu_2$, as used in Question 9 for example, would be inappropriate.

8. NSP

## Problems of Chap. 12

### 12.1

1. Under $H_0$, $\bar{X} \sim N(3, 3.5^2/50)$.
   Under $H_1$, $\bar{X} \sim N(4, 3.5^2/50)$.
   Then, with the usual notation,

$$\alpha = P(\text{Reject } H_0 | H_0 \text{ true})$$

$$= P(\bar{X} > 3.4 | \bar{X} \sim N(3, 3.5^2/50))$$

$$= 1 - P(\bar{X} \leq 3.4 | \bar{X} \sim N(3, 3.5^2/50))$$

$$= 1 - \Phi\left[\frac{(3.4 - 3)\sqrt{50}}{3.5}\right]$$

$$= 1 - \Phi(0.8081)$$

$$= 0.20952 \ (1 - \text{pnorm}(0.8081) \text{ in R}).$$

Also

$$\beta = P(\text{Accept } H_0 | H_1 \text{ true})$$

$$= P(\bar{X} \leq 3.4 | \bar{X} \sim N(4, 3.5^2/50))$$

$$= \Phi\left[\frac{(3.4 - 4)\sqrt{50}}{3.5}\right]$$

$$= \Phi(-1.2122)$$

$$= 1 - \Phi(1.2122)$$

$$= 1 - 0.88728$$

$$= 0.11272 \ (1 - \text{pnorm}(1.2122) \text{ in R}).$$

2. NSP
3. We take the hypotheses to be

$$H_0 : \mu = 1.2; \ H_1 : \mu > 1.2,$$

where $\mu$ is the mean weight of a randomly selected cabbage. Assuming that the weights are normally distributed, a suitable critical region having a 10% significance level is

$$C = \{t : t > t_{0.90}(11) = 1.363)$$

($\mathrm{qt}(0.9, \mathrm{df} = 11)$ in R) where, in the usual notation,

$$t = \frac{\bar{x} - 1.2}{s/\sqrt{12}}.$$

For the given data, $\bar{x} = 1.21$, $s = 0.031334$ and we find that $t = 1.1055$.

This t-value is not significant at the 10% level and the greengrocer should not buy the cabbages.

4. NSP
5. We consider the increases in the batting averages and use a paired comparison test.

| Batsman | 1 | 2 | 3 | 4 | 5 | 6 | 7 | 8 |
|---------|---|---|---|---|---|---|---|---|
| Increase | 1.83 | 3.74 | −1.49 | −2.84 | 6.22 | 0.29 | 1.56 | 4.05 |

Let $X$ denote the increase in the batting average of a randomly selected batsman and assume that $X \sim N(\mu, \sigma^2)$. We test the hypotheses

$$H_0 : \mu = 0; \ H_1 : \mu > 0.$$

Under $H_0$,

$$T = \frac{\bar{X}}{S/\sqrt{n}}$$

follows the Student's t-distribution with $n - 1$ degrees of freedom, where $n$ denotes the sample size and $\bar{X}$ and $S^2$ denote the sample mean and variance. A suitable critical region having significance level $\alpha$ is

$$C = \{t : t > t_{1-\alpha}(n - 1) = t_{1-\alpha}(7)),$$

since $n = 8$. From the data, $\bar{x} = 1.67$ and $s = 2.998$ so that $t = 1.58$.

Since $t_{0.9}(7) = 1.415$ ($\mathrm{qt}(0.9, \mathrm{df} = 7)$ in R) but $t_{0.95}(7) = 1.895$ ($\mathrm{qt}(0.95, \mathrm{df} = 7)$ in R), we see that the computed value is significant at the 10% level but not at the 5% level.

The winter practice is only one of several factors which might improve the batting average. For example, the second summer might be a good one with truer pitches than in the previous summer.

6. NSP

7. Let $p$ denote the proportion of wives who are happy with their marriage, and let $r$ be the probability that a response is 'Yes'.

$$r = P(\text{response is yes})$$
$$= P(\text{response is yes|heads})P(\text{heads}) + P(\text{response is yes|tails})P(\text{tails})$$
$$= \frac{3}{4} \times \frac{1}{2} + p \times \frac{1}{2}.$$

Hence $p = 2r - 0.75$.

Now $\hat{r} = 350/500 = 0.7$

and so $\hat{p} = 1.4 - 0.75 = 0.65$.

A 90% confidence interval for $r$ is given approximately by

$$\left[\hat{r} - 1.645\sqrt{\frac{\hat{r}(1-\hat{r})}{500}}, \hat{r} + 1.645\sqrt{\frac{\hat{r}(1-\hat{r})}{500}}\right],$$

i.e. $\left[0.7 - 1.645\sqrt{\frac{0.21}{500}}, 0.7 + 1.645\sqrt{\frac{0.21}{500}}\right]$ or $[0.6663, 0.7337]$.

Hence an approximate 90% confidence interval for $p$ is

$$[2 \times 0.6663 - 0.75, 2 \times 0.7337 - 0.75], \text{ i.e. } [0.5826, 0.7174].$$

Note that we cannot estimate $p$ directly as we do not know the proportion of answers 'yes' to the relevant question.

Since this confidence interval does not include the value $p = 0.8$, we reject the null hypothesis that at least 80% of wives are happy with their marriage at 10% level of significance.

8. The mean of the distribution is

$$E(X) = \int_0^\infty \frac{3\theta^3 x}{(x+\theta)^4} \, dx$$
$$= 3\theta^3 \int_\theta^\infty \frac{y-\theta}{y^4} \, dy \quad (\text{where } y = x + \theta)$$
$$= 3\theta^3 \left[-\frac{1}{2y^2} + \frac{\theta}{3y^3}\right]_\theta^\infty$$
$$= 3\theta^3 \left(\frac{1}{2\theta^2} - \frac{1}{3\theta^2}\right) = \theta/2.$$

Hence $E(X_i) = \theta/2, i = 1, 2, \ldots, n$, and $E(\hat{\theta}) = E(2\bar{X}) = \theta$.

Thus $\hat{\theta}$ is an unbiased estimator for $\theta$.

$$E(X^2) = \int_0^\infty \frac{3\theta^3 x^2}{(x+\theta)^4}\,dx = 3\theta^3 \int_\theta^\infty \frac{(y-\theta)^2}{y^4}\,dy$$

$$= 3\theta^3\left[-\frac{1}{y} + \frac{\theta}{y^2} - \frac{\theta^2}{3y^3}\right]_\theta^\infty$$

$$= \theta^2.$$

(a) So $\text{Var}(X) = \theta^2 - (\theta/2)^2 = 3\theta^2/4$.

(b) Hence $\text{Var}(\hat{\theta}) = 4\text{Var}(\bar{X}) = \frac{3\theta^2}{n}$. Therefore, standard deviation of $\hat{\theta}$ is given by $\theta\sqrt{3/n}$. Hence the required standard error is $\hat{\theta}\sqrt{3/n}$.

(c) The CLT states that

$$\hat{\theta} \sim N\left(\theta, \frac{3\theta^2}{n}\right)$$

asymptotically for large values of $n$.

(d) Suppose $P(-h \le Z \le h) = 1 - \alpha$ where $Z \sim N(0,1)$. Using the CLT, we have,

$$P\left(-h \le \frac{\hat{\theta} - \theta}{\theta\sqrt{3/n}} \le h\right) = 1 - \alpha$$

$$\Rightarrow P\left(-h\sqrt{3/n} \le \frac{\hat{\theta}}{\theta} - 1 \le h\sqrt{3/n}\right) = 1 - \alpha$$

$$\Rightarrow P\left(1 - h\sqrt{3/n} \le \frac{\hat{\theta}}{\theta} \le 1 + h\sqrt{3/n}\right) = 1 - \alpha$$

$$\Rightarrow P\left(\frac{\hat{\theta}}{1 + h\sqrt{3/n}} < \theta < \frac{\hat{\theta}}{1 - h\sqrt{3/n}}\right) = 1 - \alpha.$$

Thus, $\left(\frac{\hat{\theta}}{1+h\sqrt{3/n}}, \frac{\hat{\theta}}{1-h\sqrt{3/n}}\right)$ provides a $100(1-\alpha)\%$ confidence for $\theta$.

An alternative confidence interval $\left(\hat{\theta} \pm h\hat{\theta}\sqrt{3/n}\right)$, based on the less accurate approximation

$$\hat{\theta} \sim N\left(\theta, \frac{3\hat{\theta}^2}{n}\right),$$

will also receive some partial credit here.

(e) Here $\hat{\theta} = 2\bar{x} = 2 \times 11.25 = 22.5$. For 95% confidence interval, we take $h = 1.96$ and hence

$$\left( \frac{\hat{\theta}}{1 + h\sqrt{3/n}}, \frac{\hat{\theta}}{1 - h\sqrt{3/n}} \right) = \left( \frac{22.5}{1 + 1.96\sqrt{3/48}}, \frac{22.5}{1 - 1.96\sqrt{3/48}} \right),$$

which is given by $(15.1, 44.1)$.

Now, at 5% level of significance, we reject the null hypotheses: (i) $H_0$ : $\theta = 15$ since 15 is not contained in the 95% confidence interval. However, we fail to reject the null hypothesis, $H_0 : \theta = 20$ at 5% level of significance since the 95% confidence interval does contain the null value of $\theta = 20$.

The less accurate alternative 95% confidence interval $\left( \hat{\theta} \pm 1.96\,\hat{\theta}\sqrt{3/n} \right)$ turns out to be $22.5\left(1 \pm 1.96\sqrt{3/48}\right) = (11.47, 33.52)$, includes both the null hypotheses here.

9. NSP

10. (a) We assume that $X \sim N(\mu_x, \sigma_x^2)$. A 95% confidence interval for $\mu_x$ is

$$\bar{x} \pm qt(0.975, df = 24) \times \frac{s_x}{\sqrt{n}} = 16.6 \pm 2.06 \times \sqrt{\frac{47.6}{25}} = (13.8, 19.4).$$

(b) We assume that $X \sim N(\mu_x, \sigma_x^2)$, $Y \sim N(\mu_y, \sigma_y^2)$ independently and $\sigma_x^2 = \sigma_y^2$.

A 95% confidence interval for $\mu_x - \mu_y$ is

$$\bar{x} - \bar{y} + qt(0.975, df = n + m - 2)\, s\, \sqrt{\frac{1}{n} + \frac{1}{m}}$$

where

$$s^2 = \frac{(n-1)s_x^2 + (m-1)s_y^2}{n + m - 2} = 36.18.$$

Hence the required confidence interval is:

$$16.6 - 9.5 + 2.02\sqrt{\frac{1}{25} + \frac{1}{15}}\sqrt{36.18} = (3.1, 11.1)$$

(c) The 95% interval for the difference $\mu_x - \mu_y$ does not contain the hypothesised value 0 of the difference $\mu_x - \mu_y$. Hence there is significant evidence to reject the null hypothesis at 5% level of significance.

(d) The test statistic is:

$$t = \sqrt{\frac{nm}{n + m}} \frac{\bar{X} - \bar{Y}}{S}$$

where

$$S^2 = \frac{(n-1)S_x^2 + (m-1)S_y^2}{n+m-2}.$$

Under the null hypothesis the test statistic $t$ follows the $t$-distribution with $n+m-2$ degrees of freedom.

Here

$$t_{obs} = \sqrt{\frac{(25)(15)}{25+15}} \frac{16.6 - 9.5}{\sqrt{36.18}} = 3.61.$$

We reject the null hypothesis if $|t_{obs}| > t_{critical}$ where $t_{obs}$ is the observed value of the $t$-statistic and $t_{critical}$ is the $100\left(1-\frac{\alpha}{2}\right)\%$ percentile of the $t$ distribution with $n+m-2$ degrees of freedom.

Rejection region is the region shaded in red, where h is the critical value.

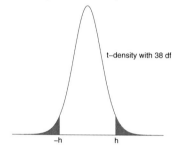

t–density with 38 df

The p-value of the test is given by

$$\begin{aligned} 2P\left(T > |t_{obs}|\right) &= 2 * \left(1 - pt\left(|t_{obs}|, df = n+m-2\right)\right) \\ &= 2 * (1 - pt(3.61, df = 38)) \\ &= 0.0009. \end{aligned}$$

## Problems of Chap. 13

### 13.1

1.(a) Here

$$\begin{aligned} M'(t) &= \tfrac{1}{4}\left(e^t - e^{-t}\right) \\ M''(t) &= \tfrac{1}{4}\left(e^t + e^{-t}\right) \end{aligned}$$

Thus

$$E(X) = M'(t)\big|_{t=0} = 0, \quad E(X^2) = M''(t)\big|_{t=0} = \tfrac{1}{2}.$$

Hence, $\text{Var}(X) = \frac{1}{2} = \sigma^2$.

(b) Standard devation of $Y = |-7|\sigma = \frac{7}{\sqrt{2}}$.

(c) Here is the pmf

| $x$ | 0 | 1 | $-1$ |
|---|---|---|---|
| $f(x)$ | $\frac{1}{2}$ | $\frac{1}{4}$ | $\frac{1}{4}$ |

2. NSP
3. The mgf is

$$M_X(t) = E(e^{tX})$$

$$= \sum_{x=1}^{\infty} e^{tX} q^{x-1} p$$

$$= \frac{p}{q} \sum_{x=1}^{\infty} (qe^t)^x$$

$$= \frac{p}{q} \{qe^t + (qe^t)^2 + (qe^t)^3 + \ldots\}$$

$$= \frac{p}{q} \frac{qe^t}{1 - qe^t} \text{ if } |qe^t| < 1 \Leftrightarrow e^t < \frac{1}{q} \Leftrightarrow t < -\log(1-p)$$

$$= \frac{pe^t}{1 - qe^t} \text{ if } t < -\log(1-p).$$

We can directly obtain the pgf similarly.

$$H(t) = E(t^X) = \frac{pt}{1 - qt} \text{ if } t < \frac{1}{q}$$

$$\frac{dH(t)}{dt} = p\frac{1 - qt + qt}{(1 - qt)^2} = \frac{p}{(1 - qt)^2}$$

$$\frac{d^2 H(t)}{dt^2} = \frac{2pq}{(1 - qt)^3}$$

Therefore $E(x) = \left.\frac{dH(t)}{dt}\right|_{t=1} = \frac{p}{(1-q)^2} = \frac{p}{p^2} = \frac{1}{p}$

$$E\{X(X-1)\} = \left.\frac{d^2 H(t)}{dt^2}\right|_{t=1} = \frac{2pq}{(1-q)^3} = \frac{2pq}{p^3} = \frac{2q}{p^2}$$

Therefore $E(X^2) - E(X) = \frac{2q}{p^2} \Rightarrow E(X^2) = \frac{2q}{p^2} + \frac{1}{p}$.

Therefore

$$var(X) = E(X^2) - \{E(X)\}^2$$

$$= \frac{2q}{p^2} + \frac{1}{p} - \frac{1}{p^2}$$

$$= \frac{2q + p - 1}{p^2}$$

$$= \frac{q + q + p - 1}{p^2}$$

$$= \frac{q}{p^2}$$

4.

$$M_X(t) = E(e^{tX})$$

$$= \int_0^\infty e^{tx} e^{-x} \, dx$$

$$= \frac{1}{1-t}, \quad t < 1$$

$$= 1 + t + t^2 + t^3 + \dots \text{ valid for } |t| < 1.$$

However

$$E(X^k) = \frac{d^k M_X(t)}{dt^k} \bigg|_{t=0}$$

$$= k! + \frac{(k+1)!}{1!} t + \frac{(k+1)!}{2!} t^2 + \dots \bigg|_{t=0}$$

$$= k!$$

(See this by taking k=1,2,3, and so on)
Alternative:

$$E(X^k) = \int_0^\infty x^k e^{-x} f(x) \, dx$$

$$= -x^k e^{-x} \bigg|_0^\infty + \int_0^\infty k x^{k-1} e^{-x} \, dx \quad \text{[Integration by parts]}$$

$$= k \int_0^\infty x^{k-1} e^{-x} \, dx$$

$$\vdots$$

$$= k! \int_0^\infty e^{-x} \, dx$$

$$= k!$$

5. NSP

6. We have

$$
\begin{aligned}
m_{aX+b}(t) &= \mathrm{E}[e^{t(aX+b)}] \\
&= \mathrm{E}[e^{taX}e^{tb}] \\
&= e^{tb}\mathrm{E}[e^{taX}] \\
&= e^{tb}m_X(at)
\end{aligned}
$$

as required.

It is known that $X \sim N(\mu, \sigma^2)$ if and only if $m_X(t) = \exp(\mu t + \sigma^2 t^2/2)$.
Applying (a)

$$
m_{aX+b}(t) = e^{(a\mu+b)t+a^2\sigma^2 t^2/2}
$$

and so $aX + b \sim N(a\mu + b, a^2\sigma^2)$. Consequently, $a = 1/\sigma$ and $b = -\mu/\sigma$ so
that $aX + b \sim N(0, 1)$.

7. NSP

8.

$$
\begin{aligned}
M_X(t) &= E(e^{tY}) \\
&= E\{e^{tX^2}\} \\
&= \int_{-\infty}^{\infty} e^{tx^2}\frac{e^{-\frac{1}{2}x^2}}{\sqrt{2\pi}}\,dx \\
&= \int_{-\infty}^{\infty} \frac{e^{-\frac{1}{2}x^2(1-2t)}}{\sqrt{2\pi}}\,dx \\
&= \int_{-\infty}^{\infty} \frac{e^{-\frac{1}{2}\frac{x^2}{b^2}}}{\sqrt{2\pi}}\,dx = b
\end{aligned}
$$

$$
\text{where } b^2 = \frac{1}{1-2t} \quad (\text{if } 1 - 2t > 0 \Rightarrow t < \frac{1}{2})
$$

$$
= \left(\frac{1}{1-2t}\right)^{\frac{1}{2}} \text{ if } t < \frac{1}{2}.
$$

Since the above is the mgf of the $\chi^2$ distribution with 1 degree of freedom we
can conclude that $Y = X^2$ follows the $\chi^2$ distribution with 1 degree of freedom.

9. It is known that $X \sim N(\mu, \sigma^2)$ if and only if $M_X(t) = e^{\mu t + \frac{\sigma^2 t^2}{2}}$
   Therefore $M_X(t) = e^{3t+t^2}$ is the mgf of $N(\mu = 3, \sigma^2 = 2)$
   Let $Y = X - \mu$
   Therefore $E\{[X - E(X)]^r\} = E(Y^r)$

But

$$M_Y(t) = e^{-t\mu}e^{\mu t + t^2}$$

$$= e^{t^2}$$

$$= \sum_{k=0}^{\infty} \frac{t^{2k}}{k!}$$

$$= \sum_{k=0}^{\infty} \frac{(2k)!}{k!} \cdot \frac{t^{2k}}{(2k)!}$$

Therefore $E(Y^r) = \begin{cases} 0 & \text{if } r \text{ is odd} \\ \frac{(2k)!}{k!} & \text{if } r = 2k \end{cases}$

10. NSP

---

## Problems of Chap. 14

### 14.1

1.(a) Since $f(x) = \frac{dF(x)}{dx}$ for a continuous random variable $X$, we have:

$$f(x) = \frac{2x}{4} = \frac{x}{2} \quad \text{if } 0 < x < 2.$$

(b) The transformation implies $x = 2 - y$ and $0 < y < 2$. We have $J = \frac{dx}{dy} = -1$. Hence the pdf of $Y$ is:

$$g(y) = \frac{2-y}{2}|J| = 1 - \frac{y}{2} \quad \text{if } 0 < y < 2.$$

2. NSP
3. The transformation is

$$y = (8x)^{\frac{1}{3}} = 2x^{\frac{1}{3}}$$

Therefore the range of $y$ is $0 < y < 2$.

Also $x = \frac{y^3}{8}$.

Therefore $\frac{dx}{dy} = \frac{1}{8} \cdot 3 \cdot y^2$.

Therefore the pdf of $Y$ if

$$g(y) = 1 \cdot \left| \frac{3}{8} y^2 \right|, \quad 0 < y < 2$$

$$= \frac{3}{8} y^2, \quad 0 < y < 2.$$

4. NSP
5. $f(x) = \frac{1}{2}(1+x), \quad -1 < x < 1.$

The transformation is $y = x^2$. It is decreasing in $-1 < x \leq 0$ and increasing in $0 < x < 1$.

Also $0 < y < 1$ and $x = \pm\sqrt{y}$

Therefore $\left| \dfrac{dx}{dy} \right| = \dfrac{1}{2\sqrt{y}}$

The pdf of $Y$ is

$$g(y) = \frac{1}{2}(1 - \sqrt{y}) \cdot \frac{1}{2\sqrt{y}} + \frac{1}{2}(1 + \sqrt{y}) \cdot \frac{1}{2\sqrt{y}}, \quad 0 < y < 1$$

$$= \frac{1}{4\sqrt{y}} \cdot 2, \quad 0 < y < 1$$

$$= \frac{1}{2\sqrt{y}}, \quad 0 < y < 1.$$

Alternative solution:

$$F(x) = \int_{-1}^{x} f(t)\, dt$$

$$= \frac{1}{2} \int_{-1}^{x} (1 + t)\, dt$$

$$= \frac{1}{2} \left[ t + \frac{t^2}{2} \right]_{-1}^{x}$$

$$= \frac{1}{2} \left\{ x + \frac{x^2}{2} + 1 - \frac{1}{2} \right\}$$

$$= \frac{1}{2} \left\{ x + \frac{x^2}{2} + \frac{1}{2} \right\}$$

$$G(y) = P(Y \leq y)$$

$$= P\{x^2 \leq y\}$$

$$= P\{-\sqrt{y} \le x \le \sqrt{y}\}$$

$$= F(\sqrt{y}) - F(-\sqrt{y})$$

$$= \frac{1}{2}\left\{\sqrt{y} + \frac{y}{2} + \frac{1}{2} + \sqrt{y} - \frac{y}{2} - \frac{1}{2}\right\}$$

$$= \sqrt{y}$$

Therefore the pdf of $Y$ is $g(y) = \dfrac{dG(y)}{dy} = \dfrac{1}{2\sqrt{y}}, \quad 0 < y < 1.$

6. NSP

7. Here the pdf of $X$ is:

$$f(x|\mu, \sigma^2) = \frac{1}{\sqrt{2\pi\sigma^2}} \exp\left\{-\frac{(x-\mu)^2}{2\sigma^2}\right\}, \quad -\infty < x < \infty.$$

Here the transformation implies,

$$x = \log(y), \text{ hence } J = \frac{dx}{dy} = \frac{1}{y},$$

and $0 \le y < \infty$. Hence, the pdf of $Y$ is,

$$
\begin{aligned}
g(y) &= \frac{1}{\sqrt{2\pi\sigma^2}} \exp\left\{-\frac{(\log(y)-\mu)^2}{2\sigma^2}\right\}\frac{1}{y} \\
&= \frac{1}{y\sqrt{2\pi\sigma^2}} \exp\left\{-\frac{(\log(y)-\mu)^2}{2\sigma^2}\right\}, \quad y \ge 0.
\end{aligned}
$$

**14.2**

1.(a) Here the Jacobian of the transformation is:

$$J = \begin{vmatrix} \frac{\partial x}{\partial w} & \frac{\partial x}{\partial z} \\ \frac{\partial y}{\partial w} & \frac{\partial y}{\partial z} \end{vmatrix} = \begin{vmatrix} 1 & 0 \\ 1 & 1 \end{vmatrix} = 1.$$

The transformation is one-to-one and $z > 0$ and $w > 0$. Hence

$$
\begin{aligned}
g(w, z) &= \frac{1}{\Gamma(m)\Gamma(n)} w^{m-1} z^{n-1} e^{-(z+w)} \\
&= \frac{1}{\Gamma(m)} w^{m-1} e^{-w} \frac{1}{\Gamma(n)} z^{n-1} e^{-z},
\end{aligned}
$$

for $w > 0$ and $z > 0$. Hence $W$ and $Z$ are independent gamma random variables since their joint pdf factorises into the product of the marginal pdfs.

(b) We know $E(Z) = n$ and $\mathrm{Var}(Z) = n$ and $E(W) = m$ and $\mathrm{Var}(W) = m$

$$E(Y) = E(W) + E(Z) = m + n, \quad \mathrm{Var}(Y) = \mathrm{Var}(W) + \mathrm{Var}(Z) = m + n.$$

2. NSP
3.(a) Here

$$E(Y) = \int_0^\infty y f(y) dy$$
$$= \int_0^\infty \frac{y}{\sqrt{2\pi}} e^{-\frac{y^2}{2}} dy$$
$$= \sqrt{\frac{2}{\pi}} \int_0^\infty e^{-u} du \quad [\text{substitute} u = \frac{y^2}{2}]$$
$$= \sqrt{\frac{2}{\pi}}.$$

$$E(Y^2) = \int_0^\infty y^2 f(y) dy$$
$$= \int_0^\infty \frac{y^2}{\sqrt{2\pi}} e^{-\frac{y^2}{2}} dy$$
$$= \sqrt{\frac{2}{\pi}} \int_0^\infty \sqrt{2u} e^{-u} du \quad [\text{substitute} u = \frac{y^2}{2}]$$
$$= \sqrt{\frac{2}{\pi}} \sqrt{2} \int_0^\infty u^{\frac{3}{2}-1} e^{-u} du$$
$$= \sqrt{\frac{2}{\pi}} \sqrt{2} \Gamma\left(\frac{3}{2}\right)$$
$$= \sqrt{\frac{2}{\pi}} \sqrt{2} \tfrac{1}{2} \sqrt{\pi}$$
$$= 1.$$

$$\text{Var}(Y) = E(Y^2) - (E(Y))^2 = 1 - \frac{2}{\pi}.$$

(b)

$$E(U) = E(X + \alpha Y)$$
$$= E(X + \alpha Y)$$
$$= E(X) + \alpha E(Y)$$
$$= \alpha \sqrt{\frac{2}{\pi}}.$$

$$\text{Var}(U) = \text{Var}(X + \alpha Y)$$
$$= \text{Var}(X) + \alpha^2 \text{Var}(Y)$$
$$= 1 + \alpha^2 \left(1 - \frac{2}{\pi}\right).$$

(c) The joint pdf of $X$ and $Y$ is:

$$f(x, y) = \frac{1}{\sqrt{2\pi}} e^{-\frac{x^2}{2}} \sqrt{\frac{2}{\pi}} e^{-\frac{y^2}{2}}, \quad -\infty < x < \infty, y > 0$$
$$= \frac{1}{\pi} e^{-\frac{x^2+y^2}{2}}.$$

Our transformation is $x = u - \alpha v$ and $y = v$. Hence the Jacobian of the transformation is:

$$J = \begin{vmatrix} \frac{\partial x}{\partial u} & \frac{\partial x}{\partial v} \\ \frac{\partial y}{\partial u} & \frac{\partial y}{\partial v} \end{vmatrix} = \begin{vmatrix} 1 & -\alpha \\ 0 & 1 \end{vmatrix} = 1.$$

We also have

$$x^2 + y^2 = (u - \alpha v)^2 + v^2$$
$$= (1 + \alpha^2)v^2 - 2\alpha u v + u^2$$
$$= u^2 + (1 + \alpha^2)\left(v^2 - 2v\frac{\alpha u}{1 + \alpha^2}\right)$$
$$= u^2 + (1 + \alpha^2)\left(v - \frac{u\alpha}{1 + \alpha^2}\right)^2 - \frac{u^2\alpha^2}{1 + \alpha^2}$$
$$= u^2\left(1 - \frac{\alpha^2}{1 + \alpha^2}\right) + (1 + \alpha^2)\left(v - \frac{u\alpha}{1 + \alpha^2}\right)^2$$
$$= \frac{u^2}{1 + \alpha^2} + (1 + \alpha^2)\left(v - \frac{u\alpha}{1 + \alpha^2}\right)^2$$

Now the joint pdf of $U$ and $V$ is:

$$g(u, v) = \frac{1}{\pi} e^{-\frac{(u-\alpha v)^2 + v^2}{2}} |J|, \quad -\infty < u < \infty, v > 0.$$

where $|J| = 1$.

(d) For the marginal pdf of $U$, we have:

$$g_U(u) = \int_0^\infty g(u, v) dv$$
$$= \int_0^\infty \frac{1}{\pi} e^{-\frac{1}{2}[(u-\alpha v)^2 + v^2]} dv$$

$$= \frac{1}{\pi} \int_0^\infty e^{-\frac{1}{2}\left[\frac{u^2}{1+\alpha^2} + (1+\alpha^2)\left(v - \frac{u\alpha}{1+\alpha^2}\right)^2\right]} dv$$

$$= \frac{1}{\pi} e^{-\frac{1}{2}\frac{u^2}{1+\alpha^2}} \int_0^\infty e^{-\frac{1}{2}(1+\alpha^2)\left(v - \frac{u\alpha}{1+\alpha^2}\right)^2} dv$$

To evaluate the last integral we substitute

$$z = \sqrt{1 + \alpha^2}\left(v - \frac{u\alpha}{1+\alpha^2}\right)$$

so that

$$dv = \frac{dz}{\sqrt{1 + \alpha^2}}$$

and when $v = 0$, $z = -\frac{u\alpha}{\sqrt{1+\alpha^2}}$. Now for $-\infty < u < \infty$,

$$g_U(u) = \frac{1}{\pi} e^{-\frac{1}{2}\frac{u^2}{1+\alpha^2}} \int_0^\infty e^{-\frac{1}{2}(1+\alpha^2)\left(v - \frac{u\alpha}{1+\alpha^2}\right)^2} dv$$

$$= \frac{1}{\pi} e^{-\frac{1}{2}\frac{u^2}{1+\alpha^2}} \int_{-\frac{u\alpha}{\sqrt{1+\alpha^2}}}^\infty e^{-\frac{1}{2}z^2} \frac{dz}{\sqrt{1 + \alpha^2}}$$

$$= \frac{1}{\pi} e^{-\frac{1}{2}\frac{u^2}{1+\alpha^2}} \frac{1}{\sqrt{1 + \alpha^2}} \sqrt{2\pi} \int_{-\frac{u\alpha}{\sqrt{1+\alpha^2}}}^\infty \frac{1}{\sqrt{2\pi}} e^{-\frac{1}{2}z^2} dz$$

$$= \sqrt{\frac{2}{\pi}} \frac{1}{\sqrt{1 + \alpha^2}} e^{-\frac{1}{2}\frac{u^2}{1+\alpha^2}} \left[1 - \Phi\left(-\frac{u\alpha}{\sqrt{1+\alpha^2}}\right)\right]$$

$$= \sqrt{\frac{2}{\pi}} \frac{1}{\sqrt{1 + \alpha^2}} e^{-\frac{1}{2}\frac{u^2}{1+\alpha^2}} \Phi\left(\frac{u\alpha}{\sqrt{1+\alpha^2}}\right),$$

as required.
4. NSP

## 14.3

1. NSP
2. Given that $X \sim$ exponential($\beta$) and $P\{X \le 1000\} = 0.75$ then the problem is to find $\beta$ since $E(X) = \beta$.
$$f(x|\beta) = \frac{1}{\beta} e^{\frac{-x}{\beta}}, \quad 0 < x < \infty$$

For any $a > 0$, $P\{x \le a\} = F(a) = \int_0^a \frac{1}{\beta} e^{-\frac{y}{\beta}} \, dy = 1 - e^{-\frac{a}{\beta}}$

$P\{X \le 1000\} = 1 - e^{-\frac{1000}{\beta}} = 0.75 \Rightarrow \beta = 721.35$

3. NSP

4. We have $f(x|\alpha, \beta) \frac{1}{B(\alpha, \beta)} x^{\alpha-1}(1-x)^{\beta-1}$

The transformation $Y = \frac{1}{X} - 1$ is decreasing and continuous if $x \in (0, 1)$ and also $0 < y < \infty$.

$$x = \frac{1}{1+y} \Rightarrow \frac{dx}{dy} = -\frac{1}{(1+y)^2}$$

Therefore the pdf of $Y$ is

$$g(y) = \frac{1}{B(\alpha, \beta)} \cdot \left(\frac{1}{1+y}\right)^{\alpha-1} \cdot \left(1 - \frac{1}{1+y}\right)^{\beta-1} \cdot \left(\frac{1}{1+y}\right)^2, \quad 0 \le y < \infty$$

$$= \frac{1}{B(\alpha, \beta)} \cdot \frac{y^{\beta-1}}{(1+y)^{\alpha+\beta}}, \quad 0 \le y < \infty$$

5. NSP

6. Here $Y = \delta X$. Hence $X = Y/\delta$ and $J = \frac{dx}{dy} = 1/\delta$. Also, $0 < x < 1$ implies $0 < y < \delta$. Now,

$$f_Y(y) = \frac{\Gamma(\alpha + \beta)}{\Gamma(\alpha)\Gamma(\beta)} \left(\frac{y}{\delta}\right)^{\alpha-1} \left(1 - \frac{y}{\delta}\right)^{\beta-1} \frac{1}{\delta}, \qquad 0 < y < \delta$$

$$= \frac{\Gamma(\alpha + \beta)}{\Gamma(\alpha)\Gamma(\beta)\delta^{\alpha+\beta-1}} y^{\alpha-1}(\delta - y)^{\beta-1}, \qquad 0 < y < \delta$$

Now the second part. We have

$$f_{X_1,X_2}(x_1, x_2) = \frac{\Gamma(\alpha_1 + \beta_1)\Gamma(\alpha_2 + \beta_2)}{\Gamma(\alpha_1)\Gamma(\beta_1)\Gamma(\alpha_2)\Gamma(\beta_2)} x_1^{\alpha_1-1}(1-x_1)^{\beta_1-1} x_2^{\alpha_2-1}(1-x_2)^{\beta_2-1},$$

$$0 < x_1 < 1, 0 < x_2 < 1.$$

Now, $y_1 = x_1 x_2$ and $y_2 = x_1(1 - x_2)$ imply $x_1 = y_1 + y_2$ and $x_2 = \frac{y_1}{y_1+y_2} = 1 - \frac{y_2}{y_1+y_2}$. These relations also imply:

(a) $0 < y_1 < 1$

(b) $0 < y_2 < 1$

(c) $0 < y_1 + y_2 < 1$

Now the Jacobian

$$J = \begin{pmatrix} \frac{\partial x_1}{\partial y_1} & \frac{\partial x_1}{\partial y_2} \\ \frac{\partial x_2}{\partial y_1} & \frac{\partial x_2}{\partial y_2} \end{pmatrix} = \begin{pmatrix} 1 & 1 \\ \frac{y_2}{(y_1+y_2)^2} & -\frac{y_1}{(y_1+y_2)^2} \end{pmatrix}$$

Hence $|J| = \frac{1}{y_1+y_2}$. Now,

$$\begin{aligned}
f_{Y_1,Y_2}(y_1, y_2) &= \frac{\Gamma(\alpha_1 + \beta_1)\Gamma(\alpha_2 + \beta_2)}{\Gamma(\alpha_1)\Gamma(\beta_1)\Gamma(\alpha_2)\Gamma(\beta_2)}(y_1 + y_2)^{\alpha_1 - 1}(1 - y_1 - y_2)^{\beta_1 - 1} \\
&\quad \left(\frac{y_1}{y_1 + y_2}\right)^{\alpha_2 - 1}\left(\frac{y_2}{y_1 + y_2}\right)^{\beta_2 - 1}\frac{1}{y_1 + y_2} \\
&= \frac{\Gamma(\alpha_1 + \beta_1)\Gamma(\alpha_2 + \beta_2)}{\Gamma(\alpha_1)\Gamma(\beta_1)\Gamma(\alpha_2)\Gamma(\beta_2)}y_1^{\alpha_2 - 1}y_2^{\beta_2 - 1}(y_1 + y_2)^{\alpha_1 - \alpha_2 - \beta_2} \\
&\quad (1 - y_1 - y_2)^{\beta_1 - 1} \\
&= \frac{\Gamma(\alpha_1 + \beta_1)\Gamma(\alpha_2 + \beta_2)}{\Gamma(\alpha_1)\Gamma(\beta_1)\Gamma(\alpha_2)\Gamma(\beta_2)}y_1^{\alpha_2 - 1}y_2^{\beta_2 - 1}(1 - y_1 - y_2)^{\beta_1 - 1},
\end{aligned}$$

since $\alpha_1 = \alpha_2 + \beta_2$,

when $0 < y_1 < 1, 0 < y_2 < 1$ and $0 < y_1 + y_2 < 1$.
Now

$$\begin{aligned}
f_{Y_1}(y_1) &= \int_0^{1-y_1} \frac{\Gamma(\alpha_1 + \beta_1)\Gamma(\alpha_2 + \beta_2)}{\Gamma(\alpha_1)\Gamma(\beta_1)\Gamma(\alpha_2)\Gamma(\beta_2)}y_1^{\alpha_2 - 1}y_2^{\beta_2 - 1}(1 - y_1 - y_2)^{\beta_1 - 1}dy_2 \\
&= \frac{\Gamma(\alpha_1 + \beta_1)\Gamma(\alpha_2 + \beta_2)}{\Gamma(\alpha_1)\Gamma(\beta_1)\Gamma(\alpha_2)\Gamma(\beta_2)}y_1^{\alpha_2 - 1}\int_0^{1-y_1} y_2^{\beta_2 - 1}(1 - y_1 - y_2)^{\beta_1 - 1}dy_2 \\
&= \frac{\Gamma(\alpha_1 + \beta_1)\Gamma(\alpha_2 + \beta_2)}{\Gamma(\alpha_1)\Gamma(\beta_1)\Gamma(\alpha_2)\Gamma(\beta_2)}y_1^{\alpha_2 - 1}\frac{\Gamma(\beta_2)\Gamma(\beta_1)}{\Gamma(\beta_1 + \beta_2)}(1 - y_1)^{\beta_1 + \beta_2 - 1} \\
&= \frac{\Gamma(\alpha_1 + \beta_1)\Gamma(\alpha_2 + \beta_2)}{\Gamma(\alpha_1)\Gamma(\alpha_2)\Gamma(\beta_1 + \beta_2)}y_1^{\alpha_2 - 1}(1 - y_1)^{\beta_1 + \beta_2 - 1} \\
&= \frac{\Gamma(\alpha_2 + \beta_1 + \beta_2)}{\Gamma(\alpha_2)\Gamma(\beta_1 + \beta_2)}y_1^{\alpha_2 - 1}(1 - y_1)^{\beta_1 + \beta_2 - 1}
\end{aligned}$$

for $0 < y_1 < 1$. This shows that $Y_1$ follows the beta distribution with parameter $\alpha_2$ and $\beta_1 + \beta_2$. Similarly, it can be shown that $Y_2$ has the beta distribution with parameters $\beta_2$ and $\alpha_2 + \beta_1$.

No, $Y_1$ and $Y_2$ are not independent.

7. NSP

## Problems of Chap. 15

### 15.1

1. First we show that the total probability is 1. We have,

$$\int_0^1 \int_0^1 f(x, y)dxdy = \int_0^1 \int_0^1 2xdxdy$$

$$= \int_0^1 dy \int_0^1 2xdx$$

$$= 2\frac{x^2}{2}\Big|_0^1$$

$$= 1,$$

as required. The marginal pdf of $X$ is:

$$f_X(x) = \int_0^1 f(x, y)dy = \int_0^1 2xdy = 2x, \quad 0 \le x \le 1.$$

The marginal pdf of $X$ is:

$$f_Y(x) = \int_0^1 f(x, y)dx = \int_0^1 2xdx = 1, \quad 0 \le y \le 1.$$

The conditional pdf of $Y|X = x$ is

$$f(Y|x) = \frac{f(x, y)}{f_X(x)} = 1, \quad 0 \le y \le 1.$$

Hence the the random variables $X$ and $Y$ are independent. Now,

$$P\left(X^2 < Y < X\right) = \int_0^1 2xdx \int_{x^2}^x dy$$

$$= \int_0^1 2x\left[y|_{x^2}^x\right]dx$$

$$= \int_0^1 2x(x - x^2)dx$$

$$= 2 \left. \frac{x^3}{3} - \frac{x^4}{4} \right|_0^1$$

$$= \frac{1}{6}$$

2. First we show that the total probability is 1. We have,

$$\int_0^1 \int_0^1 f(x, y)dxdy = \int_0^1 \int_0^1 (x + y)dxdy$$

$$= \int_0^1 \left[ \int_0^1 (x + y)dx \right] dy$$

$$= \int_0^1 \left[ \left. \left( \frac{x^2}{2} + xy \right) \right|_0^1 \right] dy$$

$$= \int_0^1 \left[ \tfrac{1}{2} + y \right] dy$$

$$= \left. \frac{y}{2} + \frac{y^2}{2} \right|_0^1$$

$$= 1,$$

as required. The marginal pdf of $X$ is:

$$f_X(x) = \int_0^1 f(x, y)dy = \int_0^1 (x + y)dy = \tfrac{1}{2} + x, \quad 0 \le< x \le 1.$$

Similarly, the marginal pdf of $Y$ is $f_Y(y) = \tfrac{1}{2} + y, \quad 0 \le< y \le 1.$ The conditional pdf of $Y|X = x$ is

$$f(y|x) = \frac{f(x, y)}{f_X(x)} = \frac{x + y}{\tfrac{1}{2} + x} = 2\frac{x + y}{1 + 2x}, \quad 0 \le y \le 1.$$

To show that $f(Y|x)$ is indeed a pdf, we have

$$\int_0^1 f(y|x)dy = \int_0^1 2\frac{x + y}{1 + 2x}dy$$

$$= \frac{2}{1 + 2x} \int_0^1 (x + y)dy$$

$$= \frac{2}{1 + 2x} \left( x + \tfrac{1}{2} \right)$$

$$= 1,$$

as required. Now we find $E(Y|X = x)$. We have,

$$E(Y|X = x) = \int_0^1 yf(y|x)dy$$

$$= \int_0^1 y2\frac{x+y}{1+2x}dy$$

$$= \frac{2}{1+2x}\int_0^1 y(x+y)dy$$

$$= \frac{2}{1+2x}\left. x\frac{y^2}{2} + \frac{y^3}{6}\right|_0^1$$

$$= \frac{2}{1+2x}\left(\frac{x}{2} + \frac{1}{3}\right)$$

$$= \frac{3}{1+2x}(3x+1), \quad 0 \le x \le 1.$$

Now we find the probability,

$$P\left(X > \sqrt{Y}\right) = \int_0^1 \left[\int_{\sqrt{y}}^1 (x+y)dx\right] dy$$

$$= \int_0^1 \left[\left. \frac{x^2}{2} + xy\right|_{\sqrt{y}}^1\right] dy$$

$$= \int_0^1 \left(\tfrac{1}{2} + y - \frac{y}{2} - y^{\frac{3}{2}}\right) dy$$

$$= \int_0^1 \left(\tfrac{1}{2} + \frac{y}{2} - y^{\frac{3}{2}}\right) dy$$

$$= \frac{7}{20}.$$

3. NSP

4.(a)

$$\int_0^1 \int_0^1 f(x, y)dxdy = \int_0^1 \int_0^1 Cdxdy$$

$$= C\int_0^1 \left[\int_0^{\sqrt{1-x^2}} dy\right] dx$$

$$= C\int_0^1 \left[\sqrt{1-x^2}\right] dx$$

$$= C \left. \frac{1}{2} x \sqrt{1 - x^2} + \frac{1}{2} \sin^{-1}(x) \right|_0^1$$

$$= C \frac{1}{2} \sin^{-1}(1)$$

$$= C \frac{1}{2} \frac{\pi}{2}$$

Hence, $\frac{C\pi}{4} = 1$ implies $C = \frac{4}{\pi}$.

(b) To find the probability, the region of integration is bounded by $0 \leq x \leq 1$, $0 \leq y \leq 1$, $0 \leq x^2 + y^2 \leq 1$ and $0 \leq x + y \leq 1$ which is given by: $0 \leq x \leq 1$, $0 \leq y \leq 1$, and $0 \leq x + y \leq 1$. Now,

$$P(X + Y \leq 1) = \int_0^1 \left[ \int_0^{1-x} \frac{4}{\pi} dy \right] dx$$

$$= \frac{4}{\pi} \int_0^1 (1 - x) dx$$

$$= \frac{4}{\pi} \frac{1}{2}$$

$$= \frac{2}{\pi}.$$

(c) NSP

(d) Here

$$E(X) = \int_0^1 x f_X(x) dx$$

$$= \int_0^1 \frac{4}{\pi} x \sqrt{1 - x^2} dx$$

$$= \frac{4}{\pi} \int_0^1 \sqrt{u} \frac{-du}{2} \quad \text{[substitute } u = 1 - x^2\text{]}$$

$$= \frac{2}{\pi} \left. \frac{u^{3/2}}{3/2} \right|_0^1$$

$$= \frac{4}{3\pi}.$$

We also evaluate:

$$E(X^2) = \int_0^1 x^2 f_X(x)dx$$

$$= \int_0^1 \frac{4}{\pi} x^2 \sqrt{1-x^2}dx$$

$$= \frac{4}{\pi} \int_0^1 u\sqrt{1-u}\frac{du}{2\sqrt{u}} \quad \text{[substitute } u = x^2\text{]}$$

$$= \frac{2}{\pi} \int_0^1 \sqrt{u}\sqrt{1-u}du$$

$$= \frac{2}{\pi} \int_0^1 u^{\frac{3}{2}-1}(1-u)^{\frac{3}{2}-1}du$$

$$= \frac{2}{\pi} B\left(\frac{3}{2},\frac{3}{2}\right)du$$

$$= \frac{2}{\pi}\frac{\left[\Gamma\left(\frac{3}{2}\right)\right]^2}{\Gamma(3)}$$

$$= \frac{2}{\pi}\frac{\left[\frac{1}{2}\Gamma\left(\frac{1}{2}\right)\right]^2}{2}$$

$$= \frac{1}{4\pi}\left[\sqrt{\pi}\right]^2$$

$$= \frac{1}{4}.$$

(e) To find $E(XY)$ we integrate subject to the constraints: $0 \leq x^2 + y^2 \leq 1$ and $0 \leq x \leq 1$ and $0 \leq y \leq 1$.

$$E(XY) = \int_0^1 \int_0^1 xy\frac{4}{\pi}dxdy$$

$$= \frac{4}{\pi} \int_0^1 x\left[\int_0^{\sqrt{1-x^2}} ydy\right] dx$$

$$= \frac{2}{\pi} \int_0^1 x(1-x^2)dx$$

$$= \frac{2}{\pi}\left(\frac{1}{2} - \frac{1}{4}\right)$$

$$= \frac{1}{2\pi}.$$

Hence,

$$\text{Var}(X) = \text{Var}(Y) = E(X^2) - (E(X))^2 = \frac{1}{4} - \frac{16}{9\pi^2}.$$

$$\text{Cov}(X, Y) = E(XY) - E(X)E(Y) = \frac{1}{2\pi} - \frac{16}{9\pi^2}.$$

Hence,

$$\text{Cor}(X, Y) = \frac{E(XY) - E(X)E(Y)}{\sqrt{\text{Var}(X)\text{Var}(Y)}}$$

$$= \frac{\frac{1}{2\pi} - \frac{16}{9\pi^2}}{\frac{1}{4} - \frac{16}{9\pi^2}}.$$

$$= \frac{18\pi - 64}{9\pi^2 - 64}.$$

(f) NSP

**15.2**

1. NSP
2. We know

$$E(X|Y = y) = \mu_x + \rho \frac{\sigma_x}{\sigma_y}(y - \mu_y)$$

$$E(Y|X = x) = \mu_y + \rho \frac{\sigma_y}{\sigma_x}(x - \mu_x)$$

$$\text{var}(Y|X = x) = \sigma_y^2(1 - \rho^2)$$

Here:

$$E(X|Y = y) = 3.7 - 0.15y$$

$$E(Y|X = x) = 0.4 - 0.6x$$

$$\text{var}(Y|X = x) = 3.64$$

Note that the coefficient of $y$ in $E(X|Y = y)$ is $\rho \frac{\sigma_x}{\sigma_y}$
Note that the coefficient of $x$ in $E(Y|X = x)$ is $\rho \frac{\sigma_y}{\sigma_x}$
Multiplying the two we get $\rho^2$.
Therefore $\rho^2 = (-0.15)(-0.6) = 0.09$

Therefore $\rho = -\sqrt{0.09} = -0.3$

Negative sign because coefficient of $y$ in $E(X|Y = y)$ is $= -0.15 = \rho\frac{\sigma_y}{\sigma_x}$ and $\sigma_x$ and $\sigma_y$ are positive.

$$\text{var}(Y|X = x) = \sigma_y^2(1 - \rho^2) = 3.64$$

$$\Rightarrow \sigma_y^2(1 - 0.09) = 3.64$$

$$\Rightarrow \sigma_y^2 = 4$$

Now

$$\frac{\rho\sigma_x}{\sigma_y} = -0.15 \Rightarrow \frac{(-0.3)\sigma_x}{2} = -0.15$$

$$\Rightarrow \sigma_x = 1$$

Now $\mu_x - 0.15(-\mu_y) = 3.7$
and $\mu_y - 0.6(-\mu_x) = 0.4$ $\Big\}$ solve for $\mu_x, \mu_y$.

Final answer: $\mu_x = 4$, $\mu_y = 2$, $\sigma_x = 1$, $\sigma_y = 2$, $\rho = -0.3$

3. NSP

4. Let $U = X + Y$, $V = X - Y$

Find the joint pdf of $U$ and $V$ where the joint pdf of $X$ and $Y$ is

$$f_{X,Y}(x, y) = \frac{1}{2\pi\sqrt{1 - \rho^2}}e^{-\frac{1}{2(1-\rho)^2}\{x^2-2\rho xy+y^2\}},$$

$$(-\infty < x < \infty, \; -\infty < y < \infty)$$

$x = \frac{u+v}{2}$; $y = \frac{u-v}{2}$

$J = \begin{vmatrix} \frac{\partial x}{\partial u} & \frac{\partial x}{\partial v} \\ \frac{\partial y}{\partial u} & \frac{\partial y}{\partial v} \end{vmatrix} = \begin{vmatrix} \frac{1}{2} & \frac{1}{2} \\ \frac{1}{2} & -\frac{1}{2} \end{vmatrix} = -\frac{1}{2}$

Now

$$x^2 - 2\rho xy + y^2$$

$$= \left(\frac{u+v}{2}\right)^2 + \left(\frac{u-v}{2}\right)^2 - 2\rho\frac{u+v}{2}\frac{u-v}{2}$$

$$= \frac{u^2}{2}(1 - \rho) + \frac{v^2}{2}(1 + \rho) \text{ (after algebra)}$$

Therefore

$$f_{U,V}(u,v) = \frac{1}{2\pi\sqrt{1-\rho^2}}\, e^{-\frac{1}{2(1-\rho)^2}\left\{\frac{u^2}{2}(1-\rho)+\frac{v^2}{2}(1-\rho)\right\}}\left|-\frac{1}{2}\right|$$

$$= \frac{1}{4\pi\sqrt{1-\rho^2}}\, e^{-\frac{u^2}{4(1+\rho)}-\frac{v^2}{4(1+\rho)}}$$

$$-\infty < u < \infty,\ -\infty < v < \infty$$

Easy to see that $U \sim N\{0, \sigma^2 = 2(1+\rho)\}$ and $V \sim N\{0, \sigma^2 = 2(1-\rho)\}$ and $U$ and $V$ are independent.

5. NSP

6.(a) Here we have

$$\rho\frac{\sigma_y}{\sigma_x} = 1.6,\ \rho\frac{\sigma_x}{\sigma_y} = 0.4$$

which implies $\rho^2 = 0.64$, ie. $\rho = 0.8$ with a positive sign. Given $\sigma_y = 2$ we obtain $\sigma_x = 1$.

Now fronm the two conditional expectations we have,

$$\mu_y - 1.6\mu_x = 3$$
$$\mu_x - 0.4\mu_y = -1.2.$$

Solving these we have $\mu_y = 3$ and $\mu_x - 0$.

(b)

$$Var(Y - bX) = Var(Y) - bCov(X,Y) + b^2Var(X)$$
$$= 4 - b(0.8)(2) + b^2$$

This is minimised at $b = 0.8$, which is equal to $\rho$.

(c) For the given parameters:

$$f(x,y) = \frac{1}{2\pi\sqrt{1-\rho^2}} \times$$
$$\exp\left[-\frac{1}{2(1-\rho^2)}\left\{(x^2 - 2\rho xy + y^2)\right\}\right],$$

for $-\infty < x < \infty$ and $-\infty < y < \infty$. Here the transformation is:

$$x = u,\ y = \rho u + v.$$

Hence the Jacobian is:

$$J = \begin{vmatrix} \frac{\partial x}{\partial u} & \frac{\partial x}{\partial v} \\ \frac{\partial y}{\partial u} & \frac{\partial y}{\partial v} \end{vmatrix} = \begin{vmatrix} 1 & 0 \\ \rho & 1 \end{vmatrix} = 1.$$

Also

$$x^2 - 2\rho xy + y^2 = u^2 + (\rho u + v)^2 - 2\rho u(\rho u + v)$$
$$= u^2(1 - \rho^2) + v^2.$$

Thus the joint density of $U$ and $V$ is:

$$g(u, v) = \frac{1}{2\pi\sqrt{1-\rho^2}} \times$$
$$\exp\left[-\frac{1}{2(1-\rho^2)}\left\{u^2(1 - \rho^2) + v^2\right\}\right],$$

for $-\infty < u < \infty$ and $-\infty < v < \infty$. Hence $U$ and $V$ are independent and $V \sim N(0, 1)$.

## 15.3

1. NSP
2.

|  | Prob. | Number |
|---|---|---|
| Freshman | $p_1 = 0.4$ | $x_1$ |
| Sophomore | $p_2 = 0.3$ | $x_2$ |
| Junior | $p_3 = 0.2$ | $x_3$ |
| Senior | $p_4 = 0.1$ | $x_4$ |

n=10. $\text{var}(X_i) = np_i(1 - p_i)$   $\text{cov}(X_i, X_j) = -np_i p_j$
$\text{var}(X_1) = 2.4$,   $\text{var}(X_2) = 2.1$,   $\text{var}(X_3) = 1.6$,   $\text{var}(X_4) = 0.9$
We can form a matrix of variances and covariances.

$$\begin{array}{c c c c c} & X_1 & X_2 & X_3 & X_4 \\ \begin{array}{c} X_1 \\ X_2 \\ X_3 \\ X_3 \end{array} & \left(\begin{array}{c} 2.4 \\ -1.2 \\ -0.8 \\ -0.4 \end{array}\right. & \begin{array}{c} -10(0.4)(0.3) \\ 2.1 \\ -0.6 \\ -0.3 \end{array} & \begin{array}{c} -10(0.4)(0.2) \\ -10(0.3)(0.2) \\ 1.6 \\ -0.2 \end{array} & \left.\begin{array}{c} -10(0.4)(0.1) \\ -10(0.3)(0.1) \\ -10(0.2)(0.1) \\ 0.9 \end{array}\right) \end{array}$$

Then we can find the correlations $\rho_{X_i, X_j} = \dfrac{\text{cov}\left(X_i, X_j\right)}{\sqrt{\text{var}\left(X_i\right)\text{var}\left(X_j\right)}}$

(a) $\begin{pmatrix} 1 & -0.534 & -0.408 & -0.272 \\ -0.534 & 1 & -0.327 & -0.218 \\ -0.408 & -0.327 & 1 & -0.167 \\ -0.272 & -0.218 & -0.167 & 1 \end{pmatrix}$ = correlation matrix

(b) $i = 1$, $j = 2$
(c) $i = 3$, $j = 4$

## Problems of Chap. 16

### 16.1

1. Here

$$
\begin{aligned}
E(Y_n) &= \tfrac{1}{n+1} \sum_{i=1}^{n} E(X_i) \\
&= \tfrac{1}{n+1} \sum_{i=1}^{n} \mu \\
&= \tfrac{n}{n+1} \mu \\
&= \mu \left( 1 - \tfrac{1}{n+1} \right) \\
&= \mu \quad \text{as } n \to \infty
\end{aligned}
$$

This proves the result.

2. NSP

3. Here $E(X_i) = \theta \le 1$ and $\mathrm{Var}(X_i) = \theta(1 - \theta) \le 0.25$ and thus both the mean and variance are finite. Moreover,

$$
E(\bar{X}_n) = \theta, \quad \mathrm{Var}(X_n) = \frac{\theta(1 - \theta)}{n}.
$$

Thus $\bar{X}_n$ is unbiased for $\theta$ and $\mathrm{Var}(X_n) \to 0$ as $n \to \infty$. Hence, $\bar{X}_n \to \theta$ in probability.

4.(a) Here $E(X) = \lambda$ and $\mathrm{Var}(X) = \lambda < \infty$. Also we have, $E(\bar{X}_n) = \lambda$ and $\mathrm{Var}(\bar{X}_n) = \tfrac{\lambda}{n}$. Because $\bar{X}_n$ is sample mean of a random sample, the CLT holds and consequently,

$$
Z_n = \frac{\sqrt{n}(\bar{X}_n - \lambda)}{\sqrt{\lambda}} \to N(0, 1)
$$

in distribution as $n \to \infty$.

(b) NSP

(c) As $\bar{X}_n$ converges to $\lambda$ in probability we claim that $T_n = \sqrt{\frac{\lambda}{\bar{X}_n}}$ converges to 1 in probability. Now,

$$
\begin{aligned}
Z_n T_n &= \frac{\sqrt{n}(\bar{X}_n - \lambda)}{\sqrt{\lambda}} \sqrt{\frac{\lambda}{\bar{X}_n}} \\
&= \frac{\sqrt{n}(\bar{X}_n - \lambda)}{\sqrt{\bar{X}_n}} \\
&\equiv Y_n.
\end{aligned}
$$

Hence using part 2 of the Slutsky's theorem, we claim that $Z_n T_n \equiv Y_n$ converges to $N(0, 1)$ in distribution as $n \to \infty$.

5. NSP

6.(a) Here

$$L(\theta; \mathbf{x}) = f(\mathbf{x}|\theta)$$
$$= \frac{1}{\theta^n} e^{-\frac{\sum_{i=1}^{n} x_i}{\theta}}$$
$$\implies \log(L(\theta; \mathbf{x})) = -n \log(\theta) - \frac{\sum_{i=1}^{n} x_i}{\theta}$$
$$\implies \frac{d \log(L(\theta; \mathbf{x}))}{d\theta} = -n\frac{1}{\theta} + \frac{\sum_{i=1}^{n} x_i}{\theta^2}$$
$$\implies \frac{d^2 \log(L(\theta; \mathbf{x}))}{d\theta^2} = n\frac{1}{\theta^2} - 2\frac{\sum_{i=1}^{n} x_i}{\theta^3}$$

Therefore, $\frac{d \log(L(\theta; \mathbf{x}))}{d\theta} = 0$ implies $\hat{\theta} = \bar{x}_n$. Also $\frac{d^2 \log(L(\theta; \mathbf{x}))}{d\theta^2}$ is negative at $\theta = \bar{x}_n$. Hence $\bar{X}_n$ is the maximum likelihood estimator of $\theta$.

(b) For the exponential distribution $E(X) = \theta$ and $\sigma^2 = \mathrm{Var}(X) = \theta^2$. Therefore, $E(\bar{X}_n) = \theta$, i.e. $\bar{X}_n$ is unbiased for $\theta$. Moreover,

$$\mathrm{Var}(\bar{X}_n) = \frac{\sigma^2}{n} = \frac{\theta}{n},$$

which goes to zero as $n \to \infty$. Hence $\bar{X}_n$ is a consistent estimator of $\theta$.

**16.2**

1. NSP

2. We have

$$u = \frac{d}{d\theta} \log f_{\mathbf{X}}(\mathbf{x}; \theta) = \frac{n}{\theta} + \sum_{i=1}^{n} \log x_i$$

and

$$E(U) = 0 \implies E\left(\frac{n}{\theta} + \sum_{i=1}^{n} \log X_i\right) = 0$$
$$\implies E\left(-\sum_{i=1}^{n} \log X_i\right) = \frac{n}{\theta}$$
$$\implies E\left(-\frac{1}{n} \sum_{i=1}^{n} \log X_i\right) = \frac{1}{\theta}.$$

Hence, $-\overline{\log X}$ is unbiased for $\theta^{-1}$. Now we can write

$$U = -n\left(-\frac{1}{n}\sum_{i=1}^{n}\log x_i - \frac{1}{\theta}\right).$$

3. We have

$$u = \frac{d}{d\theta}\log f_{\mathbf{X}}(\mathbf{x};\theta) = \frac{n}{\theta} - \frac{1}{1-\theta}\left(\sum_{i=1}^{n}x_i - n\right)$$

and

$$E(U) = 0 \implies E\left[\frac{n}{\theta} - \frac{1}{1-\theta}\left(\sum_{i=1}^{n}X_i - n\right)\right] = 0$$

$$\implies \frac{1}{1-\theta}E\left(\sum_{i=1}^{n}X_i\right) - \frac{n}{1-\theta} = \frac{n}{\theta}$$

$$\implies \frac{1}{n}E\left(\sum_{i=1}^{n}X_i\right) = \frac{1-\theta}{\theta} + 1$$

$$\implies E[\overline{X}] = \frac{1}{\theta}.$$

Hence, $\overline{X}$ is unbiased for $\theta^{-1}$. Now we can write

$$U = -\frac{n}{1-\theta}\left(\overline{X} - 1 - \frac{1-\theta}{\theta}\right) = -\frac{n}{1-\theta}\left(\overline{X} - \frac{1}{\theta}\right).$$

4. NSP
5. We have

$$f_{\mathbf{X}}(\mathbf{x};\theta) = \prod_{i=1}^{n}\frac{x_i^2}{2\theta^3}\exp\left(-\frac{x_i}{\theta}\right) = (2\theta^3)^{-n}\exp\left(-\frac{1}{\theta}\sum_{i=1}^{n}x_i\right)\prod_{i=1}^{n}x_i^2$$

$$\implies \log f_{\mathbf{X}}(\mathbf{x};\theta) = -n\log 2 - 3n\log\theta - \frac{1}{\theta}\sum_{i=1}^{n}x_i + 2\sum_{i=1}^{n}\log x_i$$

$$\implies u(\theta) = \frac{d}{d\theta}\log f_{\mathbf{X}}(\mathbf{x};\theta) = -\frac{3n}{\theta} + \frac{1}{\theta^2}\sum_{i=1}^{n}x_i$$

$$= -\frac{3n}{\theta} + \frac{n\overline{x}}{\theta^2}. \quad (2)$$

Hence,

$$E[U(\theta)] = 0 \implies E\left(-\frac{3n}{\theta} + \frac{n\overline{X}}{\theta^2}\right) = 0$$

$$\implies \frac{nE(\overline{X})}{\theta^2} = \frac{3n}{\theta}$$

$$\implies E(\overline{X}) = 3\theta.$$

The m.l.e. $\hat{\theta}$ satisfies $U(\hat{\theta}) = 0$ and hence

$$-\frac{3n}{\hat{\theta}} + \frac{n\bar{x}}{\hat{\theta}^2} = 0 \implies \hat{\theta} = \frac{\bar{x}}{3}.$$

The asymptotic distribution is $\hat{\theta} \sim N\left[\theta, \frac{1}{\mathcal{I}(\theta)}\right]$, where

$$\mathcal{I}(\theta) = E\left[-\frac{d^2}{d\theta^2}\log f_{\mathbf{X}}(\mathbf{x}; \theta)\right]$$

$$= E\left(-\frac{3n}{\theta^2} + \frac{2n\overline{X}}{\theta^3}\right), \quad \text{by differentiating (2),}$$

$$= -\frac{3n}{\theta^2} + \frac{2nE(\overline{X})}{\theta^3}$$

$$= -\frac{3n}{\theta^2} + \frac{6n\theta}{\theta^3}$$

$$= \frac{3n}{\theta^2}.$$

Hence, asymptotically, $\hat{\theta} = \overline{X}/3 \sim N\left(\theta, \frac{\theta^2}{3n}\right)$.

---

## Problems of Chap. 17

### 17.1

1.(a) We recall that

$$\hat{\beta}_1(Y, X) = \frac{\sum_{i=1}^n y_i(x_i - \bar{x})}{S_{xx}} \quad \text{and} \quad \hat{\beta}_1(Y, X) = \bar{y} - \hat{\beta}_1(Y, X)\bar{x}.$$

Now

$$\hat{\beta}_1(Z, X) = \frac{\sum_{i=1}^{n} z_i(x_i - \bar{x})}{S_{xx}}$$

$$= \frac{\sum_{i=1}^{n}(ay_i + b)(x_i - \bar{x})}{S_{xx}}$$

$$= \frac{\sum_{i=1}^{n} ay_i(x_i - \bar{x})}{S_{xx}} + \frac{\sum_{i=1}^{n} b(x_i - \bar{x})}{S_{xx}}$$

$$= a\hat{\beta}_1(Y, X) + b\frac{0}{S_{xx}}$$

$$= a\hat{\beta}_1(Y, X),$$

and

$$\hat{\beta}_0(Z, X) = \bar{z} - \hat{\beta}_1(Z, X)\bar{x}$$
$$= a\bar{y} + b - a\hat{\beta}_1(Y, X)\bar{x}$$
$$= a\hat{\beta}_1(Y, X) + b.$$

Also:

$$\hat{Z}_i = a\hat{Y}_i + b$$

$$R_i(Z) = Z_i - \hat{Z}_i$$

$$= aY_i + b - (a\hat{Y}_i + b)$$

$$= a(Y_i - \hat{Y}_i)$$

$$= aR_i(Y)$$

(b) We have

$$\bar{w} = c\bar{x} + d, \text{ and } w_i - \bar{w} = c(x_i - \bar{x}).$$

Now

$$\hat{\beta}_1(Y, W) = \frac{\sum_{i=1}^{n} y_i(w_i - \bar{w})}{S_{ww}}$$

$$= \frac{\sum_{i=1}^{n} y_i c(x_i - \bar{x})}{c^2 S_{xx}}$$

$$= \frac{1}{c}\hat{\beta}_1(Y, X)$$

and

$$\hat{\beta}_0(Y, W) = \bar{y} - \hat{\beta}_1(Y, W)\bar{w}$$

$$= \bar{y} - \frac{1}{c}\hat{\beta}_1(Y, X)(c\bar{x} + d)$$

$$= \hat{\beta}_0(Y, X) - \frac{d}{c}\hat{\beta}_1(Y, X)$$

Also:

$$\hat{Y}_i(W) = \hat{\beta}_0(Y, W) + \hat{\beta}_1(Y, W)w_i$$

$$= \hat{\beta}_0(Y, X) - \frac{d}{c}\hat{\beta}_1(Y, X) + \frac{1}{c}\hat{\beta}_1(Y, X)(cx_i + d)$$

$$= \hat{\beta}_0(Y, X) + \hat{\beta}_1(Y, X)x_i$$

$$= \hat{Y}_i(X).$$

and now $R_i(W) = y_i - \hat{Y}_i(W) = y_i - \hat{Y}_i(X) = R_i(X)$.
(c) Let

$$S = \sum_{i=1}^{n} \frac{(y_i - \beta_0 - \beta_1 x_i)^2}{1 + \beta_1^2}.$$

We have

$$\frac{\partial S}{\partial \beta_0} = \sum_{i=1}^{n} \frac{1}{1 + \beta_1^2}(-2)(y_i - \beta_0 - \beta_1 x_i).$$

Therefore $\frac{\partial S}{\partial \beta_0} = 0$ implies $\sum_{i=1}^{n}(y_i - \tilde{\beta}_0 - \tilde{\beta}_1 x_i) = 0$, that is,

$$\tilde{\beta}_0 = \bar{y} - \tilde{\beta}_1 \bar{x}.$$

Now we have:

$$\frac{\partial S}{\partial \beta_1} = \sum_{i=1}^{n} \left\{ \frac{1}{1 + \beta_1^2}(-2x_i)(y_i - \beta_0 - \beta_1 x_i) + (y_i - \beta_0 - \beta_1 x_i)^2 \frac{-2\beta_1}{(1 + \beta_1^2)^2} \right\}.$$

Now setting the above to zero and substituting the equation for $\beta_0$ we have:

$$\sum_{i=1}^{n}\left\{x_i(y_i - \beta_0 - \beta_1 x_i) + (y_i - \beta_0 - \beta_1 x_i)^2 \frac{\beta_1}{1 + \beta_1^2}\right\} = 0$$

$$\Longrightarrow \sum_{i=1}^{n}\left\{x_i(1 + \beta_1^2)(y_i - \bar{y} + \beta_1\bar{x} - \beta_1 x_i)\right.$$

$$\left. + \beta_1(y_i - \bar{y} + \beta_1\bar{x} - \beta_1 x_i)^2\right\} = 0$$

$$\Longrightarrow \beta_1^3\left\{-\sum x_i(x_i - \bar{x}) + \sum(x_i - \bar{x})^2\right\}$$

$$+ \beta_1^2\left\{\sum x_i(y_i - \bar{y}) - 2\sum(y_i - \bar{y})(x_i - \bar{x})\right\}$$

$$+ \beta_1\left\{\sum(y_i - \bar{y})^2 - \sum x_i(x_i - \bar{x})\right\} + \sum x_i(y_i - \bar{y}) = 0$$

$$\Longrightarrow \beta_1^2 S_{xy} + \beta_1\left(S_{xx} - S_{yy}\right) - S_{xy} = 0$$

Clearly, a solution of the above is:

$$\tilde{\beta}_1 = \frac{S_{yy} - S_{xx} + \sqrt{(S_{yy} - S_{xx})^2 + 4S_{xy}^2}}{2S_{xy}}.$$

a. Letting $Z_i = aY_i + b$,

$$\tilde{\beta}_1(Z, X) = \frac{S_{zz} - S_{xx} + \sqrt{(S_{zz} - S_{xx})^2 + 4S_{xz}^2}}{2S_{xz}}$$

$$= \frac{a^2 S_{yy} - S_{xx} + \sqrt{(a^2 S_{yy} - S_{xx})^2 + 4a^2 S_{xy}^2}}{2a S_{xy}}$$

b. Letting $w_i = ax_i + d$ we have

$$\tilde{\beta}_1(Z, W) = \frac{a^2 S_{yy} - S_{ww} + \sqrt{(a^2 S_{yy} - S_{ww})^2 + 4a^2 S_{wy}^2}}{2a S_{wy}}$$

$$= \frac{a^2(S_{yy} - S_{xx}) + a^2\sqrt{(S_{yy} - S_{xx})^2 + 4S_{xy}^2}}{2a^2 S_{xy}}$$

$$= \frac{(S_{yy} - S_{xx}) + \sqrt{(S_{yy} - S_{xx})^2 + 4S_{xy}^2}}{2S_{xy}}$$

$$= \tilde{\beta}_1(Y, X).$$

Now,

$$\tilde{\beta}_0(Z, W) = \bar{z} - \bar{w}\tilde{\beta}_1(Z, W)$$

$$= a\bar{y} + b - (a\bar{x} + d)\tilde{\beta}_1(Y, X)$$

$$= a\tilde{\beta}_0(Y, X) + b - d\tilde{\beta}_1(Y, X).$$

$$\hat{Z}_i = \tilde{\beta}_0(Z, W) + w_i\tilde{\beta}_1(Z, W)$$

$$= a\tilde{\beta}_0(Y, X) + b - d\tilde{\beta}_1(Y, X) + (ax_i + d)\tilde{\beta}_1(Y, X)$$

$$= a\tilde{\beta}_0(Y, X) + ax_i\tilde{\beta}_1(Y, X) + b$$

$$= a\hat{Y}_i + b,$$

as expected and

$$R_i(Z) = Z_i - \hat{Z}_i$$

$$= aY_i + b - (a\hat{Y}_i + b)$$

$$= aY_i + b - (a\hat{Y}_i + b)$$

$$= a(Y_i - \hat{Y}_i)$$

$$= aR_i(Y).$$

2. NSP
3. NSP
4.(a) The scatter plot on the left panel of Fig. 20.1 shows that the relationship between age and value is not linear so we need to consider a transformation. A suitable transformation would be a logarithmic transformation of the value variable. The scatterplot on this transformed scale can be seen on the right panel of Fig. 20.1. This scatterplot shows a linear relationship so we should fit a linear model on the transformed scale.

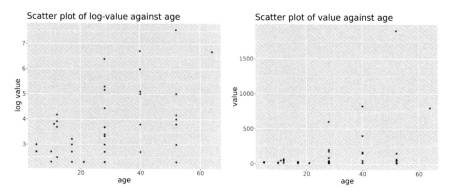

**Fig. 20.1**   Scatter plots on different scales

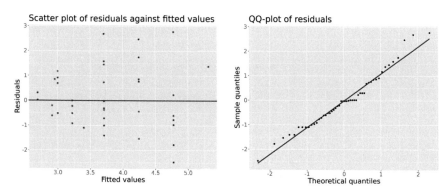

**Fig. 20.2**   Two residual plots for the linear regression model

(b) Let $Y_i$ denote the logarithm of the values of the Beanie Baby and $x_i$ the age of the Beanie Baby in the $i$th sample, where $i = 1, \ldots, 50$. We will fit the simple linear regression model using R:

$$Y_i = \beta_0 + \beta_1 x_i + \epsilon_i, \quad \epsilon_i \sim N(0, \sigma^2), \quad i = 1, \ldots, 50. \tag{20.1}$$

We examine two diagnostic plots in Fig. 20.2, the Anscombe plot and the normal probability plot. We can see some funneling in the Anscombe plot which suggests that out assumption of constant error variance may not be justified. We can also see some small departures from normality in the normal probability plot. This suggests that a linear model may not be the best fitting model for this data.

The parameter estimate table for the simple linear regression model is given by:

|             | Estimate | Std. Error | t value | Pr(>\|t\|) |
|-------------|----------|------------|---------|-----------|
| (Intercept) | 2.46806  | 0.31537    | 7.826   | 4.04e-10  |
| age         | 0.04442  | 0.01022    | 4.345   | 7.18e-05  |

When we test the hypothesis $H_0 : \beta_1 = 0$, we observe the value 4.345 for the test statistic with a P-value of less than 0.0001. This means the data provides strong evidence against the null hypothesis that there is no relationship between age and value.

A 95% confidence interval for the slope parameter, $\beta_1$, is given by $(0.04442 \pm 0.01022 \times 2.010635)$ which is $(0.0239, 0.0650)$.

The estimate of $\sigma^2$ is 1.2973 with 48 residual degrees of freedom.

(c) The fitted model on the original scale is given by:

Fitted value of Beanie Baby $= e^{2.4681} \exp(0.0444 \times \text{Age of Beanie Baby})$.

The predicted value for the logarithm of the mean value of a Beanie Baby aged 35 months is 4.0228. This means the predicted value on the original scale is $e^{4.0228} = 55.8517$. This mean value has 95% confidence interval given by : $(e^{3.6548}, e^{4.3906}) = (38.6598, 80.6888)$.

The predicted value for the logarithm of the mean value of a Beanie Baby aged 45 months is 4.4669. This means the predicted value on the original scale is $e^{4.4669} = 87.0863$. This value has 97% prediction interval given by : $(e^{1.859}, e^{7.075}) = (6.41, 1182.04)$.

---

## Problems of Chap. 18

### 18.1

1. NSP
2. We know $\hat{\beta} = (X^T X)^{-1} X^T y$.
   (i) $R = y - X\hat{\beta} = y - X(X^T X)^{-1} X^T y = (I - H)y$.
   (ii)

$$
\begin{aligned}
R^T R &= \{(I - H)y\}^T (I - H)y \\
&= y^T (I - H)^T (I - H)y \\
&= y^T (I - H)(I - H)y \quad (H \text{ is symmetric}) \\
&= y^T (I - H)y \quad (\text{idempotent}) \\
&= y^T R \\
&= R^T y \quad (\text{scalar})
\end{aligned}
$$

(iii) The general result $HX = X$ implies $H1 = 1$ since $1$ is a column of $X$. Thus $1^T H = 1^T$.

$$
\begin{aligned}
1^T R &= 1^T (I - H)y \\
&= 1^T y - 1^T H y \\
&= 1^T y - 1^T y \\
&= 0
\end{aligned}
$$

Hence the result.

(iv)

$$\sum_{i=1}^{n}(R_i - \bar{R})(y_i - \bar{y}) = \sum_{i=1}^{n} R_i(y_i - \bar{y}) \quad \text{(since } \bar{R} = 0\text{)}$$
$$= \sum_{i=1}^{n} R_i y_i - \bar{y}\sum_{i=1}^{n} R_i$$
$$= \sum_{i=1}^{n} R_i y_i$$
$$= R^T R \quad \text{(from (ii) above)}$$

(v)

$$1 - r^2 = 1 - \frac{(n-1)s_y^2 - R^T R}{(n-1)s_y^2}$$
$$= \frac{R^T R}{(n-1)s_y^2}$$

The sample correlation co-efficient between the residuals and observed values is

$$\text{Corr}(R, \mathbf{y}) = \frac{\sum_{i=1}^{n}(r_i - \bar{r})(y_i - \bar{y})}{(n-1)s_y\sqrt{\frac{1}{n-1}\sum_{i=1}^{n}(r_i - \bar{r})^2}}$$
$$= \frac{R^T R}{(n-1)s_y\sqrt{\frac{R^T R}{n-1}}}$$
$$= \sqrt{\frac{R^T R}{(n-1)s_y^2}}$$
$$= \sqrt{1 - r^2}.$$

(vi) $R = (I - H)\mathbf{y}$ and the fitted values is $H\mathbf{y}$, since $(I - H)H = 0$ by the general theorem proved in class the correlation between the residuals and the fitted values is zero.

(vii)

$$Var(R) = Var[(I - H)\mathbf{y}]$$
$$= (I - H)Var[\mathbf{y}](I - H)^T$$
$$= (I - H)\sigma^2 I(I - H)$$
$$= \sigma^2(I - H)$$

(viii) • (v) says the plot of residuals against observed values is not going to be informative.

• (vi) says that the plot of the residuals against the fitted values should be a random scatter since there is no correlation expected between them if the model was correct.

• (vii) says that the residuals are correlated and they may have less variability than the observed values.

3. NSP

4. (i) Here

$$X = \begin{pmatrix} -1 & 2 \\ 1 & 1 \\ 2 & -1 \end{pmatrix}.$$

$$X^T X = \begin{pmatrix} 6 & -3 \\ -3 & 6 \end{pmatrix}, \quad (X^T X)^{-1} = \frac{1}{9} \begin{pmatrix} 2 & 1 \\ 1 & 2 \end{pmatrix}, \quad X^T \mathbf{y} = \begin{pmatrix} -y_1 + y_2 + 2y_3 \\ 2y_1 + y_2 - y_3 \end{pmatrix}.$$

Therefore,

$$\hat{\boldsymbol{\beta}} = (X^T X)^{-1} X^T \mathbf{y} = \frac{1}{3} \begin{pmatrix} y_2 + y_3 \\ y_1 + y_2 \end{pmatrix}.$$

(ii) Here the $F$-test will be equivalent to the $t$-test since it is testing just one parameter equals to zero. We have:

$$H = X(X^T X)^{-1} X^T = \begin{pmatrix} -1 & 2 \\ 1 & 1 \\ 2 & -1 \end{pmatrix} \frac{1}{9} \begin{pmatrix} 2 & 1 \\ 1 & 2 \end{pmatrix} \begin{pmatrix} -1 & 1 & 2 \\ 2 & 1 & -1 \end{pmatrix} = \frac{1}{3} \begin{pmatrix} 2 & 1 & -1 \\ 1 & 2 & 1 \\ -1 & 1 & 2 \end{pmatrix}$$

Therefore,

$$I - H = \frac{1}{3} \begin{pmatrix} 1 & -1 & 1 \\ -1 & 1 & -1 \\ 1 & -1 & 1 \end{pmatrix}.$$

Easy to verify that:

$$\mathbf{y}^T (I - H) \mathbf{y} = \frac{1}{3}(y_1 - y_2 + y_3)^2.$$

Now

$$S^2 = \frac{1}{3-2} \mathbf{y}^T (I - H) \mathbf{y} \\ = \frac{1}{3}(y_1 - y_2 + y_3)^2.$$

To test $H_0 : a = 0$ we use the $t$-statistic (see page 26 of chapter 2):

$$\frac{\hat{\beta}_j - 0}{s[(X^T X)^{-1}]_{jj}^{1/2}} = \frac{\frac{1}{3}(y_2 + y_3)}{\sqrt{\frac{2}{9}\frac{1}{3}(y_1 - y_2 + y_3)^2}} = \frac{(y_2 + y_3)}{\sqrt{\frac{2}{3}(y_1 - y_2 + y_3)^2}}$$

Therefore, the $F$ statistic is given by:

$$F = \frac{3(y_2 + y_3)^2}{2(y_1 - y_2 + y_3)^2}$$

which follows the $F$ distribution with 1 and 1 degree of freedom under $H_0$.

(iii) Residual plots: Calculate residuals $\hat{\epsilon}_i = y_i - (X\hat{\beta})_i$.

Order and plot the residuals against the normal order statistics to check for normality. Plot should be a straight line.

Plot the residuals against the fitted values $(X\hat{\beta})_i$ to check for homoscedasticity. Plot should look like a random scatter.

5. **Analysis of the crime data**

(a) The crime data set reports the crime rate, y, (number of offenses reported to police per million population) and some related demographic statistics for 47 US states in 1960. The data were collected from the FBI's Uniform Crime Report and other government agencies to determine how the variable crime rate depends on the other variables measured in the study. This is the main objective of the study here.

There are 13 explanatory variables in the data set. Our first task here is to fit a linear model including all 13 covariates. Then we remove the variables one by one using the drop1 command in R until we cannot drop any more variable without losing significant reduction in fit. The $R^2$ value for the model with all 13 variables is reported to be 0.7692.

The command drop1 with the fitted full model as the only argument shows the $C_p$ statistics values of the full model and other models by removing one variable at a time. In our stepwise method, we sequentially drop the variable removal of which leads to the lowest $C_p$ value in each step. We stop at a point where dropping of any further variable will only increase the $C_p$ value.

In this way we drop the variables NW, LF, S, N, Ex1, M, and U1. We are then left with the model:

$$Y_i = \beta_1 + \beta_2\text{Age}_i + \beta_3\text{Ed}_i + \beta_4\text{Ex0}_i + \beta_5\text{U2}_i + \beta_6\text{W}_i$$
$$+ \beta_7\text{X}_i + \epsilon_i, \quad i = 1, \ldots, 47. \tag{20.2}$$

For this model R produces the following table of $C_p$ values.

| | Df | Sum of Sq | RSS | Cp |
|---|---|---|---|---|
| \<none\> | | | 17351.06 | 23423.93 |
| Age | 1 | 4461.00 | 21812.07 | 27017.38 |
| Ed | 1 | 6214.73 | 23565.79 | 28771.11 |
| Ex0 | 1 | 15596.51 | 32947.58 | 38152.89 |
| U2 | 1 | 1628.68 | 18979.75 | 24185.06 |
| W | 1 | 1252.58 | 18603.65 | 23808.96 |
| X | 1 | 8932.28 | 26283.34 | 31488.66 |

From the above output, we see that no further variables can be dropped without increasing the $C_p$ value. The multiple $R^2$ for this final model is 0.7478 which is not substantially lower than 0.7692 corresponding to the full model. Thus we take the above model as the final model for our subsequent analysis.

(b) The parameter estimates for the final model in Eq. (20.2) are given in the following table:

```
              Estimate      se      lower     upper
(Intercept)  -618.503  108.246  -837.275  -399.730
        Age     1.125    0.351     0.416     1.834
         Ed     1.818    0.480     0.847     2.789
        Ex0     1.051    0.175     0.697     1.405
         U2     0.828    0.427    -0.036     1.692
          W     0.160    0.094    -0.030     0.349
          X     0.824    0.181     0.457     1.190
```

The column 'Estimate' provides the least-square estimates of the regression coefficients, the standard errors of the estimates are given in the next column named se and the remaining two columns provide the 95% confidence interval estimates. At 5% level of significance, the co-efficients for U2 and W are not significant. The other regressors namely Age, Ed, Ex0 and X have positive effect on the crime rate since the corresponding co-efficients are estimated to be positive and their 95% confidence intervals do not include the point zero.

The estimates are little difficult to interpret. We first note down the definitions of the regressors, Age, Ed, Ex0 and X.

• **Age:** The number of males of age 14–24 per 1000 population.
• **Ed:** Mean number of years of schooling (times 10) for persons of age 25 or older.
• **Ex0:** 1960 per capita expenditure on police by state and local government.
• **X:** The number of families per 1000 earning below 1/2 the median income.

It is difficult to interpret the positive co-efficients for the explanatory variables Ed and Ex0. Intuitively, crime rate should go down as the population become more educated, that is, the co-efficient for education should be negative. Similarly, the co-efficient for Ex0 should be negative since additional money spent on police should decrease the time rate.

This model emphasises the need to look at the direction of the coefficients. From these coefficients, it appears that more education and police expenditures increase the crime rate. Perhaps there is another variable, a "lurking variable" not collected with these data, which causes both education and crime rate to increase together.

(c)

The residual plots are given in Fig. 20.3. The plot of the residuals against the fitted values does not show any clearly visible pattern. However, the variability

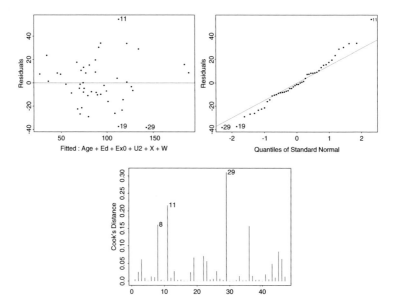

**Fig. 20.3**   Residual plots for the final model fitted to the crime data set

of the residuals is seen to be little higher near the center of the scale of the fitted values. This raises some concerns regarding the assumption of homoscedasticity (equal variance). However, the patterns shown in the plot are not strong enough and those do not suggest an alternative model for the data.

By default R flags three observations which have the highest absolute residual values. In this case the three observations are 11, 19 and 29. However, these are not necessarily outliers. Further invstigation is necessary before such a conclusion can be made.

The normal qq-plot is a plot of the residuals against corresponding quantiles of the standard normal distribution. This plot is often used to check normality of the residuals (and hence data). The plot does not show strong departure from normality, although there are residual values which are away from the straight line in the plot.

The Cook's distance plot is often used to detect outliers or influential observations in the data. Again by default S-plus flags up three most influential observations which may or may not be outliers. The most influential data point is observation number 29. Examining this in detail we see that this observation has the minimal Age and maximal Ex0 values, the value for W and X were also outside the first and third quartiles. The data value of 104.3 and the other two regressors (Ed and U2) were within the corresponding first and third quartile. Thus for this observation four out of six regressors had values outside the first and third quartile while the value of the response variable was within the first and third quartiles. This probably has caused the observation number 29 being the most influential.

This data set is a good example of what can go wrong in a regression analysis. Although the model fits reasonably well with $R^2$ values of about 75%, the interpretation suffers because of many reasons including the one which says that both education and crime rate increase together.

## Problems of Chap. 19

### 19.1

1. (a) One way ANOVA decomposition for

$$Y_{ij} = \mu + \alpha_i + \epsilon_{ij}, i = 1, \dots, k, \ j = 1, \dots, n_i.$$

where $\epsilon_{ij} \sim N(0, \sigma^2)$

$$\sum_i \sum_j (y_{ij} - \bar{y})^2 = \sum_i \sum_j (y_{ij} - \bar{y}_i)^2 + \sum_i n_i (\bar{y}_i - \bar{y})^2.$$

Let $n = \sum_i n_i$. Note that:

$$\sum_i \sum_j (y_{ij} - \bar{y})^2 = \sum_i \sum_j y_{ij}^2 - n\bar{y}^2, \quad \sum_i n_i (\bar{y}_i - \bar{y})^2 = \sum_i n_i \bar{y}_i^2 - n\bar{y}^2.$$

Here

$$\sum_i \sum_j y_{ij}^2 = 17322090 + 16116822 + 12208021 + 14929994 = 60576927$$

$$n\bar{y}^2 = \frac{1}{40}(13046 + 12592 + 10965 + 12128)^2 = 59367759.0$$

$$\sum_i n_i \bar{y}_i^2 = \frac{13046^2}{10} + \frac{12592^2}{10} + \frac{10965^2}{10} + \frac{12128^2}{10} = 59607618.9.$$

Therefore,

$$\sum_i \sum_j (y_{ij} - \bar{y})^2 = 1209168.0, \quad \sum_i n_i (\bar{y}_i - \bar{y})^2 = 239859.9,$$

$$\sum_i \sum_j (y_{ij} - \bar{y}_i)^2 = 469308.1.$$

To test the null hypothesis of no regional effect:

$$H_0 : \alpha_1 = \alpha_2 = \alpha_3 = \alpha_4 = 0,$$

we calculate

$$F = \frac{\sum_i n_i (\bar{y}_i - \bar{y})^2 / 3}{\sum_i \sum_j (y_{ij} - \bar{y}_i)^2 / 36} = \frac{79953.3}{26925.2} = 2.97.$$

Under $H_0$, $F \sim F_{3,36}$. From tables, we see that $F_{0.05}(3, 36) = 2.88$. The observed $F$ exceeds this, so we conclude there is a significant difference between regions, at the 5% level of significance.

(b) Under our model, for a future observation $y_{i0}$ in region $i$,

$$\hat{y}_{i0} = \hat{E}(y_{i0}) = \bar{y}_i,$$

$$Var\left(\hat{E}(y_{i0}) - E(y_{i0})\right) = Var(\bar{y}_i) = \frac{\sigma^2}{10}$$

which implies s.e.$(\hat{E}(y_{i0})) = \frac{s}{\sqrt{10}}$.

$$Var\left(\hat{y}_{i0} - y_{i0}\right) = Var(\bar{y}_i) + Var(y_{i0}) = \frac{\sigma^2}{10} + \sigma^2,$$

which implies s.e.$(\hat{y}_{i0}) = s\sqrt{\frac{11}{10}}$.

Here

$$s^2 = \frac{1}{\text{residual d.f.}} RSS = \frac{1}{36} \sum_i \sum_j (y_{ij} - \bar{y}_i)^2 = 26925.2,$$

hence $s = 164.1$.

Hence (i) $\hat{E}(y_{10}) = 1304.6$, 95% C.I. is $1304.6 \pm 2.03 \frac{164.1}{\sqrt{10}} = (1199.3, 1409.9)$.

(ii) $\hat{y}_{30} = 1096.5$, 95% C.I. is $1096.5 \pm 2.03 \times 164.1\sqrt{\frac{11}{10}} = (747.1, 1445.9)$.

Intervals for future observations are wider than the intervals for conditional means.

2. NSP

# Correction to: Introduction to Probability, Statistics & R

## Correction to:
**S. K. Sahu, *Introduction to Probability, Statistics & R*,**
**https://doi.org/10.1007/978-3-031-37865-2**

The original version of the book was inadvertently published with two typos in the R code. The readers trying to copy code from book to R will find out the mistakes and their code will not work. The problem is not there in the code distributed in the R package. The chapters affected due to the two typos of wrong R code are Chapter 2: Getting Started with R on pages 33, 35 (six times), 36 (two times), 37 (four times), Chapter 19: Analysis of Variance on pages 461 (two times), 462 (two times), Chapter 20: Solutions to Selected Exercises in page 471. First typo is in R code, it should be just $ instead of $. The second typo is in R code, it should be $ instead of $. In the book back matter on page 551 reference number 7 appears as Goon, A.M., Gupta, M.K., Dasgupta, B.: An Outline of Statistical Theory, vol I. World Press, Kolkatta (1994). The type setters added Kolkatta. It should be either Kolkata or Calcutta but not Kolkatta.

The correction chapters and the book have been updated with these changes.

---

The updated version of these chapters can be found at
https://doi.org/10.1007/978-3-031-37865-2_2
https://doi.org/10.1007/978-3-031-37865-2_19
https://doi.org/10.1007/978-3-031-37865-2_20

© The Author(s), under exclusive license to Springer Nature Switzerland AG 2024   C1
S. K. Sahu, *Introduction to Probability, Statistics & R*,
https://doi.org/10.1007/978-3-031-37865-2_21

# Appendix: Table of Common Probability Distributions

**A**

## A.1 Discrete Distributions

1. **Binomial** A random variable $X$ is said to follow the binomial distribution, denoted by $\text{Bin}(n, p)$, if it has the probability mass function:

$$f(x|n, p) = \binom{n}{x} p^x (1 - p)^{n-x}, \quad x = 0, 1, \ldots, n, \tag{A.1}$$

where $n$ is a positive integer and $0 < p < 1$ and

$$\binom{n}{x} = \frac{n!}{x!(n - x)!}.$$

It can be shown that $E(X) = np$ and $\text{Var}(X) = np(1 - p)$. The Bernoulli distribution is a special case when $n = 1$.

In Sect. 13.4 it has been proved that the mgf of the binomial distribution $\text{Bin}(n, p)$ is

$$M(t) = (pe^t + q)^n.$$

for any value of $t$. In Sect. 6.2, we have discussed that the binomial probabilities can be evaluated using the R commands dbinom and pbinom.

2. **Negative binomial** A random variable $X$ is said to follow the negative binomial distribution, denoted by $\text{NBin}(r, p)$, if it has the probability mass function:

$$f(x|r, p) = \frac{\Gamma(r + x)}{\Gamma(x + 1)\Gamma(r)} p^r (1 - p)^x, \quad x = 0, 1, \ldots, \tag{A.2}$$

---

The original version of this Back Matter has been revised. A correction to this Back Matter can be found at https://doi.org/10.1007/978-3-031-37865-2_21

for a positive integer $r > 0$ and $0 < p < 1$. The random variable $X$ can be interpreted as the number of failures in a sequence of independent Bernoulli trials before the occurrence of the $r$th success, where the success probability is $p$ in each trial. The **geometric** distribution is a special case when $r = 1$.

In Sect. 6.5 we have discussed the use of the R commands dnbinom and pnbinom to evaluate negative binomial probabilities. Using the sum of the negative binomial series in Sect. 6.3.2, the mgf of a random variable having the NBin$(r, p)$ distribution can be obtained as:

$$M(t) = \left( \frac{p}{1 - (1 - p)e^t} \right)^r, \text{ for } t < -\log(1 - p).$$

Section 6.5 also derives the results $E(X) = \frac{r(1-p)}{p}$ and $\text{Var}(X) = \frac{r(1-p)}{p^2}$.

3. **Poisson** A random variable $X$ is said to follow the Poisson distribution, denoted by $P(\lambda)$, if it has the probability mass function:

$$f(x|\lambda) = e^{-\lambda}\frac{\lambda^x}{x!}, \quad x = 0, 1, \ldots, \tag{A.3}$$

where $\lambda > 0$. It can be shown that $E(X) = \lambda$ and $\text{Var}(X) = \lambda$, see Sect. 6.6. The Poisson distribution is a limiting case of the Binomial distribution when $n \to \infty$, $p \to 0$ but $\lambda = np$ remains finite in the limit, see Sect. 6.6.4. The R commands dpois and ppois can be used to evaluate probabilities, see Sect. 6.6. The mgf of the Poisson distribution $P(\lambda)$ can be proved to be

$$M(t) = e^{\lambda(e^t - 1)},$$

for any value of $t$ see Example 13.16.

4. **Multinomial**

A random vector has the *multinomial distribution* with parameters $n$ and $\mathbf{p} = (p_1, \ldots, p_k)$. if it has the pmf

$$P(X_1 = x_1, \ldots, X_k = x_k) = \frac{n!}{x_1! \cdots x_k!} p_1^{x_1} \cdots p_k^{x_k},$$

where $0 < p_i < 1$ and $0 < x_i < n$ for all $i = 1, \ldots, n$ and

$$\sum_{i=1}^{k} p_i = 1, \quad \text{and} \quad \sum_{i=1}^{k} x_i = n.$$

The special case when $k = 2$ is the binomial distribution.

## A.2   Continuous Distributions

1. **Uniform** A random variable $X$ follows the uniform distribution $U(a, b)$ if it has the pdf

$$f(x|a, b) = \frac{1}{b - a}, \quad \text{when } a \leq x \leq b$$

It can be shown that $E(X) = \frac{1}{2}(a + b)$ and $\text{Var}(X) = (b - a)^2/12$.

2. **Cauchy** A random variable $X$ follows the Cauchy distribution $C(a, b)$ if it has the pdf

$$f(x|a, b) = \frac{b}{\pi \left[b^2 + (x - a)^2\right]}, \quad \text{for } -\infty < x < \infty, \ b > 0.$$

The mean and variance do not exist for this distribution. The standard Cauchy distribution is the special case of this distribution when $a = 0$ and $b = 1$. The mgf of this distribution does not exist since the required integral is not finite. As a result for this distribution the mean and variance do not exist. The R command pcauchy with the optional arguments location=a and scale=b can be used to evaluate probabilities under this distribution.

3. **Gamma** A random variable $X$ follows the gamma distribution with parameters $a > 0$ and $b > 0$, denoted by $G(a, b)$, if it has the pdf

$$f(x|a, b) = \frac{b^a}{\Gamma(a)} x^{a-1} e^{-bx}, x > 0. \tag{A.4}$$

The fact that the above is a density function implies that

$$\int_0^\infty x^{a-1} e^{-bx} dx = \frac{\Gamma(a)}{b^a}, \quad \text{for } x > 0. \tag{A.5}$$

Thus $\Gamma(a)$ is the definite integral above when $b = 1$. Using the gamma integral (A.5) it can be shown that:

$$E(X) = \frac{a}{b}, \text{ and } \text{Var}(X) = \frac{a}{b^2}.$$

We also can prove the results:

$$E\left(\frac{1}{X}\right) = \frac{b}{a - 1}, \text{ if } a > 1 \text{ and } \text{Var}\left(\frac{1}{X}\right) = \frac{b^2}{(a - 1)^2(a - 2)}, \text{ if } a > 2.$$

In fact, the distribution of the random variable $\frac{1}{X}$ is known as the inverse gamma distribution with parameters $a$ and $b$, denoted by $\text{IG}(a, b)$.

The mgf of the Gamma distribution has been derived in Example 13.3 to be:

$$M(t) = \left(\frac{b}{b-t}\right)^a.$$

The suite of R commands dgamma, pgamma and qgamma can be used to evaluate the density cumulative probabilities and to invert the cdf of the gamma distribution as discussed in Sect. 7.3.2.

The Gamma distribution has two important special cases:

a. **Exponential** When $a = 1$ the gamma distribution reduces to the exponential distribution which has pdf

$$f(x|b) = be^{-bx}, x > 0, \tag{A.6}$$

for $b > 0$.

b. $\chi^2$ When $a = \frac{n}{2}$ and $b = \frac{1}{2}$ the gamma distribution reduces to the $\chi^2$-distribution with $n$ degrees of freedom.

4. **Inverse Gamma** A random variable $X$ follows the inverse gamma distribution with parameters $a > 0$ and $b > 0$ if it has the pdf

$$f(x|a, b) = \frac{b^a}{\Gamma(a)} \frac{1}{x^{a+1}} e^{-\frac{b}{x}}, x > 0. \tag{A.7}$$

As stated above, it can be shown that $E(X) = \frac{b}{a-1}$ if $a > 1$ and $\text{Var}(X) = \frac{b^2}{(a-1)^2(a-2)}$ if $a > 2$.

5. **Beta** A random variable $X$ follows the beta distribution with parameters $a > 0$ and $b > 0$ if it has the pdf

$$f(x|a, b) = \frac{1}{B(a, b)} x^{a-1} (1 - x)^{b-1}, 0 < x < 1, \tag{A.8}$$

where

$$B(a, b) = \int_0^1 x^{a-1} (1 - x)^{b-1} dx.$$

It can be shown that $E(X) = \frac{a}{a+b}$ and $\text{Var}(X) = \frac{ab}{(a+b)^2(a+b+1)}$ and

$$B(a, b) = \frac{\Gamma(a)\Gamma(b)}{\Gamma(a + b)}, \quad a > 0, b > 0.$$

The R functions dbeta, pbeta, qbeta can be used to evaluate the density, cumulative probabilities and to invert the cdf of the beta distribution. The additional arguments are shape1=a and shape2=b.

6. **Univariate normal**: A random variable $X$ has the normal distribution, denoted by $N(\mu, \sigma^2)$, if it has the probability density function

$$f(x|\mu, \sigma^2) = \frac{1}{\sqrt{2\pi\sigma^2}} e^{-\frac{1}{2\sigma^2}(x-\mu)^2}, \quad -\infty < x < \infty, \tag{A.9}$$

where $\sigma^2 > 0$ and $\mu$ is unrestricted. It can be shown that $E(X) = \mu$ and $\text{Var}(X) = \sigma^2$. Example 13.4 and the discussion that followed showed that the mgf of the normal $N(\mu, \sigma^2)$ distribution is:

$$M(t) = e^{\mu t + \frac{\sigma^2}{2}t^2},$$

for any value of $t$. The R command pnorm calculates the cumulative probabilities and to invert the cdf we use the command qnorm.

7. **Bivariate normal** A pair of random variables $X$ and $Y$ is said to have the bivariate normal distribution $N_2(\mu_x, \mu_y, \sigma_x^2, \sigma_y^2, \rho)$ it it has the pdf

$$f(x, y) = \frac{1}{2\pi\sigma_x\sigma_y\sqrt{1-\rho^2}} \times$$
$$\exp\left[-\frac{1}{2(1-\rho^2)}\left\{\left(\frac{x-\mu_x}{\sigma_x}\right)^2 - 2\rho\left(\frac{x-\mu_x}{\sigma_x}\right)\left(\frac{y-\mu_y}{\sigma_y}\right) + \left(\frac{y-\mu_y}{\sigma_y}\right)^2\right\}\right],$$

for $-\infty < x, y < \infty$. We have the following results:

$$E(X) = \mu_x, E(Y) = \mu_y, \text{Var}(X) = \sigma_x^2, \text{Var}(Y) = \sigma_y^2, \text{Corr}(X, Y) = \rho,$$

$$E(Y|X = x) = \mu_y + \rho\frac{\sigma_y}{\sigma_x}(x - \mu_x) \text{ and } E(X|Y = y) = \mu_x + \rho\frac{\sigma_x}{\sigma_y}(y - \mu_y).$$

8. **Multivariate normal**: A $p$ dimensional random variable $\mathbf{X}$ has the multivariate normal distribution, denoted by $N_p(\boldsymbol{\mu}, \Sigma)$, if it has the probability density function

$$f(\mathbf{x}|\boldsymbol{\mu}, \Sigma) = \left(\frac{1}{2\pi}\right)^{\frac{p}{2}} |\Sigma|^{-\frac{1}{2}} e^{-\frac{1}{2}(\mathbf{x}-\boldsymbol{\mu})'\Sigma^{-1}(\mathbf{x}-\boldsymbol{\mu})}, \tag{A.10}$$

where $-\infty < x_i < \infty$, $i = 1, \ldots, p$ and $\Sigma$ is a $p \times p$ positive semi-definite matrix, $|\Sigma|$ is the determinant of the matrix $\Sigma$. It can be shown that $E(\mathbf{X}) = \boldsymbol{\mu}$ and $\text{Var}(X) = \Sigma$. The matrix $\Sigma$ is also called the covariance matrix of $X$ and the inverse matrix $\Sigma^{-1}$ is called the inverse covariance matrix. This distribution is a generalisation of the univariate normal distribution.

The conditional distribution of a subset of the random variables $\mathbf{X}$ given the other random variables. Suppose that we partition the $p$-dimensional vector $X$

into one $p_1$ and another $p_2 = p - p_1$ dimensional random variable $\mathbf{X}_1$ and $\mathbf{X}_2$. Similarly partition $\boldsymbol{\mu}$ into two parts $\boldsymbol{\mu}_1$ and $\boldsymbol{\mu}_2$ so that we have:

$$\mathbf{X} = \begin{pmatrix} \mathbf{X}_1 \\ \mathbf{X}_2 \end{pmatrix}, \quad \boldsymbol{\mu} = \begin{pmatrix} \boldsymbol{\mu}_1 \\ \boldsymbol{\mu}_2 \end{pmatrix}.$$

Partition the $p \times p$ matrix $\Sigma$ into four matrices: $\Sigma_{11}$ having dimension $p_1 \times p_1$, $\Sigma_{12}$ having dimension $p_1 \times p_2$, $\Sigma_{21} = \Sigma'_{12}$ having dimension $p_2 \times p_1$, and $\Sigma_{22}$ having dimension $p_2 \times p_2$ so that we can write

$$\Sigma = \begin{pmatrix} \Sigma_{11} & \Sigma_{12} \\ \Sigma_{21} & \Sigma_{22} \end{pmatrix}.$$

The conditional distribution of $\mathbf{X}_1 | \mathbf{X}_2 = \mathbf{x}_2$ is the following normal distribution:

$$N\left( \boldsymbol{\mu}_1 + \Sigma_{12}\Sigma_{22}^{-1}(\mathbf{x}_2 - \boldsymbol{\mu}_2), \ \Sigma_{11} - \Sigma_{12}\Sigma_{22}^{-1}\Sigma_{21} \right).$$

The marginal distribution of $\mathbf{X}_i$ is $N\left( \boldsymbol{\mu}_i, \Sigma_{ii} \right)$ for $i = 1, 2$.

A key distribution theory result that we use in the book is the distribution of a linear function of the multivariate random variable $\mathbf{X}$. Let

$$\mathbf{Y} = \mathbf{a} + B\mathbf{X}$$

where $\mathbf{a}$ is a $m$ $(< p)$ dimensional vector of constants and $B$ is an $m \times p$ matrix. The first part of the result is that

$$E(\mathbf{Y}) = \mathbf{a} + B\boldsymbol{\mu} \quad \text{and} \quad \text{Var}(\mathbf{Y}) = B\Sigma B'$$

if $E(\mathbf{X}) = \boldsymbol{\mu}$ and $\text{Var}(\mathbf{X}) = \Sigma$. If in addition we assume that $X \sim N_p(\mu, \Sigma)$ then

$$\mathbf{Y} \sim N_m(\mathbf{a} + B\boldsymbol{\mu}, B\Sigma B').$$

9. **Univariate t**: A random variable $X$ has the $t$-distribution, $t(\mu, \sigma^2, \nu)$ if it has the density function:

$$f(x|\mu, \sigma^2, \nu) = \left( 1 + \frac{(x - \mu)^2}{\nu\sigma^2} \right)^{-\frac{\nu+1}{2}}, \quad -\infty < x < \infty, \qquad (A.11)$$

when $\nu > 0$. It can be shown that

$$E(X) = \mu \text{ if } \nu > 1 \text{ and } \text{Var}(X) = \frac{\nu}{\nu - 2}\sigma^2 \text{ if } \nu > 2.$$

Also,

$$E(X^2) = \mu^2 + \sigma^2 \frac{\nu}{\nu - 2}, \text{Var}(X^2) = \frac{2\sigma^4 \nu^2 (\nu - 1)}{(\nu - 4)(\nu - 2)^2} + 8\sigma^2 \mu^2 \frac{\nu}{\nu - 2},$$

$$(A.12)$$

when $\nu > 4$. In general, the mgf does not exist for te $t$-distribution. The R command to calculate probability is pt and qt is the command to invert the cdf. Both of these require the additional argument df, which passes the number of degrees of freedom, $\nu$.

## A.3   An Important Mathematical Result

- **Differentiating an integral** The following result of calculus is useful in finding the pdf of a random variable. Suppose that $r(y), s(y)$ and $g(x, y)$ are suitable functions, then

$$\frac{d}{dy} \int_{r(y)}^{s(y)} g(x, y) dx$$
$$= \int_{r(y)}^{s(y)} \frac{\partial}{\partial y} g(x, y) dx + g(s(y), y) \frac{d}{dy} s(y) - g(r(y), y) \frac{d}{dy} r(y).$$

## A.4   Revision of Some Useful Matrix Algebra Results

1. Verify the matrix multiplication:

$$\begin{pmatrix} -1 & 2 & 3 \\ 5 & 1 & 2 \end{pmatrix} \begin{pmatrix} 1 & 2 & 0 \\ 4 & 1 & 3 \\ 3 & 1 & 2 \end{pmatrix} = \begin{pmatrix} 16 & 3 & 12 \\ 15 & 13 & 7 \end{pmatrix}$$

2. Satisfy yourself that the following inverses are correct by veryfying that $A^{-1}A = I$ where $A$ is the original matrix and $A^{-1}$ is its inverse and $I$ is the identity matrix.

   a.

$$\begin{pmatrix} b & c \\ c & d \end{pmatrix}^{-1} = \frac{1}{bd - c^2} \begin{pmatrix} d & -c \\ -c & b \end{pmatrix}$$

b.

$$
\begin{pmatrix} a\ 0\ 0 \\ 0\ b\ c \\ 0\ c\ d \end{pmatrix}^{-1} = \frac{1}{bd - c^2} \begin{pmatrix} \frac{bd-c^2}{a} & 0 & 0 \\ 0 & d & -c \\ 0 & -c & b \end{pmatrix}
$$

c. In general, when $A$ and $B$ are invertible square matrices:

$$
\begin{pmatrix} A\ \mathbf{0} \\ \mathbf{0}\ B \end{pmatrix}^{-1} = \begin{pmatrix} A^{-1} & \mathbf{0} \\ \mathbf{0} & B^{-1} \end{pmatrix}
$$

d.

$$
\begin{pmatrix} a\ 0\ 0 \\ 0\ b\ 0 \\ 0\ 0\ d \end{pmatrix}^{-1} = \begin{pmatrix} \frac{1}{a} & 0 & 0 \\ 0 & \frac{1}{b} & 0 \\ 0 & 0 & \frac{1}{d} \end{pmatrix}
$$

Thus, to invert a diagonal matrix we just invert the diagonals.

# References

1. Bal, S., Ojha, T.P.: Determination of biological maturity and effect of harvesting and drying conditions on milling quality of paddy. J. Agric. Eng. Res. **20**, 353–361 (1975)
2. Böhning, D.A., van der Heijden, P.G.M., Bunge, J.: Capture-Recapture Methods for the Social and Medical Sciences. CRC Press, Boca Raton (2018)
3. Casella, G., Berger, R.: Statistical Inference. Wadsworth Inc., Belmont (1990)
4. DeGroot, M.H., Schervish, M.J.: Probability and Statistics, 4th edn. Addison-Wesley, Reading (2012)
5. Dunstan, F.D.J., Nix, A.B.J., Reynolds, J.F., Rowlands, R.J.: Worked Examples in Probability and Distribution Theory. RND Publications, Cardiff (1981)
6. Dunstan, F.D.J., Nix, A.B.J., Reynolds, J.F., Rowlands, R.J.: Worked Examples in Statistical Inference. RND Publications, Cardiff (1984)
7. Goon, A.M., Gupta, M.K., Dasgupta, B.: An Outline of Statistical Theory, vol I. World Press, Kolkata (1994)
8. Hand, D.J., Daly, F., McConway, K., Lunn, D.E., Ostrowski, E.: A Handbook of Small Data Sets. CRC Press, Boca Raton (1993)
9. Kendall, M.G., O'Hagan, A., Forster, J.: Kendall's Advanced Theory of Statistics, volume 2, Part 2. Arnold Publishers (2004). ISBN: 9780340807521
10. Lehmann, E.L.: Theory of Point Estimation. John Wiley & Sons, New York (1983)
11. Levitsky, D., Halbmaier, C.A., Mrdjenovic, G.: The freshman weight gain: a model for the study of the epidemic of obesity. Int. J. Obes. Relat. Metab. Disord: J. Int. Assoc. Study Obes. **28**, 1435–1442 (2004)
12. Liero, H., Zwanzig, S.: Introduction to the Theory of Statistical Inference. Chapman and Hall/CRC, New York (2011)
13. Lindenmayer, D.B., Viggers, K.L., Cunningham, R.B., Donnelly, C.F.: Morphological variation among columns of the mountain brushtail possum, Trichosurus caninus Ogilby (Phalangeridae: Marsupiala). Aust. J. Zool. **43**, 449–458 (1995)
14. Nettleship, D.N.: Breeding success of the common puffin on different habitats at Great Island, Newfoundland. Ecol. Monogr. **42**, 246–252 (1972)
15. Rao, C.R.: Statistics and Truth, 2nd edn. World Scientific, Singapore (1997)
16. Ravishanker, N., Chi, Z., Dey D.K.: A First Course in Linear Model Theory. Chapman and Hall, Boca Raton (2021)
17. Ross, S.: A First Course in Probability, 8th edn. Pearson, Upper Saddle River (2010)
18. Searle, S.R.: Linear Models. John Wiley & Sons Inc., New York (1971)
19. Shaw, L.P., Shaw, L.F.: The flying bomb and the actuary. Significance Mag. **16**, 12–17 (2019)
20. Spiegelhalter, D.J.: The Art of Statistics - Learning from data. Pelican Books, London (2019)
21. Welsh, A.H.: Aspects of Statistical Inference. Wiley Interscience (1996). ISBN: 978-0471115915

© The Author(s), under exclusive license to Springer Nature Switzerland AG 2024, corrected publication 2024
S. K. Sahu, *Introduction to Probability, Statistics & R*,
https://doi.org/10.1007/978-3-031-37865-2

# Index

© The Author(s), under exclusive license to Springer Nature Switzerland AG 2024, corrected publication 2024
S. K. Sahu, *Introduction to Probability, Statistics & R*,
https://doi.org/10.1007/978-3-031-37865-2

553

Printed in the United States
by Baker & Taylor Publisher Services